Carbon Dioxide Utilisation

P. N-S.

Also of interest

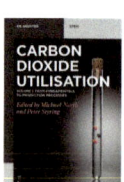
Carbon Dioxide Utilisation
Volume 1: Fundamentals
North, Styring (Eds.), 2019
ISBN 978-3-11-056309-2,
e-ISBN 978-3-11-056319-1

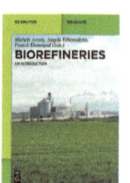
Biorefineries.
An Introduction
Aresta, Dibenedetto, Dumeignil, 2015
ISBN 978-3-11-033153-0,
e-ISBN 978-3-11-033158-5

Environmental Chemistry.
Principles and Practices
Rihana-Abdallah, Benvenuto, Roberts-Kirchhoff,
Lanigan, Evans, 2019
ISBN 978-3-11-044330-1,
e-ISBN 978-3-11-044331-8

Integrated Bioprocess Engineering.
Posten, 2018
ISBN 978-3-11-031538-7,
e-ISBN 978-3-11-031539-4

Sustainable Process Engineering.
Prospects and Opportunities
Koltuniewicz, 2014
ISBN 978-3-11-030875-4,
e-ISBN 978-3-11-030876-1

Carbon Dioxide Utilisation

Transformations

Edited by
Michael North and Peter Styring

Volume 2

DE GRUYTER

Editors

Professor Michael North
University of York
Dept. of Chemistry
Heslington
York YO10 5DD
United Kingdom
michael.north@york.ac.uk

Professor Peter Styring
The University of Sheffield
UK Centre for Carbon Dioxide Utilisation
Department of Chemical & Biological Engineering
Sir Robert Hadfield Building
Sheffield S1 3JD
United Kingdom
p.styring@sheffield.ac.uk

ISBN 978-3-11-066503-1
e-ISBN (PDF) 978-3-11-066514-7
e-ISBN (EPUB) 978-3-11-066517-8

Library of Congress Control Number: 2019947493

Bibliographic information published by the Deutsche Nationalbibliothek
The Deutsche Nationalbibliothek lists this publication in the Deutsche Nationalbibliografie;
detailed bibliographic data are available on the Internet at http://dnb.dnb.de.

© 2019 Walter de Gruyter GmbH, Berlin/Boston
Cover image: sergmam / iStock / Getty Images Plus
Typesetting: Integra Software Services Pvt. Ltd.
Printing and binding: CPI books GmbH, Leck

www.degruyter.com

MIX
Papier aus verantwor-
tungsvollen Quellen
FSC
www.fsc.org FSC® C083411

About the Editors

Michael North is professor of Green Chemistry within the Green Chemistry Centre of Excellence (GCCE) at the University of York. He obtained his first-class honours degree in chemistry from Durham University and D.Phil. from the University of Oxford. Prior to his appointment at York, he held organic chemistry academic positions at the Universities of Newcastle, London and Wales. Michael has published over 200 papers and has an h-index of 50. He was awarded the 2001 Descartes Prize by the European Commission and the 2014 Green Chemistry Award by the Royal Society of Chemistry. He is deputy director of CO2Chem, the largest international network to bring together academics, industrialists and other parties interested in developing carbon dioxide utilisation, and contributed to the 2017 Royal Society Policy Briefing Paper on "The Potential and Limitations of Carbon Dioxide Utilisation". His research interests are focussed on carbon dioxide utilisation, catalysis by the abundant metals in the Earth's crust, chemistry in green solvents and synthesis of polymers from sustainable feedstocks.

Peter Styring is professor of Chemical Engineering and Chemistry at the University of Sheffield. Based in the Department of Chemical and Biological Engineering, he is a former head of Department of Chemistry at Sheffield. He is a fellow of both the Royal Society of Chemistry (FRSC) and the Institution of Chemical Engineers (FIChemE) and is a Chartered Engineer (CEng) and Chemist (CChem). In 2006, he was awarded a prestigious EPSRC Senior Media Fellowship to work with TV, radio and the printed media for promoting chemical engineering and chemistry to the public. Peter is also an associate fellow in the Understanding of Politics at the University of Sheffield.

He is director of the UK Centre for Carbon Dioxide Utilisation and Director of the CO2Chem Network. Peter sits on the Editorial Boards of Frontier in Energy Research (CCUS) and *Journal of CO$_2$ Utilisation*. He is a member of the Scientific Advisory Board of Carbon XPRIZE and the Board of the Global CO$_2$ Initiative at the University of Michigan. Peter was awarded the 2007 Hanson Medal by the IChemE and was a member of the 2015 Royal Society MP Pairing Scheme, spending time in the Foreign & Commonwealth Office. Peter was panel lead at Mission Innovation on Accelerating CCUS in 2017 and was one of the lead authors of the report published in May 2018. He also contributed to the 2017 Royal Society Policy Briefing Paper on "The Potential and Limitations of Carbon Dioxide Utilisation" and is a member of the Royal Society Steering Group on "Synthetic Fuels".

He has been PI on a number of EU grants as UK Lead (WasteKit, SCOT, Artificial Photosynthesis, CarbonNext, Utilising Gaseous Industrial Emissions, Interreg Electrons 2 Chemicals, COZMOS) with total grants approaching 2 M€. He was a Co-I on the 4CU Programme Grant (£4.5M) and PI on numerous UKRI awards [C-Cycle, CO2Chem (three awards), Sustainable Fertilisers, Liquid Fuel and bioEnergy Supply from CO2 Reduction]. Recently, he was PI on a project funded by Costa Coffee, which successfully identified routes to the sustainable recycling of paper coffee cups. This led to him

https://doi.org/10.1515/9783110665147-202

being Co-I on a recent awards Plastics Recycling grant at Sheffield in response to the Attenborough work on single-use plastics highlighted in BBC's "Blue Planet 2". He has published over 110 scientific and numerous policy papers and has an h-index of 30. His co-authored book *Carbon Capture and Utilisation in the Green Economy* has over 25k online views and new work with the Global CO_2 Initiative has resulted in the publication of a set of extensively peer-reviewed standardisation guidelines for Life Cycle and Techno-Economic Assessment of CO_2 Utilisation.

Contents

Part VI: Photo- and plasma induced reactions of CO_2

List of contributing authors

Chapter 15
Katie Lamb
Department of Chemistry
University of York
Heslington
York YO10 5DD

Chapter 16
Antonella Pizzolante
Sarah-Elisabeth Dechent
Arjan W. Kleij
ICIQ
Institut Català
d'Investigació Química
Av. Països Catalans 16
43007 Tarragona
Spain

Christopher M. Kozak
Department of Chemistry
Memorial University of Newfoundland
St. John's NL A1B 3X7
CANADA

Chapter 17
Andreas Weilhard
Faculty of Engineering
University of Nottingham
Nottingham, NG7 2RD
UK

And

GSK Carbon Neutral Laboratory
University of Nottingham
Nottingham, NG8 2GA
UK

Jairton Dupont
Laboratory of Molecular Catalysis
Institute of Chemistry
UFRGS
Av. Bento Gonçalves, 9500
Porto Alegre 91501-970
Brazil

Victor Sans Sangorrin
Faculty of Engineering
University of Nottingham
Nottingham, NG7 2RD
UK

And

Institute of Advanced Materials (INAM)
Universitat Jaume I
Avda Sos Baynat s/n
12006, Castellón
Spain

Chapter 18
Renata Jorge da Silva
Institute of Chemistry
Federal University of Rio de Janeiro
Rua Hélio de Almeida, 40
Cidade Universitária – RJ 21941-614
Brazil
renatajs@iq.ufrj.br

Claudio J. A. Mota
Institute of Chemistry
Federal University of Rio de Janeiro
Rua Hélio de Almeida, 40
Cidade Universitária – RJ 21941-614
Brazil
cmota@iq.ufrj.br

Chapter 19
Sigrid Douven
Department of Chemical Engineering
University of Liège
B6a, Quartier Agora
Allèe du six Août, 11
4000 Liège
Belgium

Hana Benkoussas
Department of Chemical Engineering
University of Liège
B6a, Quartier Agora
Allèe du six Août, 11
4000 Liège
Belgium

https://doi.org/10.1515/9783110665147-204

Carolina Font-Palma
Department of Chemical Engineering
Thornton Science Park
University of Chester
Chester, CH2 4NU
UK
c.fontpalma@chester.ac.uk

Grégoire Léonard
Department of Chemical Engineering
University of Liège
B6a, Quartier Agora
Allèe du six Août, 11
4000 Liège
Belgium

Chapter 20
Alain Bengaouer
CEA-Grenoble
CEA-Liten
17 rue des martyrs
38054 Grenoble cedex 9
France
alain.bengaouer@cea.fr

Laurent Bedel
CEA-Grenoble
CEA-Liten
17 rue des martyrs
38054 Grenoble cedex 9
France
laurent.bedel@cea.fr

Chapter 21
James McGregor
University of Sheffield
Department of Chemical and Biological
Engineering
Sheffield S1 3JD
UK
james.mcgregor@sheffield.ac.uk

Chapter 22
Dongwei Du
School of Engineering
University of Warwick
Coventry CV4 7AL
UK

Shanwen Tao
School of Engineering
University of Warwick
Coventry CV4 7AL
UK
S.Tao.1@warwick.ac.uk

Chapter 23
Stephen M. Lyth
Platform of Inter/Transdisciplinary Energy
Research (Q-PIT)
International Institute for Carbon-Neutral
Energy Research (I2CNER)
Kyushu University, Japan
Department of Mechanical Engineering
University of Sheffield
UK
lyth@i2cner.kyushu-u.ac.jp

Chapter 24
Esperanza Ruiz Martínez
Unit for Sustainable Thermochemical
Valorization
CIEMAT
Avenida Complutense 40
28040 Madrid
Spain
esperanza.ruiz@ciemat.es

Josemaria Sanchez Hervas
Unit for Sustainable Thermochemical
Valorization
CIEMAT
Avenida Complutense 40
28040 Madrid
Spain

Chapter 25
Keerthiga Gopalram
Department of Chemical Engineering
Indian Institute of Technology Madras
India

And

Department of Chemical Engineering
SRM Institute of Science and Technology
Katankulathur
India
keerthigagopal@gmail.com

Raghuram Chetty
Department of Chemical Engineering
Indian Institute of Technology Madras
India

Chapter 26
Volodymyr Tabas
Department of Chemistry
Loughborough University
Ashby Road
Loughborough
Leicestershire LE11 3TU
UK

Benjamin R. Buckley
Department of Chemistry
Loughborough University
Ashby Road
Loughborough
Leicestershire LE11 3TU
UK
B.R.Buckley@lboro.ac.uk

Chapter 27
Jean-Marie Fontmorina
School of Engineering
Newcastle University
NE17RU Newcastle upon Tyne
UK

Paniz Izadi
School of Engineering
Newcastle University
NE17RU Newcastle upon Tyne
UK

Shahid Rasul
School of Engineering
Newcastle University
NE17RU Newcastle upon Tyne
UK

And

Faculty of Engineering and Environment
Northumbria University
NE18ST Newcastle upon Tyne
UK

Eileen H. Yu
School of Engineering
Newcastle University
NE17RU Newcastle upon Tyne
UK

Chapter 28
Annemie Bogaerts
Research group PLASMANT
Department of Chemistry
University of Antwerp
Universiteitsplein 1
BE-2610 Antwerp
Belgium
annemie.bogaerts@uantwerpen.be

Xin Tu
Department of Electrical Engineering and
Electronics
University of Liverpool
Liverpool L69 3GJ
UK
Xin.Tu@liverpool.ac.uk

Gerard van Rooij
Dutch Institute for Fundamental Energy
Research
De Zaale 20
5612 AJ Eindhoven
The Netherlands

Richard van de Sanden
Dutch Institute for Fundamental Energy
Research
De Zaale 20
5612 AJ Eindhoven
The Netherlands

Chapter 29
Andreas Kafizas
Molecular Science Research Hub
Imperial College
White City Campus
London, W12 0BZ
UK

And

The Grantham Institute
Imperial College London
South Kensington
London, SW7 2AZ
UK
a.kafizas@imperial.ac.uk

Chapter 30
Simon C. Parker
Department of Chemistry
The University of Sheffield
Sheffield, S3 7HF
UK

Andrew J. Sadler
Department of Chemistry
The University of Sheffield
Sheffield, S3 7HF
UK

James D. Shipp
Department of Chemistry
The University of Sheffield
Sheffield, S3 7HF
UK

Julia A. Weinstein
Department of Chemistry
The University of Sheffield
Sheffield, S3 7HF
UK

Part IV: **Catalytic reactions of CO$_2$**

Katie Lamb

15 Catalysts for the conversion of CO_2 to cyclic and polycarbonates

15.1 Introduction

Perhaps one of the most studied carbon dioxide utilisation (CDU) processes is the synthesis of organic carbonates. The thermodynamic stability and kinetics of CO_2, however, means that cyclic and polycarbonate formation often requires catalysts to promote the reaction and to ensure the process is economically viable. This chapter will present the significance and "greenness" of catalytically synthesising cyclic and polycarbonates from epoxides and CO_2. First, a brief introduction to the use and importance of cyclic and polycarbonates will be given, followed by a brief discussion of the reaction mechanism to form these carbonates via epoxides and CO_2. A selection of (primarily) metal-based catalytic systems reported in the literature, for the conversion of CO_2 to cyclic and polycarbonates, will then be discussed to give a general overview of this important area of CDU research.

15.1.1 Use and importance of cyclic and polycarbonates

Depending on their structure, organic carbonates can be classified as acyclic (or linear), cyclic or polycarbonates (or polymeric carbonates, Figure 15.1). Acyclic carbonates are used as gasoline additives, cosmetic thickeners and pesticides [1–3]. Cyclic carbonates are used as electrolytes in lithium-ion batteries, precursors for polymer synthesis [2] and as alternative "green" dipolar aprotic solvents, as they are non-toxic, biodegradable and non-corrosive chemicals [4, 5]. The optical transparency and impact-resistant properties of polycarbonates make them ideal for CDs, DVDs and aircraft windows [6].

Organic carbonates have been industrially important chemicals since the mid-1950s [8] and their global demand keeps growing. In 2016, worldwide production of organic carbonates reached approximately 7 million tonnes per year and is predicted to reach 20 million tonnes per year by 2030 [2]. The most industrially important organic carbonates are cyclic and polycarbonates. In 2014, global production of cyclic carbonates and polycarbonates reached 0.1 and 4 million tonnes per year, respectively [9, 10]. In efforts to satisfy this global growth in demand, sustainably and economically, numerous methods have been reported in the literature for carbonate synthesis via CDU (Figure 15.2).

Katie Lamb, Department of Chemistry, University of York, Heslington York

https://doi.org/10.1515/9783110665147-015

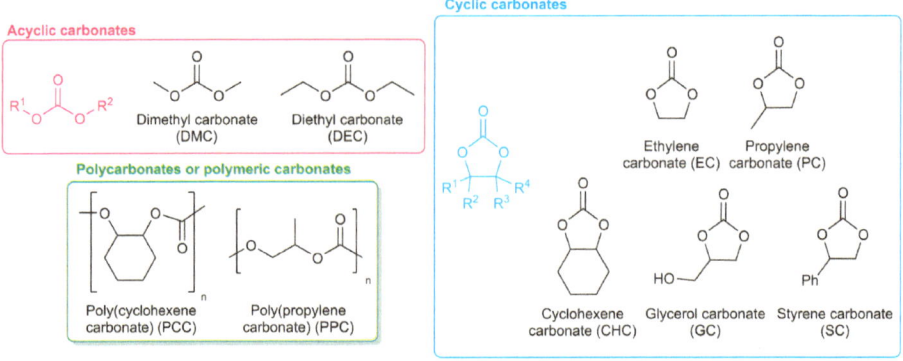

Figure 15.1: Examples of organic carbonates used in industry today [4, 7].

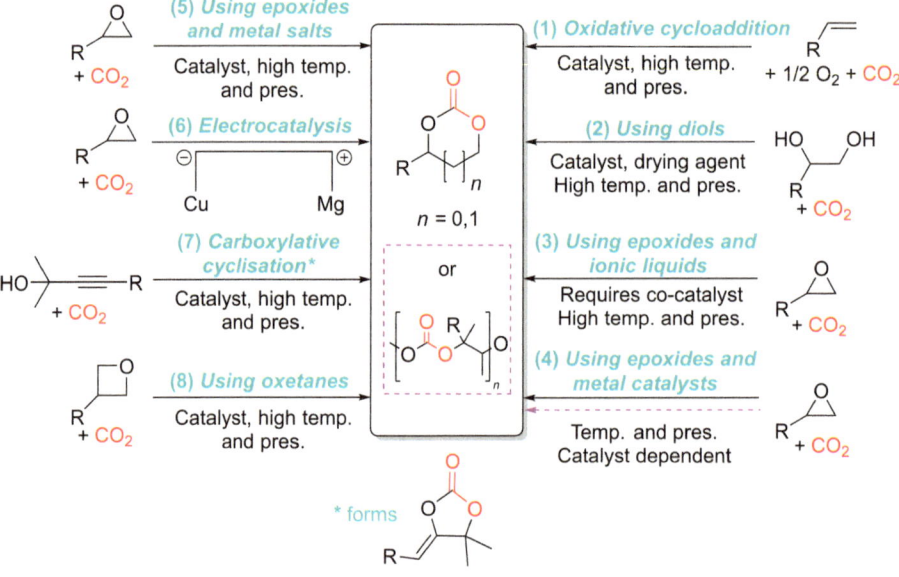

Figure 15.2: Cyclic (black solid line) and polycarbonate (dashed purple line) synthesis via CDU [1, 11–14].

The most popular substrates to use for cyclic and polycarbonate synthesis via CDU are epoxides (routes 3–6, Figure 15.2), due to the synthetic versatility of epoxide conversion. Carbonate synthesis via epoxides usually requires a catalyst due to the thermodynamic stability and kinetics of CO_2. The reported success of using metal catalysts to convert epoxides under mild conditions, combined with 100% atom economy, is also ideal in terms of Green Chemistry (*vide infra*). Cyclic and polycarbonate synthesis from epoxides shall therefore be the focus throughout the rest of this chapter. There are excellent reviews for those interested in other reaction processes [7, 15, 16].

15.1.2 Why is cyclic and polycarbonate synthesis via CDU green?

Since the establishment of Green Chemistry in the 1990s, there has been a paradigm shift towards promoting greener and more sustainable chemical processes globally. A real emphasis towards green, sustainable and energy-efficient CDU methodology has therefore been promoted in recent literature, in an effort to ensure CO_2 emissions are depleted without simultaneously affecting the environment. Not only are cyclic and polycarbonates economically viable products, but they can also be produced via Green Chemistry principles. One way to consider if a reaction pathway is "green" is to determine if any of the "Twelve Principles of Green Chemistry" are satisfied (Figure 15.3) [17]. Other important factors to consider are (but are not limited to) green metrics and Life Cycle Analysis (LCA).

1. Prevent waste
2. Atom economy
3. Use less hazardous synthesis
4. Design benign synthesis
5. Use benign solvents and auxiliaries
6. Improve energy efficiency
7. Renewable feedstocks
8. Reduce derivatives
9. Catalytic versus stoichiometric
10. Design for degradation
11. Real-time analysis for pollution prevention
12. Inherently benign chemistry for accident prevention

Figure 15.3: A simplistic view of the "Twelve Principles of Green Chemistry" [17, 18].

Cyclic carbonate and polycarbonate syntheses in conjunction with metal catalysis and CDU can satisfy many Green Chemistry principles (Figure 15.3). These reactions can occur with 100% atom economy (Principle 2), rarely require extremely toxic reagents or reaction conditions (Principle 3) and in the presence of a catalyst (Principle 9) can be performed under ambient or near-ambient conditions (Principle 6). The formation of cyclic and polycarbonates via renewable feedstocks (Principle 7) is a current hot topic of research (see Chapter 16). All CDU reactions also arguably prevent waste by consuming "waste" CO_2 gas (Principle 1). Synthesising cyclic carbonates and polycarbonates via CDU also provides an alternative and greener synthetic route to the traditional methods of using phosgene and bis(phenol)A, respectively. These carbonates also have many green applications. For example, cyclic carbonates can be used as alternative green solvents [19, 20]. Polycarbonate synthesis via CDU could also substitute synthetic polymers prepared from non-renewable petroleum resources, which in 2011 equated to approximately 7% of worldwide oil and gas consumption [21, 22]. Synthesising organic carbonates via CDU therefore provides a huge opportunity to reduce CO_2 emissions while encouraging sustainable chemistry.

15.2 Reaction mechanisms for cyclic and polycarbonate formation from CO_2 and epoxides

Both cyclic and polycarbonates can be formed from CO_2 insertion into epoxides and follow the same reaction mechanism (in the initial stages). A vital difference in the reaction profile is that cyclic carbonates are the thermodynamically favoured product, whereas polycarbonates are the kinetically favoured product [21, 23]. As a result, reaction conditions can be tailored to favour which pathway occurs. Numerous methods exist in the literature, but perhaps the most studied catalytic system is the combination of a Lewis acidic metal complex and a Lewis basic co-catalyst. The Lewis acid initiates the reaction by activating the epoxide and/or CO_2 via electrophilic activation, then the Lewis base acts as a nucleophile and ring-opens the epoxide (Scheme 15.1) [12, 21]. It is generally accepted this catalytic system occurs via a co-operative mechanism with (currently) four possible mechanisms reported in the literature [21]. These include:

1) a monometallic mechanism involving one nucleophile (the most commonly accepted mechanism, Scheme 15.1),
2) a monometallic mechanism involving two nucleophiles,
3) a bimetallic mechanism involving two different metal complexes, and
4) a bimetallic mechanism involving two metal centres from the same complex.

In the case of monometallic mechanism (Scheme 15.1) if ring-closure or "back-biting" can occur at the metal carbonate stage (ring-closure is favoured, the nucleophile is a good leaving group or high temperatures are used), the reaction mechanism will favour cyclic carbonate formation (Pathway 1, Scheme 15.1). Alternatively, if ring-opening further equivalents of epoxides and repeating CO_2 addition is favoured at the metal carbonate stage (ring-closure is unfavourable, nucleophile is a poor leaving group or low temperatures are used), polymerisation will occur (Pathway 2, Scheme 15.1) [21].

In polycarbonate synthesis, the formation of the metal carbonate is termed as the initiation step (Scheme 15.2). The vital difference is that instead of the metal carbonate intermediate undergoing ring-closure (Pathway 1, Scheme 1) propagation occurs (Pathway 2, Scheme 15.1), as the metal carbonate ring-opens another epoxide molecule via nucleophilic substitution. The polymer will continue to propagate until termination

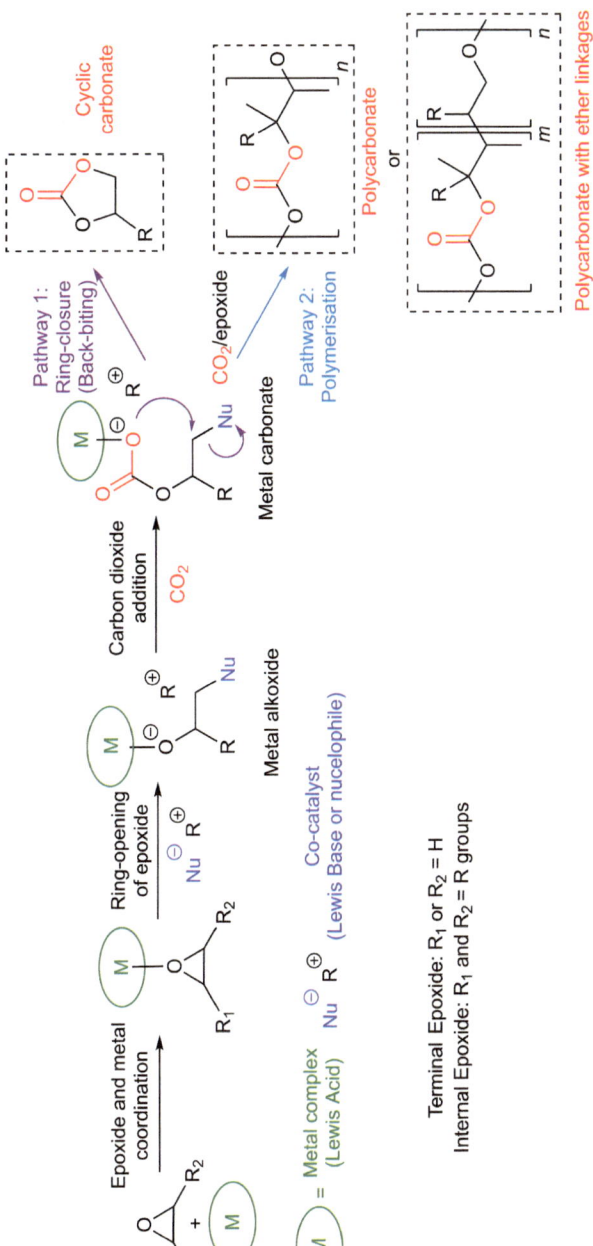

Scheme 15.1: The generally accepted co-operative mechanism for cyclic carbonate (Pathway 1) and/or polycarbonate synthesis (Pathway 2), using epoxides and CO_2 with a metal catalyst (Lewis acid) and co-catalyst (Lewis base) [12, 21].

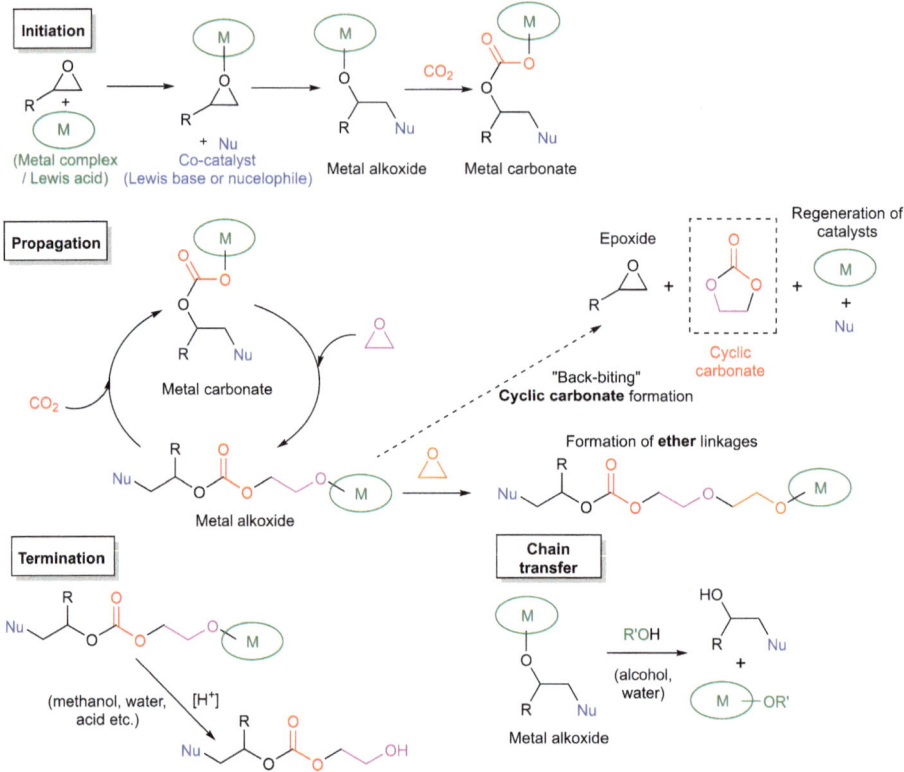

Scheme 15.2: Polycarbonate synthesis using epoxides and CO_2 [24].

is initiated via hydrolysis of the polymer (often acid hydrolysis in methanol), thus forming a polymer chain end-capped with a hydroxyl group (Scheme 15.2) [24].

Other reactions can occur and interfere with the propagation step, thus preventing 100% carbonate linkages in the copolymer and can even terminate the reaction. These include chain transfer reactions with adventitious water, alcohols or acids, propagation of epoxides into the polymer to form ether linkages, and intramolecular "back-biting" or ring-closure to form cyclic carbonates (Scheme 15.2) [22]. Polycarbonate syntheses are thus performed under anhydrous conditions, with low epoxide/catalyst concentrations, low temperatures (polycarbonates are kinetically favoured) and specially designed catalysts that disfavour "back-biting" of the metal carbonate. Further information on polycarbonate formation mechanisms and physicochemical properties can be found in Perscamona's 2014 review [25].

As both cyclic and polycarbonate syntheses follow similar reaction mechanisms, it can be difficult to create catalytic systems that selectively form cyclic or polycarbonates, or those with switchable favourability towards either carbonate. When also considering terminal epoxides, for example propylene oxide, the difference in activation energies for cyclic carbonate synthesis versus copolymer formation is often minimal [23]. Therefore, not all systems reported in the literature can selectively form cyclic or polycarbonates, and very few can alternate which carbonate product is selectively formed by simply changing the reaction conditions (*vide infra*).

15.3 Catalysts for cyclic and polycarbonate formation

Many synthetic methods have been reported for CO_2 insertion into terminal and internal (di-substituted) epoxides to form five-membered cyclic carbonates [1, 11–13, 15]. Non-metal-based examples are discussed in Section 15.3.1, and homo- and heterogeneous metal-based catalysts are discussed in Sections 15.3.2–15.3.4. For other possible reaction processes, there are some excellent articles in the literature [15, 26, 27]. Those interested in the use of sustainable metals for cyclic carbonate synthesis should consult North's 2015 review [13]. A review into synthesising cyclic carbonates using alternative reagents to epoxides was recently written by Zhang et al. [28].

The number of metal complexes reported in the literature is extremely vast, thus only some will be discussed in this chapter (Sections 15.3.2–15.3.4). More comprehensive reviews on cyclic [1, 29] and polycarbonate syntheses [22, 30, 31] are available in the literature. Comparisons between different catalytic systems developed for the synthesis of styrene carbonate (SC), or poly(cyclohexane carbonate, PCC) and poly(propylene carbonate, PPC), will be made when possible.

Metal-based catalysts are often used as the Lewis acid, as they enable the reaction to proceed under mild (low temperatures and CO_2 pressure) and thus more energy-efficient conditions. Intrinsic properties of the metal catalyst, such as metal acidity, ligand structure, solubility and flexibility, as well as co-catalyst properties, will influence which reaction pathway is favoured. Epoxide conversion is possible in the sole presence of simple nucleophiles, which can ring-open the epoxide without a metal catalyst, but often requires harsh conditions. Tetraalkylammonium salts, phosphonium salts and nitrogen-based compounds, such as 4-(dimethylamino)-pyridine (DMAP), are often used as the Lewis base/nucleophile/co-catalyst, due to their low cost, basicity, nucleophilicity and global abundance [1, 12, 21].

15.3.1 Non-metal (and electrochemical) systems for cyclic carbonate formation

There are a few notable examples in the literature in which cyclic carbonate synthesis occurs without a metal catalyst, but these are rare. In 2018, a polymeric salicylimine system was reported to convert epoxides into cyclic carbonates in high yields, and could be recycled without any significant loss in conversion. However, the absence of a metal-based catalyst meant that high temperatures (>100 °C) were required and this system was only screened against a few terminal epoxides (Scheme 15.3a) [32]. In 2016, Hirose reported a simple binary system consisting of pyridine reagents and tetrabutylammonium (TBA) salts, which required no solvent and achieved sufficient conversions under ambient conditions (Scheme 15.3b). High catalytic loadings were, however, required for good conversions, and the catalyst was inefficient at ring-opening internal epoxides [33]. In 2011, an intricate system by Buckley et al. was devised to electrochemically synthesise cyclic carbonates in a single compartment cell with a Cu cathode and Mg anode. The conversion of more challenging internal epoxides was not reported (Scheme 15.3c) [14].

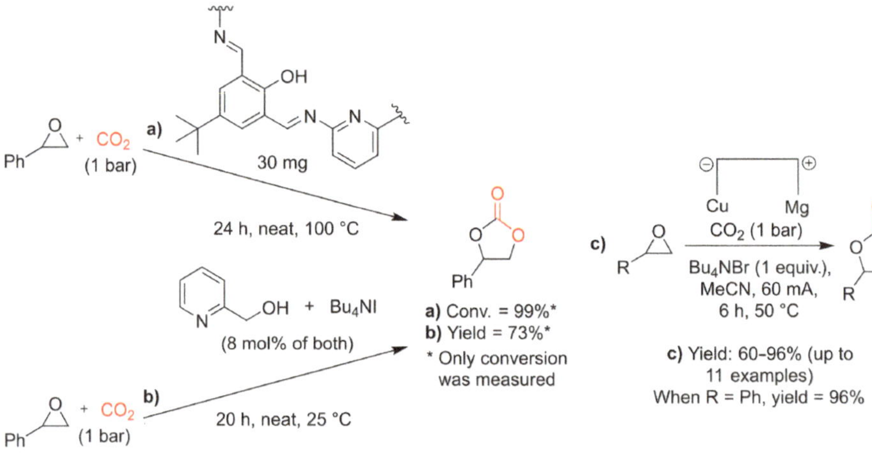

Scheme 15.3: Non-metal (and electrochemical) systems used for cyclic carbonate synthesis [14, 32, 33].

15.3.2 Metal catalysts for cyclic carbonate formation

Some of the most active metal complexes reported in the literature in cyclic carbonate formation are acen, amino-tris(phenolate), scorpionate, salen and salophen (or salphen) complexes. The metals employed in these complexes include (but are not limited to) Na, Mg, Al, Ca, Cr, Mn, Fe, Co and Zn [1, 12, 13]. During the 1990s, Darensbourg et al. extensively studied Co(III) and Cr(III) salen complexes primarily for polycarbonate synthesis but interestingly discovered that certain Cr(III) salophen complexes promoted cyclic carbonate formation from cyclohexane oxide, rather than polymerisation (**Cat 1–6**, Scheme 15.4) [34]. These intriguing results encouraged further investigation into these complexes by North et al. (*vide infra*).

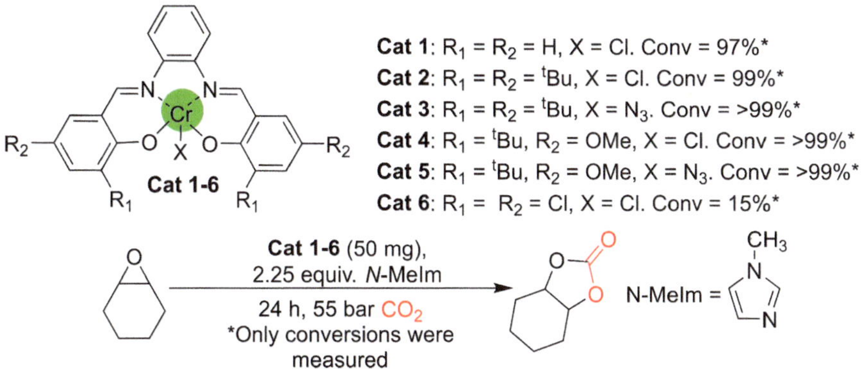

Cat 1: $R_1 = R_2 = H$, $X = Cl$. Conv = 97%*
Cat 2: $R_1 = R_2 = {}^tBu$, $X = Cl$. Conv = 99%*
Cat 3: $R_1 = R_2 = {}^tBu$, $X = N_3$. Conv = >99%*
Cat 4: $R_1 = {}^tBu$, $R_2 = OMe$, $X = Cl$. Conv = >99%*
Cat 5: $R_1 = {}^tBu$, $R_2 = OMe$, $X = N_3$. Conv = >99%*
Cat 6: $R_1 = R_2 = Cl$, $X = Cl$. Conv = 15%*

Cat 1-6 (50 mg),
2.25 equiv. *N*-MeIm

24 h, 55 bar CO_2
*Only conversions were
measured

N-MeIm =

Scheme 15.4: Cr(III) salophen catalysts reported by Darensbourg in the synthesis of cyclic carbonates from epoxides and CO_2 [34].

In 2007, North et al. synthesised exceptionally active bimetallic Al(III) salen complexes, which could ring-open terminal epoxides with tetrabutylammonium bromide under neat and mild conditions [35]. The most active complex was **Cat 7**, and this catalyst could even work without a co-catalyst at elevated temperatures and pressures, which is an extremely rare feat for these Lewis acidic compounds (**Cat 7**, Scheme 15.5) [36]. Between 2016 and 2017, North's research group also studied Al(III) and Cr(III) salophen complexes,[37, 38] and discovered that these catalysts were exceptionally active in ring-opening internal epoxides under mild conditions (**Cat 8** and **9**, Scheme 15.5). In 2011, North tested the compatibility of a supported bimetallic Al(III) salen catalyst (**Cat 10**, Scheme 15.5) with CO_2 from flue gas. Despite these "unclean" conditions, the catalytic ability of **Cat 10** was only slightly affected after exposure to flue gas from coal burning and could be regenerated. This research highlighted the extraordinary potential of Al(III) salen catalysts to be implemented in industrial plants to promote large-scale CDU (and large-scale Carbon Capture and Storage (CCS)) [39].

Many metal catalysts can ring-open terminal epoxides but cannot ring-open internal epoxides, due to the steric hindrance of functional groups on internal epoxides blocking any nucleophile from ring-opening the epoxide. Nevertheless, there are some notable catalysts in the literature that can form cyclic carbonates from internal epoxides (e.g. North's Al(III) and Cr(III) salophen catalysts, Scheme 15.5). During 2015–2017, Kleij et al. developed an interesting strategy to form tri-substituted cyclic carbonates from hydroxyl-substituted cyclic epoxides, using numerous Al(III) and Fe(III) tris(phenolate) complexes (**Cat 10–13**) at slightly elevated temperatures and pressures [40]. These catalysts were also capable of ring-opening oxetanes (four-membered cyclic epoxides) to form six-membered cyclic carbonates. Kleij took the conversion of epoxides to cyclic carbonates even further by illustrating the effectiveness of this reaction, using

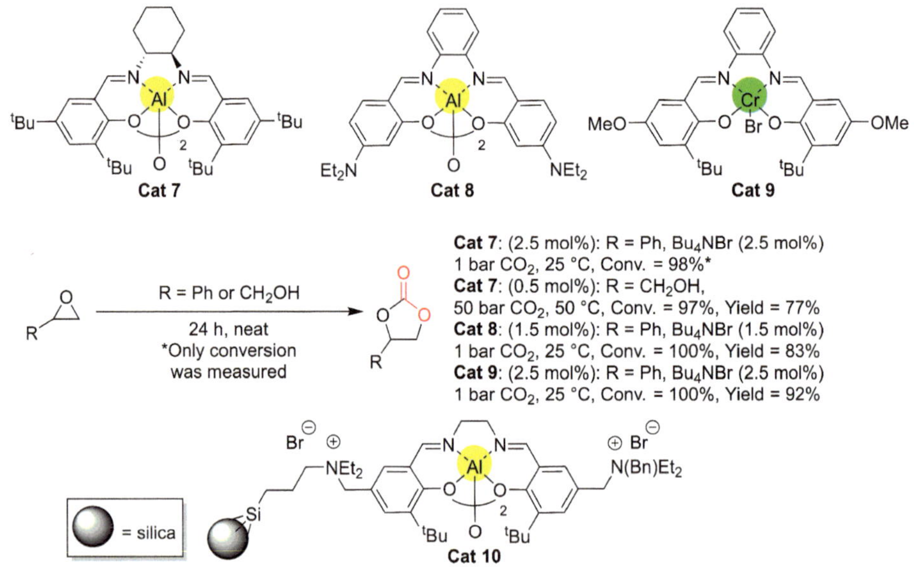

Scheme 15.5: Al(III) and Cr(III) catalysts reported by North in the synthesis of cyclic carbonates from epoxides and CO_2 [35–39].

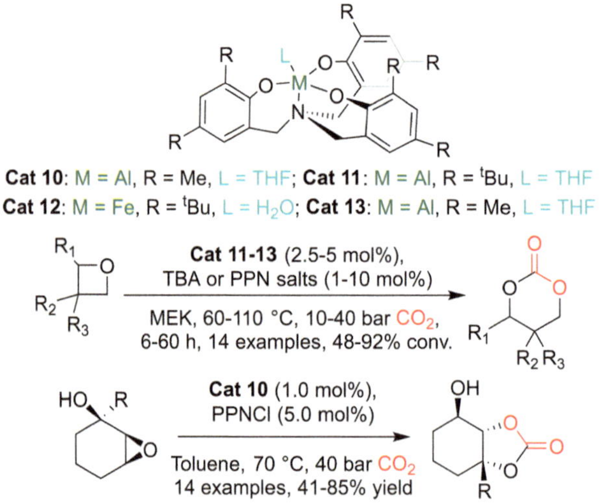

Scheme 15.6: Kleij's synthesis of cyclic carbonates from epoxides and CO_2 using tris(phenolate) metal complexes [40, 41].

Cat 10, to synthesise functionalised cyclic compounds, which can be used as precursors for medicinally important compounds (Scheme 15.6) [41].

15.3.3 Metal catalysts for polycarbonate formation

The synthesis of polycarbonates via CDU occurs from Ring-Opening COPolymerisation (ROCOP) of epoxides and CO_2, via homo- or heterogeneous catalysis [24]. In general, homogeneous catalysts deliver a much greater uptake of CO_2 into the polymer, which results in balanced epoxide–CO_2 enchainment and produces aliphatic polycarbonates. By contrast, heterogeneous catalysts require considerably higher pressures and result in lower levels of CO_2 incorporation, thereby producing polyether carbonates, where the ether linkages result from sequential epoxide enchainment [42]. This reaction can be used to form (poly)hydroxyl-terminated "polyols," which themselves are important for polyurethane formation and polyesters. This research, however, is beyond the scope of this chapter and only the conversion of epoxides with CO_2 will be discussed. For recent advances in using renewable feedstocks, see Chapter 16 and Darensbourg's 2017 review [43].

Not only must catalysts be designed to selectively form polymers with decent yields, but must also synthesise polymers with uniform monomer distributions and high percentage of carbonate linkages. The thermophysical factors of the synthesised polymer are therefore essential, as these properties determine the polymer's flexibility, rigidness, glass-transition temperature (T_g), crystallinity, purity and even biodegradability. Variables such as "number average molecular weight" (M_N), "weight average molecular weight" (M_W) and polydispersity (PDI or Đ) provide vital information about polymer chain lengths and hence the distribution of monomers (repeat units) in the polymer. The closer the PDI value is to 1, the greater the monodispersity of the polymer. All of these variables are catalyst dependent and are critically important in controlling the composition, properties and thus future applications of the polymer. Other important variables to consider include the stereochemistry and tacticity (configuration) of functional groups in the backbone of the polymer units, of which the former determines polymer flexibility. Further information on the physical properties of polymers and their analysis are in Pescarmona's 2014 review [25].

Inoue et al. were the first to report the copolymerisation of epoxides and CO_2 in 1969, with diethylzinc (ZnEt$_2$), in a 1:1 mixture with water, forming PPC from propylene oxide and CO_2 [44]. Intrigued by these findings, Inoue also reported in 1978 the successful formation of PPC from propylene oxide and CO_2, using Al(III) and Zn(II) tetraphenylporphyrin complexes [45]. These catalysts, however, required long reaction times and illustrated extremely poor control of molecular weight, producing polymers with a broad molecular weight distribution and low percentage of carbonate linkages.

These Zn(II) complexes were the first homogenous catalysts reported in the literature and this discovery encouraged others to investigate similar catalysts for polycarbonate synthesis. In 2001, Coates et al. studied Zn(II) β-diiminate complexes (e.g. **Cat 14–16**, Scheme 15.7) for the alternating copolymerisation of cyclohexene oxide (CHO) and CO_2, reporting the most active set of catalysts for this reaction at that time [46]. In 2017, Williams et al. synthesised numerous di-zinc-aryl complexes

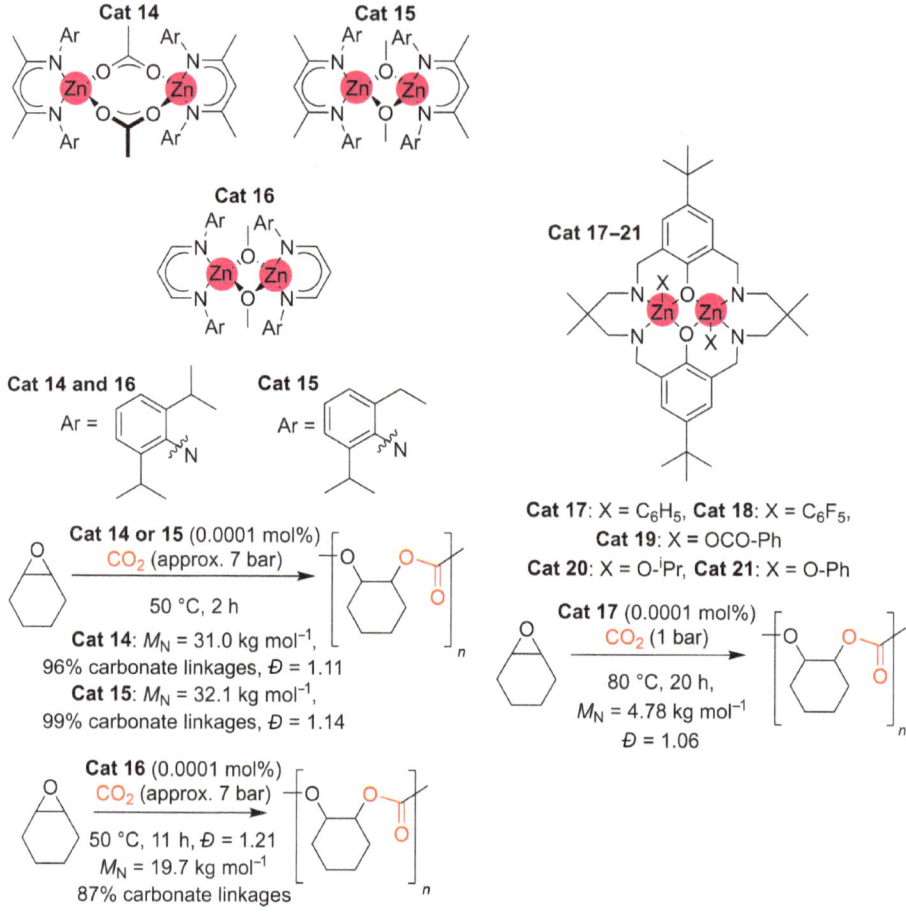

Scheme 15.7: Zn(II) complexes reported by Coates and Williams et al. in the ROCOP of CHO and CO_2 [46, 47].

(e.g. **Cat 17–21**, Scheme 15.7), with **Cat 17** exhibiting moderate activity for the ROCOP of CHO and CO_2, as well as the polymerisation of phthalic anhydride with CHO (Scheme 15.7) [47].

Numerous Co(III) salen catalysts have been investigated due to their promising favourability towards polymerisation versus cyclic carbonate formation, especially by Darensbourg as described in his *Chem. Soc. Rev.* review [23]. In 2005, Coates et al. were the first to report a regioselective lone catalytic system, a Co(III) salen catalyst (**Cat 22**, Scheme 15.8), for CO_2 and propylene oxide copolymerisation. This approach produced PPC with >99% selectivity and with 90–99% carbonate linkages at ambient temperature, but required high CO_2 pressures (55 bar) to obtain satisfactory results [48]. In 2009, Lu et al. developed a Co(III) salen catalyst with onium salts or organic

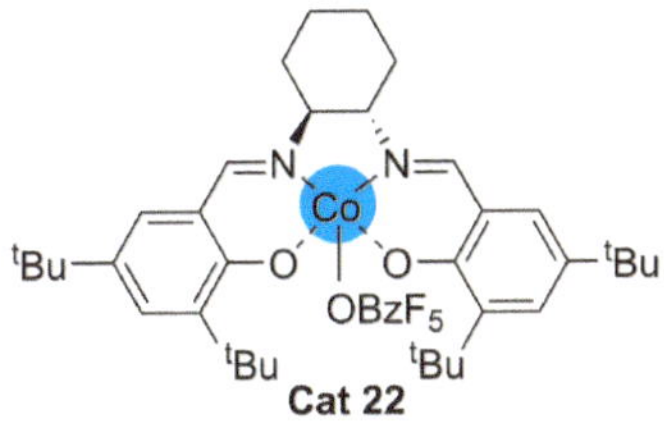

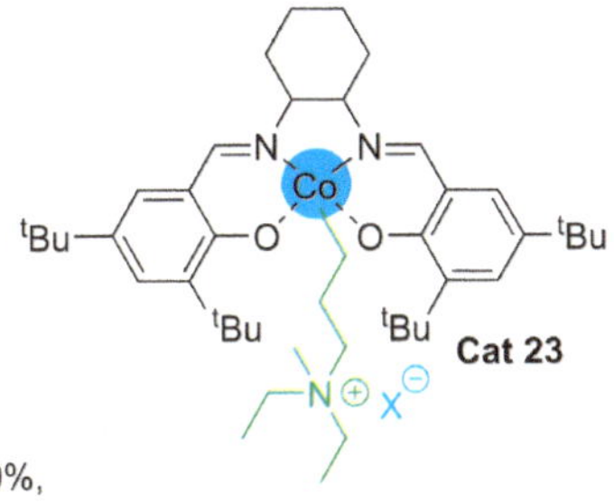

Cat 22: (0.0005 mol%): PPNCl (0.0005 mol%),
R = (S)-Me, 0.5 h, 14 bar CO$_2$, 22 °C, %PPC = >99%,
M_N = 22.2 kg mol^{-1}, Đ = 1.15
Cat 22: (0.0005 mol%): PPNCl (0.0005 mol%),
R = (R)-Me, 2 h, 14 bar CO$_2$, 22 °C, %PPC = 98%,
M_N = 19.1 kg mol^{-1}, Đ = 1.12
Cat 23: (0.0005 mol%): Bu$_4$NBr (0.0005 mol%),
R = rac-Me, 4 h, 15 bar CO$_2$, 25 °C, %PPC = >99%,
M_N = 30.4 kg mol^{-1}, Đ = 1.10

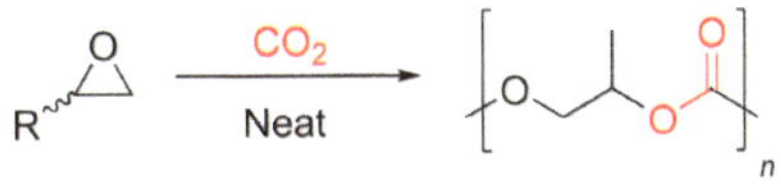

Scheme 15.8: Co(III) salen complexes developed by Lu for the copolymerisation of CO$_2$ and propylene oxide [48, 49].

bases attached to the catalyst via alkyl linkers (**Cat 23**, Scheme 15.8) [49]. This catalyst had 100% selectivity for the formation of PCC and was the first complex to achieve selective copolymerisation at high temperatures (around 100 °C). Lu hypothesised that this occurred due to the intermediate anionic polymer remaining in close contact with the metal centre via electrostatic interactions, when displaced by the epoxide substrate in the rate-determining step. The formed ion-pair interaction consequently reduces the probability of the free anionic polymer chain undergoing backbiting, thus forming a cyclic carbonate, and also stabilises the Co(III) active species at high temperatures [49].

In 2018, Darensbourg et al. reported the synthesis of CO$_2$-based polycarbonates, using a Co(III) salen complex with a PPN-TFA salt (**Cat 24**), which were capable of self-healing after being cut into half with a sharp blade, and could reconnect by simply placing the two separate pieces back together. By employing a three-step tandem synthesis, incorporating epoxide hydrolysis, CO$_2$/epoxide copolymerisation and thio-ene click reactions consecutively, polymers with desired ratios of "stiff" PPC and "soft" poly(allylglycidyl ether carbonate, PAC) units could be synthesised (Scheme 15.9) [50]. These polymers are the first autonomic CO$_2$ polymers to be reported in the literature that can self-heal. Although this work is only an initial study, it highlights the achievements that can be obtained by combining different monomer units into the polymer and will hopefully encourage others to create more diverse CO$_2$-based polymers.

In 2018, Williams et al. developed the first indium catalysts ever reported for the selective formation of polycarbonates from CO$_2$ and epoxides, and illustrated high activity even under mild pressures (**Cat 25–27**, Scheme 15.10) [51]. Interestingly, these

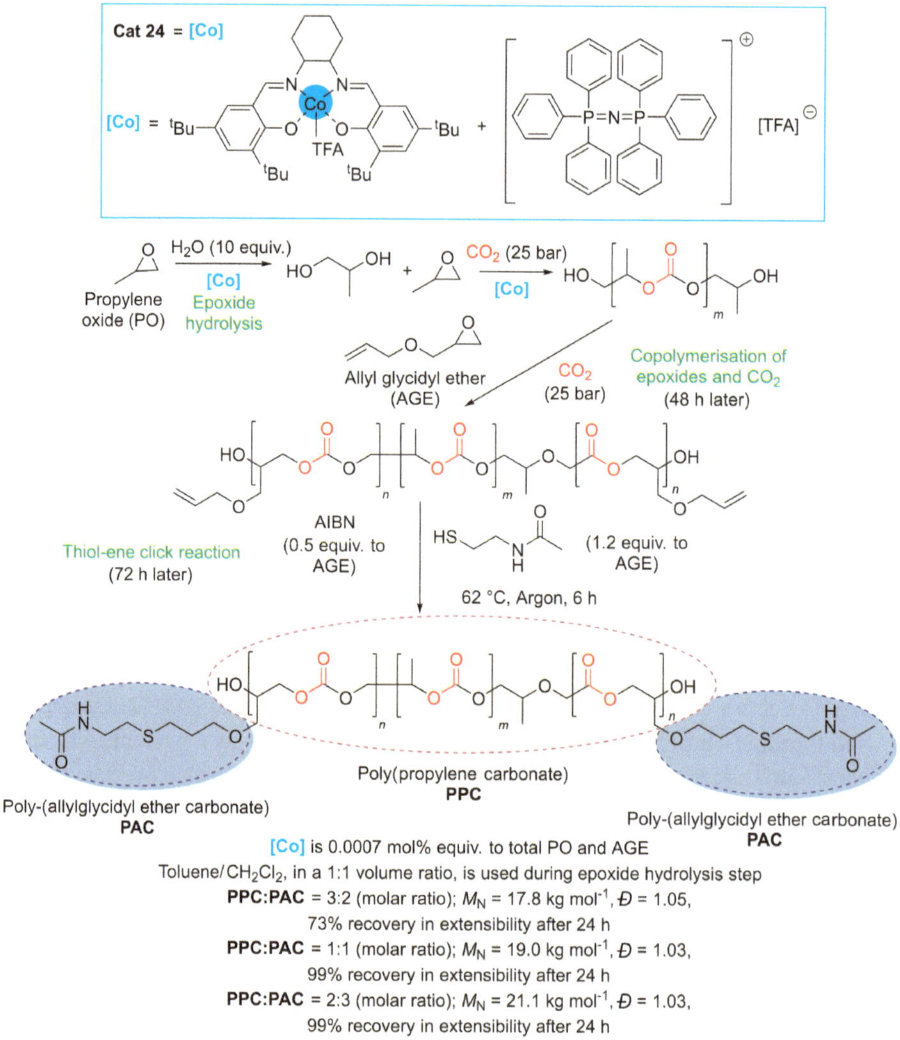

Scheme 15.9: The synthesis of "self-healing" CO_2-based polymers as reported by Darensbourg [50].

complexes only formed polycarbonates when phosphasalen ligands were used in the catalyst structure, whereas salen ligands formed polyethers. The combination of co-catalysts DMAP or PPNCl hindered the catalyst or promoted cyclic carbonate formation, respectively. By performing a detailed structural analysis of these complexes, **Cat 25** was found to be the most active and selective catalyst for polycarbonate formation. Interestingly, the reaction mechanism takes an unexpected mononuclear pathway (i.e. has a first-order dependence on catalyst concentration), compared to the bimetallic or bicomponent pathway usually reported for

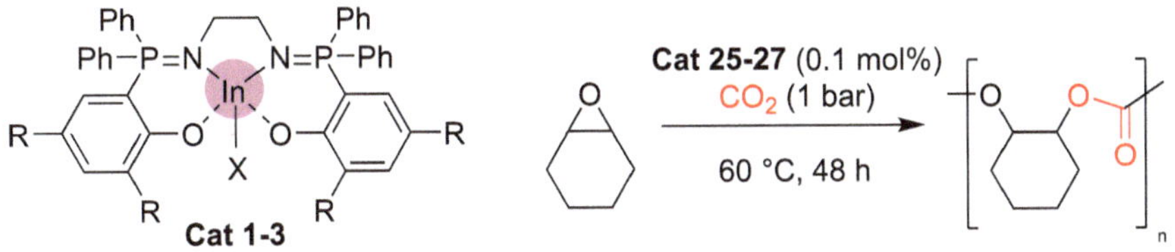

Cat 25-27 (0.1 mol%)
CO_2 (1 bar)
60 °C, 48 h

Cat 1-3

Cat 25: R = ^tBu, X = O^tBu
Cat 26: R = Amyl, X = O^tBu
Cat 27: R = Cumyl, X = O^tBu

Cat 25: %PPC >99%, M_N = 1.41 kg mol^{-1}, Đ = 1.23
Cat 26: %PPC >99%, M_N = 1.60 kg mol^{-1}, Đ = 1.23 (80 °C)
Cat 27: %PPC >99%, M_N = 2.90 kg mol^{-1}, Đ = 1.16

Scheme 15.10: In(III) phosphasalen complexes reported by Williams in the ROCOP of CHO and CO_2 [51].

salen metal systems. This was hypothesised to be due to the larger ionic radius of In(III) compared to traditional salen metals (r_{ionic}(pm): In(III) = 80, Cr(III) = 62, Co (III) = 55, Al(III) = 54), which in theory could accommodate the cis-coordination of an epoxide during the reaction mechanism. This large ionic radius could consequently reduce the energy barrier of insertion and thus promote mononuclear polymerisation pathways. This hypothesis was further supported by Lewis bases hindering the catalysts, as Lewis base coordination directly competes with metal–epoxide coordination [51].

Numerous catalysts can convert CO_2 into polycarbonates, but there is an urgent need for these systems to work with captured CO_2 or flue gas, rather than pure CO_2 gas. Some studies have performed these experiments, with a notable example reported by Williams et al. in 2015, when dinuclear Mg(II) and Zn(II) catalysts **Cat 28–30** (Scheme 15.11) converted CHO into PCC using CO_2 directly taken from the Ferrybridge CCS power plant in the UK (Scheme 15.11). The recyclability and effectiveness of the catalysts were diminished by the presence of H_2O, SO_x and NO_x gases, but could still form polycarbonates with reasonable M_N values and with narrow polydispersities, even in the presence of 400 equivalents of H_2O [52].

15.3.4 Metal catalysts for cyclic and polycarbonate formation

Very few catalytic systems have been designed that can effectively "switch" favourability towards cyclic or polycarbonate synthesis, and maintain high selectively and yields. One of the most intriguing, and rare, catalytic systems to achieve this were the Fe(III) amino triphenolate catalysts developed in 2013 by Pescarmona et al. (**Cat 31–33**, Scheme 15.12). These complexes could change favourability towards cyclic or polycarbonate formation by simply changing the reaction conditions, for example, co-catalytic loading and temperature [53]. Creating more catalysts that are as versatile as Pescarmona's is still currently an issue.

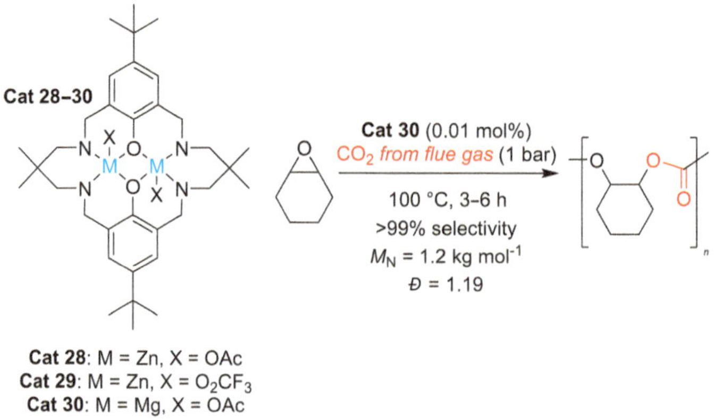

Cat 28–30

Cat 28: M = Zn, X = OAc
Cat 29: M = Zn, X = O_2CF_3
Cat 30: M = Mg, X = OAc

Scheme 15.11: Mg(II) and Zn(II) complexes reported by Williams et al. in the ROCOP of CHO and CO_2 from flue gas [52].

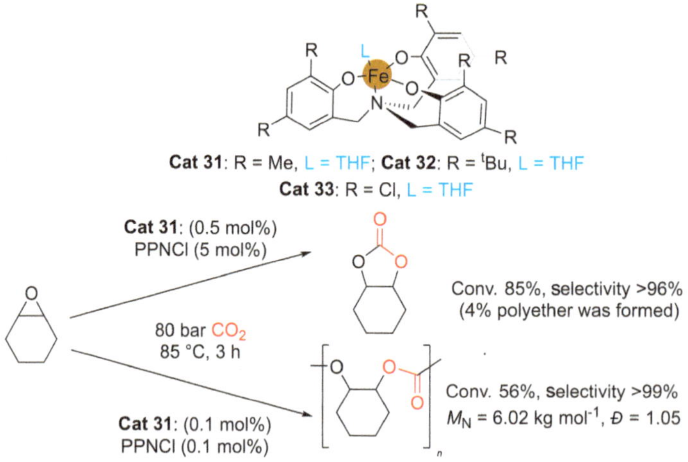

Cat 31: R = Me, L = THF; **Cat 32**: R = ^{t}Bu, L = THF
Cat 33: R = Cl, L = THF

Scheme 15.12: Fe(III) complexes reported by Pescarmona et al. for cyclic and polycarbonate synthesis from CO_2 [53].

15.4 Summary and concluding remarks

By carefully tailoring reaction conditions and catalytic structures, numerous homo- and heterogeneous catalysts have been successfully designed to convert epoxides into cyclic or polycarbonates via CDU. This reaction not only provides a "green" alternative to synthesising commercially valuable products, but also provides a sustainable method for utilising "waste" CO_2 and reducing CO_2 emissions. The ever-increasing importance and acceptance of Green Chemistry has also led to further success in performing these reactions under mild, energy-efficient and green reaction conditions.

Perhaps one of the important hurdles to still overcome is the conversion of di- and tri-substituted cyclic carbonates under energy-efficient conditions. Recent research, however, indicates that more and more progress is being achieved in this field. For polycarbonate synthesis, most studies have focused on utilising simple terminal epoxides, such as propylene oxide, or only polymerising one epoxide throughout the entire reaction. More emphasis therefore needs to be directed towards utilising more complex monomers, and using different mixtures and ratios of epoxides to create intriguing and diverse polymers. Tackling the conversion of more challenging substrates, such as bio-based and natural epoxides, also needs to be addressed. A greater focus towards creating catalysts that can go beyond research in the lab is crucial, and more catalytic systems must be tested under flue gas conditions. Investigating the ability of catalysts to be employed in an industrial setting is essential to determine if these catalysts can realistically synthesise cyclic and/or polycarbonates via CDU on a global scale, and ultimately provide a realistic opportunity to deplete global CO$_2$ emissions via CDU.

References

[1]   North M, Pasquale R, Young C. "Synthesis of cyclic carbonates from epoxides and CO$_2$". Green Chem. 2010;12:1514–39.
[2]   Aresta M, Dibenedetto A, Quaranta E. "State of the art and perspectives in catalytic processes for CO$_2$ conversion into chemicals and fuels: The distinctive contribution of chemical catalysis and biotechnology". J. Catal. 2016;343:2–45.
[3]   Aresta M, Dibenedetto A, Angelini A, Papai I. "Reaction mechanisms in the direct carboxylation of alcohols for the synthesis of acyclic carbonates". Top. Catal. 2015;58:2–14.
[4]   Schaffner B, Schaffner F, Verevkin SP, Borner A. "Organic carbonates as solvents in synthesis and catalysis". Chem. Rev. 2010;110:4554–81.
[5]   Parker HL, Sherwood J, Hunt AJ, Clark JH. "Cyclic carbonates as green alternative solvents for the heck reaction". ACS Sustainable Chem. Eng. 2014;2:1739–42.
[6]   Besse V, Camara F, Voirin C, Auvergne R, Caillol S, Boutevin B. "Synthesis and applications of unsaturated cyclocarbonates". Polym. Chem. 2013;4:4545–61.
[7]   Miao C-X, Wang J-Q, He L-N. Catalytic processes for Chemical Conversion of Carbon dioxide into Cyclic carbonates and Polycarbonates. Open Org. Chem. J. 2008;2:68–82.
[8]   Clements JH. "Reactive applications of cyclic alkylene carbonates". Ind. Eng. Chem. Res. 2003;42:663–74.
[9]   Dibenedetto A, Angelini A. Synthesis of Cyclic Carbonates. In: Aresta M, Van Eldik R, eds. Advances in Inorganic Chemistry: CO$_2$ Chemistry. 1st ed. Oxford: Elsevier; 2014:25–81.
[10] Aresta M, Dibenedetto A, Angelini A. "Catalysis for the valorization of exhaust carbon: from co$_2$ to chemicals, materials, and fuels". Technological Use of CO$_2$. Chem. Rev. 2014;114: 1709–42.

[11] Maeda C, Miyazaki Y, Ema T. "Recent progress in catalytic conversions of carbon dioxide". Catal. Sci. Technol. 2014;4:1482–97.

[12] Martin C, Fiorani G, Kleij AW. "Recent advances in the catalytic preparation of cyclic organic carbonates". ACS Catal. 2015;5:1353–70.

[13] Comerford JW, Ingram IDV, North M, Wu X. "Sustainable metal-based catalysts for the synthesis of cyclic carbonates containing five-membered rings". Green Chem. 2015;17: 1966–87.

[14] Buckley BR, Patel AP, Wijayantha KGU. "Electrosynthesis of cyclic carbonates from epoxides and atmospheric pressure carbon dioxide". Chem. Commun. 2011;47:11888–90.

[15] Fiorani G, Guo WS, Kleij AW. "Sustainable conversion of carbon dioxide: the advent of organocatalysis". Green Chem. 2015;17:1375–89.

[16] Sun JM, Fujita S, Arai M. "Development in the green synthesis of cyclic carbonate from carbon dioxide using ionic liquids". J. Organomet. Chem. 2005;690:3490–7.

[17] Anastas PT, Warner JC. Green Chemistry: Theory and Practice. Oxford: Oxford University Press; 1998.

[18] Green Chemistry Pocket Guides. American Chemical Society, 2017. (Accessed 18/ 10/2018,at https://www.acs.org/content/acs/en/greenchemistry/principles/12-principles-of-green-chemistry.html)

[19] Lawrenson SB, Arav R, North M. "The greening of peptide synthesis". Green Chem. 2017;19: 1685–91.

[20] Alder CM, Hayler JD, Henderson RK, et al. "Updating and further expanding GSK's solvent sustainability guide". Green Chem. 2016;18:3879–90.

[21] Pescarmona PP, Taherimehr M. "Challenges in the catalytic synthesis of cyclic and polymeric carbonates from epoxides and CO_2". Catal. Sci. Technol. 2012;2:2169–87.

[22] Kember MR, Buchard A, Williams CK. "Catalysts for CO_2/epoxide copolymerisation". Chem. Commun. 2011;47:141–63.

[23] Lu XB, Darensbourg DJ. "Cobalt catalysts for the coupling of CO_2 and epoxides to provide polycarbonates and cyclic carbonates". Chem. Soc. Rev. 2012;41:1462–84.

[24] Trott G, Saini PK, Williams CK. "Catalysts for CO_2/epoxide ring-opening copolymerization". Philos. Trans. R. Soc., A 2016;374:1–19.

[25] Taherimehr M, Pescarmona PP. "Green Polycarbonates Prepared by the Copolymerization of CO_2 with Epoxides". J. Appl. Polym. Sci. 2014;131,1–17.

[26] Lan DH, Fan N, Wang Y, et al. "Recent advances in metal-free catalysts for the synthesis of cyclic carbonates from CO_2 and epoxides". Chin. J. Catal. 2016;37:826–45.

[27] Xu BH, Wang JQ, Sun J, et al. "Fixation of CO_2 into cyclic carbonates catalyzed by ionic liquids: a multi-scale approach". Green Chem. 2015;17:108–22.

[28] Riduan SN, Zhang YG. "Recent developments in carbon dioxide utilization under mild conditions". Dalton. Trans. 2010;39:3347–57.

[29] Martinez J, Castro-Osma JA, Earlam A, et al. "Synthesis of cyclic carbonates catalysed by aluminium heteroscorpionate complexes". Chem. – Eur. J. 2015;21:9850–62.

[30] Darensbourg DJ, Yeung AD. "A concise review of computational studies of the carbon dioxide-epoxide copolymerization reactions". Polym. Chem. 2014;5:3949–62.

[31] Darensbourg DJ, Mackiewicz RM, Phelps AL, Billodeaux DR. "Copolymerization of CO_2 and epoxides catalyzed by metal salen complexes". Acc. Chem. Res. 2004;37:836–44.

[32] Subramanian S, Park J, Byun J, Jung Y, Yavuz CT. "Highly efficient catalytic cyclic carbonate formation by pyridyl salicylimines". ACS Appl. Mater. Interfaces 2018;10: 9478–84.

[33] Wang L, Zhang GY, Kodamaa K, Hirose T. "An efficient metal- and solvent-free organocatalytic system for chemical fixation of CO$_2$ into cyclic carbonates under mild conditions". Green Chem. 2016;18:1229–33.

[34] Darensbourg DJ, Mackiewicz RM, Rodgers JL, Fang CC, Billodeaux DR, Reibenspies JH. "Cyclohexene oxide/CO$_2$ copolymerization catalyzed by chromium(III) salen complexes and N-methylimidazole: Effects of varying salen ligand substituents and relative cocatalyst loading". Inorg. Chem. 2004;43:6024–34.

[35] Melendez J, North M, Pasquale R. "Synthesis of cyclic carbonates from atmospheric pressure carbon dioxide using exceptionally active aluminium(salen) complexes as catalysts". Eur. J. Inorg. Chem. 2007:3323–6.

[36] Castro-Osma JA, North M, Wu X. "Development of a halide-free aluminium-based catalyst for the synthesis of cyclic carbonates from epoxides and carbon dioxide". Chem. – Eur. J. 2014;20:15005–8.

[37] Castro-Osma JA, Lamb KJ, North M. "Cr(salophen) complex catalyzed cyclic carbonate synthesis at ambient temperature and pressure". ACS Catal. 2016;6:5012–25.

[38] Castro-Osma JA, North M, Wu X. "Synthesis of cyclic carbonates catalysed by chromium and aluminium salphen complexes". Chem. – Eur. J. 2016;22:2100–7.

[39] North M, Wang BD, Young C. "Influence of flue gas on the catalytic activity of an immobilized aluminium (salen) complex for cyclic carbonate synthesis". Energy Environ. Sci. 2011;4: 4163–70.

[40] Laserna V, Martin E, Escudero-Adan EC, Kleij AW. "Substrate-triggered stereoselective preparation of highly substituted organic carbonates". ACS Catal. 2017;7:5478–82.

[41] Rintjema J, Guo WS, Martin E, Escudero-Adan EC, Kleij AW. "Highly chemoselective catalytic coupling of substituted oxetanes and carbon dioxide". Chem. – Eur. J. 2015;21:10754–62.

[42] Zhu YQ, Romain C, Williams CK. "Sustainable polymers from renewable resources". Nature 2016;540:354–62.

[43] Poland SJ, Darensbourg DJ. "A quest for polycarbonates provided via sustainable epoxide/ CO$_2$ copolymerization processes". Green Chem. 2017;19:4990–5011.

[44] Inoue S, Koinuma H, Tsuruta T. "Copolymerization of carbon dioxide and epoxide". J. Polym. Sci., Part B: Polym. Lett. 1969;7:287–292.

[45] Takeda N, Inoue S. "Polymerization of 1,2-epoxypropane and co-polymerization with carbon dioxide catalyzed by metalloporphyrins". Macromol. Chem. Phys. 1978;179:1377–81.

[46] Cheng M, Moore DR, Reczek JJ, Chamberlain BM, Lobkovsky EB, Coates GW. "Single-site beta-diiminate zinc catalysts for the alternating copolymerization of CO$_2$ and epoxides: Catalyst synthesis and unprecedented polymerization activity". J. Am. Chem. Soc. 2001;123:8738–49.

[47] Romain C, Garden JA, Trott G, Buchard A, White AJP, Williams CK. "Di-Zinc-Aryl Complexes: CO$_2$ Insertions and Applications in Polymerisation Catalysis". Chem. – Eur. J. 2017;23: 7367–76.

[48] Cohen CT, Chu T, Coates GW. "Cobalt catalysts for the alternating copolymerization of propylene oxide and carbon dioxide: Combining high activity and selectivity". J. Am. Chem. Soc. 2005;127:10869–78.

[49] Ren WM, Liu ZW, Wen YQ, Zhang R, Lu XB. "Mechanistic aspects of the copolymerization of co$_2$ with epoxides using a thermally stable single-site cobalt(III) catalyst". J. Am. Chem. Soc. 2009;131:11509–18.

[50] Yang GW, Zhang YY, Wang Y, Wu GP, Xu ZK, Darensbourg DJ. "Construction of autonomic self-healing co$_2$-based polycarbonates via one-pot tandem synthetic strategy". Macromolecules 2018;51:1308–13.

[51] Thevenon A, Cyriac A, Myers D, White AJP, Durr CB, Williams CK. "Indium catalysts for low-pressure co_2/epoxide ring-opening copolymerization: evidence for a mononuclear mechanism?" J. Am. Chem. Soc. 2018;140:6893–903.

[52] Chapman AM, Keyworth C, Kember MR, Lennox AJJ, Williams CK. "Adding value to power station captured CO_2: Tolerant Zn and Mg homogeneous catalysts for polycarbonate polyol production". ACS Catal. 2015;5:1581–8.

[53] Taherimehr M, Al-Amsyar SM, Whiteoak CJ, Kleij AW, Pescarmona PP. "High activity and switchable selectivity in the synthesis of cyclic and polymeric cyclohexene carbonates with iron amino triphenolate catalysts". Green Chem. 2013;15:3083–90.

Antonella Pizzolante, Sarah-Elisabeth Dechent, Arjan W. Kleij
and Christopher M. Kozak

16 Sustainable feedstock for conversion of CO_2 to cyclic and polycarbonates

16.1 General introduction

The use of carbon dioxide in chemical synthesis has witnessed substantial growth in the last decade [1]. This renewable carbon feedstock combines a number of advantageous properties in synthetic chemistry such as availability and low cost [2]. Despite these attractive features, the high stability of carbon dioxide limits large-scale application as high-energy reactants and highly efficient catalysts are required to provide efficient turnover kinetics. The progress made in the area of CO_2 catalysis has, however, paved the way towards a wide variety of organic molecules and materials with value beyond academic interest. The formation of organic carbonates is probably the most intensely studied process in CO_2 valorisation using readily available epoxides and CO_2 as partner reagents [3]. More recently, a shift towards the use of renewable feedstock in these CO_2-based conversions has taken place. The use of low-cost, environmentally benign starting materials such as sugar residues, fatty acids and terpenes has created a new momentum in organic carbonate catalysis providing new functional synthons and material precursors. This chapter will focus on two subclasses of organic carbonates (cyclic and polycarbonates) that are based on the aforementioned three categories of renewable monomers. Further specific details on cyclic and polycarbonate synthesis from petrochemically derived feedstock are described in Chapter 15.

16.2 Biobased cyclic carbonates

16.2.1 Cyclic carbonates derived from terpene precursors

Terpene-based structures represent renewable resources and typical examples include limonene, camphene, vitamin A, steroids and natural rubber (Figure 16.1). Recently, terpenes have increasingly found applications in polymer science and sustainable materials as renewable feedstock upon combining with carbon dioxide. In particular, limonene has received a great deal of attention because of its availability and post-synthetic potential.

Limonene is a major constituent of citrus fruit with a content of up to 90 wt% [4]. This terpene contains two synthetically useful olefin bonds, one *endo*- and the other being *exo*-cyclic. Both the mono- and diepoxide of limonene (LO and LDO,

https://doi.org/10.1515/9783110665147-016

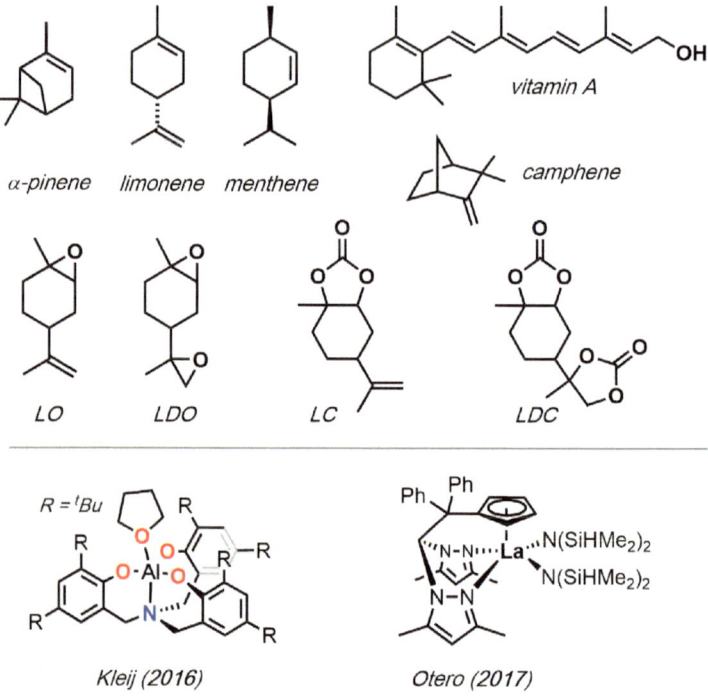

Figure 16.1: Selected terpene scaffolds and prominent examples of efficient catalysts for their conversion to cyclic carbonate products such as limonene carbonate (LC) and limonene dicarbonate (LDC). The stereochemistry in the products is omitted for clarity.

respectively; Figure 16.1) have been probed towards the synthesis of cyclic carbonates. In 2012, Müllhaupt and co-workers achieved the catalytic formation of limonene dicarbonate (LDC) from LDO and CO_2 at 140 °C and 30 bar pressure in the presence of a quaternary ammonium salt [5]. The LDC is a versatile monomer since both carbonate rings are reactive towards aminolysis and offer a more sustainable way to construct non-isocyanate-based polyurethanes (NIPUs). Unfortunately, the formation of LDC suffered from various side reactions such as (oligo)ether formation at pressures lower than 30 bar.

Kleij and co-workers investigated a wide scope of trisubstituted terpene monooxides [6]. A homogenous binary catalyst comprising an Al(III) aminotriphenolate complex (Figure 16.1) and a nucleophilic additive in methyl ethyl ketone was found to be successful towards the formation of various terpene carbonates at 85 °C and 10 bar of CO_2. The coupling reactions that involved other bicyclic terpene oxides including carvone oxide, limonene diepoxide and menthene oxide showed similar reactivity, typically giving yields in the range of 42–52% with high chemoselectivity. Terpene carbonates based on acyclic precursors, however, showed poorer outcome in terms of chemoselectivity (despite the rather similar isolated yields) because of the rather slow reaction rates of these cycloaddition reactions that involve sterically crowded oxiranes. In 2017, a new rare-earth metal-based heteroscorpionate complex (Figure 16.1) was developed for the synthesis of cyclic carbonates including examples based on terpene oxides (LO and LDO) derived from limonene [7]. These reactions were carried out at 100 °C under 10 bar CO$_2$ pressure in the absence of solvent and also required a chloride nucleophile. Similar order of magnitude yields of both limonene carbonates were achieved under these conditions.

Though not strictly terpene based, 1,4-cyclohexadiene (CHD) is a common by-product obtained from oleochemical olefin metathesis reactions and offers an alternative synthon with double bond functionalities reminiscent of many terpene scaffolds. [8] Its oxide (1,4-cyclohexadiene oxide, CHDO) can be easily obtained from CHD by standard oxidation methods, and various catalysts have shown potential to convert CHDO into its corresponding cyclic carbonate typically using binary catalysts comprising a Lewis acid and the accompanying nucleophile [9, 10].

Calcium-based catalysts have been hardly investigated for cyclic carbonate formation despite calcium being abundant, inexpensive and environmentally benign. Recently, however, a Ca(II)-based catalyst comprising a crown ether and a triphenylphosphine additive was successfully used to convert various biosourced terpene oxides at 75 °C and 50 bar [11]. Although cyclic terpene oxides have been proved to be rewarding substrates giving high yields of their corresponding carbonate upon coupling with CO$_2$, under similar reaction conditions acyclic substrates gave poor yield (<25%) of the cyclic carbonate. Nonetheless, this Ca-based catalyst stands out as an easily assembled, multicomponent structure based on simple and accessible building blocks. However, the coupling of terpene oxides (which typically contain tri-substituted oxirane groups) remains a challenge, and new catalyst designs are warranted to expand on the scope of carbonates under more sustainable conditions.

16.2.2 Cyclic carbonates derived from sugars

Natural sugars represent a promising, inexpensive and renewable alternative to fossil fuel-based feedstock, as they are highly abundant, have low toxicity and feature attractive structural and stereochemical diversity. The rigid cyclic structure, which is adopted by many sugars, can further make them suitable for polymer applications since the imparted stiffness in the polymeric backbone has potential to enhance the material properties (e.g. glass transition temperature, T_g). Therefore, polymerisation reactions of sugar-based cyclic carbonates through ring-opening polymerisation (ROP) lead to biobased aliphatic polycarbonates (APCs) [12–15].

ROP of sugar-derived cyclic carbonates is a very promising non-toxic alternative for the preparation of well-defined, high-molecular-weight polycarbonates while using CO_2 as an abundant and cheap feedstock. These bioderived polymers show, due to their biodegradability and biocompatibility, high potential in several applications such as tissue engineering scaffolds and vehicles for drug delivery [12–15].

To date, sugar-derived cyclic carbonates used as monomers for ROP mainly include those based on D-glucose [16–18] and D-xylose [19]. Just recently, it was shown that the carboxylation of D-mannose (Figure 16.2) can be readily performed in a two-step synthesis with 1-O-methyl-α-D-mannose as a commercially available reagent and CO_2 as C1 synthon. First, the hydroxyl groups at the 2- and 3-positions undergo isopropylidene protection (part in blue) followed by the insertion of CO_2, predominantly into the less sterically hindered primary hydroxyl group mediated by 1,8-diazabicyclo[5.4.0]-undec-7-ene (DBU) in order to form an ionic salt. Then, consecutive addition of the tosyl chloride (TsCl) in the presence of NEt_3 allows for cyclic carbonate formation via a nucleophilic addition–elimination pathway with retention of the original stereochemistry.

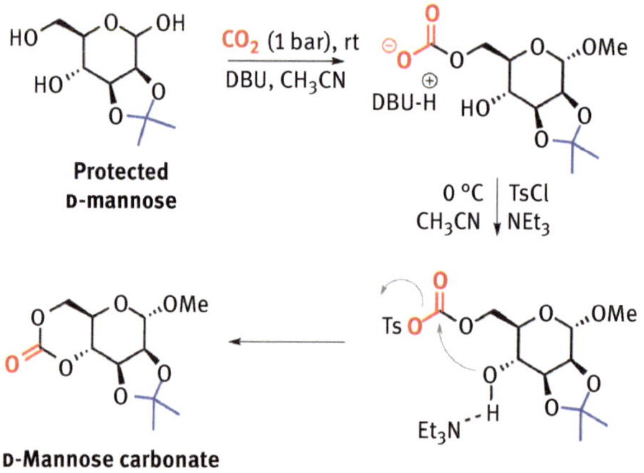

Figure 16.2: Two-step approach towards the carboxylation of D-mannose.

Later on, it turned out that this synthetic approach is not applicable to all sugar-related feedstock. DNA-based 2′-deoxyribonucleosides have shown high potential for the application in advanced polymeric materials due to their supramolecular base-pairing abilities. However, these sugar precursors turned out not to be able to form cyclic carbonates neither using the pathway presented in Figure 16.2 nor via traditional phosgene-based routes. To overcome these difficulties, a slightly different approach was demonstrated to be more successful and included the conversion of methylated thymidine 2′-deoxyribonucleoside into its carbonate analogue. Similar to the applied strategy in Figure 16.2, first the CO_2 insertion into the

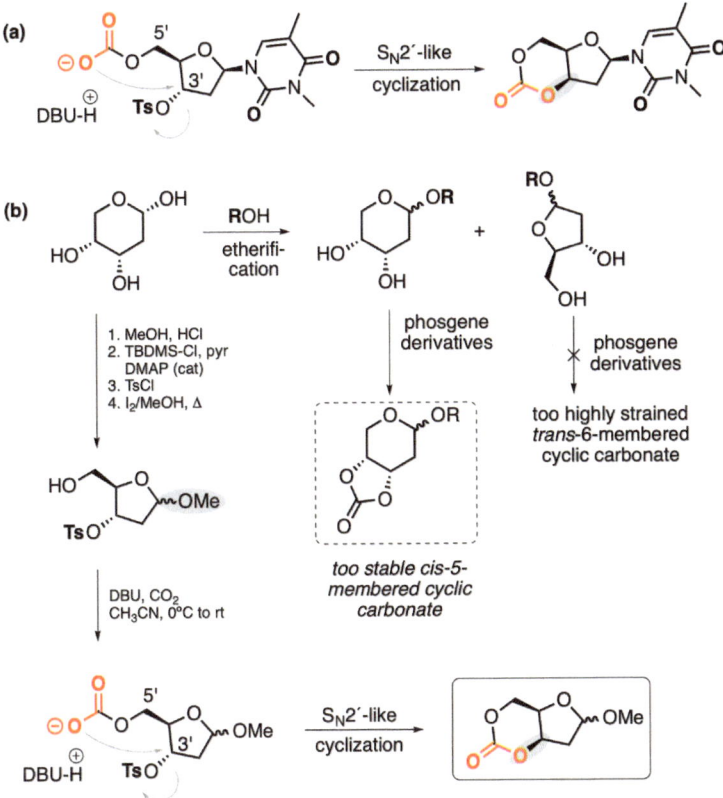

Figure 16.3: (a) Synthesis of the cyclic carbonate of methylated thymidine 2′-deoxyribonucleoside with stereochemical inversion at the 3′-position. (b) Challenges in the conversion of cyclic carbonate ROP monomers from 2-deoxy-D-ribose and its synthesis in several steps via a stereochemical inversion at the 3′-position.

primary alcohol group was facilitated by DBU. Then, contrary to the synthesis of D-mannose carbonate, an intramolecular $S_N2′$-like displacement of the tosyl group results in a stereochemical inversion at the 3′-position (Figure 16.3a) [15].

In a rather similar fashion 2-deoxy-D-ribose, which also presents an abundant and simple pentose-like building block, could be converted into a sugar-based cyclic carbonate suitable for use as a monomer for ROP to form APCs (Figure 16.3b). Thereto, the thermodynamically less favoured furanose form was kinetically trapped by methylation of the anomeric hydroxyl group and then tosylated via several protection steps in order to facilitate the pathway described in Figure 16.3a. The stereochemical inversion in the cyclic carbonate product plays a crucial role in relation to its potential for ROP. Previous reactions that were carried out between phosgene derivatives and the pyranose showed the formation of stable *cis*-five-membered cyclic carbonates, which in contrast to their *trans*-fused analogues do not readily undergo ROP. Additionally, the direct carboxylation of the furanose form in the presence of (tri)phosgene results in the formation of an unstable *trans*-six-membered cyclic carbonate. These observations led to the idea of the reaction pathway described in Figure 16.3b, and underlines the subtleties of designing productive routes towards ROP monomers derived from sugar precursors.

16.2.3 Cyclic carbonates derived from fatty acids

Within the range of biobased raw materials, (epoxidised) vegetable oils and their fatty acid derivatives play a crucial role in order to form new, highly innovative green materials. Like other bioderivable feedstock, these compounds are typically available in relatively high quantities at low-to-moderate cost, renewable and environmentally benign [20–24]. Cyclic carbonates derived from oleochemical compounds and CO_2 allow for variation of several physicochemical properties and consequently have emerged as useful synthons for a wide scope of applications including their use as emollients [25], fuel additives [26, 27], polymer precursors [28, 29], (industrial) lubricants and chemical solvents [26, 29].

A special interest has been devoted to fatty acid esters, which have demonstrated promising features to function as plasticisers [30], as fuel additives or as components of biomedical devices [31]. Additionally, cyclic carbonate-based vegetable oils have been reported as attractive synthons for the inexpensive and green production of NIPUs, avoiding the requirement for toxic isocyanates and use of phosgene reagent [23, 28, 29, 32–34]. These formal poly(hydroxyurethane)s (PHUs) have, due to the presence of hydroxyl groups throughout the polymer backbone, very specific properties and are to date the most promising green substitutes for conventional polyurethanes widely applied in paints, in coatings and in biomaterials [28, 35].

The requisite epoxidation of the olefinic double bonds of oleochemicals can be easily carried out at larger scale by the use of peracetic or performic acids in combination with H_2O_2 as oxidising agent, or on a lab scale with *meta*-chloroperoxybenzoic acid (*m*CPBA). The epoxidised and typically ester-protected fatty acid derivatives can be coupled with CO_2 to form their cyclic carbonates via methods discussed in Chapter 15, and a broad range of mostly metal-based catalytic systems have been developed [11, 24, 36–38]. Generally, these types of carboxylation reactions require harsher conditions (>100 °C, $p(CO_2) > 25$ bar) and long reaction times to facilitate high conversion of the sterically demanding internal epoxides into their cyclic carbonate analogues [11, 38, 39].

Pioneering work in this context was realised by Wilkes et al. [29] by converting epoxidised soybean oil (ESBO) into carbonated soybean oil (CSBO) in the presence of tetrabutylammonium bromide (TBAB) though under rather harsh reaction conditions (Figure 16.4). This triple epoxide conversion turned out to be crucially dependent on the CO_2 pressure, on the reaction temperature and on the applied amount

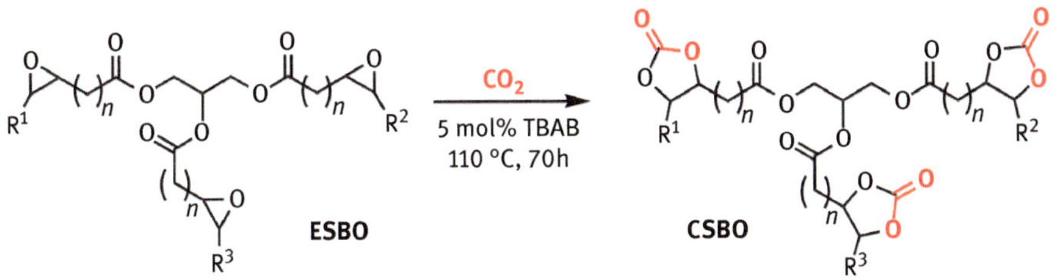

Figure 16.4: Conversion of epoxidised soybean oil (ESBO) into carbonated soybean oil (CSBO).

of catalyst [32]. The use of supercritical CO_2 was highly beneficial in order to improve the yield and the kinetics of the carboxylation reactions. In addition, these conditions also proved effective towards the conversion of various other epoxidised fatty acid methyl esters (EFAMEs) into their corresponding carbonated fatty acid methyl esters [40, 41].

To facilitate separation of the product and catalyst, the TBAB catalyst was heterogenised by immobilising it onto a silica surface, though mass transfer limitations were observed resulting in a significant drop in activity [38]. Despite this drawback, Mülhaupt demonstrated that such a catalytic system can still be applied for the carboxylation of epoxidised linseed and soybean oil by simply applying longer reaction times [23]. Further optimisation of the reactivity was reported by Leitner et al. [36] through the introduction of polyoxometalates (POM) as co-catalysts. Upon combination of the metal-substituted silicotungstate POM (THA-Cr-Si-POM), a significant fourfold increase in activity and an increase in the *cis/trans* selectivity were observed. A synergistic mechanism accelerating the double inversion pathway process was suggested (Figure 16.5).

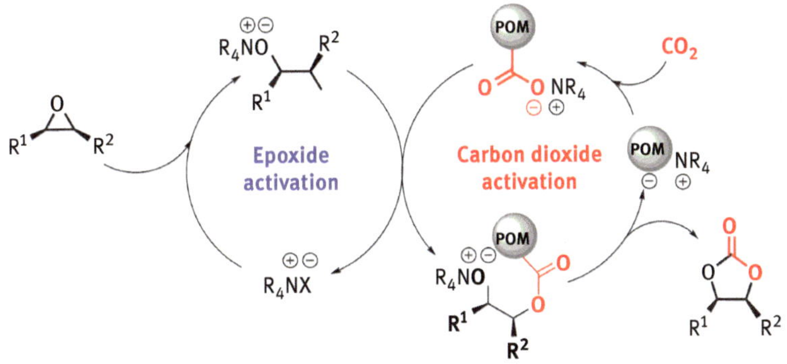

Figure 16.5: Synergistic pathway towards the formation of fatty acid-based cyclic carbonate.

In this manifold, TBAB acts as a nucleophile mediating the ring opening of the epoxide, and consecutively transfers the resulting alkoxide to the POM-activated CO_2 improving overall kinetics. Although a broad range of mono-epoxy oleochemical derivatives could be carboxylated, the POM-based binary catalyst was not chemoselective for poly-epoxidised compounds as these easily underwent side reactions. These side reactions could at a later stage be minimised while allowing for virtual complete stereospecific conversions through the use of an Al(III) aminotriphenolate/PPNCl (PPN = bis(triphenylphosphine)iminium) binary catalyst, which to date provides the only example of complete stereocontrol for the synthesis of mono-, di- and even tri-cyclic carbonate-based fatty acids [42].

A metal-free approach for the enhancement of a halide catalyst was studied by Tassaing and coworkers [35], and epoxidised linseed oil (cf. ESBO, Figure 16.4) was transformed into its tricarbonate using various nucleophiles combined with a hydrogen-bond activator (i.e. an organocatalytic binary catalyst). A screening was carried out to identify the most suitable halide salt or ionic liquid combined with commercially available (multi)phenolic compounds or fluorinated alcohols to activate the oxirane units via OH···O hydrogen bond (HB) interactions. 1,3-Bis (2-hydroxyhexafluoroisopropyl)benzene was identified as the most effective HB activator with about a twofold increase in epoxide conversion compared to the same reaction in the absence of the H-bond donor. Despite the fact that this organocatalytic approach indeed proved to be feasible, the reaction conditions needed to convert the internal epoxide fragments into their respective cyclic carbonates still remained rather harsh (100 °C, 100 bar).

Further improvement of this HB approach was reported by Werner et al. [24], applying a screening that focused on a detailed structure–activity relationship using one-component bifunctional phosphorus-based organocatalysts incorporating both a phosphonium salt and phenolic moiety (Figure 16.6). The phenolic phosphonium salts indeed have high efficiencies, and the most efficient example

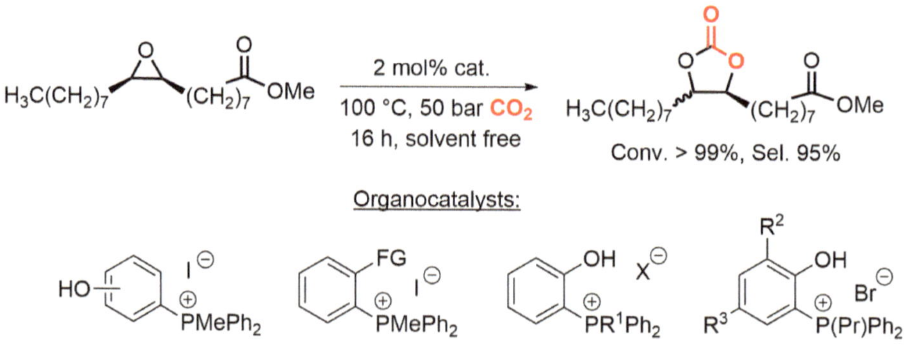

Figure 16.6: Screening conditions applied for the carboxylation of methyl oleate applying bifunctional phosphorus-based organocatalysts.

showed favourable stability towards air and moisture. Excellent conversions were obtained for a range of fatty acid methyl ester carboxylations under comparatively mild conditions (80 °C, 25 bar) in a solvent and metal-free medium.

Apart from organohalides, simple alkaline- and alkaline-earth metal halides have proven to be efficient catalysts for the conversion of oleochemicals into their carbonate derivatives following a phase transfer catalysis approach in order to achieve sufficient solubility [10, 38]. Rokicki and co-workers converted ESBO (Figure 16.4) in the presence of KI/18-crown-6 at 60 bar CO$_2$ pressure and 130 °C [43]. This procedure was also effective for EFAMEs conversion by applying a variety of salts assisted by crown ethers under slightly different reaction conditions resulting in high conversions and good selectivities, with further addition of glycols being required to improve the economics of the catalyst systems.

Other recently reported crown-ether-inspired catalysts include a system based on dicyclohexyl-functionalised 18-crown-6 ether and calcium iodide that is able to mediate the conversion of fatty acid esters and oils into their carbonate derivatives under mild reaction conditions (60 °C, 20 bar) [11]. The addition of triphenylphosphine significantly improved the catalyst activity and showed comparable performance as additive compared to DBU and 1,5,7-triazabicyclo[4.4.0] dec-5-ene (TBD) but was less expensive and easier to use. Various (transition) metal complexes/salts in combination with phosphonium salts as catalytic systems for the carboxylation of oleochemical derivatives have been recently developed [24, 39]. Similar studies have been performed on Fe catalysts as a sustainable alternative with Fe being an abundant and inexpensive metal with low toxicity. These Fe-based catalysts [39] displayed similar catalytic features in terms of structure/reactivity relationships and levels of stereospecificity as their Mo analogues [44].

16.2.4 Cyclic carbonates derived from other biobased compounds

Apart from terpenes, sugars and fatty acids, there are some other feedstock that may serve as alternatives for fossil fuel-based cyclic carbonates, including glycerol. Glycerol is a polyol compound industrially derived from triglycerides via saponification, hydrolysis and transesterification and is available on a large scale. Its main applications are in the food industry as a sweetener and in pharmaceutical formulations. Glycerol has been investigated as a green substrate for the synthesis of glycerol carbonate (GC), which is of considerable interest towards greener polyurethanes or solvents, cosmetic products, eco-composites and as electrolytes for lithium batteries. Recently, different catalytic systems have been developed for the synthesis of GC or diglycerol dicarbonate (DGDC). The majority of the catalytic systems employ carbonylating reagents such as CO$_2$, urea, diethyl carbonate or dimethyl carbonate (DMC).

For instance, homogeneous and heterogeneous CO_2-masked *N*-heterocyclic carbene-based organocatalysts were shown to be very efficient towards the formation of DGDC in the presence of DMC under mild and solvent-free conditions [45]. This catalyst system also tolerated the use of crude, industrial glycerol containing impurities such as soaps, water and esters though the catalyst activity was generally lower requiring longer reaction times. Other biobased diols containing 1,2-, 2,3- and 1,3-diol fragments also proved to be feasible substrates yielding five- and six-membered cyclic carbonate products, though 1,3-diols were significantly more reluctant to undergo selective cyclisation.

Three different glycidyl esters of glycerol, pentaerythritol and trimethylolpropane were converted into their corresponding cyclic carbonates GGC, PEC and TMC by direct conversion with CO_2 and TBAB as catalyst at 120 °C and 30 bar. The reaction showed full conversion (96%) of the epoxy moieties resulting in a carbonate content of around 31 wt% and 22 wt% of "fixed CO_2" [46].

Direct use of carbon dioxide to produce cyclic carbonates from (biobased) diols has also been reported, and can be achieved via a carboxylation/hydration cascade catalyst composed of cerium oxide (CeO_2) and 2-cyanopyridine in the synthesis of DMC from methanol and CO_2 [47]. Various 1,2-diols and 1,3-diols can be converted into five-membered and six-membered cyclic carbonates in high yields and selectivities but require the presence of the dehydrating agent 2-cyanopyridine. Recent work has also demonstrated that bioderived 1,2- and 1,3-diols can be converted into their cyclic carbonates using a combination of NEt_3, TsCl and DBU in a stoichiometric approach [48].

Lignin is an amorphous cross-linked polymer that gives structural integrity to plants, making up 25–35% of woody biomass. Lignin is among the most abundant feedstock for aromatic and aliphatic compounds bearing hydroxyl, methoxyl, carboxyl and carbonyl moieties, and offers access to monomers for cyclic carbonate synthesis via modification of aromatic lignin components to prepare "green" and stiff NIPUs. Several examples of carboxylated lignin derivatives have been reported and investigated as potentially green monomers. The presence of cyclic carbonates in these phenolic and aromatic lignin substrates has opened up a new sustainable branch in the field of the synthesis of polyesters and polyurethanes.

In a typical example, two technical lignin derivatives named soda pulping herbaceous lignin and a hardwood kraft lignin were converted into cyclic carbonates via a binary-type catalyst comprising 1-allyl-3-methylimidazolium bromide and $CoCl_2$ [49]. In particular, the involvement of both catalyst components is believed to promote the reaction by dual activation of both the epoxide and CO_2. The reaction was studied in detail converting epoxidised model compounds (guaiacol; 2,6-dimethoxyphenol) and vanillic acid at 20 bar and 80 °C. Other simple and inexpensive binary halide-derived catalysts have been reported for the efficient conversion of the most abundant lignin-derived monomers vanillin and creosol, either using CO$_2$ or DMC as a carbonylating agent [50, 51].

16.3 Polycarbonates derived from renewable compounds

The copolymerisation of CO$_2$ has been studied for a large variety of epoxides, and many of which are discussed in Chapter 15. This section discusses the use of epoxides that can be obtained from renewable sources and is subdivided into the origin of the epoxide monomers. Sustainable polymers from renewable resources [52], including renewable epoxides [53], have been recently reviewed. Poland and Darensbourg categorise renewably sourced epoxides as being derived from a "metabolite" or a direct "precursor" [53]. Epoxides derived from direct precursors include those of pinene and limonene (Figure 16.1). A "metabolite"-derived epoxide is one where the epoxide is generated via multiple reactions from the product of biological metabolism, and examples of such epoxides are given in Figure 16.7. An

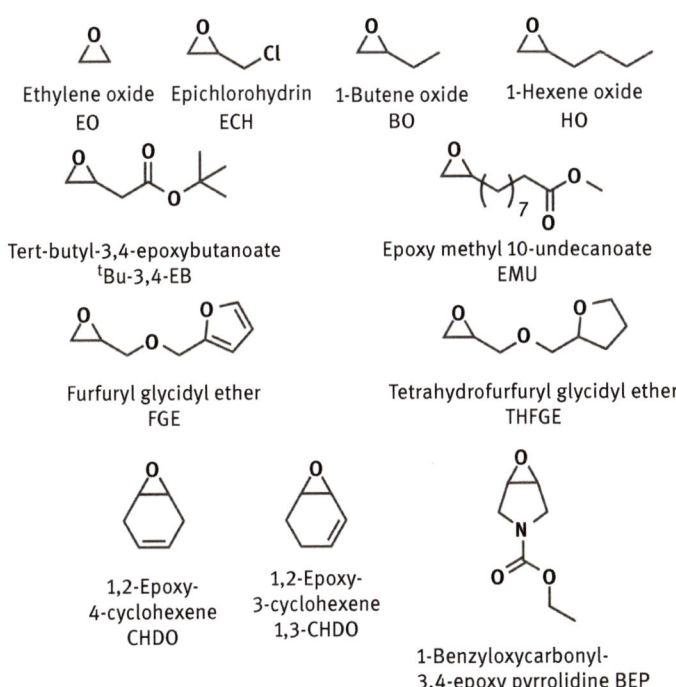

Figure 16.7: Epoxides derived from renewable metabolites.

example is ethanol, which can be metabolically generated from renewable starting materials but must be dehydrated to give ethylene, which is then epoxidised to ethylene oxide. The number of steps required and their efficiency influence the overall sustainability of the resulting epoxide monomer. A precursor-derived epoxide is one whereby the olefin precursor (the compound that is converted into the epoxide) can be directly extracted from the renewable raw material, such as limonene and pinene. Thus, the olefin can be converted to an epoxide theoretically via a single, efficient step. Of course, this step too can vary greatly in terms of its "greenness" depending on the oxidising agent (e.g. mCPBA compared to H_2O_2).

16.3.1 From alkylene oxide monomers

There are numerous alkylene oxides (epoxides of olefinic precursors) that can be sourced from renewable metabolites. Ethylene oxide, although primarily derived from ethylene obtained from petroleum sources, can be obtained from bioethylene obtained from bioethanol, which is produced globally at ca. 100 billion litres per year. Bioethylene can be obtained via catalytic dehydration and oxidised to form ethylene oxide. The source of the bioethanol must be considered, however, as sugarcane and corn are common biomass feedstocks that compete with food production. Lignocellulosic biomass, however, offers an alternative using waste products of forestry and agriculture.

Ethylene oxide is extremely toxic and flammable and this has limited its academic study compared to propylene oxide (PO). Some examples of its use in copolymerisation reactions with CO_2 to give poly(ethylene carbonate) have, however, been reported. Yamada and co-workers described di(ketoiminato) cobalt complexes (Figure 16.8) for the copolymerisation of CO_2 with ethylene oxide in the presence of PPNCl to provide completely alternating poly(ethylene carbonate) [54]. Copolymerisation of 1-butene oxide (BO) and 1-hexene oxide (HO) (which can be potentially metabolite derived) with CO_2 could be achieved using bifunctional salenCo(III) compounds. Molar masses over 30,000 g mol^{-1} could be obtained with narrow dispersities by Nozaki and co-workers using a piperidinium arm containing complex (Figure 16.8) [55]. Lee and co-workers reported a related salenCo(III) catalyst possessing four quaternary ammonium salt groups, which exhibits very high activity leading to polycarbonates with high molar masses over 200,000 g mol^{-1} (Figure 16.8) [56]. This catalyst also yielded terpolymers of BO and HO with PO.

Tert-butyl-3,4-epoxybutanoate (^{t}Bu-3,4-EB) can be obtained from (S)-3,4-dihydroxybutyric acid (a normal human urinary metabolite) via protection of the acid functionality and conversion of the 3,4-diol to an epoxide. Copolymerisation of ^{t}Bu-3,4-EB with CO_2 could be achieved by the group of Darensbourg using a salenCo(III) catalyst with an ionic dicyclohexylmethylpropyl ammonium group (Figure 16.9) [57]. The copolymer showed 100% head-to-tail regioselectivity with

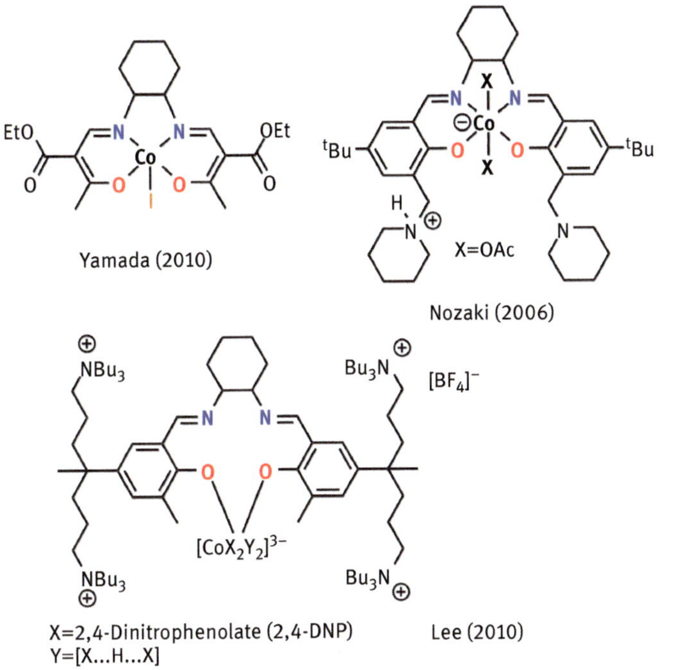

Figure 16.8: Catalysts employed for ethylene oxide/CO$_2$ copolymerisation.

ring opening of the epoxide occurring at the least substituted methylene carbon of the epoxide. Deprotection of the *tert*-butyl groups can be performed leading to carboxylic acid groups, which can be used for coupling with amino acids such as aspartic acid to give water-soluble and biodegradable polycarbonates.

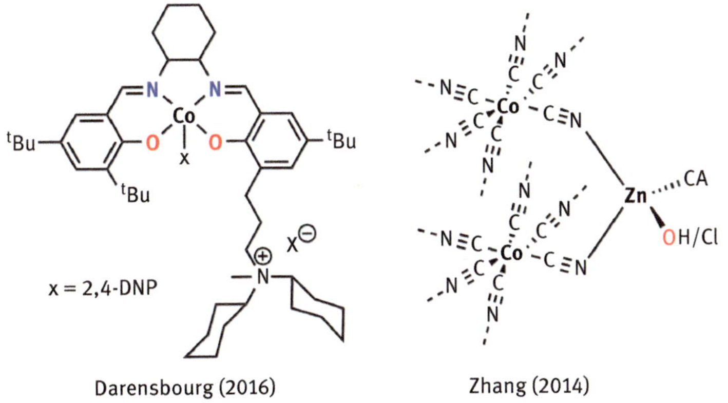

Figure 16.9: Catalysts capable of *tert*-butyl-3,4-epoxybutanoate and 2-oxoundecanoic acid methyl ester copolymerisation with CO$_2$.

2-Oxoundecanoic acid methyl ester, also known as epoxy methyl 10-undecanoate (EMU), can be sourced from ricinoleic acid, which is in turn a major component of castor oil. The copolymerisation of EMU with CO_2 using heterogeneous Zn-Co(III) double metal cyanide (DMC) catalyst (Figure 16.9) gives polycarbonate with >99% CO_2 incorporation [58]. The glass transition temperatures, T_g, of the resulting polymer were very low, between −38 and −44 °C, and molar masses between 3700 and 18,600 g mol^{-1} with Đ between 2.1 and 2.9. The content of ether linkages increased when reactions were performed above 80 °C.

Epichlorohydrin, while commonly produced from petrochemically sourced propene, can also be obtained from glycerol, which is in turn a byproduct of biodiesel synthesis. Perfectly alternating poly(chloropropylene carbonate) (PCPC) can be obtained at low temperatures using binary or bifunctional cobalt salen complexes with TBD anchored to the ligand framework, or bearing an appended quaternary ammonium salt (Figure 16.10) [59]. Regioregular, isotactic PCPC can be obtained using a bulky adamantyl group containing bifunctional catalyst [60]. Regioirregular PCPC (commonly abbreviated to PECHC) is amorphous with a T_g of approximately 30 °C, whereas isotactic poly(R-chloropropylene carbonate) possesses a T_g of 42 °C and a melting point of 108 °C.

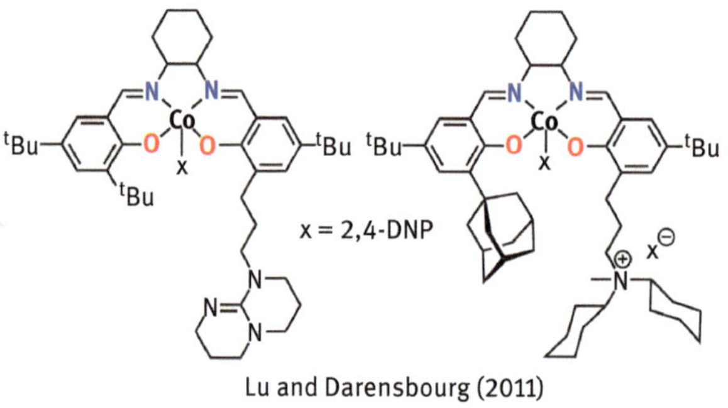

Lu and Darensbourg (2011)

Figure 16.10: Bifunctional salenCo(III) catalysts used in poly(epichlorohydrin carbonate) formation.

Furfuryl glycidyl ether (FGE) can be obtained from an etherification reaction of epichlorohydrin with furfuryl alcohol. Furfuryl alcohol can be obtained from the reduction of furfural, which is a bioplatform molecule obtained from hemicellulose-rich feedstock. Copolymerisation of FGE and CO_2 can be achieved using a ternary yttrium(III) trichloroacetate/diethylzinc/glycerine catalyst system [61], or a binary salenCo(III) catalyst with PPNCl (Figure 16.11) [62]. The unsaturated furan group in the polycarbonate leads to instability to oxidation and cross-linking, resulting in discoloration. The polymer can be stabilised, however, by addition of antioxidants or by Diels–Alder reaction with N-phenylmaleimide. Alternatively, the saturated

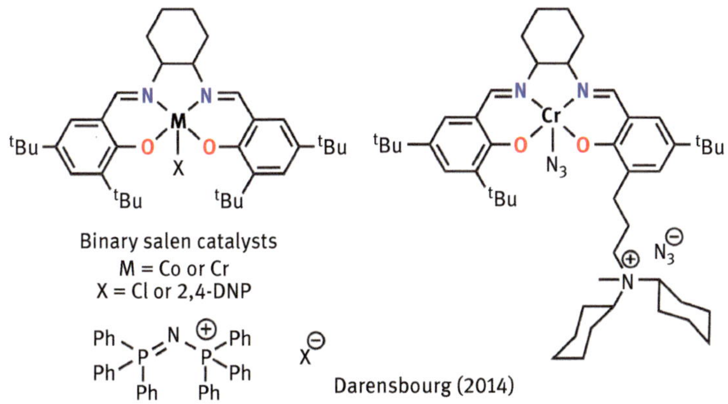

Figure 16.11: Binary cobalt and chromium and bifunctional chromium catalysts used for CO$_2$/1,2-epoxy-4-cyclohexene copolymerisation.

analogue tetrahydrofurfuryl glycidyl ether can be copolymerised with CO$_2$ to give a polycarbonate that is more stable to discoloration.

CHD is a common by-product of oleochemical olefin metathesis reactions of, for example, unsaturated fatty acids derived from plant oils. Epoxidation of CHD by *m*CPBA or oxone (potassium peroxymonosulfate) results in 1,2-epoxy-4-cyclohexene, CHDO. Hydrogenation of CHDO gives cyclohexene oxide (CHO), which is the predominant epoxide monomer used in CO$_2$ coupling and copolymerisation reactions. The use of CHO for these reactions is described in Chapter 17 but CHDO/CO$_2$ copolymerisation will be discussed here. There are several examples of catalysts that give poly(cyclohexadiene carbonate) (PCHDC) from CHDO and CO$_2$ [8, 63, 64], of which the best activities are shown by binary salenCoX/PPNX systems (where X = 2,4-dinitrophenolate or chloride, Figure 16.11) [65]. At temperatures of up to 40 °C and pressures of 20 bar, polymer selectivity was effectively quantitative, giving narrow dispersity polymers ($Đ$ = 1.10 – 1.30) of moderately high molecular weight. Binary salenCrN$_3$/PPNN$_3$ and a bifunctional salenCrN$_3$ system having an ionic dicyclohexylmethylpropyl ammonium azide group (Figure 16.11) show lower activities and temperature-dependant polymer selectivities [65].

Terpolymers of CHDO, CHO and CO$_2$ can be formed using a salenCoCl/PPNCl catalyst system [8]. CHO incorporation proceeded at a faster rate than CHDO with respective conversions of 72% and 22% into polycarbonate. The resulting copolymer had a molar mass of 11,500 g mol^{-1} and $Đ$ = 1.12. The double bond in PCHDC or its terpolymers allows further post-polymerisation modifications, such as AIBN-initiated alkene hydrothiolation (AIBN = Azobisisobutyronitrile)[63]. The functionalisation of PCHDC to give, for example, polar tags (and therefore hydrophilic polycarbonate), bromination of the alkene site or incorporation of other monomers has a significant effect on T_g values, which range from 36 to 128 °C [53]. Isomerisation of 1,4-cyclohexadiene to

1,3-cyclohexadiene is possible, and its epoxidation gives 1,3-cyclohexadiene oxide. Copolymerisation of this epoxide with CO_2 gives poly(1,3-cyclohexadiene carbonate), which under identical conditions proceeds much faster than for 1,4-CHDO [63]. The T_g of the 1,3-polycarbonate was reported as 104–108 °C, which is slightly lower than the 1,4-copolymer.

16.3.2 From terpene-based monomers

Limonene oxide is the predominant terpene-sourced epoxide from which copolymerisation with CO_2 has been achieved. The first report of polycarbonate formation from limonene oxide and CO_2 was from the group of Coates in 2004 using a β-diiminate (BDI) zinc acetate catalyst [66]. The [(BDI)ZnOAc] catalyst (Figure 16.12) employs mild conditions to produce poly(limonene carbonate) (PLC) at 25 °C and 6.9 bar CO_2 pressure. Catalytic activity of 32 TO h^{-1} resulted in polymer M_n = 9300 g mol^{-1} with Đ = 1.13 and no evidence of ether linkages. PLC formation is very sensitive to the presence of chain transfer agents (CTA), such as water or hydroxyl-functionalised molecules. Even in small amounts, CTAs cause the formation of low molar mass polymers; therefore, rigorous purification of LO to remove trace hydroxyl-containing compounds is needed to obtain high molar masses. Greiner showed that PLC can be obtained on kilogram scales and with molar masses over 100,000 g mol^{-1} using [(BDI)ZnOAc] if the hydroxyl-containing impurities are O-methylated [67]. Activities can be improved by employing electron-withdrawing CF_3 groups on the BDI backbone (up to 310 TO h^{-1}) [68], and replacing the acetate ligand with hydrolysable $N(SiMe_3)_2$ groups allows for hydroxyl-terminated polycarbonate diols to be obtained [69].

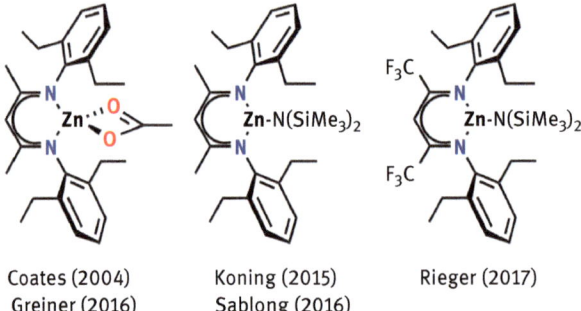

Coates (2004)   Koning (2015)   Rieger (2017)
Greiner (2016)   Sablong (2016)

Figure 16.12: Zinc BDI complexes used in limonene oxide/CO_2 copolymerisation.

A report by Greiner and co-workers describes the many possible modifications of PLC prepared by BDI zinc catalysts [70]. For example, hydrophilicity of the PLC can be improved by adding polar side chains. In addition to LO, the diepoxide LDO has been copolymerised using [(BDI)ZnN(SiMe$_3$)$_2$] to give poly(limonene-8,9-oxide carbonate) (PLOC), possessing epoxide functionalities that can be post-synthetically modified [71, 72]. Examples of modification include reactions with thiols, carboxylic acids or CO_2, which were found to influence glass transition temperatures, T_g. As discussed earlier, LDO can be coupled with CO_2 to give LDC, which can be used to synthesise NIPU by reaction with diamines [73].

The aminotriphenolate aluminium complex of Kleij (Figure 16.1 with R = H or Me) in the presence of added halide nucleophile from PPNX (X = Cl or Br) copolymerised LO and CO_2 [74]. Polymer molar masses of up to 10,600 g mol^{-1} and moderate control of dispersity ($Đ$ = 1.43) were obtained at 10 bar CO_2 pressure. Interestingly, this catalyst system was found to be more selective for reaction with *cis*-(R)-LO. This catalyst was subsequently used to prepare terpolymers of LO, CHO and CO_2 [75]. The degree of LO incorporation and thus the number of olefin units in the polymer could be controlled. Post-modification through thiol-ene chemistry provided cross-linked materials with improved thermal properties exhibiting higher T_g values than their terpolymer precursors (Figure 16.13). Modulation of glass

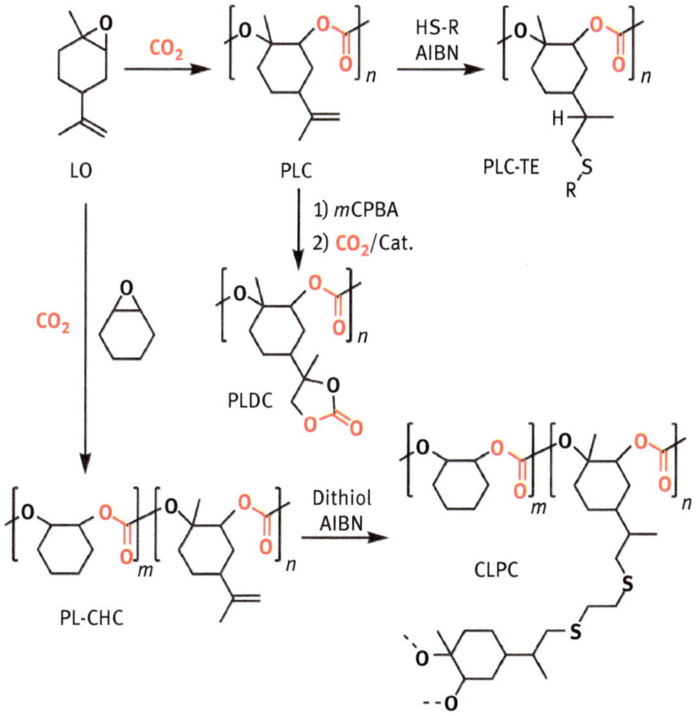

Figure 16.13: Synthesis of PLC, PLDC, PL-CHC terpolymer and cross-linked PLC polymers.

transition temperatures in limonene-derived polycarbonates was achieved by epoxidation of the alkene units using mCPBA to give PLOC and subsequent reaction with CO_2 to give poly(limonene dicarbonate)s (PLDC) (Figure 16.13) [76]. PLOC exhibits a T_g of 135 °C, slightly higher than that of PLC, and PLDC has an even higher T_g of 180 °C.

The decomposition pathways of PLC have been studied. Products obtained from thermal decomposition include limonene, limonene oxide, and alcohols and diols obtained from LO [68, 69]. Unlike other poly(alkylene carbonate)s that exhibit backbiting pathways to form cyclic carbonates (see Chapter 17), cyclic limonene carbonate is typically not observed from depolymerisation of PLC unless maleic anhydride end groups are used. The unique depolymerisation of PLC was proposed to be due to PLC's tetrasubstituted cyclohexane rings [77]. Due to steric hindrance and conformationally inaccessible intermediates required for carbonate cyclisation, carbon dioxide release and epoxide reformation is preferred [78]. An equilibrium exists between PLC and LO/CO_2 monomers, which is temperature dependent [68].

The physical properties of PLC vary depending on the report, and enantiopure PLC is amorphous. Co-precipitation of poly(R-limonene carbonate) and poly (S-limonene carbonate) yields a crystalline racemic stereocomplex [79]. Glass transition temperatures are wide ranging depending on the polymer molar masses and dispersities, with T_g = 70–130 °C and decomposition temperatures, T_d, as low as 180 °C [53]. Although the constituent monomers of PLC are biologically sourced, PLC has shown some resistance to biodegradation. It is believed that due to its hydrophobicity, PLC displays no evidence of degradation in active compost, unlike poly(L-lactide), which disintegrated within 1 week under these conditions [70]. It was thought that increasing the hydrophilicity as discussed earlier would lead to better degradability, but results to date were unsuccessful.

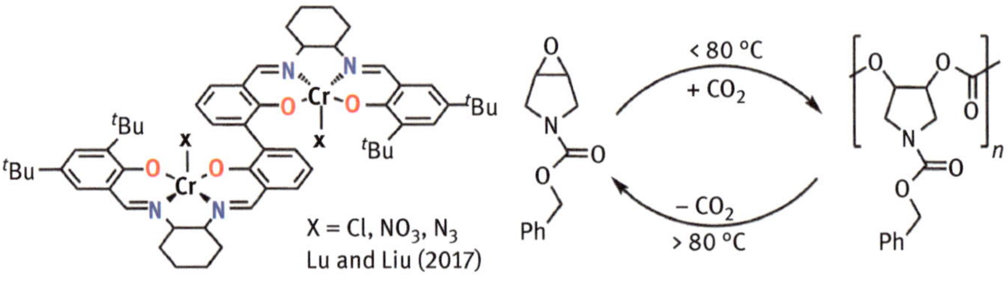

Figure 16.14: Dinuclear salenCrX complexes for temperature-dependent BEP/CO_2 copolymerisation and depolymerisation.

Liu, Lu and co-workers described the copolymerisation of 1-benzyloxycarbonyl-3,4-epoxy pyrrolidine (BEP) and CO_2 to give a new polycarbonate using dinuclear salenCrX complexes in the presence of PPNF (Figure 16.14).[80] BEP can be produced from furfural via a multistep process, and thus can be considered to be a renewable epoxide. Once copolymerised with CO_2, the polycarbonate can completely depolymerise to its monomeric form when heated above 100 °C. These polycarbonates possessed molecular weights ranging from 2.7 to 10.2 kg mol^{-1} with moderate dispersities averaging $Đ$ = 1.32.

16.3.3 From lipids

The cyclic carbonates derived from fatty acids and lipids described earlier are useful synthons for the preparation of NIPUs. CSBO was shown to easily react with di- or tri-primary amines to give the corresponding NIPUs [29]. Reaction between CSBO and n-butylamine, 1,2-ethylene diamine (EDA), 1,6-hexamethylene diamine (HMD) and tris(2-aminoethyl)amine (TA) at room temperature to 60 °C showed the effective ring opening of five-membered cyclic carbonate moieties in the triglyceride molecules by the amine to form β-hydroxyurethane systems. Samples made from triamine TA showed the highest T_g of 43 °C. In contrast, samples containing HMD exhibited the lowest T_g (ca. 18 °C). NIPUs prepared at room temperature showed no evidence of amide formation by reaction at the ester functionality. A subsequent report investigated the reaction of ED, HMD and 1,4-butane diamine (BDA) at curing temperatures of 70 to 100 °C [32]. Amide formation and ester group cleavage was observed in all samples prepared at these temperatures. Furthermore, stoichiometric carbonate-to-amine ratios were found to give the highest T_g values, hardness and tensile strengths. T_g variations in samples prepared from the three diamines were not large, but hardness and tensile strength were highest for those with ED and lowest for HMD.

Internal and terminal carbonated fatty acid diesters obtained from the reaction of CO_2 with their epoxide precursors were reacted with diamines to give polyurethanes of molar masses up to 13,500 g mol^{-1} [34]. These polyurethanes had T_g values of −15 °C. Side reactions may occur at the ester functionality giving amide linkages when EDA is used. Amide linkages were not observed, however, when isophorone diamine (IPDA, a secondary diamine) was used as the comonomer.

Linseed and soybean oil-based polyurethanes could be prepared from their carbonated oils and different amines [23]. The linseed oil carbonates with 20.2 to 26.8 wt% carbonate content were cured with EDA, BDA and IPDA. Using IPDA for curing carbonated linseed oil, the T_g increased from 20 to 60 °C compared to EDA. The stiffness of IPDA-cured polyurethane, as reflected by the Young's modulus, increased three orders of magnitude at the expense of elongation at break. Most conventional CSBO-based polymers are soft and flexible; thus, this approach indicates that NIPUs can be developed with higher dimensional stability and stiffness to meet the demands of engineering applications.

Fatty acid-based bis-cyclic five-membered carbonates containing amide linkages were prepared from methyl 10-undecenoate. The reaction of these biobased carbonates with BDA, IPDA, Priamine 1075 and Jeffamine 400 (a polyether amine) produced poly(hydroxyurethane amide)s with molar masses up to 31,000 g mol^{-1}. These PHUs exhibited amorphous to semicrystalline features depending on the monomers used [28].

Biobased polyurethanes were synthesised from cyclocarbonated broccoli seed oil and different di- or triamines [33]. The broccoli seed oil was first epoxidised, then carbonated with CO_2 at 50 bar to form cyclic carbonates. Polyurethanes were prepared through reaction of the cyclocarbonated oil with BDA, *m*-xylene diamine and bis(hexamethylene triamine). The T_g of the different polyurethanes depended on the amine used and the amine-to-carbonate ratios, but were relatively similar ranging between 15 and 23 °C.

16.4 Applications of biobased polycarbonates

For a process to be considered sustainable, several criteria must be considered. Societal, ecological and economic factors must be addressed for a system/process to be truly sustainable. In terms of coupling and copolymerisation reactions of CO_2 and epoxides, these reactions continue to advance in terms of high catalytic activity, product selectivity, regio- and stereoselectivity, diversity in terms of epoxide substrate and tolerance to use impure sources of CO_2 as well as lower pressures of CO_2.

Cyclic carbonates are finding uses in electrolyte mixtures for lithium ion batteries and synthesis of fine chemicals and polymers. Polycarbonate polyols have found application in polyurethane synthesis, and cyclic carbonates can be used as reagents in non-isocyanate polyurethane synthesis. The technology has led to the birth of new companies specialising in exploiting this chemistry, as well as considerable interest from multinational corporations as part of their R&D portfolios. Life cycle analyses comparing polycarbonate polyols with the traditional polyether polyols used in polyurethane production have been performed [81]. While perfectly alternating polycarbonate polyols can contain up to 50 wt% CO_2, their production still generates CO_2 (2.65–2.86 kg CO_2 per 1.0 kg of polycarbonate), hence they cannot be considered entirely carbon neutral. However, if polyol production is required, the utilisation of CO_2 creates significant impact reductions compared to the synthesis of conventional polyether polyols.

The excellent material properties of PLC have led to much excitement in its development. While it has a promising life cycle assessment, it suffers from potential supply chain issues, given the volatility of limonene availability arising from global citrus production. PCHDC, on the other hand, may have greater potential if its synthesis from waste fatty acids and plant oils can be further improved, as it has a CO_2 incorporation ability and can be functionalised to address task-specific requirements.

16.5 Conclusions and outlook

The promise of cyclic carbonates and polycarbonates from renewables will require much development in terms of not only their application-driven properties and efficiency of synthesis, but also in terms of production of their sustainable feedstock. Better use must be made of waste products of agriculture, forestry and other industries to produce suitable monomers/reagents for reaction with CO_2. At the other end of the spectrum, the biodegradability of carbonate materials, especially polymers, is a concern. Some evidence for degradation of the most industrially developed CO_2-derived polymer, poly(propylene carbonate), in compost has been reported, whereas PLC has shown resistance to biodegradation. Improving hydrophilicity may allow improved hydrolysis of the carbonate unit.

References

[1]   Liu Q, Wu L, Jackstell R, Beller M. "Using carbon dioxide as a building block in organic synthesis". Nat. Commun. 2015, 6, 5933.
[2]   Klankermayer J, Wesselbaum S, Beydoun K, Leitner W. "Selective catalytic synthesis using the combination of carbon dioxide and hydrogen: catalytic chess at the interface of energy and chemistry". Angew. Chem. Int. Ed. 2016, 55, 7296–343.
[3]   Comerford JW, Ingram IDV, North M, Wu X. "Sustainable metal-based catalysts for the synthesis of cyclic carbonates containing five-membered rings". Green Chem. 2015, 17, 1966–87.
[4]   Parrino F, Fidalgo A, Palmisano L, Ilharco LM, Pagliaro M, Ciriminna R. "Polymers of limonene oxide and carbon dioxide: polycarbonates of the solar economy". ACS Omega. 2018, 3, 4884–90.
[5]   Bähr M, Bitto A, Mulhaupt R. "Cyclic limonene dicarbonate as a new monomer for non-isocyanate oligo- and polyurethanes (NIPU) based upon terpenes". Green Chem. 2012, 14, 1447–54.
[6]   Fiorani G, Stuck M, Martin C, et al. "Catalytic coupling of carbon dioxide with terpene scaffolds: access to challenging bio-based organic carbonates". ChemSusChem 2016, 9, 1304–11.
[7]   Martinez J, Fernandez-Baeza J, Sanchez-Barba LF, Castro-Osma JA, Lara-Sanchez A, Otero A. "An efficient and versatile lanthanum heteroscorpionate catalyst for carbon dioxide fixation into cyclic carbonates". ChemSusChem 2017, 10, 2886–90.
[8]   Winkler M, Romain C, Meier MAR, Williams CK. "Renewable polycarbonates and polyesters from 1,4-cyclohexadiene". Green Chem. 2015, 17, 300–6.
[9]   Laserna V, Fiorani G, Whiteoak CJ, Martin E, Escudero-Adan E, Kleij AW. "Carbon dioxide as a protecting group: highly efficient and selective catalytic access to cyclic cis-diol scaffolds". Angew Chem. Int. Ed. 2014, 53, 10416–9.
[10] Steinbauer J, Werner T. "Poly(ethylene glycol)s as ligands in calcium-catalyzed cyclic carbonate synthesis". ChemSusChem 2017, 10, 3025–9.
[11] Longwitz L, Steinhauer J, Spannenberg A, Werner T. "Calcium-based catalytic system for the synthesis of bio-derived cyclic carbonates under mild conditions". ACS Catal. 2018, 8, 665–72.

[12] Gregory GL, Jenisch LM, Charles B, Kociok-Kohn G, Buchard A. "Polymers from sugars and CO_2: synthesis and polymerization of a d-mannose-based cyclic carbonate". Macromolecules 2016, 49, 7165–9.

[13] Gregory GL, Kociok-Kohn G, Buchard A. "Polymers from sugars and CO_2: ring-opening polymerisation and copolymerisation of cyclic carbonates derived from 2-deoxy-D-ribose". Polym. Chem. 2017, 8, 2093–104.

[14] Gregory GL, Lopez-Vidal EM, Buchard A. "Polymers from sugars: cyclic monomer synthesis, ring-opening polymerisation, material properties and applications". Chem. Commun. 2017, 53, 2198–217.

[15] Gregory GL, Hierons EM, Kociok-Kohn G, Sharma RI, Buchard A. "CO_2-Driven stereochemical inversion of sugars to create thymidine-based polycarbonates by ring-opening polymerisation". Polym. Chem. 2017, 8, 1714–21.

[16] Haba O, Tomizuka H, Endo T. "Anionic ring-opening polymerization of methyl 4,6-O-benzylidene-2,3-O-carbonyl-alpha-D-glucopyranoside: A first example of anionic ring-opening polymerization of five-membered cyclic carbonate without elimination of CO_2". Macromolecules 2005, 38, 3562–3.

[17] Kumar R, Gao W, Gross RA. "Functionalized polylactides: Preparation and characterization of L-lactide-co-pentofuranose". Macromolecules 2002, 35, 6835–44.

[18] Mikami K, Lonnecker AT, Gustafson TP, et al. "Polycarbonates derived from glucose via an organocatalytic approach". J. Am. Chem. Soc. 2013, 135, 6826–9.

[19] Shen YQ, Chen XH, Gross RA. "Polycarbonates from sugars: Ring-opening polymerization of 1,2-O-isopropylidene-D-xylofuranose-3,5-cyclic carbonate (IPXTC)". Macromolecules 1999, 32, 2799–802.

[20] Gobin M, Loulergue P, Audic JL, Lemiegre L. "Synthesis and characterisation of bio-based polyester materials from vegetable oil and short to long chain dicarboxylic acids". Ind. Crops. Prod. 2015, 70, 213–20.

[21] Liu ZS, Shah SN, Evangelista RL, Isbell TA. "Polymerization of euphorbia oil with Lewis acid in carbon dioxide media". Ind. Crops. Prod. 2013, 41, 10–6.

[22] Wang Z, Zhang X, Wang RG, et al. "Synthesis and characterization of novel soybean-oil-based elastomers with favorable processability and tunable properties". Macromolecules 2012, 45, 9010–9.

[23] Bähr M, Mülhaupt R. Linseed and soybean oil-based polyurethanes prepared via the non-isocyanate route and catalytic carbon dioxide conversion. Green Chem 2012, 14, 483–9.

[24] Büttner H, Steinbauer J, Wulf C, Dindaroglu M, Schmalz HG, Werner T. "Organocatalyzed synthesis of oleochemical carbonates from CO_2 and renewables". ChemSusChem 2017, 10, 1076–9.

[25] Dierker M. "Oleochemical carbonates-an overview". Lipid Technol. 2004, 16, 130–3.

[26] Kenar JA. "Current perspectives on oleochemical carbonates". Inform 2004, 15, 580–2.

[27] Kenar JA, Knothe G, Dunn RO, Ryan TW, Matheaus A. "Physical properties of oleochemical carbonates". J. Am. Oil. Chem. Soc. 2005, 82, 201–5.

[28] Maisonneuve L, More AS, Foltran S, et al. "Novel green fatty acid-based bis-cyclic carbonates for the synthesis of isocyanate-free poly(hydroxyurethane amide)s". RSC Adv. 2014, 4, 25795–803.

[29] Tamami B, Sohn S, Wilkes GL. "Incorporation of carbon dioxide into soybean oil and subsequent preparation and studies of nonisocyanate polyurethane networks". J. Appl. Polym. Sci. 2004, 92, 883–91.

[30] Riedeman WL, inventor. Rohm & Haas Co., assignee. Carbonato esters of fatty acids for use as plasticizers for vinyl resins patent US2858286. 1958.

[31] Bender AD, Berkoff CE, Groves WG, et al. "Synthesis and biological properties of some novel heterocyclic homoprostanoids". J. Med. Chem. 1975, 18, 1094–8.

[32] Javni I, Hong DP, Petrović ZS. "Soy-based polyurethanes by nonisocyanate route". J. Appl. Polym. Sci. 2008, 108, 3867–75.

[33] Loulergue P, Amela-Cortes M, Cordier S, Molard Y, Lemiegre L, Audic JL. "Polyurethanes prepared from cyclocarbonated broccoli seed oil (PUcc): New biobased organic matrices for incorporation of phosphorescent metal nanocluster". J. Appl. Polym. Sci. 2017, 134. J. Appl. Polym. Sci. 2017, 134, 45339.

[34] Boyer A, Cloutet E, Tassaing T, Gadenne B, Alfos C, Cramail H. "Solubility in CO$_2$ and carbonation studies of epoxidized fatty acid diesters: towards novel precursors for polyurethane synthesis". Green Chem. 2010, 12, 2205–13.

[35] Alves M, Grignard B, Gennen S, Detrembleur C, Jerome C, Tassaing T. "Organocatalytic synthesis of bio-based cyclic carbonates from CO$_2$ and vegetable oils". RSC Adv. 2015, 5, 53629–36.

[36] Langanke J, Greiner L, Leitner W. "Substrate dependent synergetic and antagonistic interaction of ammonium halide and polyoxometalate catalysts in the synthesis of cyclic carbonates from oleochemical epoxides and CO$_2$". Green Chem. 2013, 15, 1173–82.

[37] Maisonneuve L, Wirotius AL, Alfos C, Grau E, Cramail H. "Fatty acid-based (bis) 6-membered cyclic carbonates as efficient isocyanate free poly(hydroxyurethane) precursors". Polym. Chem. 2014, 5, 6142–7.

[38] Schäffner B, Blug M, Kruse D, et al. "Synthesis and application of carbonated fatty acid esters from carbon dioxide including a life cycle analysis". ChemSusChem 2014, 7, 1133–9.

[39] Büttner H, Grimmer C, Steinbauer J, Werner T. "Iron-based binary catalytic system for the valorization of CO$_2$ into biobased cyclic carbonates". ACS Sustain. Chem. Eng. 2016, 4, 4805–14.

[40] Doll KM, Erhan SZ. "Synthesis of carbonated fatty methyl esters using supercritical carbon dioxide". J. Agric. Food. Chem. 2005, 53, 9608–14.

[41] Doll KM, Erhan SZ. "The improved synthesis of carbonated soybean oil using supercritical carbon dioxide at a reduced reaction time". Green Chem. 2005, 7, 849–54.

[42] Peña Carrodeguas L, Cristòfol À, Fraile JM, et al. "Fatty acid based biocarbonates: Al-mediated stereoselective preparation of mono-, di- and tricarbonates under mild and solventless conditions". Green Chem. 2017, 19, 3535–41.

[43] Parzuchowski PG, Jurczyk-Kowalska M, Ryszkowska J, Rokicki G. "Epoxy resin modified with soybean oil containing cyclic carbonate groups". J. Appl. Polym. Sci. 2006, 102, 2904–14.

[44] Tenhumberg N, Büttner H, Schäffner B, Kruse D, Blumenstein M, Werner T. "Cooperative catalyst system for the synthesis of oleochemical cyclic carbonates from CO$_2$ and renewables". Green Chem. 2016, 18, 3775–88.

[45] Stewart JA, Drexel R, Arstad B, Reubsaet E, Weckhuysen BM, Bruijnincx PCA. "Homogeneous and heterogenised masked N-heterocyclic carbenes for bio-based cyclic carbonate synthesis". Green Chem. 2016, 18, 1605–18.

[46] Fleischer M, Blattmann H, Mulhaupt R. "Glycerol-, pentaerythritol- and trimethylolpropane-based polyurethanes and their cellulose carbonate composites prepared via the non-isocyanate route with catalytic carbon dioxide fixation". Green Chem. 2013, 15, 934–42.

[47] Honda M, Tamura M, Nakao K, Suzuki K, Nakagawa Y, Tomishige K. "Direct cyclic carbonate synthesis from CO$_2$ and diol over carboxylation/hydration cascade catalyst of CeO$_2$ with 2-cyanopyridine". ACS Catal. 2014, 4, 1893–6.

[48] Gregory GL, Ulmann M, Buchard A. "Synthesis of 6-membered cyclic carbonates from 1,3-diols and low CO_2 pressure: a novel mild strategy to replace phosgene reagents". RSC Adv. 2015, 5, 39404–8.

[49] Fache M, Darroman E, Besse V, Auvergne R, Caillol S, Boutevin B. "Vanillin, a promising biobased building-block for monomer synthesis". Green Chem. 2014, 16, 1987–98.

[50] Chen Q, Gao KK, Peng C, Xie HB, Zhao ZK, Bao M. "Preparation of lignin/glycerol-based bis (cyclic carbonate) for the synthesis of polyurethanes". Green Chem. 2015, 17, 4546–51.

[51] Kühnel I, Saake B, Lehnen R. "A new environmentally friendly approach to lignin-based cyclic carbonates". Macromol. Chem. Phys. 2018, 219, 1700613.

[52] Zhu YQ, Romain C, Williams CK. "Sustainable polymers from renewable resources". Nature 2016, 540, 354–62.

[53] Poland SJ, Darensbourg DJ. "A quest for polycarbonates provided via sustainable epoxide/ CO_2 copolymerization processes". Green Chem. 2017, 19, 4990–5011.

[54] Okada A, Kikuchi S, Nakano K, Nishioka K, Nozaki K, Yamada T. "New class of catalysts for alternating copolymerization of alkylene oxide and carbon dioxide". Chem. Lett. 2010, 39, 1066–8.

[55] Nakano K, Kamada T, Nozaki K. "Selective formation of polycarbonate over cyclic carbonate: copolymerization of epoxides with carbon dioxide catalyzed by a cobalt(III) complex with a piperidinium end-capping arm". Angew Chem. Int. Ed. 2006, 45, 7274–7.

[56] Seong JE, Na SJ, Cyriac A, Kim BW, Lee BY. "Terpolymerizations of CO_2, propylene oxide, and various epoxides using a cobalt(III) complex of salen-type ligand tethered by four quaternary ammonium salts". Macromolecules 2010, 43, 903–8.

[57] Tsai FT, Wang YY, Darensbourg DJ. "Environmentally benign CO_2-based copolymers: degradable polycarbonates derived from dihydroxybutyric acid and their platinum-polymer conjugates". J. Am. Chem. Soc. 2016, 138, 4626–33.

[58] Zhang YY, Zhang XH, Wei RJ, Du BY, Fan ZQ, Qi GR. "Synthesis of fully alternating polycarbonate with low T-g from carbon dioxide and bio-based fatty acid". RSC Adv. 2014, 4, 36183–8.

[59] Wu GP, Wei SH, Ren WM, Lu XB, Xu TQ, Darensbourg DJ. "Perfectly alternating copolymerization of CO_2 and epichlorohydrin using cobalt(III)-based catalyst systems". J. Am. Chem. Soc. 2011, 133, 15191–9.

[60] Wu GP, Xu PX, Lu XB, et al. "Crystalline CO_2 copolymer from epichlorohydrin via Co(III)-complex-mediated stereospecific polymerization". Macromolecules 2013, 46, 2128–33.

[61] Hu Y, Qiao L, Qin Y, et al. "Synthesis and stabilization of novel aliphatic polycarbonate from renewable resource". Macromolecules 2009, 42, 9251–4.

[62] Hilf J, Scharfenberg M, Poon J, Moers C, Frey H. "Aliphatic polycarbonates based on carbon dioxide, furfuryl glycidyl ether, and glycidyl methyl ether: reversible functionalization and cross-linking". Macromol. Rapid. Commun. 2015, 36, 174–9.

[63] Darensbourg DJ, Chung WC, Yeung AD, Luna M. "Dramatic behavioral differences of the copolymerization reactions of 1,4-cyclohexadiene and 1,3-cyclohexadiene oxides with carbon dioxide". Macromolecules 2015, 48, 1679–87.

[64] Honda S, Mori T, Goto H, Sugimoto H. "Carbon-dioxide-derived unsaturated alicyclic polycarbonate: Synthesis, characterization, and post-polymerization modification". Polymer 2014, 55, 4832–6.

[65] Darensbourg DJ, Chung WC, Arp CJ, Tsai FT, Kyran SJ. "Copolymerization and cycloaddition products derived from coupling reactions of 1,2-epoxy-4-cyclohexene and carbon dioxide". Postpolymerization Functionalization via Thiol-Ene Click Reactions. Macromolecules 2014, 47, 7347–53.

[66] Byrne CM, Allen SD, Lobkovsky EB, Coates GW. "Alternating copolymerization of limonene oxide and carbon dioxide". J. Am. Chem. Soc. 2004, 126, 11404–5.

[67] Hauenstein O, Reiter M, Agarwal S, Rieger B, Greiner A. "Bio-based polycarbonate from limonene oxide and CO$_2$ with high molecular weight, excellent thermal resistance, hardness and transparency". Green Chem. 2016, 18, 760–70.

[68] Reiter M, Vagin S, Kronast A, Jandl C, Rieger B. "A Lewis acid beta-diiminato-zinc-complex as all-rounder for co- and terpolymerisation of various epoxides with carbon dioxide". Chem Sci. 2017, 8, 1876–82.

[69] Li CL, Sablong RJ, Koning CE. "Synthesis and characterization of fully-biobased alpha, omega-dihydroxyl poly(limonene carbonate)s and their initial evaluation in coating applications". Eur. Polym. J. 2015, 67, 449–58.

[70] Hauenstein O, Agarwal S, Greiner A. "Bio-based polycarbonate as synthetic toolbox". Nat. Commun. 2016, 7, 11862.

[71] Li CL, Sablong RJ, Koning CE. "Chemoselective alternating copolymerization of limonene dioxide and carbon dioxide: a new highly functional aliphatic epoxy polycarbonate". Angew. Chem. Int. Ed. 2016, 55, 11572–6.

[72] Li CL, van Berkel S, Sablong RJ, Koning CE. "Post-functionalization of fully biobased poly (limonene carbonate)s: Synthesis, characterization and coating evaluation". Eur. Polym. J. 2016, 85, 466–77.

[73] Schimpf V, Ritter BS, Weis P, Parison K, Mülhaupt R. "High purity limonene dicarbonate as versatile building block for sustainable non-isocyanate polyhydroxyurethane thermosets and thermoplastics". Macromolecules 2017, 50, 944–55.

[74] Peña Carrodeguas L, González-Fabra J, Castro-Gómez F, Bo C, Kleij AW. "Al-III-catalysed formation of poly(limonene)carbonate: DFT analysis of the origin of stereoregularity". Chem. Eur. J. 2015, 21, 6115–22.

[75] Martin C, Kleij AW. "Terpolymers derived from limonene oxide and carbon dioxide: access to cross-linked polycarbonates with improved thermal properties". Macromolecules 2016, 49, 6285–95.

[76] Kindermann N, Cristofol A, Kleij AW. "Access to biorenewable polycarbonates with unusual glass transition temperature (T$_g$) modulation". ACS Catal. 2017, 7, 3860–3.

[77] Li CL, Sablong RJ, van Benthem R, Koning CE. "Unique base-initiated depolymerization of limonene-derived polycarbonates". ACS Macro. Letters. 2017, 6, 684–8.

[78] Auriemma F, De Rosa C, Di Caprio MR, Di Girolamo R, Coates GW. "Crystallization of alternating limonene oxide/carbon dioxide copolymers: determination of the crystal structure of stereocomplex poly(limonene carbonate)". Macromolecules 2015, 48, 2534–50.

[79] Auriemma F, De Rosa C, Di Caprio MR, Di Girolamo R, Ellis WC, Coates GW. "Stereocomplexed poly(limonene carbonate): a unique example of the cocrystallization of amorphous enantiomeric polymers". Angew. Chem. Int. Ed. 2015, 54, 1215–8.

[80] Liu Y, Zhou H, Guo JZ, Ren WM, Lu XB. "Completely recyclable monomers and polycarbonate: approach to sustainable polymers". Angew. Chem. Int. Ed. 2017, 56, 4862–6.

[81] von der Assen N, Bardow A. "Life cycle assessment of polyols for polyurethane production using CO$_2$ as feedstock: insights from an industrial case study". Green Chem. 2014, 16, 3272–80.

Andreas Weilhard, Jairton Dupont and Victor Sans

17 Carbon dioxide hydrogenation to formic acid

17.1 Introduction

Formic acid (FA) was first discovered in the strongly corroding venom of red ants due to its low pK_a [1]. It is the smallest carboxylic acid with a high boiling point of 100.8 °C and a low pK_a of 3.74. Physically, it is a colourless liquid with high pungent odour, and is soluble in organic solvents such as alcohols, ethers and acetone. Its main applications include animal feed, silage preservatives, cleaning agents and leather production.1 With its application as a direct product, some 10% of the worldwide FA has been transformed from organic synthesis into fine chemicals and pharmaceuticals. For instance, FA can be used as a reducing agent [2]. Notably, here is the reduction of imines (Schiff bases) and the formylation of primary amines from ammonia, ketones and FA (Leuckart reaction) [2]. Another important transformation is into branched carboxylic acids in the Koch carboxylic acid synthesis [2, 3]. The global market for FA was valued $517M in 2016 and is expected to be valued $879M by the end of 2,027, mostly due to rapid market growth in Asia [3].[1] This rapid market growth has attracted interdisciplinary interest both in academia and industry for the synthesis of FA.

The main producers of FA today are BASF SE, BP and Eastman (formerly Taminco/Kemira). The main synthetic pathways include the formal methanol carbonylation, as done by BASF SE, BP and Eastman [4, 5]. In the first step, methanol is carbonylated with CO in the presence of catalytic amounts of sodium methanolate to yield methylformate, which is then hydrolysed in the presence of excess amounts of water to form methanol and FA [1]. The purification of the FA synthesised by this method is tedious as FA forms a positive azeotrope with water (bp: 107.6 °C, 22.4 wt% water) [1].

Other methods for the formation of FA involve the carbonylation of hydroxides, followed by the acidification of the formate salts formed, the hydrolysis of formamides and it can be separated as a side product during the synthesis of acetic acid by oxidation of hydrocabons [1]. Recently, it was proposed that FA can be synthesised

---

1 Data extracted from: https://www.marketresearchfuture.com/reports/formic-acid-market-1132. Last retrieved 21 July 2018.

Andreas Weilhard, Faculty of Engineering, University of Nottingham, Nottingham, UK; GSK Carbon Neutral Laboratory, University of Nottingham, Nottingham, UK
Jairton Dupont, Laboratory of Molecular Catalysis, Institute of Chemistry, Brazil
Victor Sans, Institute of Advanced Materials (INAM), Universitat Jaume I, Avda Sos Baynat s/n 12006, Castellón Spain

https://doi.org/10.1515/9783110665147-017

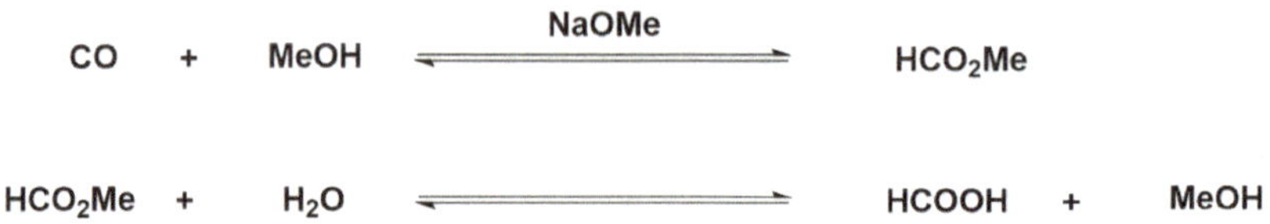

$$CO \quad + \quad MeOH \quad \overset{NaOMe}{\rightleftharpoons} \quad HCO_2Me$$

$$HCO_2Me \quad + \quad H_2O \quad \rightleftharpoons \quad HCOOH \quad + \quad MeOH$$

Figure 17.1: Formal water carbonylation for the synthesis of FA [12].

by the catalytic transformation of biomass with Keggin-type polyoxometallates or vanadium oxides [6–8]. It is important to mention that the CO used for the carbonylation of methanol or hydroxides is formed by the oxidation of methane to CO, thus it is directly linked to the employment of natural gases as a resource [9, 10].

The synthesis of FA from CO_2 and hydrogen would be advantageous, since it represents a virtually unlimited carbon source. Furthermore, if it is efficiently transformed to FA, it could be introduced easily into existing synthetic value chains. The direct hydrogenation of CO_2 to FA would be completely atom efficient, which is very interesting from a green chemistry perspective [11]. However, it is important to understand the main limitations, which can lead to increased costs and higher CO_2 emissions than the amount that is transformed. One notorious issue is hydrogen generation. In order to be viable, the H_2 used in the hydrogenation of CO_2 should be generated from renewable sources. Doing so would constitute a step towards decarbonising the industrial synthetic processes. Consequently, the hydrogenation of CO_2 is closely linked to emerging technologies such as water splitting.

In general, the reactivity of FA can be summarised in the FA triangle (Figure 17.2) [12]. FA can be synthesised from the direct hydrogenation of CO_2 or from the reaction of CO and H_2O. FA can be regarded as a safe method to store H_2 and CO. Its dehydrogenation to form CO_2 and H_2, can be subsequently employed to perform hydrogenation reactions or in fuel cells [13–17]. Alternatively, FA might be considered as a feedstock for carbon monoxide, which can be liberated on demand under formation of water as the side product [18]. This would avoid the storage of high amounts of toxic carbon monoxide replacing carbon monoxide by a relatively non-toxic liquid, ultimately easing the storage demands [1]. Most recently, FA has been discussed as

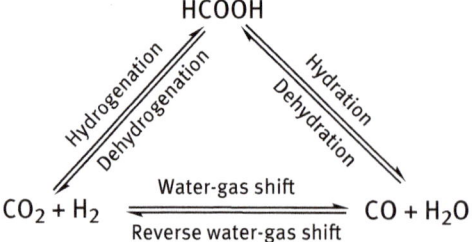

Figure 17.2: Main reactivity pathways of HCOOH described via the FA triangle. Adapted from reference 12.

a hydrogen storage material. As mentioned earlier, FA can be dehydrogenated to liberate H_2 on demand, without contamination with undesired side products such as CO. This makes the path especially attractive for the fuel cell technology, where the technology is known to be sensitive towards the CO contaminations [13, 19–22].

It is crucial to understand the thermodynamic and kinetic challenges faced in the hydrogenation of CO_2 to realise the above-mentioned applications. In this chapter, the general aspects of CO_2 hydrogenation to FA in terms of kinetics and thermodynamics will be discussed. After giving the reader a fundamental understanding of those challenges, examples of catalytic systems will be outlined. In the final part of this chapter, some process schemes shall be discussed, where FA is synthesised from CO_2 and H_2.

17.2 General aspects of CO_2 hydrogenation to FA

17.2.1 Thermodynamic considerations

In the hydrogenation of CO_2 to FA, formally one molecule of H_2 is reacted with one molecule of CO_2 to give HCOOH in 100% atomic yield. However, both reagents are gaseous under standard conditions, while the FA is a liquid at room temperature. Thus, the reaction is entropically (ΔS) highly unfavoured, rendering the process endergonic (Scheme 17.1a). In order to shift the thermodynamics towards more favourable reaction conditions, the entropic barrier has to be lowered, or be overcompensated by the enthalpic (ΔH) contribution. In other words, the thermodynamic equilibrium has to be disturbed, for example, by a consecutive reaction. This is typically achieved by the addition of bases, yielding formate adducts and salts. Here the formed FA reacts exothermically with the base, thus increasing the enthalpic contribution (ΔH). Consequently, the thermodynamic equilibrium (K_{eq}) is shifted towards the product side (Scheme 17.1b). Unfortunately, the energy shift in the formation of formate adducts and formate salts has to be overcome later during downstream processing to generate free FA. Alternatively, the reaction can be carried out in solvents, which lowers the entropic barrier (ΔS), allowing to shift the equilibrium towards the product side. Indeed, under these conditions the thermodynamic equilibrium (K_{eq}) is shifted slightly to the product side, giving slightly negative free Gibbs energies (ΔG), or in other terms the reaction becomes exergonic (Scheme 17.1c) [4, 12, 23–25].

As shown above, the free Gibbs energy is only slightly negative. In order to achieve suitable concentrations, the reactions have to be carried out near the thermodynamic equilibrium. The concentration of FA achieved during the hydrogenation can further be improved by increasing the amount of reagents (CO_2 and H_2) in the solution (principle of Le Chatelier). This can also be expressed by the thermodynamic equilibrium in relationship with the Gibbs free energy:

$$CO_2 \text{ (g)} + H_2 \text{ (g)} \longrightarrow HCOOH \text{ (l)}$$

$$\left.\begin{array}{l} \Delta H^0 = -31.5 \text{ kJ mol}^{-1} \\ \Delta S^0 = -250 \text{ J mol}^{-1} \text{K}^{-1} \end{array}\right\} \Delta G^0_{298\,K} = 43 \text{ kJ mol}^{-1}$$

$$NH_3 \text{ (aq)} + CO_2 \text{ (g)} + H_2 \text{ (g)} \longrightarrow HCOO^- NH_4^+ \text{ (aq)}$$

$$\left.\begin{array}{l} \Delta H^0 = -84.3 \text{ kJ mol}^{-1} \\ \Delta S^0 = -250 \text{ J mol}^{-1} \text{K}^{-1} \end{array}\right\} \Delta G^0_{29\,8K} = -9.8 \text{ kJ mol}^{-1}$$

Scheme 17.1: Influence of base addition on the thermodynamics of CO_2 hydrogenation to FA. Values of enthalpies and entropy extracted from reference 12.

$$\Delta G = -RT \ln K_{eq} \qquad (17.1)$$

The equilibrium constant is directly linked to the concentration of reagents (in this case, CO_2 and H_2) in solution, as described in the following equation:

$$K_{eq} = \frac{[HCOOH]}{[CO_2]_1[H_2]_1} \qquad (17.2)$$

It is evident that, if the equilibrium constant must be constant, increasing the gas pressure must increase the concentration of FA achieved at constant temperatures. The relationship between gas pressure and gas concentration in solution can be expressed by Henry's law, see eq. (17.3) (C is the solubility of a gas at a fixed temperature in a particular solvent, k is the Henry constant, P_x is the partial pressure of gas x):

$$C = kP_x \qquad (17.3)$$

Furthermore, the Gibbs free energy is directly linked to entropy and enthalpy as follows:

$$\Delta G = \Delta H - T\Delta S \qquad (17.4)$$

Combining eq. (17.4) with eq. (17.1) gives the van't Hoff equation, which can be used to calculate the enthalpy and entropy of a reaction; see eq. (17.5) [26]:

$$\ln K_{eq} = -\frac{\Delta H}{RT} + \frac{\Delta S}{R} \qquad (17.5)$$

Inserting eq. (17.2) into eq. (17.5) and resolving the natural logarithm gives the following equation:

$$\frac{[HCOOH]}{[CO_2]_1[H_2]_1} = \exp\left(-\frac{\Delta H}{RT} + \frac{\Delta S}{R}\right) \qquad (17.6)$$

From these equations, it can easily be observed that reactions yielding FA must be carried out at relatively low temperatures (the reaction is exothermic) and high pressures (principle of Le Chatelier), see eq. (17.6). Furthermore, increasing the temperature will favour the formation of CO via the reverse water gas shift (rWGS) reaction [27, 28]. The rWGS can be regarded as a related reaction to FA synthesis. In both cases, the oxidation state of the carbon atom is reduced by 2. However, the hydrogenation of CO_2 to HCOOH is exothermic while the rWGS is highly endothermic and is therefore carried out at elevated temperatures [28].

17.2.2 Kinetic considerations

Carbon dioxide is an unreactive gas, with two slightly nucleophilic oxygens and a slightly positively polarised carbon [29, 30]. In order to transfer a nucleophile onto the aforementioned carbon, an electron-rich nucleophile is needed. Hence, in the bespoke hydrogenation of CO_2 to FA an electron-rich hydride is needed [30]. It is well understood that early transition metals form strong hydrides, thus facilitating the insertion of CO_2 into the metal hydride bond (M–H). Unfortunately, early transition metals also form strong metal oxygen bonds, thus the dissociation of formate from the metal centre becomes challenging. Consequently, most catalytic systems consist of metals from groups 8 and 9 (group 8: Fe, Ru; group 9: Co, Rh, Ir). A balance between hydride donation and M–O bond strength is key for the development of efficient catalysts [12, 23, 31–33].

The hydrogenation of CO_2 to FA can be separated into two main steps: (1) the insertion of CO_2 into an M–H bond and (2) the hydride recovery.

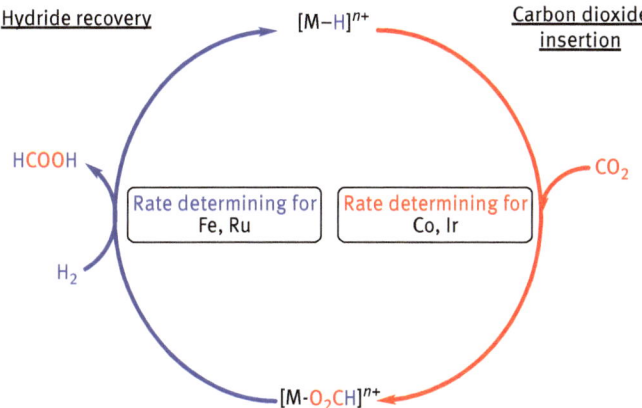

Figure 17.3: Main synthetic steps in the hydrogenation of CO_2 to FA/formate. Carbon dioxide insertion (red line) is typically rate determining for Co- and Ir-based catalysts. Hydride recovery (blue line) is typically rate determining for Fe- and Ru-based catalysts.

Unsurprisingly, the CO_2 insertion is mostly found to be the rate determining step for the group 9 metals Co and Ir. This can be explained by the weaker M–H bond and weaker M–O bond, when compared with metals from group 8 (Ru, Fe). In fact, most catalytic systems containing Ru or Fe as the active species have the hydride recovery as rate determining step [12, 23, 31–37]. In order to give the reader a better understanding of both steps, electronic effects determining the mechanistic aspects will be discussed in the following separately.

1) CO_2 insertion

As mentioned previously, the insertion of CO_2 is rate determining for late transition metal catalysts, for example, most famously Co and Ir. In this step, formally a hydride is transferred from a metal onto CO_2. Mechanistically, two paths can be determined, the inner sphere and outer sphere mechanism. Most commonly CO_2 inserts via an outer sphere mechanism into a metal hydride. Here, two transition states and an ionic intermediate – which has been isolated in an Ar matrix – are present [38]. The inner sphere mechanism is considered faster and is preferred with strong nucleophiles and/or when enough space is generated around the active metal catalyst. In the inner sphere mechanism, only one transition state has to be overcome [39].

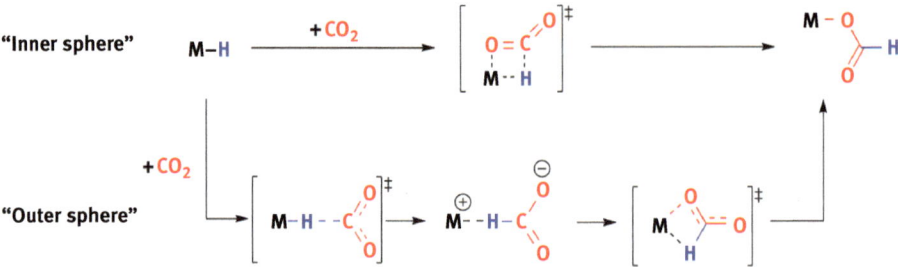

Figure 17.4: CO_2 insertion via the inner and outer sphere mechanisms into a metal hydride bond [39]. Reprinted with permission from ACS.

The thermodynamic parameters affecting the CO_2 insertion can be determined by electrochemical measurements. For instance, it has been shown that there is a relationship between the reduction potential of parent complex and its hydricity (represented as ΔG^0_{H-} in Figure 17.5a) [40, 41]. The hydricity represents the ability of a molecule to accept a hydride from another molecule. In this way, a higher hydricity means a lowered capability of a molecule to transfer a hydride. Thus, by measuring the reduction potential of a parent complex, the change in free energy for the insertion of CO_2 into an M–H bond can be calculated as the difference between the hydricity of the metal complex and formate. The hydricity of formate has been added to Figure 17.5b, red indicates a regime where the CO_2 insertion becomes endergonic (thermodynamically uphill) and green the exergonic region (thermodynamically downhill) [41].

A clear trend for a broad range of complexes can be observed. For instance, all Ni and Co complexes have higher hydricity than formate, while for all rhodium

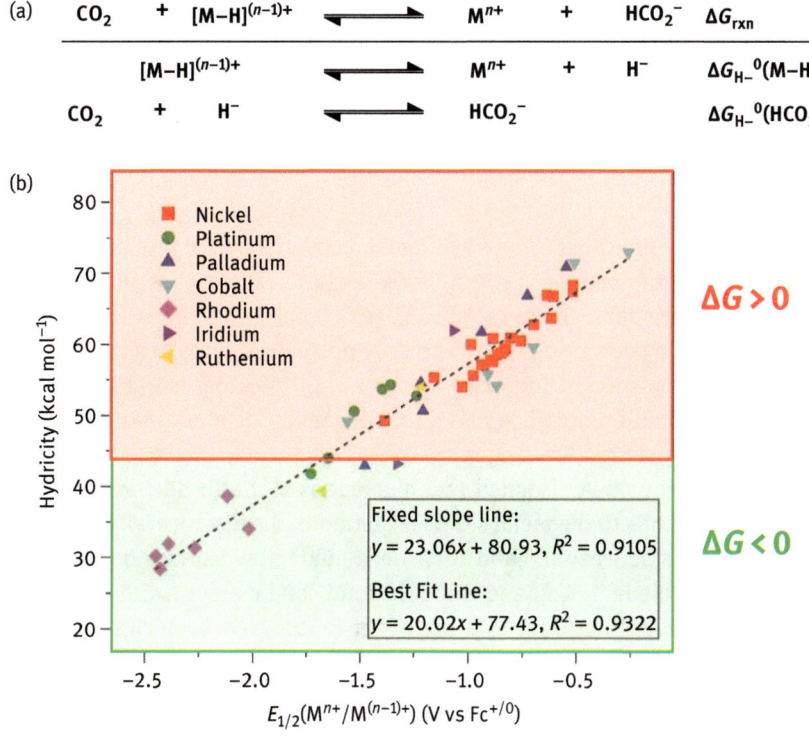

(a)

$$CO_2 \; + \; [M-H]^{(n-1)+} \rightleftharpoons M^{n+} \; + \; HCO_2^- \quad \Delta G_{rxn}$$

$$[M-H]^{(n-1)+} \rightleftharpoons M^{n+} \; + \; H^- \quad \Delta G_{H-}^{\,0}(M-H)$$

$$CO_2 \; + \; H^- \rightleftharpoons HCO_2^- \quad \Delta G_{H-}^{\,0}(HCO_2^-)$$

Figure 17.5: a) Reaction scheme to determine the thermochemical cycle and b) hydricity of various metal complexes against the reduction potential of the parent complex [41].

complexes the hydricity is lower than that of formate. For iridium and ruthenium, the values are close to those of formate, hence the most active catalysts consist of iridium or ruthenium complexes [12, 31–33]. If the hydricity of the metal hydride is close to that of formate the hydride recovery is near thermoneutral, that is, none of both steps in the catalytic transformation are strongly endergonic and nearly thermoneutral, thus easy to be operated.

The relationship between catalytic activity and hydricity has been employed to experimentally examine a wide range of Co-based catalysts. Here, a lowered hydricity led to an increased catalytic activity culminating in a time of flight (TOF) of 6,400 h^{-1} at 1.8 atm of CO_2/H_2 (1:1) at 21 °C with [Co(1,2-bis(dimethylphosphino) ethane)$_2$H] in the presence of the strong base, 2,8,9-triisopropyl-2,5,8,9-tetraaza-1-phosphabicyclo- [3,3,3] undecane, (Verkade's base; Vkd's base). The increased rate can be related to facilitated transfer of a hydride from the metal to CO_2, which is rate determining for all Co-based catalysts. Upon decreasing the hydricity of the M−H, the reaction becomes more thermoneutral. In contrast the relationship between hydricity and catalytic activity cannot be found for the analogous Rh catalyst

[42]. This can be explained by the already very low hydricity of Rh catalysts, which makes this step exergonic. However, an increased reaction rate can be obtained when the addition of dihydrogen onto the metal centre is facilitated. This step represents the rate determining step for most Rh catalysts and falls already into the hydride recovery [42].

2) Hydride recovery

The hydride recovery can be split into two different steps, the addition of dihydrogen onto the metal centre and the heterolytic splitting of H_2. In order to understand both steps, it is useful to investigate the metal dihydrogen complex (σ-M–H_2). Historically, σ-M–H_2 complexes have been assumed to be highly unstable and they were assumed to exist only as transient species. Indeed, it was not until 1983 that the first dihydrogen complex was isolated by Kubas [43, 44]. Independently, Saillard and Hoffmann were able to calculate the first dihydrogen complex [45]. Since then, a plethora of dihydrogen complexes have been isolated and characterised. In all dihydrogen complexes, the H_2 ligand binds to the metal side. Furthermore, the distance between both hydrogen atoms ranges between 0.8 and 1.6 Å. As a guide: the distance between the H atoms in a H_2 molecule is 0.74 Å, and the H–H bond can be assumed to be broken when the distance is more than 1.6 Å. The varying distances have been classified into four subclasses: the true H_2 complexes (0.8 Å to 0.9 Å), the elongated H_2 complexes (1.0 Å to 1.2 Å), compressed dihydrides (1.3 Å to 1.6 Å) and dihydrides (>1.6 Å). Note that in dihydrides the H–H bond is effectively broken, hence cannot be assigned as σ-M–H_2 complex but are best described as an M(H)$_2$ complex [43, 46].

The breaking of the H–H bond can be regarded as a continuum. On the one hand, the H_2 ligand is stabilised by π-backdonation (BD) from the metal to H_2 ligand, while on the other hand σ-donation from the H_2 ligand to the metal further contributes to the stability of the σ-M–H_2 complex. Changing the electronic parameters of a parent complex will change the local minimum for the H_2 complex on this continuum. Indeed, stable σ-M–H_2 complexes are obtained when both electronic donations are relatively strong. For instance, increasing the BD from a nucleophilic metal centre will favour the homolytic splitting of the H_2, resulting in a dihydride [47, 48]. In contrast, decreased BD from electrophilic metal centres will prefer the heterolytic cleavage of H_2 resulting in an M–H and a proton. It is important to note that cationic electrophilic complexes can have a pK_a as low as –6 and therefore can even protonate organic solvents [49].

Electrophilic metal complexes would be beneficial for the heterolytic splitting of H_2. Unfortunately, electrophilic complexes generally create weak hydride donors, hence are not suitable for the hydrogenation of CO_2 [40, 50]. Consequently, most catalytic systems require strong bases such as Vkd's base, KOH, trimethylamine and so on to recover the active hydride and to deprotonate the σ-M–H_2 complex [51, 52]. The utilisation of strong bases further shifts the thermodynamic equilibrium to the product side. The strength of the base needed to deprotonate the σ-M–H_2 complex can be determined electrochemically in a similar way as the determination of the hydricity described in Figure 17.5a. The hydricity of the parent complex can be used to estimate the strength of the base needed to facilitate an exergonic deprotonation of the σ-M–H_2 complex. When the free energy for the heterolytic splitting of molecular hydrogen and the pK_a of the conjugated form of the base are known, the change in free energy can be calculated according to the equations presented in Figure 17.6 [41, 53].

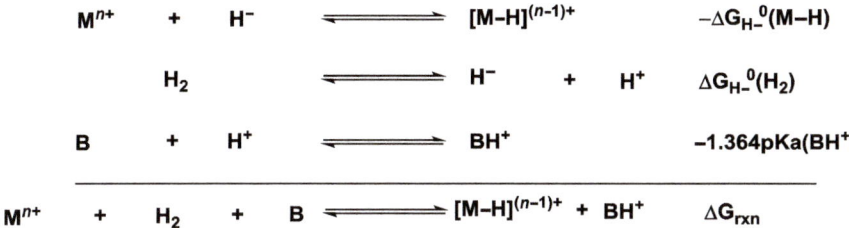

Figure 17.6: Reaction scheme for the determination of base strength needed for deprotonation [53]. Source: Reprinted with permission from ACS from reference [53].

Most catalytic systems utilise stoichiometric amounts of bases to shift the thermodynamic equilibrium to the product side, to deprotonate the intermediate σ-M–H_2 and to prevent the deactivation of the active hydride via protonation.

17.3 Development of catalytic systems

In the presence of bases, a plethora of catalysts have been shown to be active in the hydrogenation of CO_2 to formate salts and adducts. Those catalysts can be classified by the ligand architecture, into phosphine – pincer based – N-heterocyclic carbene and half sandwich catalysts [54–67]. Furthermore, the catalysts can be classified by their active metal. However, as described in the previous paragraph, the most active catalysts utilise metals from groups 8 and 9. Below, literature examples of catalysts active in the hydrogenation of CO_2 in the presence of base will be discussed based on their ligand architecture. The aim is to highlight catalyst development and cannot give a full picture of catalysts active in the hydrogenation of CO_2. A complete overview of the field can be found in recent reviews [12, 22, 31].

17.3.1 Catalysts with phospine-based ligand architecture

Pioneering work in the hydrogenation was conducted as early as 1976 by Inoue et al. using triphenylphosphine (PPh_3) complexes of Ru, Ir and Rh [68]. These catalysts showed activity in the hydrogenation of CO_2 in benzene in the presence of triethylamine (NEt_3) as a base [68]. Furthermore, small amounts of water effectively accelerated the catalytic transformation of CO_2 to formate [63, 69]. Later, this concept was further elaborated and extended to more polar solvents such as DMSO [70].

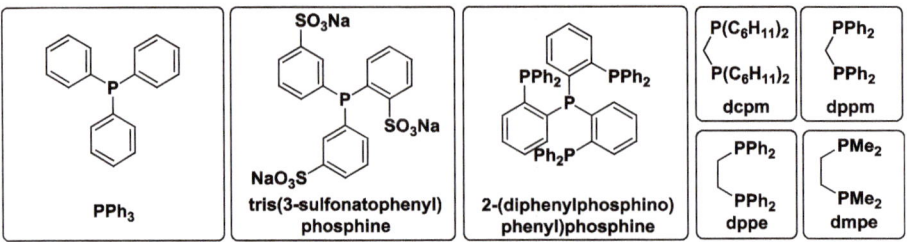

Figure 17.7: Phosphine compounds used as ligands for the synthesis of catalysts applied to the hydrogenation of CO_2 to FA [31]. Reprinted with permission from ACS.

In 1993, Leitner et al. reported the first water-soluble phosphine complex active in the hydrogenation of CO_2. They examined various catalysts and found that RhCl (tppts)$_3$ (tppts: tris(3-sulfonatophenyl)phosphine) (see Figure 17.9) was the most active catalyst at 4 MPa H_2/CO_2 achieving a TON of 3,440 at room temperature in a water:amine mixture [64, 66, 71]. Beller, Laurenczy et al. reported that [RuCl$_2$ (C$_6$H$_6$)] was active in the hydrogenation of CO_2 in the presence of aqueous NaHCO$_3$ in combination with a series of phosphine ligands. These phosphine ligands included PPh$_3$, 1,1-bis-(dicyclohexylphosphino)methane, 1,2-bis-(diphenylphosphino) ethane and 1,1-bis-(diphenylphosphino)methane (dppm). The highest catalytic activity was achieved when dppm was added, showing a TOF of 1,200 h^{-1} after 2 h at 70 °C under 8.5 MPa of H_2/CO_2 (5/3.5), even though the catalyst deactivated after few hours [72]. Non-precious phosphine-based catalysts have also been reported to be active in the hydrogenation of CO_2 or bicarbonates to FA [73, 74]. For instance, iron(II)fluorotris [2-(diphenylphosphino)phenyl)phosphino]-tetrafluoroborate complexes showed a TON of 7,500 at 100 °C under 6 MPa H_2 [74]. Linehan et al. reported a series of phosphine-based (Figure 17.8) cobalt catalysts achieving TOF values ranging between 3,400 and 74,000 h^{-1}, by varying pressures from 1 to 20 atm with equal partial pressures of CO_2/H_2. However, a very strong base (Verkade's base) is needed to deprotonate the Co–H$_2$ interemediate [42]. Some phosphine ligands reported here are depicted in Figure 17.9.

Another strategy is the in situ capturing of CO_2 as carbamate using polyethylenimine and the consecutive hydrogenation of the carbamate to FA in the presence of RhCl$_3$•3H$_2$O with a monodentate phosphine ligand [75]. The sequential hydrogenation of CO_2 achieved a maximum TON of 852. Immobilised bidentate iminophosphine ligands in combination with Ir have also been found to be active under relatively mild conditions (60 °C, 4 MPa, (p_{H2}:p_{CO2}) (1:1), 20 h). Here, a TON of 2,800 after 20 h was reported [76].

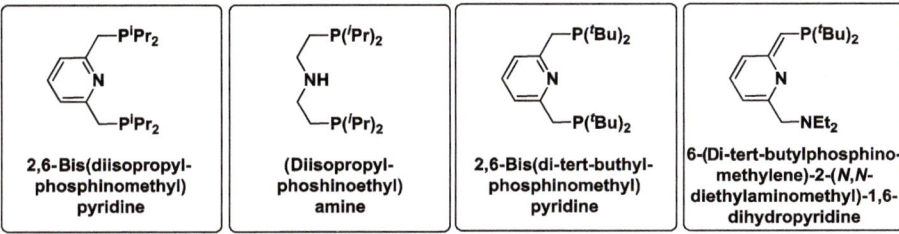

Figure 17.8: Pincer ligands employed as ligands for Ir, Rh, Ru, Fe complexes applied for the CO_2 hydrogenation [31]. Reprinted with permission from ACS.

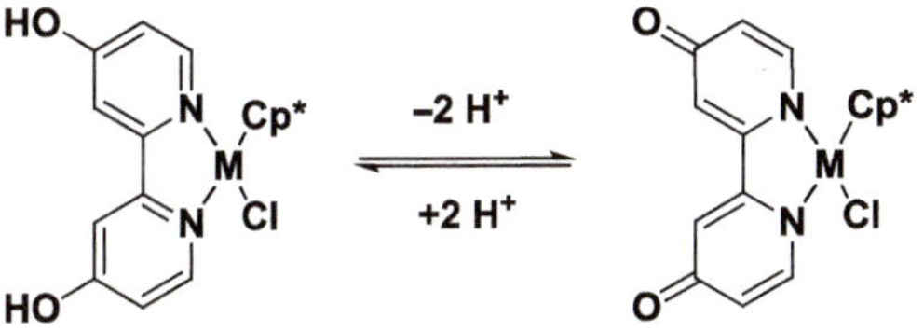

Figure 17.9: Keto resonance structure of 4,4'-dihydroxy-2,2'-bipyridine) [31, 83]. Reprinted with permission from ACS.

17.3.2 Catalysts with pincer-based ligand architecture

The highest catalytic activity reported to date was achieved with the pincer-type ligand architecture, where a TOF value of 150,000 h^{-1} at 200 °C and a TON value of 3,500,000 at 120 °C were reported in water/THF (5/1) (v/v) in the presence of KOH under 8.5 MPa H_2/CO_2 (1/1) with [IrH$_3$(2,6-bis(diisopropylphosphinomethyl)pyridine)] (see Figure 10 for the ligand structure) [59]. With the aliphatic [IrH$_3$((diisopropylphoshinoethyl) amine)] TOF values of up to 18,780 h^{-1} and TON values of 348,000 were reported [77]. In both cases, the secondary coordination sphere plays a crucial role in achieving high catalytic activities, for example, in the case of [IrH$_3$((diisopropylphoshinoethyl)amine)] a proton bridge is formed between the N–H and the formate ligand [77]. while with [IrH$_3$(2,6-bis(diisopropylphosphinomethyl)pyridine)] the α-position is deprotonated under de-aromatisation of the pyridine ring, releasing the formate from the metal centre [59]. Upon addition of H_2 the ligand is re-protonated/re-aromatised [59, 77]. High activity was also reported by Milstein et al. for trans-[FeH$_2$(CO)(2,6-bis(di-tert-butylphosphinomethyl)pyridine)]. This iron pincer complex achieves a TOF of 160 h^{-1} and a TON of up to 780 at 60 °C under 1 MPa in H$_2$O/THF (10/1) (v/v) in the hydrogenation of CO_2 to FA [78]. The de-aromatised Ru derivative of trans-[FeH$_2$(CO)(2,6-bis(di-tert-butylphosphinomethyl)pyridine)] was shown to possess even higher activity than the Fe analogue [79, 80]. Similarly, [RuH(CO)(6-(di-tertbutylphosphinomethylene)-2-(N,

N-diethylaminomethyl)-1,6-dihydropyridine)] is also active in the hydrogenation of CO_2 under basic conditions [80]. Pidko et al. reported [RuH(Cl)(CO)(2,6-bis(di-tert-butylphosphinomethyl)pyridine)] to achieve a TOF of 1,100,000 h^{-1} in dimethylforma-mide (DMF) with 1,8-Diazabicyclo[5.4.0]undec-7-ene (DBU) as a base under 4 MPa H_2/CO_2 at 120 °C. In contrast to previous studies, it was demonstrated via Density Functional Theory (DFT) calculations that the de-aromatised intermediate is inactive, while the dihydride represents the active species [60, 61, 81].

17.3.3 Catalysts with N-heterocyclic-carbene-based ligand architecture

Peris and collaborators explored various NHC-based catalysts for the hydrogenation of CO_2. Best results were obtained with [IrI_2(AcO)(bis-NHC)] as catalyst at 200 °C under 6 MPa H_2/CO_2 (1/1) [56, 57]. Here a TON of 190,000 was achieved after 75 h. NHC-carbenes show high thermal stability and lead generally to high catalytic activity, due to the high electron donating character of the ligand. Interestingly, NHC-carbene complexes were employed in the transfer hydrogenation of CO_2. For instance, when isopropanol was used as H_2 source, a TON of 1,000 was achieved [82]. Employing isopropanol as a H_2 source is safer than employing pressurised H_2. The lowered activity compared to reactions carried out with pressurised H_2 can be attributed to the difficult formation of the active metal hydride from isopropanol [56, 57, 82].

17.3.4 Half sandwich complexes as catalyst for the hydrogenation of CO_2

An interesting relationship between the electron donation of the ligand and the cat-alytic activity has been demonstrated for half sandwich complex. For instance, the complex [Cp*Ir(4,4′-dihydroxy-2,2′-bipyridine)Cl]Cl showed more than a 1,000-fold increase in its initial TOF, achieving a value of 5,100 h^{-1}, when compared to its un-substituted derivative [Cp*Ir(2,2′-bipyridine)Cl]Cl. This effect is attributed to elec-tron donation from the oxyanion to the metal centre and the consequent "keto" resonance structure [83], as shown in Figure 17.9.

The catalytic activity can be correlated to the Hammett constant σ_p^+ and could be used in future for the development of new catalysts to achieve potentially even higher catalytic activity than reported so far. However, it is noteworthy that upon switching the initial hydroxo group from the para position to the ortho position, the catalytic activity was significantly increased, reaching a TOF of 8,050 h^{-1}. As the electron donation in para and ortho positions is almost the same, the increased cat-alytic activity must be attributed to the facilitated formation of the active hydride from the M–H_2 complex. Here, the oxyanion in ortho position functions as a base, thus the proton transfer from H_2 is eased [54, 55, 58, 84–86].

17.4 Conclusions

The hydrogenation of CO_2 to FA is a vibrant research field, which has seen an increased attention in industry and academia in the last three decades. However, an industrial process remains economically challenging, mainly due to the thermodynamic limitations inherent to this reaction, that is, stoichiometric amounts of bases have to be used. This leads to energy-intensive purification steps during isolation, hindering the competitiveness of bespoke processes. As a consequence, attention has been shifted towards base-free system, although conversions and catalytic activities remain sluggish compared to system utilising bases. In this regard, the utilisation of buffering ionic liquids (ILs) can present an interesting alternative. Here, the formation of FA is favoured via weak interactions; hence, the separation of FA from the reaction media should be easier to achieve.

Bibliography

[1] Reutemann, W.; Kieczka, H. In *Ullmann's Encyclopedia of Industrial Chemistry*; Wiley-VCH Verlag GmbH & Co. KGaA: 2000.
[2] Zassinovich, G.; Mestroni, G.; Gladiali, S. Chem. Rev. 1992, *92*, 1051.
[3] Aresta, M.; Dibenedetto, A.; Angelini, A. Chem. Rev. 2014, *114*, 1709.
[4] Schaub, T.; Paciello, R. A. Angew. Chem. Int. Ed. 2011, *50*, 7278.
[5] Arpe, H.-J. *Industrial Organic Chemistry*; Wiley-VCH: Weinheim, 2010; Vol. 5.
[6] Jiang, L.; Dao-Jun, D.; Li, D.; Qing-Xiang, G.; Yao, F. *ChemSusChem* 2012, *5*, 1313.
[7] Wolfel, R.; Taccardi, N.; Bosmann, A.; Wasserscheid, P. Green Chem. 2011, *13*, 2759.
[8] Albert, J.; Wolfel, R.; Bosmann, A.; Wasserscheid, P. Energy Environ. Sci. 2012, *5*, 7956.
[9] Baufumé, S.; Grüger, F.; Grube, T.; Krieg, D.; Linssen, J.; Weber, M.; Hake, J.-F.; Stolten, D. Int. J. Hydrog. Energ. 2013, *38*, 3813.
[10] Chen, L.; Lu, Y.; Hong, Q.; Lin, J.; Dautzenberg, F. M. Appl. Catal. A-Gen. 2005, *292*, 295.
[11] Clark, J., Macquarrie, D., Gronnow, M. and Budarin, V In *Process Intensification for Green Chemistry* 2013
[12] Klankermayer, J.; Wesselbaum, S.; Beydoun, K.; Leitner, W. Angew. Chem. Int. Ed. 2016, *55*, 7296.
[13] Björn, L.; Albert, B.; Henrik, J.; Matthias, B. Angew. Chem.Int. Ed. 2008, *47*, 3962.
[14] Boddien, A.; Loges, B.; Junge, H.; Beller, M. *ChemSusChem* 2008, *1*, 751.
[15] Boddien, A.; Loges, B.; Junge, H.; Gärtner, F.; Noyes, J. R.; Beller, M. Adv. Synth. Catal. 2009, *351*, 2517.
[16] Kawanami, H.; Himeda, Y.; Laurenczy, G. In *Advances in Inorganic Chemistry*; van Eldik, R., Hubbard, C. D., Eds.; Academic Press: 2017; Vol. 70, p 395.
[17] Yuranov, I.; Autissier, N.; Sordakis, K.; Dalebrook, A. F.; Grasemann, M.; Orava, V.; Cendula, P.; Gubler, L.; Laurenczy, G. ACS Sustain. Chem. Eng. 2018, *6*, 6635.
[18] Supronowicz, W.; Ignatyev, I. A.; Lolli, G.; Wolf, A.; Zhao, L.; Mleczko, L. Green Chem. 2015, *17*, 2904.
[19] Yu, X.; Pickup, P. G. J. Power. Sources. 2008, *182*, 124.
[20] Williams, R.; Crandall, R. S.; Bloom, A. Appl. Phys. Lett. 1978, *33*, 381.
[21] Loges, B.; Boddien, A.; Gärtner, F.; Junge, H.; Beller, M. Top Catal. 2010, *53*, 902.

[22] Sordakis, K.; Tang, C.; Vogt, L. K.; Junge, H.; Dyson, P. J.; Beller, M.; Laurenczy, G. Chem. Rev. 2018, *118*, 372.

[23] Leitner, W. Angew. Chem. Int. Ed. 1995, *34*, 2207.

[24] Jessop, P. G. *The Handbook of Homogeneous Hydrogenation* Wiley-VCH: Weinheim, 2007.

[25] W. Leitner, E. D., F.Gaßner *AqueousPhase Organometallic Catalysis Concepts and Applications*; Wiley-VCH: Weinheim, 1998.

[26] Weilhard, A.; Qadir, M. I.; Sans, V.; Dupont, J. ACS Catal. 2018, *8*, 1628.

[27] Qadir, M. I.; Weilhard, A.; Fernandes, J. A.; de Pedro, I.; Vieira, B. J. C.; Waerenborgh, J. C.; Dupont, J. ACS Catal. 2018, *8*, 1621.

[28] Prieto, G. *ChemSusChem* 2017, *10*, 1056.

[29] Walther, D. Coordin. Chem. Rev. 1987, *79*, 135.

[30] Leitner, W. Coordin Chem Rev. 1996, *153*, 257.

[31] Wang, W.-H.; Himeda, Y.; Muckerman, J. T.; Manbeck, G. F.; Fujita, E. Chem. Rev. 2015, *115*, 12936.

[32] Dong, K.; Razzaq, R.; Hu, Y.; Ding, K. Top. Curr. Chem. 2017, *375*, 23.

[33] Walter, L. Angew. Chem. Int. Ed. Engl. 1995, *34*, 2207.

[34] Hayashi, H.; Ogo, S.; Fukuzumi, S. Chem. Comm. 2004, 2714.

[35] Ogo, S.; Kabe, R.; Hayashi, H.; Harada, R.; Fukuzumi, S. Dalton Trans. 2006, 4657.

[36] Mondal, B.; Neese, F.; Ye, S. Inorg. Chem. 2016, *55*, 5438.

[37] Mondal, B.; Neese, F.; Ye, S. Inorg. Chem. 2015, *54*, 7192.

[38] Li-Xue, J.; Chongyang, Z.; Xiao-Na, L.; Hui, C.; Sheng-Gui, H. Angew. Chem. Int. Ed. Engl. 2017, *56*, 4187.

[39] Hazari, N.; Heimann, J. E. Inorg. Chem. 2017, *56*, 13655.

[40] Wiedner, E. S.; Chambers, M. B.; Pitman, C. L.; Bullock, R. M.; Miller, A. J. M.; Appel, A. M. Chem. Rev. 2016, *116*, 8655.

[41] Waldie, K. M.; Ostericher, A. L.; Reineke, M. H.; Sasayama, A. F.; Kubiak, C. P. ACS Catal. 2018, *8*, 1313.

[42] Jeletic, M. S.; Hulley, E. B.; Helm, M. L.; Mock, M. T.; Appel, A. M.; Wiedner, E. S.; Linehan, J. C. ACS Catal. 2017, *7*, 6008.

[43] Kubas, G. J. Acc. Chem. Res. 1988, *21*, 120.

[44] Kubas, G. J.; Ryan, R. R.; Swanson, B. I.; Vergamini, P. J.; Wasserman, H. J. J. Am. Chem. Soc. 1984, *106*, 451.

[45] Saillard, J. Y.; Hoffmann, R. J. Am. Chem. Soc. 1984, *106*, 2006.

[46] Kubas, G. *Metal-dihydrogen and σ-bond coordination: The consummate extension of the Dewar-Chatt-Duncanson model for metal-olefin π bonding*, 2001; Vol. 635.

[47] Triguero, L.; Föhlisch, A.; Väterlein, P.; Hasselström, J.; Weinelt, M.; Pettersson, L. G. M.; Luo, Y.; Ågren, H.; Nilsson, A. J. Am. Chem. Soc. 2000, *122*, 12310.

[48] Hasegawa, T.; Li, Z.; Parkin, S.; Hope, H.; McMullan, R. K.; Koetzle, T. F.; Taube, H. J. Am. Chem. Soc. 1994, *116*, 4352.

[49] Morris, R. H. Chem. Rev. 2016, *116*, 8588.

[50] Pitman, C. L.; Brereton, K. R.; Miller, A. J. M. J. Am. Chem. Soc. 2016, *138*, 2252.

[51] Osadchuk, I.; Tamm, T.; Ahlquist, M. S. G. *Organometallics* 2015, *34*, 4932.

[52] Lilio, A. M.; Reineke, M. H.; Moore, C. E.; Rheingold, A. L.; Takase, M. K.; Kubiak, C. P. J. Am. Chem. Soc. 2015, *137*, 8251.

[53] Waldie, K. M.; Brunner, F. M.; Kubiak, C. P. ACS Sustain. Chem. Eng. **2018**, *6*, 6841.

[54] Hull, J. F.; Himeda, Y.; Wang, W.-H.; Hashiguchi, B.; Periana, R.; Szalda, D. J.; Muckerman, J. T.; Fujita, E. Nat. Chem. 2012, *4*, 383.

[55] Himeda, Y.; Miyazawa, S.; Hirose, T. *ChemSusChem* 2011, *4*, 487.

[56] Sanz, S.; Azua, A.; Peris, E. Dalton Trans. 2010, *39*, 6339.

[57] Azua, A.; Sanz, S.; Peris, E. Chem. Eur. J. 2011, *17*, 3963.
[58] Himeda, Y.; Onozawa-Komatsuzaki, N.; Sugihara, H.; Kasuga, K. *Organometallics* 2007, *26*, 702.
[59] Tanaka, R.; Yamashita, M.; Nozaki, K. J. Am. Chem. Soc. 2009, *131*, 14168.
[60] Filonenko, G. A.; Conley, M. P.; Copéret, C.; Lutz, M.; Hensen, E. J. M.; Pidko, E. A. ACS Catal. 2013, *3*, 2522.
[61] Filonenko, G. A.; van Putten, R.; Schulpen, E. N.; Hensen, E. J. M.; Pidko, E. A. *ChemCatChem* 2014, *6*, 1526.
[62] Filonenko, G. A.; Smykowski, D.; Szyja, B. M.; Li, G.; Szczygieł, J.; Hensen, E. J. M.; Pidko, E. A. ACS Catal. 2015, *5*, 1145.
[63] Tsai, J. C.; Nicholas, K. M. J. Am. Chem. Soc. 1992, *114*, 5117.
[64] Fornika, R.; Gorls, H.; Seemann, B.; Leitner, W. J. Chem. Soc., Chem. Commun. 1995, 1479.
[65] Hutschka, F.; Dedieu, A.; Eichberger, M.; Fornika, R.; Leitner, W. J. Am. Chem. Soc. 1997, *119*, 4432.
[66] Graf, E.; Leitner, W. J. Chem. Soc., Chem. Commun. 1992, 623.
[67] Elek, J.; Nádasdi, L.; Papp, G.; Laurenczy, G.; Joó, F. Appl Catal A-Gen 2003, *255*, 59.
[68] Inoue, Y.; Izumida, H.; Sasaki, Y.; Hashimoto, H. Chem. Lett. 1976, *5*, 863.
[69] Munshi, P.; Main, A. D.; Linehan, J. C.; Tai, C.-C.; Jessop, P. G. J. Am. Chem. Soc. 2002, *124*, 7963.
[70] Ezhova, N. N.; Kolesnichenko, N. V.; Bulygin, A. V.; Slivinskii, E. V.; Han, S. Russ. Chem. Bull. 2002, *51*, 2165.
[71] Gassner, F.; Leitner, W. J. Chem. Soc., Chem. Commun. 1993, 1465.
[72] Christopher, F.; Ralf, J.; Albert, B.; Gabor, L.; Matthias, B. *ChemSusChem* 2010, *3*, 1048.
[73] Christopher, F.; Carolin, Z.; Ralf, J.; Wolfgang, B.; Matthias, B. Chem-Eur. J. 2012, *18*, 72.
[74] Ziebart, C.; Federsel, C.; Anbarasan, P.; Jackstell, R.; Baumann, W.; Spannenberg, A.; Beller, M. J. Am. Chem. Soc. 2012, *134*, 20701.
[75] Li, Y.-N.; He, L.-N.; Liu, A.-H.; Lang, X.-D.; Yang, Z.-Z.; Yu, B.; Luan, C.-R. Green Chem. 2013, *15*, 2825.
[76] Zheng, X.; D., M. N.; T., N. G.; F., S. W.; C., H. J. *ChemCatChem* 2013, *5*, 1769.
[77] Schmeier, T. J.; Dobereiner, G. E.; Crabtree, R. H.; Hazari, N. J. Am. Chem. Soc. 2011, *133*, 9274.
[78] Robert, L.; Yael, D. P.; Gregory, L.; W., S. L. J.; Yehoshoa, B. D.; David, M. Angew. Chem. Int. Ed. 2011, *50*, 9948.
[79] Matthias, V.; Moti, G.; A., I. M.; Yael, D.-P.; Yehoshoa, B.-D.; David, M. Chem. Eur. J 2012, *18*, 9194.
[80] Huff, C. A.; Kampf, J. W.; Sanford, M. S. *Organometallics* 2012, *31*, 4643.
[81] Filonenko, G. A.; Hensen, E. J. M.; Pidko, E. A. Catal. Sci.Technol. 2014, *4*, 3474.
[82] Sanz, S.; Benítez, M.; Peris, E. *Organometallics* 2010, *29*, 275.
[83] Hansch, C.; Leo, A.; Taft, R. W. Chem. Rev. 1991, *91*, 165.
[84] Wang, W.-H.; Muckerman, J. T.; Fujita, E.; Himeda, Y. ACS Catal. 2013, *3*, 856.
[85] Ertem, M. Z.; Himeda, Y.; Fujita, E.; Muckerman, J. T. ACS Catal. 2016, *6*, 600.
[86] Onishi, N.; Xu, S.; Manaka, Y.; Suna, Y.; Wang, W.-H.; Muckerman, J. T.; Fujita, E.; Himeda, Y. **Inorg**. Chem. 2015, *54*, 5114.

Renata Jorge da Silva and Claudio J. A. Mota

18 CO_2 Hydrogenation to Methanol and Dimethyl Ether

18.1 Introduction

Global warming is one of the main problems society is facing at the present time. Carbon dioxide (CO_2), associated with the burning of fossil fuels, is the main greenhouse gas (GHG) in the atmosphere and is responsible for the climate changes the world is experiencing in the first decades of the twenty-first century. Despite the global policy to support the reduction of GHG emissions, humanity is still dependent on fossil fuels, derived from petroleum, coal and natural gas, as the main source of energy. Therefore, CO_2 utilisation, comprising capture and conversion of this gas into fuels and chemicals, is gaining importance in recent years.

CO_2 is considered to be easily available, non-toxic and with great economic potential, since it may be considered a waste of the combustion process. Nevertheless, there are thermodynamic and kinetic considerations that mean CO_2 utilisation is still limited to a few products and processes, especially at commercial scale. From the thermodynamic point of view, the CO_2 molecule is highly stable, with an enthalpy of formation of −94 kcal/mol. This implies that most CO_2 conversions must involve significant energy input. Although CO_2 has acidic character, it is poorly reactive towards many traditional reductants, usually requiring a catalyst to speed up the rate of the reaction to reasonable levels. Thus, any process for converting CO_2 to fuels and chemicals must address the thermodynamic and kinetic constraints.

Considering that the demand of energy by society will increase in the forthcoming decades and that fossil-derived fuels will still share a major part of the world's energy matrix, recycling the CO_2 emitted to fuels and chemicals will grow in importance, especially using renewable energy sources.

Methanol may be produced from CO_2 hydrogenation; it can be used as a fuel or feedstock for the chemical industry in the synthesis of many products. Additionally, dimethyl ether (DME), which can be obtained from methanol dehydration or directly from CO_2, is used as a raw material for the production of formaldehyde and olefins (propene, ethene, for example), besides being a substitute for diesel in compression engines, having a high cetane number and producing a cleaner combustion, with no emission of particulate matter or sulphur.

https://doi.org/10.1515/9783110665147-018

Nowadays, methanol and DME are industrially produced from synthesis gas, a mixture of CO and H_2, usually obtained through the chemical conversion of natural gas or coal. Thus, the application of a synthetic route to methanol and DME from recyclable resource, which enables sustainable development, is extremely positive in a circular economy.

According to the concept of "Methanol Economy", suggested by Olah et al. [1], CO_2 would be captured from some natural or industrial sources and further transformed into methanol, which in turn, may be converted to DME and many other products, such as fuels and chemicals (Figure 18.1). The chemical recycling of CO_2 to methanol and DME provides a renewable and abundant source of carbon, contributing to the reduction of GHG emissions into the atmosphere. The energy input to generate H_2 must come from renewable sources, such as solar, wind or geothermal, in order to achieve a neutral carbon cycle and provide sustainability to the entire cycle.

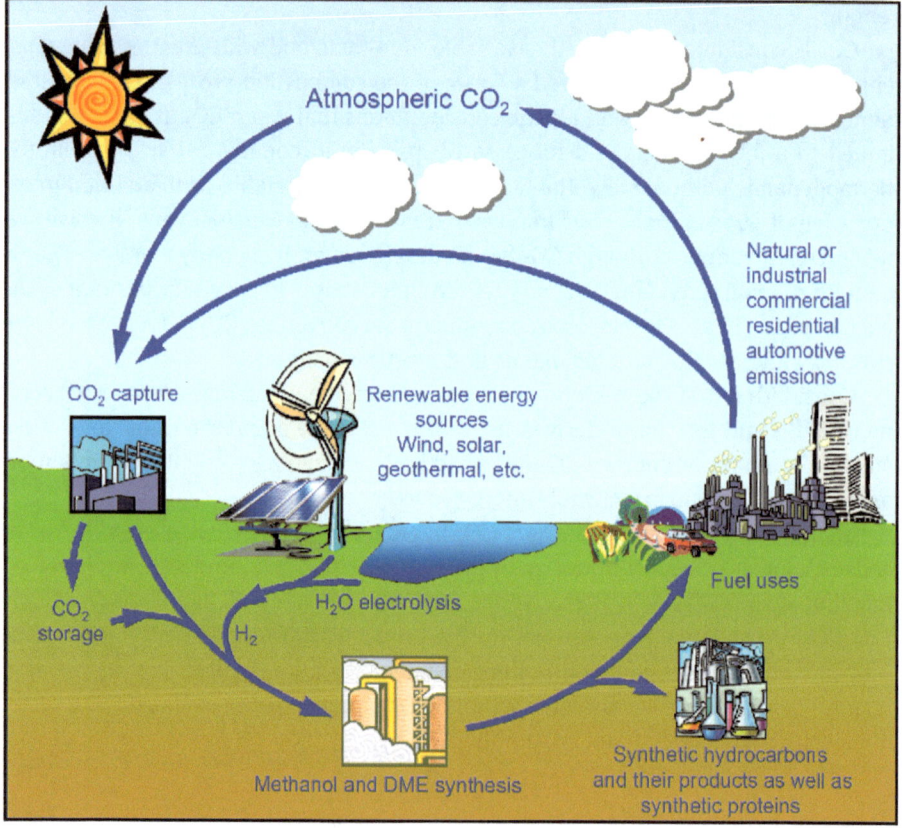

Figure 18.1: Anthropogenic carbon cycle through the production of methanol. Reprinted with permission of the American Chemical Society.

Today, there are extensive investigations to produce methanol from CO$_2$ hydrogenation. Although an industrial plant is in operation in Iceland, producing hydrogen from geothermal energy, there are still many improvements to be made to the catalyst, which is basically the same used in the traditional process from syngas. Therefore, it is clear that there is a need of a more sustainable and economically viable methanol production process that can be applied on a large scale, replacing the current technology based on fossil sources.

18.2 Importance and main uses of methanol and dimethyl ether

Currently, the main industrial route for the production of methanol and DME involves syngas, a mixture of CO and H$_2$, usually produced from natural gas. In fact, syngas may contain many components, including small amounts of CO$_2$; however, H$_2$ and carbon monoxide are the major constituents. It is usually produced through the reforming of coal or natural gas using nickel-based heterogeneous catalysts. When coal is used as feedstock, CO$_2$ is formed in higher amounts as by-product [2–5]. Alternatively, methanol and DME can also be produced from the catalytic hydrolysis of chloromethane using metal-exchanged zeolites [6–8], avoiding the need for production of syngas, which demands a great amount of energy (Scheme 18.1).

$$CH_4 + H_2O \longrightarrow CO + 3H_2 \quad \Delta H = +49 \text{ kcal mol}^{-1}$$

Scheme 18.1: Natural gas reforming to produce syngas.

The production of methanol from syngas is exothermic and occurs over a heterogeneous catalyst based on copper (Cu) and zinc oxide (ZnO), with other metal oxides as promoters (Scheme 18.2). In catalysis, promoters are components that, alone, present no or negligible catalytic activity, but when used together with an active catalyst significantly improve the activity or selectivity to a desired product. The reaction is carried out at temperatures ranging from 230 to 290 °C and at 50 to 150 bar of pressure, in a continuous flow, fixed bed reactor. Under these conditions, conversion ranges from 20 to 30% and the non-converted syngas is recycled to the reactor.

$$CO + 2H_2 \rightleftharpoons CH_3OH \quad \Delta H = -22 \text{ kcal mol}^{-1}$$

Scheme 18.2: Production of methanol from syngas.

DME may be obtained by two routes: from the bimolecular dehydration of methanol in the presence of an acidic catalyst (Scheme 18.3) or directly from syngas through the

$$2CH_3OH \xrightarrow{\text{``H}^+\text{''}} H_3COCH_3 + H_2O$$

Scheme 18.3: Production of DME from methanol dehydration.

use of a bifunctional catalyst, comprising a metallic and an acidic phase. The first route is more conventional and has already been implemented at industrial scale. Alumina or zeolites are the preferred heterogeneous acid catalysts for methanol dehydration. The direct route usually comprises a metallic catalyst, for methanol synthesis, in the presence of, or impregnated on, an acid support, to dehydrate the methanol as it is formed. The main advantage of the direct route is the possibility of shifting the equilibrium of methanol synthesis, through its further transformation into DME. The main drawback is the difficulty of tuning the experimental conditions to maximise both reactions, as well as the possibility of secondary reactions [4, 5].

Methanol is widely used as a raw material in the chemical industry, with a production of about 40 million tonnes/year [9, 10]. It is an intermediate in the production of formaldehyde and acetic acid [9, 10], which are used in the manufacture of resins, plastics, adhesives and paints, among other uses. Methanol can also be directly used as a fuel and in the production of biodiesel [11]. Although it has only half of the energy density of gasoline and diesel, it presents a higher octane rating, as well as being less flammable than gasoline and, therefore, safer. Methanol has been used in race cars for many decades. Fuel cells can also use methanol to produce electricity [3, 5].

DME is a colourless, non-toxic, non-corrosive, non-carcinogenic, environmentally friend, easily handled gas. Additionally, it can be stored in liquid form under moderate compression. The use of DME has grown considerably in Asian countries, as a substitute for diesel due to its high cetane number. In addition, it is considered as a cleaner fuel, free of sulphur and aromatics and emitting less CO, NO_x and particulate matter. It is also considered as an ideal substitute for electric power generation [4].

Methanol and DME can also be used as precursors for the production of ethylene, propylene and other hydrocarbons, replacing traditional routes based on petroleum and natural gas. The process known as methanol to gasoline (MTG) [12] was initially commercialised in the 1980s. It was further adapted to the production of light olefins, such as ethene and propene, and named methanol to olefins (MTO) [13, 14], due to the higher added value of the olefins compared to gasoline. Figure 18.2 illustrates the versatility of methanol and DME in many applications.

18.3 Thermodynamic and kinetic considerations

The hydrogenation of CO_2 to methanol is exothermic, but limited by equilibrium, being favoured at high pressures and low temperatures (Scheme 18.4) [15]. In contrast

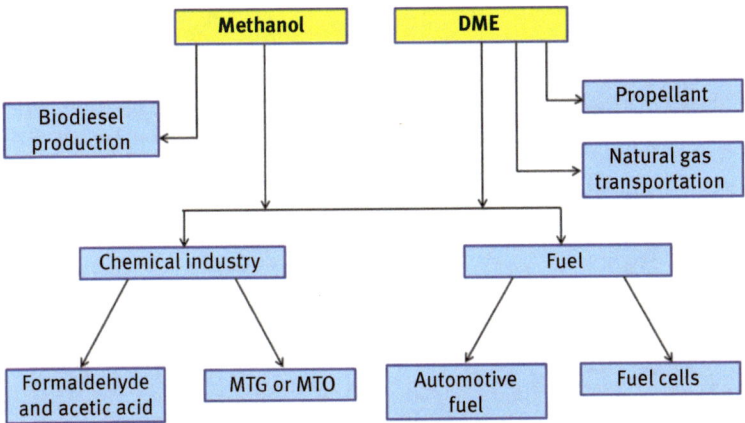

Figure 18.2: Possible applications of methanol and dimethyl ether.

$$CO_2 + 3H_2 \rightleftharpoons CH_3OH + H_2O \quad \Delta H = -12\,\text{kcal mol}^{-1}$$

Scheme 18.4: Hydrogenation of CO$_2$ to methanol.

$$CO_2 + H_2 \rightleftharpoons CO + H_2O \quad \Delta H = +10\,\text{kcal mol}^{-1}$$

Scheme 18.5: Reverse water gas shift (RWGS).

to production from syngas, where methanol is the only product, in the synthesis from CO$_2$, water is formed as a by-product. The reverse water gas shift (RWGS) reaction (Scheme 18.5), which involves the combination of CO$_2$ and H$_2$ to produce CO and water, is a possible secondary reaction that decreases the selectivity to methanol. It is mostly favoured at higher temperatures. More extensive hydrogenation of CO$_2$ to methane is also possible, but rarely observed using the traditional Cu.ZnO.Al$_2$O$_3$ catalyst under the normal reaction conditions. Formation of water accelerates the sintering of Cu and ZnO particles, resulting in faster deactivation [16].

Mota and Monteiro [17] reported that over a traditional Cu.ZnO.Al$_2$O$_3$ catalyst (50/40/10 mol%), the yield of methanol is close to the equilibrium value at 270 °C and 50 bar of pressure. Nevertheless, the yield would considerably increase at lower temperatures, according to thermodynamics. Thus, methanol synthesis from CO$_2$ must involve a catalytic system that works at the thermodynamic limits at temperatures lower than 270 °C. In addition, the increase in temperature favours the RWGS, which will produce CO and decrease the selectivity to the desired product.

Figure 18.3 shows the equilibrium methanol yield as a function of temperature and pressure of CO$_2$ hydrogenation. Higher pressures and lower temperatures favour the thermodynamic equilibrium towards methanol. Therefore, the main

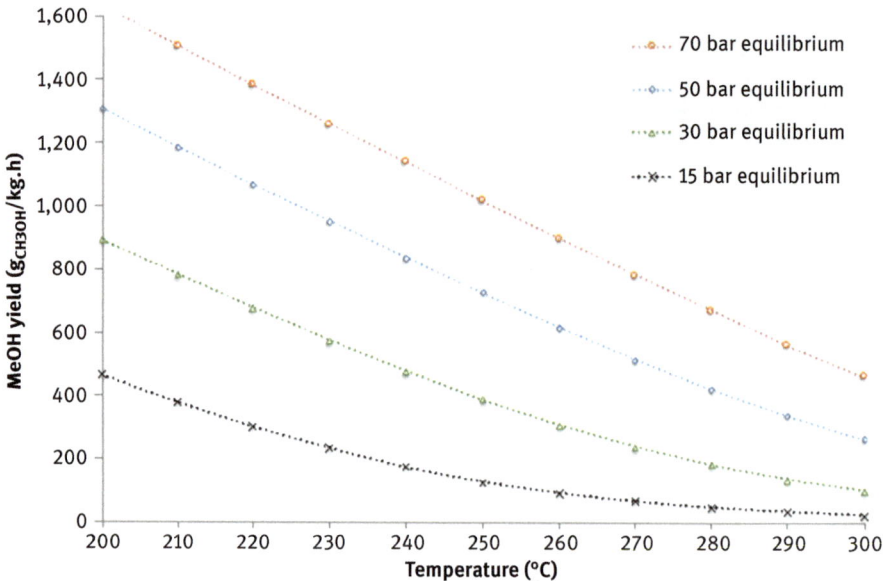

Figure 18.3: Methanol equilibrium yield as a function of reaction pressure and temperature of the CO_2 hydrogenation.

challenge is the development of highly active and selective catalytic systems, able to operate at the thermodynamic limits at lower temperatures.

18.4 Main catalysts for CO_2 hydrogenation

In general, the industrial production of methanol from syngas is carried out on Cu. ZnO catalysts, with alumina as a textural promoter, at temperatures around 220–300 °C and pressures of 50–100 bar. This methanol production process is well established industrially and many companies such as BASF, Lurgi, Haldor-Topsoe and Mitsubishi commercialised this technology [18].

Prior to 1965, methanol synthesis was performed on ZnO–Cr_2O_3-based catalysts, at 250–260 bar and a maximum temperature of 390 °C. Under these conditions, methanol could be produced with selectivity in the range of 90–94%. After 1965, the scenario was modified by the development of Cu.ZnO-based catalysts, which allowed a reduction of the pressure, to 100–150 bar, while operating at temperatures around 250 °C with 99% selectivity. Today, the most traditional catalyst for methanol synthesis is based on $Cu.ZnO.Al_2O_3$; this later component acts as a textural promoter, to increase the surface area [2, 18].

Figure 18.4 shows the most common compositions of the heterogeneous catalysts used in the hydrogenation of CO_2 to methanol. Most are based on the

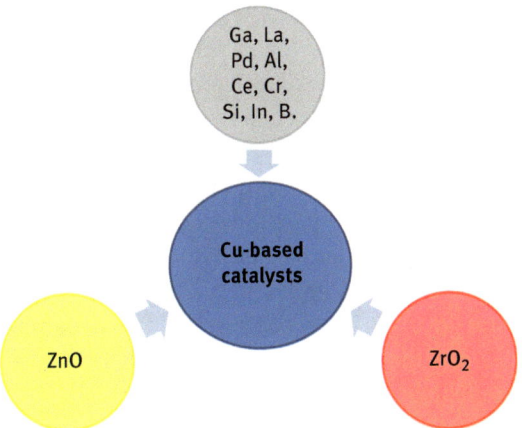

Figure 18.4: Elements and compounds used in the formulation of Cu-based catalysts for hydrogenation of CO$_2$ to methanol.

traditional Cu.ZnO catalyst, with other promoters, which may include ZrO$_2$ and other metals or metal oxides. The aim of including such promoters is to improve the activity, selectivity and stability of the catalyst [19–21].

The commercial catalyst for methanol synthesis consists of 50–70% CuO, 20–50% ZnO and 5–20% Al$_2$O$_3$. Other oxides such as chromium, aluminium, rare earth elements, palladium, gallium or zirconium can be used together or replacing Al$_2$O$_3$ in the catalyst formulations [22–31]. The catalyst must be reduced in a hydrogen-enriched gas stream, under controlled temperature conditions, prior to reaction. This procedure reduces the CuO phase to Cu^0 particles, which are the active sites for hydrogen adsorption.

The catalyst comprises Cu and ZnO nanoparticles arranged alternately, in order to guarantee the spacing and stability of the Cu particles. This minimises sintering, which reduces the surface area and leads to deactivation. The ZnO phase can also contribute to the adsorption of CO$_2$, and some models predict that reaction mostly occurs at the interface of the metallic and metal oxide phases; whereas the Cu metal adsorbs the hydrogen atoms, the metal oxide mostly adsorbs the CO$_2$, through an acid–base reaction [32].

An ideal catalyst must present high activity and selectivity to the desired product and be capable of operating for long times without appreciable loss of the catalytic properties. Thus, deactivation by sintering of the metal particles, coke formation or poisoning of impurities in the feed must be avoided.

Some strategies for the development of improved catalysts involve: (i) use of other metals as promoters, especially to favour hydrogen adsorption; (ii) modification of conventional Cu.ZnO.Al$_2$O$_3$ catalysts with other metal to achieve higher dispersion of the Cu particles to avoid sintering.

ZrO_2 is one of the main promoters of methanol synthesis catalysts. It may contribute to the improvement of the thermal stability of the catalyst, minimising the sintering of the Cu particles. Addition of lanthanum oxide can increase the Cu surface area. Other strategies to obtain high CO_2 conversion and selectivity to methanol are the use of nanocrystalline ZrO_2 [2].

Cu.ZnO modified with ZrO_2 is the most studied system due to two factors: ZrO_2 promotes a high Cu dispersion and has lower affinity to water compared to Al_2O_3. The promoter is responsible to control the texture and adsorption properties of the catalyst. ZrO_2 shows better synergy with Cu.ZnO than other promoters, such as CeO_2 and Al_2O_3. This is an important factor to increase the catalytic activity, because a synergy between the metal particles and the basic metal oxide phase is believed to play an important role in the mechanistic reaction path [24].

Typical metals employed on hydrogenation, such as Pd, Pt, Ni and Ga supported on metal oxides or nanostructured carbon materials, appear to be a growing trend among alternatives to Cu.ZnO-based catalysts. Indium oxide (In_2O_3) is another material that has received great attention in the recent years; it can be employed either as a catalyst or as a promoter. Catalytic systems using In_2O_3 have shown high activity and selectivity for CO_2 hydrogenation into methanol. The creation of oxygen vacancies has been claimed to be important during the catalytic cycle. It is possible to increase the number of active vacancies using ZrO_2 as a promoter of In_2O_3 catalysts. The In_2O_3/ZrO_2 catalytic system showed high stability and could be scaled up without losing its catalytic properties [33].

The synthesis method is an important parameter in determining the characteristics of the catalyst. The chosen technique must consider the type of target material, bulk or supported, and will directly influence the surface area, particle size, pore volume, crystalline phase and consequently the activity and selectivity of the catalyst.

The most traditional techniques for the synthesis of Cu.ZnO-based catalysts is co-precipitation, where the precursor metal oxides are precipitated from salt solutions of the metals, upon variation of the pH. Other techniques, such as impregnation, sol–gel, deposition–precipitation and combustion, can also be used.

18.5 Possible Intermediates and reaction pathway

Through the use of isotopic labelling and theoretical calculations, it has been proposed that, even in the process based on syngas, the most favourable energetic pathway to methanol synthesis involves the CO_2 molecule. Regarding the traditional Cu.ZnO-based catalyst, it is believed that reaction mostly occurs at the interface between the metal and metal oxide phase. Figure 18.5 shows a pictorial view of these processes. The copper nanoparticles are responsible for the adsorption of the hydrogen molecule, decomposing it to chemisorbed hydrides. On the other hand, the ZnO phase may interact with the CO_2 molecule to yield carbonate species.

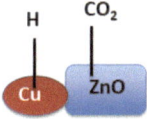

Figure 18.5: Pictorial view of the adsorption of hydrogen on the Cu particles and the CO$_2$ molecule on the ZnO phase.

The calculated potential energy surface of methanol synthesis via CO$_2$ hydrogenation is shown in Figure 18.6 [34].

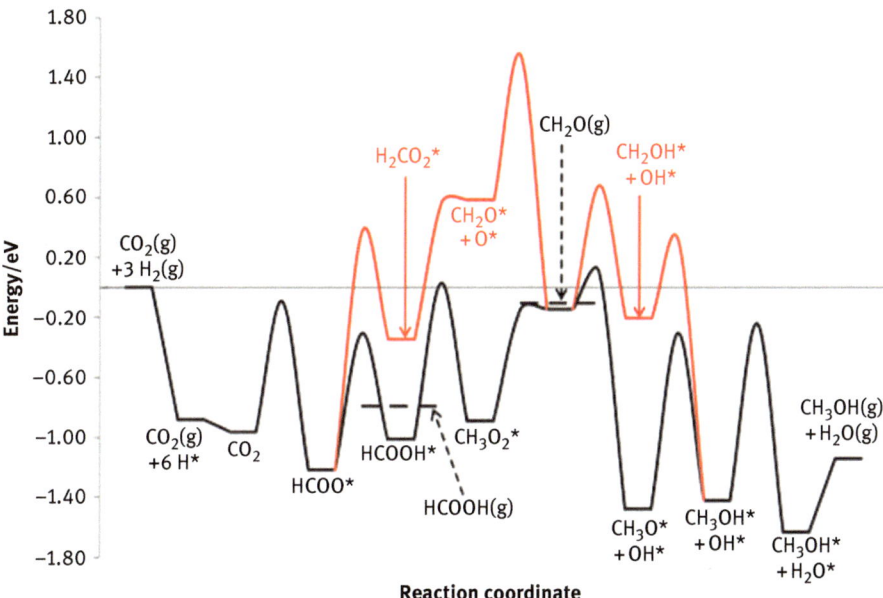

Figure 18.6: Potential energy surface of methanol synthesis via CO$_2$ hydrogenation. Reprinted with permission of the American Chemical Society.

Adsorbed CO$_2$ is transformed into adsorbed formate (HCOO*) upon reaction with an adsorbed hydrogen atom (H*), which may be further converted to adsorbed formic acid and then adsorbed hydroxyl methoxy intermediate (CH$_2$OHO*). In the sequence, there then occurs the formation of adsorbed formaldehyde (CH$_2$O*) releasing an adsorbed hydroxy group (OH*). The subsequent reaction with adsorbed hydrogen yields an adsorbed methoxy group (CH$_3$O*), which ultimately leads to adsorbed methanol (CH$_3$OH*). Water is formed through the interaction of the adsorbed hydroxy group with an adsorbed hydrogen atom [34]. Figure 18.7 shows the elementary steps.

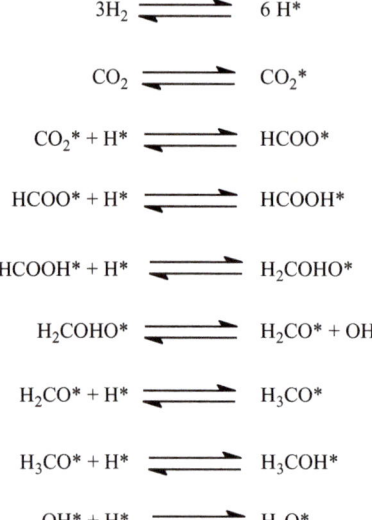

Figure 18.7: Possible mechanistic pathway of CO_2 hydrogenation to methanol. * denotes an adsorbed species.

18.6 Bifunctional catalysts for coupling methanol synthesis from CO_2 and the subsequent dehydration to dimethyl ether

The production of DME from methanol can be accomplished with the use of solid acid catalysts, at temperatures ranging from 250 to 400 °C and pressures above 18 bar [35]. This is the traditional route, with commercial units installed all over the world.

Dehydration of methanol to DME is exothermic by 23.5 kJ/mol. Ethylene and coke are formed as by-products at high temperatures. Acidic materials are used as catalysts in methanol dehydration. Among them, alumina, modified alumina, zeolites and mixed metal oxides are the most important. The acidity must be controlled to avoid side reactions, especially formation of hydrocarbons. Water plays a role in deactivating the catalyst, through adsorption on the acid sites.

On the other hand, DME can also be produced directly from syngas or CO_2 hydrogenation, coupling the methanol synthesis catalyst with an acidic material, capable of dehydrating the formed methanol molecule under the same reaction conditions. Most of the studies about DME production from syngas or CO_2 are conducted in the temperature range of 200–300 °C and pressures up to 70 bar [36]. However, no industrial process has been developed so far, either from syngas or CO_2. For the concomitant production of methanol and DME, a bifunctional catalytic system is necessary, to carry out, in a single step, the hydrogenation of CO_2 to methanol followed by dehydration to DME (Scheme 18.6).

$$2CO_2 + 6\,H_2 \;\; \overset{Cu.ZnO}{\rightleftharpoons} \;\; 2\,H_2O + 2\,CH_3OH \;\; \overset{\text{``H}^+\text{''}}{\longrightarrow} \;\; H_3COCH_3 + H_2O$$

Scheme 18.6: Direct DME formation from CO$_2$ hydrogenation using bifunctional catalytic system.

There are few studies about the direct conversion of CO$_2$ into DME compared with its direct production from syngas. The most traditional catalytic system for direct DME production involves Cu.ZnO-based catalysts in the presence of different solid acids, either as supports or in physical mixtures [37, 38]. Zeolite ZSM-5 or γ-Al$_2$O$_3$ is usually employed in methanol dehydration to DME [28, 29, 36, 39]. Other solid acid catalysts have also been exploited, including modified alumina with silica, TiO$_2$-ZrO$_2$, clays, ion exchange resins and zeolite, such as Y, Mordenite and Beta [36, 39–43].

In the case of bifunctional catalysts, the acidity of the solid acid support is not the only significant factor to maximise the production of DME. The choice of the preparation method also plays an important role. A study of bifunctional Cu.ZnO catalysts supported on Al$_2$O$_3$ or Nb$_2$O$_5$, for the direct hydrogenation of CO$_2$ to methanol, indicated that the method of preparation also affects the performance. Catalysts prepared by wet impregnation were more active than catalysts, of same composition, prepared by co-precipitation of the metals over the support [44]. This has been explained by the close contact between the metal precursors during the impregnation procedure.

The majority of the catalytic systems for the direct DME production from CO$_2$ are based on a physical mixture of the methanol synthesis catalysts, based on Cu. ZnO, and a solid acid.

Other catalyst systems have been studied for DME production from CO$_2$. An example is a catalyst of Pd supported on Ga$_2$O$_3$, which at mild reaction conditions showed conversion in the range of 5–14% and selectivity to DME varying from 4 to 53%. Nevertheless, the nature of the Ga$_2$O$_3$ phase has a great influence on the selectivity to DME and must be controlled during catalyst synthesis [45]. The utilisation of ZrO$_2$ as support is also reported, but there are disadvantages such as low surface area and the catalyst performance is influenced by its phase transformation. Zirconia presents three phases and the transition between them might change its catalytic properties [36].

Considering the thermodynamic of the reactions, the utilisation of a bifunctional catalyst can be more efficient when associated with a decrease in temperature and increase in pressure. Modifying these parameters may favour the synthesis of methanol and consecutively convert the produced methanol to DME, shifting the equilibrium of the reaction [2, 28, 29, 36].

The development of an efficient, bifunctional catalyst for the direct production of DME from CO$_2$ is still a matter of challenge, because of the need to control both catalytic functions, hydrogenation and dehydration. To improve the CO$_2$ conversion into methanol it is important to have a well-dispersed metal phase of high activity. On the other hand, the acid sites must be efficient in the dehydration of the formed methanol at the same reaction conditions, thus shifting the equilibrium of the methanol synthesis and providing the highest yield of DME. However, this acidity needs to be controlled to avoid possible by-products and coking of the active phase. Other important point to be considered is the separation of the formed DME from the remaining methanol in the effluent. This separation is not easy and could be solved using a distillation system, which would increase the costs of the process.

18.7 Industrial plant for CO_2 hydrogenation to methanol in Iceland

There is an industrial unit for methanol production by CO_2 hydrogenation, located in Iceland, which started operations in 2011. This plant was named after George Olah, and belongs to Carbon Recycling International (CRI) (Figure 18.8). The H_2 needed in the process is generated from the electrolysis of water using geothermal energy. In 2015, the capacity was expanded from 1.3 million litres per year to an annual capacity of more than 5 million litres of "Renewable Methanol". Actually, the plant recycles 5.5 tonnes of CO_2 per year, which would be released into the atmosphere if not processed to methanol. All the energy used in the plant is generated from hydropower and geothermal energy. The plant uses electricity to produce hydrogen, which is converted to methanol in a catalytic reaction with CO_2. CO_2 is captured from the flue gas released by a geothermal power plant located next to the CRI facilities [46].

Figure 18.8: George Olah renewable methanol plant in Iceland. Reprinted with permission by CRI.

According to the CRI, the expansion of new plants would be the solution to the European Union's demand for renewable fuels, since "Renewable Methanol" has the advantages of being a biodegradable fuel and less flammable than gasoline, being cleaner and safer [46].

Mitsui Corporation also built a demonstration plant in Japan, but it is no longer operational. This project was known as "The Chemical CO$_2$ Fixation and Utilisation" technology. The process included CO$_2$ capture from waste streams of industrial operations and methanol production from the hydrogenation of CO$_2$. The H$_2$ required for hydrogenation was produced by photocatalytic water splitting. A pilot-scale demonstration plant was built in 2008, in Osaka, Japan, producing 100 tonnes/year of methanol. It used a Cu.ZnO-based catalyst promoted with different metal oxides, in a fixed-bed reactor. Due to difficulties in the production of H$_2$ from water splitting through photocatalysis, the project was discontinued and did not proceed to full industrial scale [47].

A major challenge for the production of methanol from CO$_2$ is the need of sustainable and low cost-effective H$_2$ production. Hydrogen is usually obtained from natural gas by steam reforming, a process highly intensive in energy and based on fossil sources. Therefore, for the sustainable CO$_2$ conversion to methanol, it is necessary to use a H$_2$ source obtained from an environmentally friendly process, such as solar, wind and hydropower [5, 31].

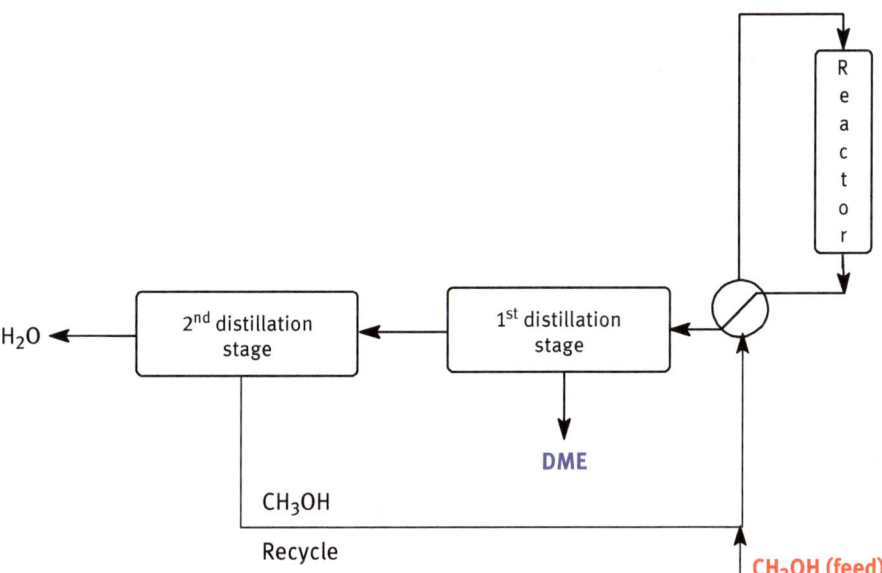

Figure 18.9: Schematic diagram of the process of methanol dehydration to DME over a fixed-bed reactor system.

There is no industrial plant for the direct DME, either from syngas or CO_2. All the plants are based on the dehydration of the methanol over acidic catalysts. Haldor-Topsoe, Lurgi, Mitsubishi Gas Chemical, Toyo Engineering Corporation, Johnson Matthey and Uhde are some of the companies with technology to convert methanol to DME and some industrial plants are in operation in the world, especially in Asia. Most of the technologies to produce DME from methanol use fixed-bed reactors, but reactive distillation can also be used. Figure 18.9 shows a simplified scheme of the process of methanol dehydration over a fixed-bed reactor system.

Although there is no commercial plant for the direct production of DME from syngas or CO_2, companies like Haldor-Topsoe, JFE and Korea Gas Corporation have developed technologies for this process. Slurry phase reactors present advantages due to lower costs and better temperature control, which is of prime importance to avoid run-away, since both reactions, methanol synthesis and methanol dehydration, are exothermic.

Bibliography

[1] Olah, G. A., Surya Prakash, G. K., Goeppert, A. Anthropogenic Chemical Carbon Cycle for a Sustainable Future. Journal of the American Chemical Society, v. 133, pp. 12881–12898, 2011.

[2] Goeppert, A.; Czaun, M.; Jones, J.-P.; Prakash, G. K. S.; Olah, G. A. Recycling of carbon dioxide to methanol and derived products – closing the loop. Chemical Society Reviews, v. 43, pp. 7995–8048, 2014.

[3] Olah, G. A. Beyond Oil and Gas: The Methanol Economy. Angewandte Chemie International Edition, v. 44, pp. 2636–2639, 2005.

[4] Olah, G. A.; Goeppert, A.; Prakash, G. K. S. Beyond Oil and Gas: The Methanol Economy, 2nd ed., Wiley-VCH, 2009.

[5] Olah, G. A.; Goeppert, A.; Prakash, G. K. S. Chemical Recycling of Carbon Dioxide to Methanol and Dimethyl Ether: From Greenhouse Gas to Renewable, Environmentally Carbon Neutral Fuels and Synthetic Hydrocarbons. Journal of Organic Chemistry, v. 74, pp. 487–498, 2009.

[6] Fernandes, D. R.; Rosenbach, N. JR.; Mota, C. J. A. Catalytic conversion of chloromethane to methanol and dimethyl ether over metal-exchanged zeolite Y. Applied Catalysis A: General, v. 367, pp. 108–112, 2009.

[7] Fernandes, D. R.; Carvalho, T. L.; MOTA, C. J. A. Catalytic Conversion of Chloromethane to Methanol and Dimethylether over Two Catalytic Beds: A Study of Acid Strength. Brazilian Journal of Petroleum and Gas, v. 4, pp. 83–89, 2010.

[8] Khaleel, A.; Shehadi, I.; Al-Marzouqi, A. Catalytic conversion of chloromethane to methanol and dimethyl ether over mesoporous γ−alumina. Fuel Processing Technology, v. 92, pp. 1783–1789, 2011.

[9] Cheng, W.-H.; Kung, H. H. Methanol Production and Use, Marcel Dekker, New York, 2016.

[10] Weissermel, K.; Arpe, H. J. Industrial Organic Chemistry, 4th ed., Wiley-VCH, Weinheim, 2003.

[11] Pinto, C. A.; Guarieiro, L. N.; Resende, M. J. C.; Ribeiro, N. M.; Torres, E. A.; Lopes, W. A.; Pereira, P. A. D.; Andrade, J.B. Journal of the Brazilian Chemical Society, v. 16, p. 1313, 2005.

[12] Tabak, S. A.; Krambeck, F. J.; Garwood, W. E. Conversion of Propylene and Butylene over ZSM-5 Catalyst. American Institute of Chemical Engineers Journal, v. 32, p. 1526, 1986.

[13] Keil, F. J. Methanol-to-hydrocarbons: process technology. Microporous and Mesoporous Materials, v. 29, p. 49–66, 1999.

[14] Stöcker, M. Methanol-to-hydrocarbons: catalytic materials and their behavior. Microporous and Mesoporous Materials, v. 29, pp. 3–48, 1999.

[15] Shen, W.-J.; Jun, K. W.; Choi, H. S.; Lee, K. W. Thermodynamic Investigation of Methanol and Dimethyl Ether Synthesis from CO$_2$ Hydrogenation. Korean Journal of Chemical Engineering, v. 17, pp. 210–216, 2000.

[16] Liu, X.-M.; Lu, G. Q.; Yan, Z.-F.; Beltramini, J. Recent Advances in Catalysts for Methanol Synthesis via Hydrogenation of CO and CO$_2$. Industrial & Engineering Chemistry Research, v. 42, pp. 6518–6530, 2003.

[17] Mota, C. J. A.; Monteiro, R. S. Química e sustentabilidade: novas fronteiras em biocombustíveis. Química Nova, v. 36, n° 10, pp. 1483–1490, 2013.

[18] Lloyd, L. Handbook of Industrial Catalysts. Capítulo 10, Methanol Synthesis, pp. 421–435. Editora Springer, 2011.

[19] Fujitani, T.; Nakamura, J. The chemical modification seen in the Cu/ZnO methanol synthesis catalysts. Applied Catalysis A: General, v. 191, pp. 111–129, 2000.

[20] Choi, Y.; Futagami, K.; Fujitani, T.; Nakamura, J. The role of ZnO in Cu/ZnO methanol synthesis catalysts: morphology effect or active site model? Applied Catalysis A: General, v. 208, pp. 163–167, 2001.

[21] Nakamura, J.; Uchijima, T.; Kanai, Y.; Fujitani, T. The role of ZnO in Cu/ZnO methanol synthesis catalysts. Catalysis Today, v. 28, pp. 223–230, 1996.

[22] Razali, N. A. M.; Lee, K. T.; Bhatia, S., Mohamed, A. R. Heterogeneous catalysts for production of chemicals using carbon dioxide as raw material: A review. Renewable and Sustainable Energy Reviews, v. 16, pp. 4951–4964, 2012.

[23] Saito, M.; Fujitan, T.; Takahara, I.; Watanabe, T.; Takeuch, M.; Kana, Y.; Moriya, K.; Kakumoto, T. Development of Cu/ZnO-Based High Performance Catalysts for Methanol Synthesis by CO$_2$ Hydrogenation, v. 36, n° 6–9, pp. 577–580, 1995.

[24] Arena, F.; Barbera, K.; Italiano, G.; Bonura, G.; Spadaro, L.; Frusteri, F. Synthesis, characterization and activity pattern of Cu–ZnO/ZrO$_2$ catalysts in the hydrogenation of carbon dioxide to methanol. Journal of Catalysis, v. 249, pp. 185–194, 2007.

[25] Mota, C. J. A.; Monteiro, R. S.; Maia, E. B. V.; Pimentel, A. F.; Miranda, J. L.; Alves, R. M. B.; Coutinho, P. L. A. O Dióxido de Carbono como Matéria-Prima para a Indústria Química. Produção do Metanol Verde. Revista Virtual de Química, v. 6, n° 1, pp. 44–59, 2013.

[27] Ganesh, I. Conversion of carbon dioxide into methanol–a potential liquid fuel: Fundamental challenges and opportunities (a review). Renewable and Sustainable Energy Reviews, v. 31, pp. 221–257, 2014.

[28] Wang, S.; Mao, D.; Guo, X.; Wu, G.; Lu, G Dimethyl ether synthesis via CO$_2$ hydrogenation over CuO-TiO$_2$-ZrO$_2$/HZSM-5 bifunctional catalysts. Catalysis Communications, v. 10, pp. 1367–1370, 2009.

[29] Wang, W.; Wang, S.; Ma, X.; Gong, J. Recent advances in catalytic hydrogenation of carbon dioxide. Chemical Society Reviews, v. 40, pp. 3703–3727, 2011.

[30] Saeidi, S.; Amin, N. A. S.; Rahimpour, M. R. Hydrogenation of CO$_2$ to value-added products – A review and potential future developments. Journal of CO$_2$ Utilization, v. 5, pp. 66–81, 2014.

[31] Jadhav, S. G.; Vaidya, P. D.; Bhanage, B. M.; Joshi, J. B. Catalytic carbon dioxide hydrogenation to methanol: A review of recent studies. Chemical Engineering Research and Design, v. 92, p. 2557–2567, 2014.

[32] Behrens, M.; Schlögl, R. How to prepare a good Cu/ZnO catalyst or the role of solid state. Zeitschrift für Anorganische und Allgemeine Chemie, v. 639, pp. 2683–2695, 2013.

[33] Martin, O.; Mart, A.J.; Mondelli, C.; Mitchell, S.; Segawa, T.F.; Hauert, R.; Drouilly, C.; Ferr, D.C.; Ramirez J.P. Indium Oxide as a Superior Catalyst for Methanol Synthesis by CO2 Hydrogenation Angew. Chem. Int., v.55, pp. 6261–6265, 2016.

[34] Grabow, L. C. and Mavrikakis, M. Mechanism of Methanol Synthesis on Cu through CO2 and CO Hydrogenation. ACS Catalysis, v.1, pp. 365–384, 2011.

[35] Yaripour, F.; Baghaei, F.; Schmidt, I.; Perregaard, J. Catalytic dehydration of methanol to dimethyl ether (DME) over solid-acid catalysts. Catalysis Communications, v. 6, pp. 147–152, 2005.

[36] Azizi, Z; Rezaeimanesh M.; Tohidian T.; Rahimpour, M.R. Dimethyl ether: A review of technologies and production challenges. Chemical Engineering and Processing, 82, 150–172, 2014.

[37] Ereña, J.; Garoña, R.; Arandes, J. M.; Aguayo, A. T.; Bilbao, J. Effect of operating conditions on the synthesis of dimethyl ether over a CuO-ZnO-Al$_2$O$_3$/NaHZSM-5 bifunctional catalyst. Catalysis Today, v. 107–108 p. 467–473, 2005.

[38] Sun, K.; Lu, W.; Qiu, F.; Liu, S.; XU, X. Low-temperature synthesis of DME from CO$_2$/H$_2$ over Pd-modified CuO–ZnO–Al$_2$O$_3$–ZrO$_2$/HZSM-5 catalysts. Catalysis Communications, v. 5, pp. 367–370, 2004.

[39] Naik, S. P.; Ryu, T.; BUI, V.; Miller, J. D.; Drinnan, N.B.; Zmierczak, W. Synthesis of DME from CO$_2$/H2 gas mixture. Chemical Engineering Journal, v. 167, pp. 362–368, 2011.

[40] Bonura, G.; Cannilla, C.; Frusteri, L.; Frusteri, F. The influence of different promoter oxides on the functionality of hybrid CuZn-ferrierite systems for the production of DME from CO2-H2 mixtures. Applied Catalysis A, 544, 21–29, 2017.

[41] Frusteri, F.; Migliori, M.; Cannilla, C.; Frusteri, L.; Catizzone, E.; Aloise, A.; Giordano G.; Bonura, G. Direct CO2-to-DME hydrogenation reaction: New evidences of a superior behaviour of FER-based hybrid systems to obtain high DME yield. Journal of CO2 Utilization.18, 353–361, 2017.

[42] Kattel, S.; Ramírez, P. J.; Chen, J. G.; Rodriguez, J. A.; Liu, P. Active sites for CO2 hydrogenation to methanol on Cu/ZnO catalysts, Science, 355, 1296–1299, 2017.

[43] Phienluphon R.; Pinkaew, K.; Yang, G.; Li, J.; Wei, Q.; Yoneyama, Y. Vitidsant, T.; Tsubaki, N. Designing core (Cu/ZnO/Al2O3)–shell (SAPO-11) zeolite capsule catalyst with a facile physical way for dimethyl ether direct synthesis from syngas. Chemical Engineering Journal. 270, 607–611, 2015.

[44] Da Silva, R. J.; Pimentel, A. F.; Monteiro, R. S.; Mota, C. J. A. Synthesis of methanol and dimethyl ether from the CO2 hydrogenation over Cu.ZnO supported on Al2O3 and Nb2O5. Journal of CO2 Utilization, 15, 83–88, 2016.

[45] Oyola-Rivera, O.; Baltanás, M. A.; Cardona-Martínez, N. CO$_2$ hydrogenation to methanol and dimethyl ether by Pd–Pd$_2$Ga catalysts supported over Ga$_2$O$_3$ polymorphs. Journal of CO$_2$ Utilization, v. 9, p. 8–15, 2015.

[46] World's Largest CO2 Methanol Plant. Carbon Recycling International. Available in: <http://www.carbonrecycling.is/projects/>. Accessed in July 2018.

[47] Mitsui Chemicals to Establish a Pilot Facility to Study a Methanol Synthesis Process from CO2. Mitsui Chemicals. Available in: <https://www.mitsuichem.com/en/release/2008/080825e.htm> Accessed in July 2018.

Sigrid Douven, Hana Benkoussas, Carolina Font-Palma
and Grégoire Léonard

19 Towards sustainable methanol from industrial CO$_2$ sources

19.1 Introduction

Worldwide methanol production was 70 Mt in 2015 and is projected to increase to 95 Mt by 2020 [1]. This fast growing consumption offers potential for the production of methanol via CO$_2$ re-use. In addition, methanol could become a key chemical in the energy transition, provided that the energy needed for methanol synthesis is supplied by renewable energies in the form of hydrogen resulting from water electrolysis. Methanol, with a molecular formula of CH$_3$OH, is the simplest alcohol. Methanol is a colourless liquid, soluble in water and with a mild alcoholic odour. Table 19.1 shows properties of methanol, including technical specifications of methanol purity.

Methanol is mostly consumed in Northeast Asia as shown in Figure 19.1, and its main use is as feedstock for the production of other chemicals such as formaldehyde or acetic acid. China is increasingly using methanol to produce olefins, the so-called methanol to olefins (MTO), where nearly 1 in 5 tonnes of methanol are predicted to be used for MTO production by 2021 to meet the growing Chinese chemical demand [4]. It may also be used to make fuel additives methyl-tert-butyl ether/methyl tert-amyl ether (MTBE/TAME) or other diesel-like fuels such as dimethyl ether. Even though methanol has a lower energy content than gasoline (46.5 MJ/kg), it also has a long history as a fuel in racing cars due to its power and safety properties relative to gasoline: harder ignition, burns more slowly and does not emit black smoke [5]. Methanol is a cleaner burning fuel, which makes it also a good candidate as a gasoline blending component. Studies using methanol blends in vehicles started in the late 1970s and 1980s, after the crude oil price shock of the 1970s. As a result, large vehicle fleet tests (approx. 1000 vehicles each) using methanol blends of up to 15 vol.% (M15) have been successfully run by automakers or oil companies in Sweden, Germany, New Zealand and China [6]. Blends with 85 vol.% (M85) were tested mostly in California, USA, showing the same or better performance than gasoline counterparts, where up to 15,000 vehicles and 100 refuelling stations were installed. However, despite the environmental benefits, automakers and refineries convinced the U.S. Environmental Protection Agency (EPA) that reformulated gasoline could meet emission standards, and by mid-1990s other fuels such as ethanol and natural gas stopped the interest in methanol [5].

Sigrid Douven, Hana Benkoussas, Grégoire Léonard Department of Chemical Engineering, University of Liège, Liège, Belgium
Carolina Font-Palma, Department of Chemical Engineering, Thornton Science Park, University of Chester, Chester, United Kingdom

https://doi.org/10.1515/9783110665147-019

Table 19.1: Methanol technical data sheet [2].

Physico-chemical properties		Usual composition specifications	
Molecular weight	32.04 g/mol	Purity	99.85 wt% min
Specific gravity (20°C)	0.791–0.793	Water (impurity)	0.10 wt% max
Boiling point	64.6°C	Acetone (impurity)	30 mg/kg max
Freezing point	−97.8°C	Ethanol (impurity)	50 mg/kg max
Energy content [3]	726.3 kJ/mol, 22.7 MJ/kg	Chloride (impurity)	0.5 mg/kg max
Energy of vapourisation [3]	38.5 kJ/mol		

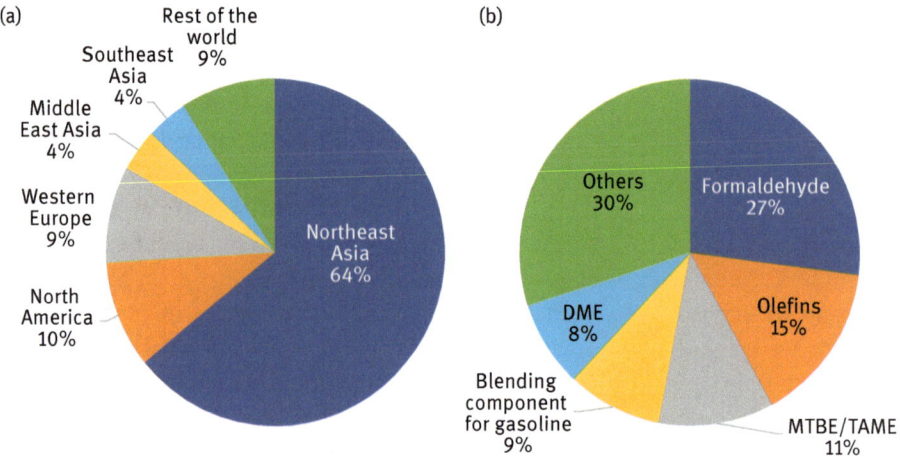

Figure 19.1: (a) World consumption of methanol, and (b) its uses (IHS, 2017b) [7].

Due to methanol's characteristics, Nobel laureate George Olah envisioned the "methanol economy", where methanol substitutes fossil fuels as raw material to synthesise hydrocarbons, as ground transportation fuel and/or for energy storage [3]. Should this methanol economy become a reality, the methanol demand would significantly increase, and its production from CO_2 re-use and renewable energies would become key towards a sustainable future.

19.2 Processes for methanol synthesis

19.2.1 State of the art of conventional methanol synthesis

Conventional industrial methanol synthesis involves reactions where the synthesis gas (also called syngas comprising a mixture of CO, CO_2 and H_2) is catalytically converted [8]. The catalytic reactions occur with the aid of the commercial catalyst $Cu/ZnO/Al_2O_3$ at conditions in the range of 250–300 °C and 50–100 bar [9]:

$$CO + 2H_2 \leftrightarrow CH_3OH \quad \Delta H_{25°C} = -90.5kJ/mol \tag{19.1}$$

$$CO_2 + 3H_2 \leftrightarrow CH_3OH + H_2O \quad \Delta H_{25°C} = -49.5kJ/mol \tag{19.2}$$

The reactions are exothermic, and heat removal is essential to keep the equilibrium towards the products. The conventional process of methanol synthesis involves three main stages: (i) the production of synthesis gas derives typically from natural gas by steam reforming, autothermal reforming or partial oxidation, but it could also come from different carbonaceous fuels, such as light naphtha, crude oil, heavy oils or biomass [10]; (ii) the synthesis of methanol starts by compressing the syngas to the desired operating pressure using compressors with intercooling, then the pressurised syngas is heated and sent to the reactor to produce methanol; (iii) the processing of methanol as observed in reaction (2) produces water. The crude methanol is sent to a separator and flashed and then is sent to a distillation column to remove by-products, for example, water and other low and high boiling components [9].

Different reactor designs exist for methanol synthesis. One design is the multistage adiabatic reactor with interstage cooling to quench the reaction either by heat removal or cold feed injection. Another design is shell and tube reactors, in which the reaction takes place in tubes that are cooled by water flowing in a shell, so the reactor is quasi-isothermal [9]. More information about reactor designs for methanol production can be found in a recent state-of-the-art review [11].

Methanol production typically emits in the range of 0.3–0.7 t CO_2/t methanol for steam reforming and up to 1.4 t CO_2/t methanol for partial oxidation of oil [12]. The CO_2 emissions originate from direct process emissions and indirect emissions coming from the electricity supply of fossil fuel power plants.

19.2.2 Innovations in methanol synthesis process

Efforts to mitigate CO_2 emissions have led to a growing interest in the conversion of CO_2 into methanol. Among the alternative options are conversion by: (a) direct hydrogenation of CO_2, (b) two-stage catalytic route consisting of CO_2 conversion into CO through a reverse water gas shift (RWGS) reaction and then hydrogenation of CO; (c) an electrolytic cell for CO_2 reduction, which produces oxygen as a by-product; and (d) a photo-electrochemical cell for CO_2 reduction with water using solar energy [13].

For the catalytic routes, the catalyst $Cu/ZnO/Al_2O_3$ typically used for CO_2-rich feedstocks presents activity issues. Therefore, $Cu–ZnO/ZrO_2$ catalysts have been proposed as more active, selective and with longer life for CO/CO_2 hydrogenation [14]. At low pressure the conventional catalyst produces more CO, while Ni–Ga catalysts have shown to reduce CO production and suppress the RWGS reaction at atmospheric pressure [15]. The two catalytic routes have been compared through simulation work, where direct hydrogenation of CO_2 presents higher economical and energetic efficiency. Nonetheless, the technology selection may greatly depend on the plant location with respect to availability of cheap electricity and/or environmental constraints [16].

For the electrochemical conversion of CO_2 to methanol, two competing reactions occur; one called hydrogen evolution reaction, where hydrogen is produced by water electrolysis and then is combined with CO_2 to form methanol, and the other is the direct reduction of CO_2 into methanol inside the electrolysis cell. For high selectivity of the direct reduction of CO_2, copper and oxidised Cu-based electrodes are promising options. However, there are several technical issues that need overcoming to make this option suitable for commercial application. The challenges include the development of electrodes with the required stability, activity and selectivity, and a better understanding of the reaction mechanisms that would lead to the development of new catalysts [17].

The design of reactors for methanol synthesis from CO_2 is also one of the upcoming challenges. For instance, efficient heat transfer is a key issue due to the exothermicity of the methanol synthesis reactions. Several studies have recently addressed this challenge by proposing the use of highly conductive structured catalysts for efficient heat transfer and low pressure drop [18, 19]. A promising reactor design has been experimentally demonstrated in two studies [20, 21]. The authors observed condensation of the reaction products (water and methanol), while unreacted gases were recycled internally within the reactor. As a result, the thermodynamic equilibrium limitation can be bypassed, leading to a practical conversion of almost 100%. Such a reactor design would avoid the need for a recycling loop and thus allow once-through operation.

Furthermore, important challenges are also upcoming regarding the origin of the reactants for methanol synthesis. Currently, fossil fuel resources are usually employed to produce the syngas needed for methanol synthesis. Instead, biomass or CO_2 could serve as carbon sources for methanol synthesis. Different industrial CO_2 sources may be identified as discussed further in Section 19.3. An even larger challenge consists of supplying the hydrogen needed for catalytic routes. This should be produced from renewables, such as electrolysis of water via carbon-free electricity or biologically, in order to make methanol synthesis an environmentally friendly option.

There are a few industrial methanol plants that can be found using exclusively CO_2 hydrogenation. The George Olah Renewable Methanol Plant in Svartsengi, Iceland, is a power-to-methanol facility that produces methanol by using CO_2 from a geothermal power plant and hydrogen produced by electrolysis of water using renewables such as hydro, geothermal and wind. It is operated by Carbon Recycling International (CRI) since 2012; the methanol is traded as Vulcanol™ and the plant expanded its capacity to 4,000 tonnes (5 ML) per year in 2015, using 5,500 tonnes/year of CO_2 [22]. Producers reported that their technology could reduce carbon emissions by 85% compared with plants using syngas from fossil fuels. Plans for a first

CO_2-to-renewable methanol plant in Canada by Blue Fuel Energy (BFE) anticipate a capacity of 2.5 ML/day [23]. These cases will be further discussed in Section 19.3.2. In addition, waste-to-methanol has been proposed to reduce municipal solid waste and produce renewable chemicals [24]. Enerkem, a Canadian company for waste to biofuels, begun operations of its first full-scale commercial facility for refuse-derived fuel (RdF) at Enerkem Alberta Biofuels. This plant started converting RdF into syngas and then into methanol in 2015 and ethanol in 2017. The company claims a capacity of 38 ML/year [25].

In the last decade, companies such as Carbon Engineering in Canada or Climeworks in Switzerland have developed technologies based on temperature-swing adsorption to capture CO_2 directly from the atmosphere, while other technologies consider humidity-swing adsorption [26]. Since the CO_2 concentration in air is much lower than in industrial off-gases (0.04 vol.% vs about 12 vol.% from a typical coal power plant), the challenges of direct air capture lie in reducing the required energy consumption. Then, methanol produced with CO_2 captured from air could be considered as a sink for CO_2 and thus be associated with negative carbon emissions. Kothandaraman et al. [27] proposed a method that uses a soluble polyamine and a ruthenium-based catalyst for the capture and direct CO_2-to-methanol conversion in a homogeneous ethereal solvent system operating at 125–165°C. With such a system, up to 79% of the CO_2 captured from air converts to methanol, opening the way to promising air-to-fuel processes.

19.3 CO_2 utilisation from industrial sources

The opportunity of a short-term development of CO_2 re-use technologies is dependent on the CO_2 availability and purity. In addition to the recent launching of several large-scale CO_2 capture units dedicated to the power sector, applications separating CO_2 in large amounts are already in operation in industrial processes such as natural gas processing, and hydrogen and ammonia production, where the separation of CO_2 is part of the production process. Indeed, such industrial processes produce CO_2 as a by-product, usually with high purity and in large quantities. Consequently, these processes would be suitable for an early development of CO_2 re-use technologies as only simple water removal and compression steps are required in addition to the existing process, making separation less expensive. Other large-scale sources of CO_2 such as power plants, cement, steel or ethanol industries have little experience in CO_2 capture, but the techno-economic feasibility of CO_2 capture and re-use to methanol feedstock is becoming a topic of considerable interest. Indeed, methanol can be seen as a versatile chemical that could become a key building block in CO_2 re-use technologies. In this section, we present a detailed case study for ammonia plants and will then consider other sources such as iron and steel manufacture, and ethanol plants.

19.3.1 Case of CO_2 re-use from ammonia plant: Process design and energy integration

Ammonia production is a large industrial source of high purity CO_2. The capture of CO_2 in ammonia plants is a mature technology since the separation of CO_2 and H_2 is required within the process. As a consequence, CO_2 is a by-product of ammonia production. Depending on the hydrogen source used for the production of ammonia, two main technologies are used: (i) steam reforming of light hydrocarbons (methane and naphtha), and (ii) partial oxidation of heavier hydrocarbons (heavy oils and coal). Even if processes can somewhat differ, the main steps can be summarised as follows: (i) the steam reforming of methane to produce synthesis gas, that is, CO and H_2, (ii) the shift conversion of CO and H_2O into CO_2 and H_2, (iii) the CO_2 removal, (iv) the methanation (removal of residual carbon oxides) and (v) the ammonia synthesis. If coal or naphtha is used, extra processes are required to gasify them before the reforming.

The selected feedstock and technology determines the amount of CO_2 produced. Typically, a modern plant based on natural gas leads to 1.7 tonnes CO_2/tonne NH_3 [13], whereas an existing plant based on partial oxidation of coal can reach 4.6 tonnes CO_2/tonne NH_3 [28]. However, ammonia plants are often integrated with urea plants as the main reactants of urea production are ammonia and CO_2. In 2017, the ammonia production was 174 Mt NH_3 [29] with around 25% produced from coal [30], mainly in China, while urea production reached 170 Mt in 2017 [29]. Assuming a complete conversion of CO_2 to urea, 0.73 tonnes CO_2/tonne urea are required. With a low CO_2-emitting plant, that is, 1.7 tonnes CO_2/tonne NH_3, and if the production of urea that is not integrated to ammonia plants is neglected (only a few plants in the world), 171.7 tonnes CO_2 are still emitted from worldwide ammonia synthesis, after urea production.

Often large amounts of high-purity CO_2 that are not used for the conversion of ammonia into urea could be available for synthesis of CO_2-sourced materials, such as methanol. Co-production of ammonia and methanol is not new as it has been demonstrated at industrial scale since 1993, but not as a CO_2 re-use technology [31]. Indeed, starting from the natural gas reforming step, several plants have been built that allow part of the syngas to produce methanol instead of ammonia [32]. Incentives for this technology are the product diversification, product flexibility and better marketing perspectives. In 2014, the process licensor, Haldor Topsøe, started the construction of a new methanol ammonia co-production plant for the Russian company Shchekinoazot with a planned capacity of 1,350 tonnes/day of methanol and 415 tonnes/day of ammonia.

In the following section, a model for the production of methanol starting from the CO_2 removal unit of an ammonia plant is developed. This study differs from previous industrial projects by combining methanol and ammonia synthesis with water electrolysis using renewable energy as the source for hydrogen used to produce methanol instead of the hydrogen present in the syngas. The global model takes into account the three submodels: (i) CO_2 capture, (ii) water/CO_2 co-electrolysis and (iii) methanol synthesis including its purification. A focus is made on the process integration, that is, the estimation of the energy requirements of the whole process and the optimisation of the energy consumption. Modelling assumptions are detailed in the following sections.

19.3.1.1 Model description

This study considers the particular case of methanol production from CO_2 effluents of the capture unit of an ammonia plant. A model has been developed in Aspen Plus V8.8 starting from the CO_2 capture step and is based on a plant designed to treat 208,000 Nm^3/h of flue gas at 26.4 bar and 37 °C, containing 60.2 mol% H_2, 21.7 mol% N_2, 17.9 mol% CO_2 and traces of H_2O, as in Alvis et al. [33]. Figure 19.2 shows the global simplified process flowsheet.

CO_2 capture
In the ammonia synthesis process, CO_2 is separated from the N_2/H_2 mixture using an amine solvent; in this case, MDEA (*N*-methyldiethanolamine) is promoted by piperazine (45 wt% MDEA/5 wt% PZ). In brief, this unit consists mainly of an absorber where CO_2 reacts with the liquid amine solvent. The solvent is then regenerated in the stripper and recycled back to the absorber as in Figure 19.2. Table 19.2 shows the CO_2 content of the treated gas at the outlet of the absorber maintained at 500 ppm, while the purity of the CO_2 after the stripper reaches 98.5%. This CO_2, considered as a by-product of ammonia production, is the feedstock for the methanol synthesis. This stream is the inlet of the subsequent step, the co-electrolysis. Finally, the ENRTL-RK method is used for evaluating the thermodynamic properties. This model is based on the electrolyte version of the Non-Random Two-Liquid thermodynamic model for the liquid phase combined with the Redlich-Kwong model for the gas phase.

Table 19.2: CO_2 capture: composition of the flue gas, the outlet of the absorber and the outlet of the stripper.

	Molar composition (%)				Molar flow rate (kmol/h)	T (°C)
	H_2O	H_2	CO_2	N_2		
Flue gas	0.25	60.2	17.9	21.7	9,217	37
Absorber outlet	0.45	73.1	0.05	26.3	7,564	51
Stripper outlet	1.44	0.01	98.5	0	1,480	30

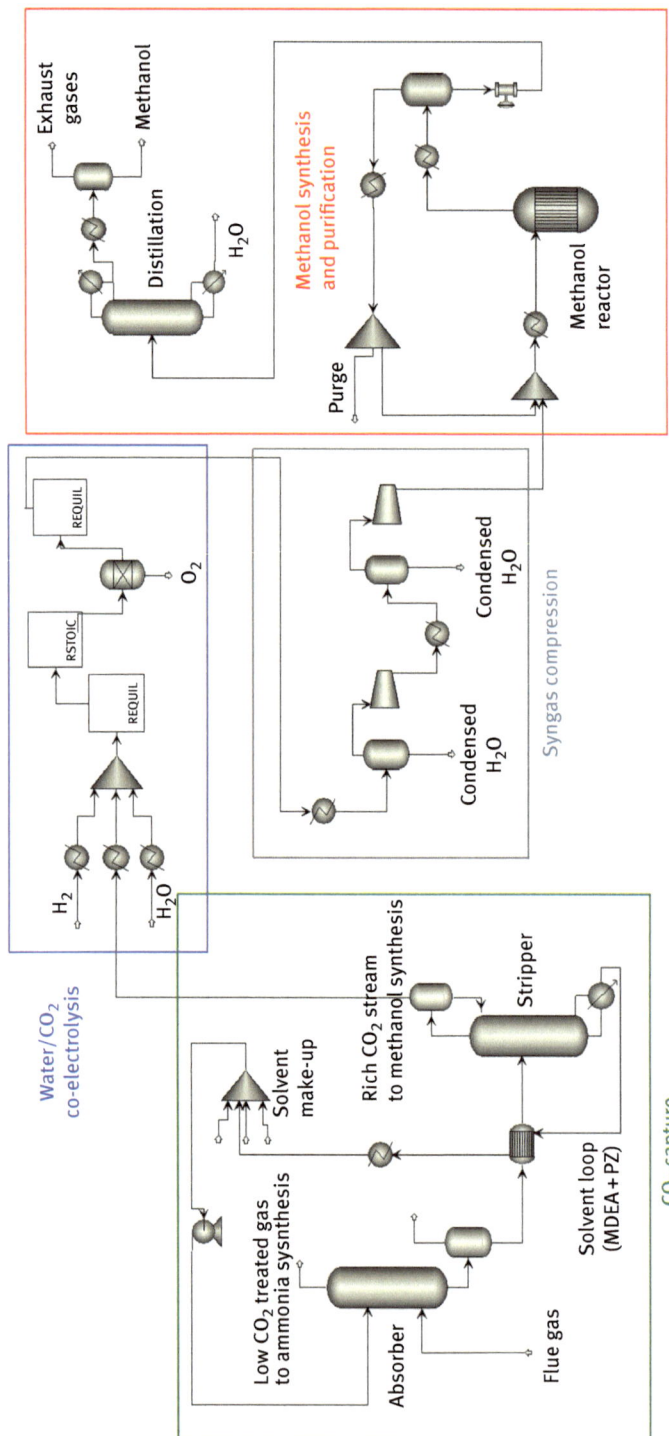

Figure 19.2: Global flowsheet of the process, including CO_2 capture, water/CO_2 co-electrolysis, methanol synthesis and purification.

Water/CO_2 co-electrolysis

The H_2O/CO_2 co-electrolysis is performed in a Solid Oxide Electrolysis Cell (SOEC) operating at 850°C and atmospheric pressure. This technology is still at the R&D stage but it has been chosen for its promising potential. Indeed, the high operating temperature of the SOEC is thermodynamically and kinetically advantageous [34, 35]. Thermodynamically, high temperatures are favourable to the endothermic H_2O and CO_2 splitting reactions. It also results in faster reaction kinetics and lower internal resistance, thus decreasing by 20-30% the electrical energy demand to reduce, respectively, carbon dioxide and water [36] and enabling potentially high efficiency. Operating at high temperature, part of the energy required for the electrolysis can be obtained from the Joule heat produced inside the cell. Furthermore, the co-electrolysis of both CO_2 and H_2O can be achieved at high temperature, while this is not the case in conventional water electrolysis (acid or alkaline).

The RK-Aspen thermodynamic method is used and is based on the Redlich-Kwong equation of state. The conversion of CO_2 and H_2O is fixed to 70% during electrolysis, in agreement with experimental results from the literature [34]. The sub-process is modelled in three stages [37]:

1. Equilibrium between reactants according to the RWGS reaction: $H_2 + CO_2 \leftrightarrow H_2O + CO$;
2. CO_2 and H_2O decomposition at 850 °C: $H_2O \rightarrow H_2 + \frac{1}{2} O_2$ and $CO_2 \rightarrow CO + \frac{1}{2} O_2$;
3. Similar equilibrium as stage 1 between reaction products.

Between stages 2 and 3, the oxygen is separated from other components by a separator simulating the anodic outlet. The oxygen, seen as a by-product of water/CO_2 electrolysis, is in fact valuable and could be sold in order to make the process more profitable as discussed in Section 19.4.

The inlet water flow rate was adjusted in order to reach a molar ratio of 100/45/10 for H_2O/CO_2/H_2. The aim is to reach a H_2-to-carbon ratio close to the ideal ratio of 3:1 at the outlet of the electrolysis, more suitable for methanol synthesis according to reaction (19.2). The ratio is actually closer to 2 according to Table 19.3, but this does not prevent the global process from reaching high efficiency and conversion. In addition, hydrogen at the inlet is used to prevent coking of the electrodes. Taking into account the three modelled stages, conversion for H_2O is 69% and the conversion for CO_2 is

Table 19.3: Co-electrolysis: composition of the electrolyser inlet stream and cathode outlet stream.

	Molar composition (%)				Molar flow rate (kmol/h)	T (°C)
	H_2O	H_2	CO	CO_2		
Inlet	64.5	6.4	0	29	5,364	850
Outlet	20.2	50.7	21.2	7.8	5,364	850

73% (Table 19.3). Those values are in agreement with the assumption of a conversion of 70% for both components for the second stage only.

Methanol synthesis

The methanol synthesis is modelled on the basis of a condensation reactor [21], so that no recompression step is needed for the recycle loop as the recycling occurs within the reactor, thanks to a density gradient. This kind of reactor is not commercially mature yet, but in the present work, we make the assumption to anticipate the development of such new technologies. The reactor operates at 250 °C and 50 bar, and is assumed at equilibrium in a first approximation. The formation of methanol from CO_2 and the water gas shift reaction are considered without side reactions, as by-products were reported by Bos and Brilman [21] to be present only in trace concentration. At the reactor outlet, the unreacted gases are separated from methanol by condensation at 25 °C and 50 bar. The recycling loop for unreacted gases allows increasing the conversion. As high selectivity towards methanol is reached, the only purification step is the removal of water done in a distillation column. A final selectivity of 98% towards methanol is reached (Table 19.4) and corresponds to a productivity of 38.7 tonne/h. The RK-Aspen model is used, except for the distillation column, where the ELECNRTL model is used due to the polar mixture in the liquid phase. The condensation of methanol in the reactor allows reaching high conversion rates, larger than 99.5% in agreement with Bos and Brilman [21].

Table 19.4: Methanol synthesis and purification: composition of (i) the inlet and (ii) outlet streams of the reactor and (iii) of the methanol final stream (after purification).

	Molar composition (%)					Molar flow rate (kmol/h)	T (°C)
	H_2O	H_2	CO	CO_2	CH_3OH		
(i) Compressor outlet	0.3	63.4	26.5	9.8	0	4,292	267
(ii) Reactor outlet	9.1	0.1	0.1	13.8	76.9	1,624	25
(iii) Methanol outlet	0.1	0	0	2.2	97.7	1,199	25

19.3.1.2 Energy integration

Heat integration was performed using the pinch method as described by Douglas [38]. The most energy consuming blocks of the global process are the stripper for the recovery of CO_2 (in capture) and the electrolysis unit, the latter representing around 50% of the total energy demand of the process. However, as energy for the electrolyser is provided as electricity (operation under thermo-neutral voltage), it is not considered in the heat integration. For the other units, the principle is to exchange heat between hot streams that need to be cooled down and cold streams that need to be warmed up. A minimum temperature difference of 20 K in the heat exchangers was assumed. As shown in Figure 19.3, the overlap between hot and cold composite curves represents the maximum heat recovery possible within the process, that is, 141 MW exchanged between hot and cold streams, corresponding to a decrease close to 60% of the energy requirement. This is a maximum value without any economic consideration.

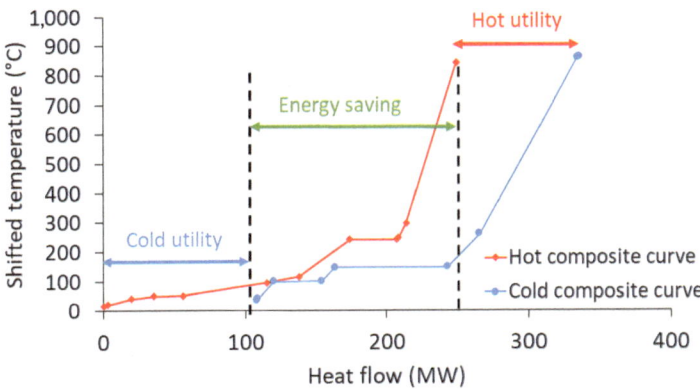

Figure 19.3: Pinch analysis of the global process.

A heat exchanger network can be established to recover the maximum heat and 11 heat exchangers are required in the optimal case. Of course, the cost of heat exchangers is an important factor not taken into account here but that should be considered in further steps. When looking at the heat exchanger network, 48 MW from the hot syngas at the outlet of the electrolyser can be used to warm up the water inlet of the electrolyser, and 33 MW of the heat generated by the methanol synthesis can supply almost all the heat needed at the reboiler of the methanol distillation column. These two exchangers already represent 57% of the maximum heat that would be recovered in a perfect heat exchanger network. In addition to the heat exchangers, cold and hot utilities are still required, but the use of heat integration significantly reduces the energy requirements from 477 to 194 MW.

19.3.1.3 Efficiency calculation

The efficiency of the global process is defined as

$$\eta = \frac{\dot{M}_{\text{Methanol}}.\text{LHV}_{\text{Methanol}}}{\dot{M}_{\text{H2}}.\text{LHV}_{\text{H2}} + Q_{\text{In}} + P_{\text{In}}}$$

where $\text{LHV}_{\text{Methanol}}$ and LHV_{H2} are the lower heating values of the methanol produced and hydrogen fed to the electrolyser (to prevent coking on the electrodes),

$\dot{M}_{Methanol}$ and $\dot{M}_{H2}$ are their respective flow rates. P_{In} and Q_{In} are the power and heat inputs (cold utilities are neglected in first approximation). Table 19.5 shows that the conversion efficiency increases from 42% to 58% by the use of heat integration.

As mentioned in Section 19.2, CRI produces green methanol using geothermal energy. Little information is available but, in the news published by the company in 2012, they claimed an efficiency for the conversion of electricity to methanol of 58%. Kauw et al. [39] determined a value of 54.1%. In our study, we reach a theoretical value of 58%, which is in good agreement with mentioned data, taking into account the model assumptions and the maximum heat recovery.

Table 19.5: Conversion efficiency of the global process (i) without and (ii) with heat integration.

Parameter	$\dot{M}_{Methanol}.LHV_{Methanol}$	$\dot{M}_{H2}.LHV_{H2}$	P_{In}	Q_{In}	η
Unit	MW	MW	MW	MW	%
(i) Without heat integration	220	23	272	228	42
(ii) With heat integration	220	23	272	87	58

In conclusion, this work presents a CO_2-to-methanol process starting from ammonia CO_2-rich flue gas. The model takes into account the different steps (CO_2 capture, co-electrolysis, methanol synthesis and purification) and this study also includes heat integration in order to improve the heat consumption of the global process. Thanks to this first approach optimisation, the process efficiency is increased by 37%, from 42% to 58%. However, the economic relevance of the CO_2 re-use into methanol still has to be qualified by an economic study, and experimental demonstration of the integrated process is needed to validate the modelling assumptions.

19.3.2 Case of CO_2 re-use from other industries

Besides ammonia plants, other industries emit carbon dioxide directly, either through fossil fuel combustion or due to the processing of their raw materials. The present section gives an overview of some important industrial processes that could be potential sources for CO_2 conversion into methanol. These sources may be subdivided into further high purity CO_2 sources (ethanol, natural gas and ethylene) and dilute sources (steel mills, cement plants and power industry).

19.3.2.1 Ethanol industry

The world's largest producer of ethanol is the United States, accounting for nearly 57% of global output with a production of about 44 Mt. The second largest producer is Brazil accounting for nearly 28%; the European Union follows with 5% and the remaining 10% is produced by the rest of the world in 2015 [40]. Ethanol can be produced from a fermentation process using renewable sources such as agricultural crops (sugarcane and sugar beets) or starch. It can also be produced from non-renewable sources like coal and gas. A large part of the CO$_2$ released during the fermentation process is highly concentrated as purities up to almost 100% can indeed be reached [41]. Thus, the CO$_2$ resulting from fermentation represents a large and high purity stream that can be seen as a promising value stream for the ethanol industry when considering CO$_2$ re-use technologies.

Only few projects exist so far to re-use the CO$_2$ emitted from ethanol plants. Most of them address re-use of CO$_2$ for Enhanced Oil Recovery (EOR), which consists of injecting the captured CO$_2$ into an oil reservoir to recover more oil. For instance, about 0.27 Mt of CO$_2$ is annually supplied by the ethanol industry for EOR in Texas and Kansas [42]. More recent work proposed to re-use the CO$_2$ emitted for the cultivation of microalgae [43]. However, no project has been identified for the re-use of CO$_2$ into methanol. Nevertheless, this technology may represent a strategic opportunity for the ethanol industry to both increase its revenue and reduce its carbon emissions. Moreover, the produced methanol would have a net negative CO$_2$ footprint if the initial carbon source is biomass.

19.3.2.2 Natural gas industry

Only one project so far intends to produce methanol based on CO$_2$ emitted from the natural gas industry. The BFE project was initiated in 2008 with plans to capture CO$_2$ during the sweetening of natural gas from a reservoir located in Canada and to combine it with water electrolysis. In addition, methanol would also be synthesised from natural gas, following the conventional process. The combination of both carbon sources (natural gas for 97% and CO$_2$ for 3%) would decrease the environmental impact of the produced methanol. The project aims to use methanol as an intermediate for the production of gasoline fuels [23]. The target is the production of 2.5 ML/day of gasoline. However, at the present day, little progress has been achieved in the project as stated in a letter addressed by the company to the British Columbia government [44].

19.3.2.3 Ethylene industry

One project has been reported in Japan for the synthesis of methanol from CO$_2$ captured in the ethylene production process, with no indication about the purity of the CO$_2$ source. The captured CO$_2$ is reacted with hydrogen coming from a photolysis process to produce methanol [45]. A pilot plant was constructed in 2009, with a capacity of about 100 tonnes/year of methanol. The objective was to study a process that could be scaled-up to produce methanol as an intermediate for olefins and aromatics synthesis. However, no recent communication about this project or possible scale-up could be retrieved.

19.3.2.4 Steel mills

The steel industry is facing a major challenge to reduce CO_2 emissions to mitigate climate change. According to the International Energy Agency [46], the iron and steel sector is the first industrial emitter of CO_2, accounting for about 7% of the worldwide emissions. According to the same source, about 1.4 tonnes of CO_2 are emitted for every tonne of crude steel produced in current average blast furnaces.

The most used and dominant process of steelmaking is blast furnace-basic oxygen furnace. Some typical gas compositions emitted from these processes are presented in Table 19.6. As these exhaust gases still have some calorific value, they are usually burned to produce heat or electricity that is then used in the manufacturing process. However, this results in the production of even more CO_2. Moreover, as the composition of the emitted gases is quite similar to the composition of syngas, it may be more suitable to re-use these gases to produce base chemicals and fertilisers instead of just generating electricity and heat.

Table 19.6: Typical gas composition of steel mill plant [47].

Gases (vol.%)	Blast furnace gas	Basic oxygen furnace gas	Coke oven gas
H_2	1–8	2–10	36–62
CO	19–27	55–80	3–6
CH_4	–	–	16–27
C_xH_y	–	–	1–2
CO_2	16–26	10–18	1–5
N_2 + Ar	44–58	8–26	2–6

Several European initiatives have started to develop dedicated CO_2 capture and re-use technologies for the steel industry. Regarding the CO_2 capture part, one can mention the Ultra-Low CO_2 Steelmaking (ULCOS) project, and more recently the STEPWISE project that aims to reduce CO_2 emissions of the iron and steel industry. For instance, the STEPWISE project was launched in 2015 with the objective to study the decarbonisation of the top gas blast furnace based on the sorption-enhanced water gas shift technology that combines CO_2 adsorption with the water gas shift reaction [48]. This technology is carried out in multiple reactors that are filled with catalytically active CO_2 adsorbent. When the carbon-rich gas from the blast furnace is fed into the reactor, the CO_2 is removed. Thus, the water gas shift reaction is moved to the right side, allowing total CO conversion and maximal H_2 production. Selection of pathways for CO_2 re-use was, however, not the focus of these projects.

In 2016, the FReSMe project was launched for 4 years to re-use the CO_2 captured from an industrial blast furnace at SSAB in Luleå (Sweden) and produce methanol. Methanol would then be used as a fuel in the ship transport sector. The nominal production rate is expected to reach 50 kg/h of methanol for a blast furnace gas feed of 800 m^3/h [49].

Furthermore, Harp et al. [50] presented an application of the CRI technology using steelworks off-gases. As presented in Table 19.6, the steelworks off-gases offer a large amount of CO_2 from the blast furnace and hydrogen from the coke oven. The process produces methanol by combining the captured CO_2 and hydrogen recovered by pressure swing adsorption with additional hydrogen produced by electrolysis. In addition, Lundgren et al. [51] reported that it is possible to integrate biomass gasification with steelworks off-gases to increase the production of methanol and at the same time to decrease the emission of CO_2 from the biomass.

Finally, it is worth mentioning that another way to re-use CO_2 emitted from the steel plant is leading to the production of ethanol instead of methanol. For this purpose, the Steelanol project [52] based on the LanzaTech technology proposes to recycle carbon-rich waste gases of the ArcelorMittal Ghent plant (Belgium) by fermentation into bio-ethanol for use in the transport sector [53]. The project started in 2015 and is still in progress.

19.3.2.5 Cement industry

Another important source of CO_2 is the cement industry. This contributes to approximately 5–8% of the total global CO_2 emissions [54] and its outlet flue gases contain 16–39% of CO_2, 6–11% of O_2 and 9–22% of H_2O. Currently, there are several on-going research projects to evaluate the feasibility of CO_2 capture and storage. Different capture technologies are studied such as post-combustion and oxy-combustion [55]. However, no industrial project to re-use CO_2 has been identified so far.

19.3.2.6 Power industry

In 2015, about two-thirds of the world electricity generation was still supplied by fossil fuels, significantly contributing to the CO_2 emissions [56]. Coal alone accounted for close to 40% of the electricity generation, despite being the most polluting fossil fuel. Table 19.7 gives the CO_2 emissions per type of fossil fuel in the world [57]. To avoid these emissions, carbon capture technologies are widely studied and are being implemented in power plants, as evidenced by the two first large-scale

Table 19.7: CO_2 emissions from electricity generation and heat sector [57].

Fuel	CO_2 worldwide emissions (Mt CO_2)
Coal	9,729
Natural gas	2,844
Oil	844
Other (include industrial and non-renewable municipal wastes)	124

projects operating: the Boundary Dam (2014) and the Petra Nova (2017) plant, which have a CO_2 capture capacity of 1.0 and 1.4 Mt CO_2 per year, respectively.

Several articles study the feasibility of re-using CO_2 emissions from the power sector to convert them into methanol. For instance, Pérez-Fortes et al. [10] performed a techno-economic study of a CO_2-to-methanol plant producing 440 kt methanol per year by re-using about one-fifth of the emissions of a 550 MW_{el} coal-fired power plant. Bellotti et al. [58] also consider a similar case, capturing 140 kg/h CO_2 from a coal power plant (about 1.1 Mt per year assuming 8,000 h of operation per year) to produce 97 kg/h methanol. So far, only one demonstration project has been initiated by an industrial consortium at the Lünen power plant in Germany. This 11 M€ project (MefCO$_2$) is supported by the Horizon2020 research fund of the European Commission. The objective is to build and demonstrate the technology with a production of 1 tonne of methanol per day [59] based on CO_2 from a coal power plant and hydrogen supplied by water electrolysis. Following the experience gained in their plant in Iceland, the company CRI is part of the project consortium, along with several academic and industrial partners.

In conclusion, various cases of CO_2 re-use were discussed earlier. It has been shown that many industries are interested in the concept of CO_2 re-uses. Thus, the interest to mitigate climate change was highlighted. In addition, the integration of CO_2 re-use technologies to produce low carbon products may allow industries to increase their profits, but market maturity still has to be discussed, which is the topic of the next section.

19.4 Economic study of methanol from CO_2

In the previous sections, the synthesis of methanol as a re-use pathway for CO_2 originating from different industrial sources has been discussed. In most cases, despite a promising potential for CO_2 re-use, only a few projects have been identified, and methanol production from CO_2 is still marginal. Methanol synthesis from pure CO_2 (instead of syngas) is considered at a technology readiness level of 6–7 [10]. This means that commercial demonstration plants are available, for instance, in Japan or

Iceland as mentioned earlier, with production of 100–5,000 tonnes/year, and that the next step for technology development is the transition to full-scale commercial plants. As discussed previously, methanol production from CO$_2$ is composed of three main parts: CO$_2$ capture, hydrogen production and methanol synthesis. For each of these parts, different options exist and the best choice depends on local conditions.

Several sources for CO$_2$ have been discussed in the previous section. Assuming near-100% conversion of the captured CO$_2$ into methanol as a first approach following reaction (19.2), it requires about 1.4 tonnes of CO$_2$ to produce 1 tonne of methanol. A recent review has evaluated the cost of CO$_2$ coming from different industrial sources, including iron and steel industries, cement plants, pulp and paper plants, oil refineries, as well as high-purity sources such as ammonia or ethanol plants [60]. Different CO$_2$ capture technologies are considered, ranging from conventional solvent-based capture to oxyfuel combustion and more innovative technologies such as calcium looping. Based on an extensive literature review, the authors present costs for CO$_2$ avoidance in USD with 2013 as reference year. Although the cost for avoided CO$_2$ may differ from the cost of captured CO$_2$, this study already gives a very good first indication of costs to compare different industrial sources. For instance, the average cost of CO$_2$ from high-purity sources equals 26.6 $/tonne, with values varying between 3.9 and 74 $/tonne. CO$_2$ resulting from ethanol production is evaluated to 12.3 $/tonne, while CO$_2$ from ammonia plants costs about 22 $/tonne. Furthermore, the cost of CO$_2$ in the cement industry equals about 74.2 $/tonne, and 66.2 $/tonne in the iron and steel industry.

The second step is the hydrogen production. Although about 96% of hydrogen is still produced from fossil resources due to lower cost [61], the focus of CO$_2$ re-use towards methanol is rather set on water electrolysis. Indeed, in order to make methanol a sustainable product, the large amounts of energy needed must come from renewable energy and the best way to include them in the production process is via water electrolysis. The disparity of the cost estimations for this step is very large, as the cost of hydrogen production is largely influenced by the cost of electricity, and the later varies a lot depending on the ability of the process to operate in flexible regimes and on the local electricity supplier. Depending on the electrolysis technology (alkaline, proton exchange membrane or solid oxide) the efficiency of the electrolysis step varies, but as a first approach the value of about 50 kWh/kg H$_2$ can be assumed. About 0.2 tonnes of hydrogen is required per tonne of methanol, so the electricity would be of 10 MWh/tonne of methanol. With an electricity price of 112 €/MWh (average European electricity price for non-household consumers in 2016 [62]), the electrolysis-induced electricity cost would already reach 1,120 € per tonne of methanol. However, much lower electricity prices can be expected if the process is directly connected to a source of renewable energy rather than to the grid, as the average cost for wind and solar electricity production in Europe is expected to be in the range of 43–51 €/MWh in 2020 [63]. Besides operating costs, the capital cost for the electrolyser is also not negligible and lies among the largest contributors to the final methanol cost. According to Atsonios et al. [64], the capital cost amounts to about 30% of the methanol cost, and two-thirds of it is due to the electrolyser cost.

The methanol synthesis from CO_2 and hydrogen itself is rather a minor cost in the methanol production chain. The capital and operative expenses of this step account in total for about 15% of the methanol cost [64]. Another study [65] assumes that about 20% of the methanol cost is related to the methanol synthesis itself.

One of the first studies to assess the feasibility of methanol production from captured CO_2 was published as early as 2003 [66]. In that work, CO_2 was captured from a coal-fired power plant, and alkaline electrolysis was considered at different pressures (1.5 and 30 bar). The conversion efficiency of renewable electricity into methanol (kW_{th}/kW_{el}) appeared to be in the range of 51–68%, depending on the option selected (for instance, higher conversion is achieved when a "free" source of waste heat is available, which is not included in the efficiency calculation). When methanol is the only product, it appears that the break-even price for methanol (the price that drives the net present value to zero at the end of the plant lifetime) was equal to 480 €/tonne, while the average European bulk market price for methanol reached 225 €/tonne in 2003. Different economic scenarios for the availability of electricity were studied and the possibility of selling oxygen as a by-product of the electrolysis was also considered. With oxygen as a by-product, the break-even price decreases by about 40% in the option considered (electrolysis at 1.5 bar), down to 288 €/tonne. However, the authors showed that the process could be more economically profitable if methanol is to be sold as a transportation fuel rather than on the bulk market. In this case, a positive business plan can be identified, provided a favourable tax rate (lower than 30%) is applied to methanol fuel, so that it becomes competitive with gasoline. In 2007, Galindo Cifre and Badr [67] found that the cost of methanol produced as a re-use pathway for captured CO_2 would reach about 500–600 €/tonne, and would thus be more expensive than the methanol produced from biomass (300–400 €/tonne) or using the conventional pathway from natural gas (100–200 €/tonne). However, the uncertainty about these costs is large as only few plants produce methanol from CO_2. In a literature review published in 2013 [68], the cost of methanol production from CO_2 was estimated to range between 510 and 900 €/tonne, being the most expensive production processes among those studied, but also one of the most sustainable ones.

Furthermore, three different works have recently studied the costs of the CO_2-to-methanol process and present a detailed repartition of its investment and operational costs. First, Bellotti et al. (58) assess the rentability of the process over a lifetime of 10 years for different plant sizes. Besides the potential of economies of scale, they show that a medium-size plant (10 kt/year) needs to sell methanol at a price of 600 €/tonne to reach a net present value of zero. They confirm that selling oxygen as a by-product could make the case more profitable as the selling price for methanol would decrease down to about 500 €/tonne. Then, Pérez-Fortes et al. [10] model and optimise a large-scale CO$_2$-to-methanol plant producing 440 kt/year of methanol. Investment costs for this plant reached a total amount of about 200 M€ (fixed capital investment). However, the revenues are lower than the operating costs of the base case conditions, so the authors conclude that the project is not financially attractive unless some key parameters are changed. For instance, having a carbon tax as high as 222 €/tonne CO$_2$, doubling the selling price of methanol compared to the bulk market price, or decreasing the cost of hydrogen by 2.5 would each lead the process to reach a break-even point over 20 years. Moreover, according to the authors, the CO$_2$ re-use potential is present, but rather limited. Indeed, the current European demand for methanol accounts for about 7.6 Mt/year. Assuming that the new CO$_2$-to-methanol technology would provide methanol to fully replace current importations (4.7 Mt/year), only about 11 of such plants would be needed in Europe. This figure would increase to about 18 plants if the whole European market is to be supplied. However, the possibility of a rapid methanol market growth where methanol would act as a sustainable energy carrier leading to the so-called methanol economy is not considered here. Finally, besides the large-scale approach, Atsonios et al. [64] also consider the case of decentralised hydrogen production occurring close to distributed renewable systems. However, the large logistic costs induced by the need to store hydrogen to achieve a constant supply to the methanol plant make the process non-financially profitable. When a constant supply is achieved (electricity supplied by the grid), the electricity price is the dominating factor. If electricity is to be used only during off-peak hours, the critical cost then becomes the electrolyser cost. Although none of the studied options is profitable, the least bad choice from a financial point seems to be the maximisation of the capacity factor of the equipment, and thus the use of grid electricity.

19.5 Conclusions

In this chapter, we discuss the opportunity of using CO$_2$ from industrial sources to produce sustainable methanol. Some important industrial sectors that could be seen as potential sources of CO$_2$ are reviewed: ammonia, steel, ethanol, ethylene, natural gas, cement and power industries. In most cases, despite a promising potential for CO$_2$ re-use, only few projects have been identified and methanol production from CO$_2$ is still marginal. A model for the CO$_2$-to-methanol process is presented based on CO$_2$-rich gas coming from ammonia production process. This model takes into account the different steps from the CO$_2$ capture to the methanol purification, and heat integration is performed in order to determine the reduction of heat consumption achievable for the global process. Even if the economic relevance of the CO$_2$ re-use into methanol still has to be qualified, it offers an estimation of the process efficiency.

Given the current market assumptions, a methanol plant would not have a positive return rate on investment. Indeed, the production cost for methanol would be equal to about 1.7 times larger than the revenue [10]. However, a positive business case could be identified if a carbon tax is introduced, if the operating costs can be reduced, or if methanol can be sold at a higher price. These elements not only depend on political will, but also on technology development. For instance, the cost of gasoline in Europe amounts to about 0.137 €/kWh (assuming 1.3 €/L and 9.5 kWh/L). In comparison, the same energy price would lead to a methanol price of 750 €/tonne of methanol, which is clearly higher than the current techno-economical estimations mentioned earlier, rather ranging between 500 and 600 €/tonne. If this is combined to a carbon tax and to forecasted technology evolution that would decrease the CAPEX and OPEX, methanol may be a competitive and sustainable alternative fuel for the future.

Finally, in order to make methanol synthesis an environmentally friendly option, hydrogen required for the synthesis needs to be produced from renewable energy. According to the currently available data, a life cycle assessment observed that the production of methanol from CO_2 and renewable energy would be associated with – 1.2 to –1.3 kg CO_2 equivalent per kg of CH_3OH (kg CO_2eq/kg CH_3OH) in comparison to 0.7 to 1.1 kg CO_2e/kg CH_3OH for the conventional fossil value chain [69]. Even if still challenging, CO_2-to-methanol conversion presents a clear benefit for the environment compared to current production from fossil fuel and, when considering the numerous uses of methanol, this process may contribute to the deployment of the so-called methanol economy.

Bibliography

[1] Alvarado M, 2016. Methanol. (Accessed July, 2018, at http://www.methanol.org/wp-content/uploads/2016/07/Marc-Alvarado-Global-Methanol-February-2016-IMPCA-for-upload-to-web site.pdf).

[2] MI, 2016a. Methanol technical data sheet for produced methanol, Methanol Institute. (Accessed July, 2018, at http://www.methanol.org/wp-content/uploads/2016/06/Methanol-Technical-Data-Sheet.pdf).

[3] Olah GA, Goeppert A, Surya Prakash GA. Beyond Oil and Gas: The Methanol Economy. Wiley-VCH, 2006.

[4] IHS, 2017a. 'Global Methanol Demand Growth Driven by Methanol to Olefins as Chinese Thirst for Chemical Supply Grows, IHS Markit Says'. (Accessed September, 2018, at https://news.ihsmarkit.com/press-release/country-industry-forecasting-media/global-methanol-de mand-growth-driven-methanol-olefi).

[5] Bechtold RL, Goodman MB, Timbario TA. 2007. Use of Methanol as a Transportation Fuel. (Accessed July, 2018, at http://methanolfuels.org/wp-content/uploads/2013/05/Bechtold-ATS-Methanol-Use-in-Transportation.pdf).

[6]   MI, 2016b. Methanol Blending -Technical Product Bulletin, Methanol Institute. (Accessed July, 2018, at http://www.methanol.org/wp-content/uploads/2016/06/Blending-Handling-Bulletin-Final.pdf).

[7]   IHS, 2017b. Methanol – Chemical Economics Handbook. (Accessed July, 2018, at https://ihs markit.com/products/methanol-chemical-economics-handbook.html).

[8]   Van-Dal ÉS, Bouallou C. Design and simulation of a methanol production plant from CO_2 hydrogenation. Journal of Cleaner Production 2013, 57, 38–45.

[9]   Ott J, Gronemann V, Pontzen F, Fiedler E, Grossmann G, Burkhard Kersebohm D, Weiss G, Witte C. Methanol. In: Ullmann's Encyclopedia of Industrial Chemistry. Wiley-VCH, 2012.

[10] Pérez-Fortes M, Schöneberger JC, Boulamanti A, Tzimas E. Methanol synthesis using captured CO_2 as raw material: Techno-economic and environmental assessment. Applied Energy 2016, 161, 718–32.

[11] Palma V, Meloni E, Ruocco C, Martino M, Ricca A. State of the Art of Conventional Reactors for Methanol Production. In: Methanol, Science and Engineering, Amsterdam, Elsevier, 2018, 29–51.

[12] Harnisch J, Jubb C, Nakhutin A, Sena Cianci VC, Lanza R, Martinsen T, Mohammad AKW, Santos MMO, McCulloch A, Mader BT. 2006. '2006 IPCC Guidelines for national greenhouse gas inventories.' In Volume 3: Industrial Processes and Product Use.

[13] Pérez-Fortes M, Tzimas E. 2016. Techno-economic and environmental evaluation of CO_2 utilisation for fuel production. Synthesis of methanol and formic acid. In: JRC Science for Policy Report.

[14] Yang C, Ma Z, Zhao N, Wei W, Hu T, Sun Y. Methanol synthesis from CO_2-rich syngas over a ZrO_2 doped CuZnO catalyst. Catalysis Today 2006, 115, 222–27.

[15] Studt F, Sharafutdinov I, Abild-Pedersen F, Elkjær CF, Hummelshøj JS, Dahl S, Chorkendorff I, Nørskov JK. Discovery of a Ni-Ga catalyst for carbon dioxide reduction to methanol. Nature Chemistry 2014, 6, 320–4.

[16] Anicic B, Trop P, Goricanec D. Comparison between two methods of methanol production from carbon dioxide. Energy 2014, 77, 279–89.

[17] Albo J, Alvarez-Guerra M, Castaño P, Irabien A. Towards the electrochemical conversion of carbon dioxide into methanol. Green Chemistry 2015, 17, 2304–24.

[18] Montebelli A, Visconti CG, Groppi G, Tronconi E, Ferreira C, Kohler S. Enabling small-scale methanol synthesis reactors through the adoption of highly conductive structured catalysts. Catalysis Today 2013, 215, 176–85.

[19] Arab S, Commenge J-M, Portha J-F, Falk L. Methanol synthesis from CO_2 and H_2 in multi-tubular fixed-bed reactor and multi-tubular reactor filled with monoliths. Chemical Engineering Research and Design 2014, 92, 2598–608.

[20] van Bennekom JG, Venderbosch RH, Winkelman JGM, Wilbers E, Assink D, Lemmens KPJ, Heeres HJ. Methanol synthesis beyond chemical equilibrium. Chemical Engineering Science 2013, 87, 204–8.

[21] Bos MJ, Brilman DWF. A novel condensation reactor for efficient CO_2 to methanol conversion for storage of renewable electric energy. Chemical Engineering Journal 2015, 278, 527–32.

[22] CRI, 2018. Carbon Recycling International (CRI). (Accessed March, 2018, at http://carbonrecy cling.is/).

[23] BFE, 2018a. Blue Fuel Energy. (Accessed September, 2018 at http://bluefuelenergy.com/)

[24] Iaquaniello G, Centi G, Salladini A, Palo E, Perathoner S, Spadaccini L. Waste-to-methanol: Process and economics assessment. Bioresource Technology 2017, 243, 611–9.

[25] Enerkem, 2018. Enerkem Alberta Biofuels. (Accessed July, 2018 at https://enerkem.com/facili ties/enerkem-alberta-biofuels/).

[26] Sanz-Pérez ES, Murdock CR, Didas SA, Jones CW. Direct Capture of CO_2 from Ambient Air. Chemical Reviews 2016, 116, 11840–76.

[27] Kothandaraman J, Goeppert A, Czaun M, Olah GA, Surya Prakash GK. Conversion of CO_2 from Air into Methanol Using a Polyamine and a Homogeneous Ruthenium Catalyst. Journal of the American Chemical Society 2016,138, 778–81.

[28] Zhou W, Zhu B, Li Q, Ma T, Hu S, Griffy-Brown C. CO_2 emissions and mitigation potential in China's ammonia industry. Energy Policy 2010, 38, 3701–9.

[29] IFA, 2017. Short-Term Fertilizer Outlook 2017 – 2018. Strategic Forum, 14–15 November 2017, Zürich, PIT and Agriculture Services, IFA.

[30] Bennett S. CCS in industrial applications. A workshop of the CCUS Action Group in preparation for the 4th Clean Energy Ministerial. OECD/IEA Report, London, UK, 2013.

[31] Holm-Larsen H. Co-production of methanol in ammonia plants. 1994 IFA Technical Conference, Amman, Jordan (Accessed September, 2018, at https://www.fertilizer.org/im ages/Library_Downloads/1994_ifa_amman_holm.pdf).

[32] Syed Othman Abu Bakar. Economics of co-production of methanol in ammonia/urea complex. 1998 IFA Technical conference, Marrakech, Morocco (Accessed September, 2018 at https:// www.fertilizer.org/images/Library_Downloads/1998_ifa_marrakech_othman.pdf).

[33] Alvis RS, Hatcher NA, Weiland RH. CO_2 removal form syngas using piperazine-activated MDEA and potassium dimethyl glycinate. Nitrogen+Syngas 2012, Athens, Greece.

[34] Sun X, Chen M, Jensen SH, Ebbesen SD, Graves C, Mogensen M. Thermodynamic analysis of synthetic hydrocarbon fuel production in pressurized solid oxide electrolysis cells. International Journal of Hydrogen Energy 2012, 37, 17101–10.

[35] Graves C, Ebbesen SD, Mogensen M. Co-electrolysis of CO_2 and H_2O in solid oxide cells: performance and durability. Solid State Ionics 2011, 192, 398–403.

[36] Rivera-Tinoco R, Mansilla C, Bouallou C. Competitiveness of hydrogen production by High Temperature Electrolysis: Impact of the heat source and identification of key parameters to achieve low production costs. Energy Conversion and Management 2010, 51, 2623–34.

[37] Redissi Y, Bouallou C. Valorization of Carbon Dioxide by Co-Electrolysis of CO_2/H_2O at High Temperature for Syngas Production. Energy Procedia 2013, 37, 6667–78.

[38] Douglas JM. Conceptual Design of Chemical Processes, International Edition, McGrawHill Book Company, 1988.

[39] Kauw M, Benders RMJ, Visser C. Green methanol from hydrogen and carbon dioxide using geothermal energy and/or hydropower in Iceland or excess renewable electricity in Germany. Energy 2015, 90, 208–17.

[40] RFA, 2016. 2016 Ethanol industry outlook. Renewable Fuels Association.

[41] Faaij A. Modern biomass conversion technologies. Mitigation and Adaptation Strategies for Global Change 2006, 11, 343–75.

[42] Fry M, Ellsworth S, Ahmad F, Crabtree B. 2017. Capturing and Utilizing CO_2 from Ethanol: Adding Economic Value and Jobs to Rural Economies and Communities While Reducing Emissions. State CO_2-EOR Deployment Work Group.

[43] Holanda LR, Ramos FS. Reuse of Waste Sugarcane Agribusiness and Green Power Generation. Journal of Clean Energy Technologies 2016, 4, 341–5.

[44] BFE, 2018b. Blue Fuel Energy. British Columbia (BC) Low Carbon Fuels Compliance Pathway Assessment. (Accessed August, 2018, at https://www2.gov.bc.ca/assets/gov/farming-natu ral-resources-and-industry/electricity-alternative-energy/transportation/renewable-low-car bon-fuels/blue_fuel_energy_corporation_-_response_to_bc_lcf_compliance_pathway_assess ment_2017pdf_717_kb.pdf).

[45] Mitsui, 2008. Mitsui Chemicals to establish a pilot facility to study a methanol synthesis process from CO_2. Press release, Mitsui Chemicals Inc.

[46] IEA, 2017a. Global Iron & Steel Technology Roadmap. Landolina S, Fernandez A. Kick-off Workshop presentation. International Energy Agency, 20 November 2017, Paris.

[47] Caillat S. Burners in the steel industry: utilization of by-product combustion gases in reheating furnaces and annealing lines. Energy Procedia 2017, 120, 20–7.

[48] van Dijk HAJ, Cobden PD, Lundqvist M, Cormos CC, Watson MJ, Manzolini G, van der Veer G, Mancuso L, Johns J, Sundelin B. Cost effective CO_2 reduction in the Iron & Steel Industry by means of the SEWGS technology: STEPWISE project. Energy Procedia 2017, 114, 6256–65.

[49] FReSMe, 2017. From residual steel gases to methanol. (Accessed August, 2018 at http://www.fresme.eu/).

[50] Harp G, Tran KC, Bergins Chr, Buddenberg T, Drach I, Koytsoumpa EI, Sigurbjornsson O. Application of Power to Methanol Technology to Integrated Steelworks for Profitability, Conversion Efficiency, and CO_2 Reduction. 2nd European Steel Technology and Application Days, 2015, Düsseldorf, Germany.

[51] Lundgren J, Ekbom T, Hulteberg C, Larsson M, Grip C-E, Nilsson L, Tunå P. Methanol production from steel-work off-gases and biomass based synthesis gas. Applied Energy 2013, 112, 431–9.

[52] Steelanol project, 2018. (Accessed September, 2018, at www.steelanol.eu/en).

[53] Van Der Maren O, 2016. Energy & climate – défis et opportunités du changement climatique. (Accessed September, 2018, at https://www.vbo-feb.be/en/business-issues/energy-mobil ity–environment/energie-mobilite–environnement/energy–climate–defis-et-opportunites-du-changement-climatique_2016-11-15/).

[54] Kajaste R, Hurme M. Cement industry greenhouse gas emissions – management options and abatement cost. Journal of Cleaner Production 2016, 112, 4041–52.

[55] Davison J, 2016. CCS in the Cement Industry, IEA Greenhouse Gas R&D Programme. CCS in Industry workshop, 28th April 2014, Vienna.

[56] IEA, 2017b. Key world energy statistics 2017. International Energy Agency.

[57] IEA, 2017c. CO_2 emissions from fuel combustion – Highlights 2017. International Energy Agency.

[58] Bellotti D, Rivarolo M, Magistri L, Massardo AF. Feasibility study of methanol production plant from hydrogen and captured carbon dioxide. Journal of CO_2 utilization 2017, 21, 132–8.

[59] PEI, 2015. Coal plant provides CO_2 for methanol production. Power Engineering International. (Accessed August, 2018 at https://www.powerengineeringint.com/articles/2015/06/coal-plant-provides-co2-for-methanol-production.html).

[60] Leeson D, Mac Dowell N, Shah N, Petit C, Fennell P. A Techno-economic analysis and systematic review of carbon capture and storage (CCS) applied to the iron and steel, cement, oil refining and pulp and paper industries, as well as other high purity sources. International Journal of Greenhouse Gas Control 2017, 61, 71–84.

[61] Nikolaidis P, Poullikkas A. A comparative overview of hydrogen production processes. Renewable and Sustainable Energy Reviews 2017, 67, 597–611.

[62] Eurostat, May 2018. Electricity price statistics. Electricity prices for non-household consumers. ISSN 2443-8219. (Accessed July, 2018, at http://ec.europa.eu/eurostat/statis tics-explained/index.php/Electricity_price_statistics).

[63] Irena, 2018. Renewable Power Generation Costs in 2017. International Renewable Energy Agency, Abu Dhabi. ISBN 978-92-9260-040-2.

[64] Atsonios K, Panopoulos KD, Kakaras E. Investigation of technical and economic aspects for methanol production through CO_2 hydrogenation. International Journal of hydrogen energy 2016, 41, 2202–14.

[65] ENEA, 2016. The potential of power-to-gas. Technology review and economic potential assessment. ENEA consulting.

[66] Mignard D, Sahibzada M, Duthie JM, Whittington H. Methanol synthesis from flue-gas CO_2 and renewable electricity: a feasibility study. International Journal of Hydrogen Energy 2003, 28, 45–64.

[67] Galindo Cifre P, Badr O. Renewable hydrogen utilisation for the production of methanol. Energy Conversion and Management 2007, 48, 519–27.

[68] IEA-ETSAP and IRENA, 2013. Production of Bio-methanol. Technology Brief 7. International Energy Agency – Energy Technology Systems Analysis Programme and International Renewable Energy Agency.

[69] Artz J, Müller TE, Thenert K, Kleinekorte J, Meys R, Sternberg A, Bardow A, Leitner W. Sustainable conversion of carbon dioxide: an integrated review of catalysis and life cycle assessment. Chemical Reviews 2018, 118, 434–504.

Alain Bengaouer and Laurent Bedel

20 CO$_2$ hydrogenation to methane

20.1 Introduction

The energy transition consists of the production of energy from non-fossil sources in order to reduce CO$_2$ emissions. In the electricity sector, the share of wind turbine and photovoltaic renewable sources has grown to represent in 2017 nearly 15% of the European production [1]. The European 2030 climate and energy framework has set a binding target at the EU level to boost the share of renewables to at least 27% of EU energy consumption by 2030 [2]. This increasing share of intermittent wind and solar electricity will lead to temporal and geographic mismatch between production and consumption. Furthermore, a characteristic of these intermittent sources of energy is their low capacity factor compared to conventional energy sources such as coal, gas and nuclear. Indeed the capacity factor of off-shore wind turbine, on-shore wind turbine and photovoltaic is respectively nearly 40%, 20–25% and 15–20% against 70–80% for conventional sources. Then for the substitution of each GW of conventional energy sources between 2 and 6 GW of intermittent sources should be implemented to ensure the same electricity production per year. This will entail large surpluses and curtailment of energy produced during sunny and windy atmospheric conditions. Therefore, large capacity of energy storage should be deployed to ensure electricity production during periods of deficit.

Among the different energy storage technologies, power-to-gas consists of the conversion of electricity into gas: hydrogen or methane. Electricity is converted into hydrogen by water electrolysis and can be further converted into methane by CO$_2$ hydrogenation known as the Sabatier reaction. Power-to-gas appears as a promising solution for long-term and large-scale energy storage, thanks to the available natural gas grid with more than 2 million km of transmission and distribution network and nearly 800 TWh [3] of underground storage across Europe.

In this chapter, a general overview of the power-to-gas concept is presented, with the description of its main components, namely the water electrolysis, the CO$_2$ capture unit and the methanation unit. The main configurations and requirements are discussed.

Then, the methanation reaction thermodynamics, catalysts and kinetics are described. The main reactor technologies used to carry on the reaction are given and some considerations on reactor modelling and optimisation are proposed. The conclusion of this chapter is devoted to a review of the main demonstration projects ongoing in Europe. These projects aim at providing a technical feedback of different

Alain Bengaouer, Laurent Bedel, CEA-Grenoble; CEA-Liten France

https://doi.org/10.1515/9783110665147-020

methanation concepts against the specific requirements for smaller scale and more compact units, and unsteady operation.

20.2 Power-to-gas systems

Power-to-gas is the link of the energetic systems allowing the conversion of electricity into gas (Figure 20.1). One of the main advantages of power-to-gas is that the gas stored can be used for (i) the production of electricity by gas combustion power plants or by fuel cells, (ii) transport using either hydrogen or CNG fuels, (iii) heating (iv) and chemistry.

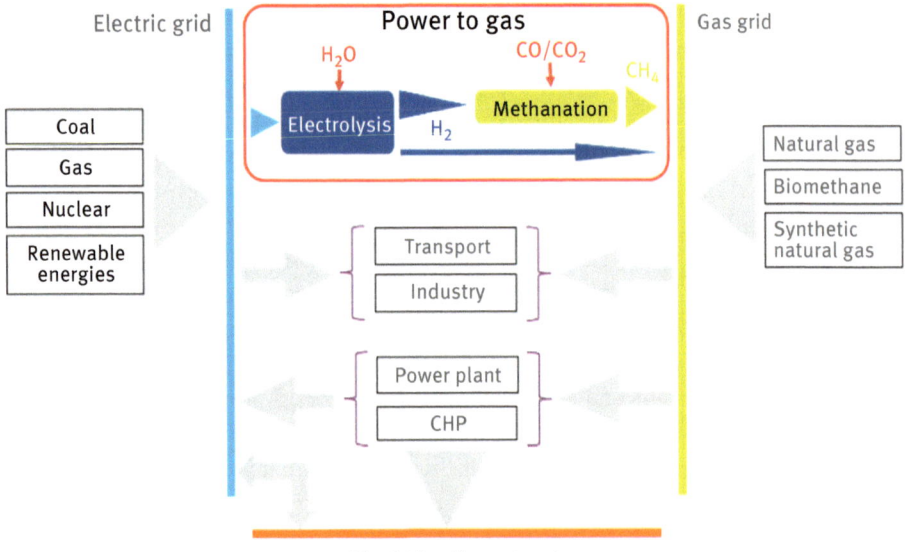

Figure 20.1: Complementarity of networks and sector coupling.

Sector coupling and the complementarity of networks: electric, gas and heat, should be essential in the future energy system.

20.3 Conversion chains of power-to-gas

a) Power to hydrogen:

The simplest power-to-gas conversion chain consists of converting electricity into hydrogen by water electrolysis. The hydrogen produced is then either injected into the gas grid or stored (Figure 20. 2).

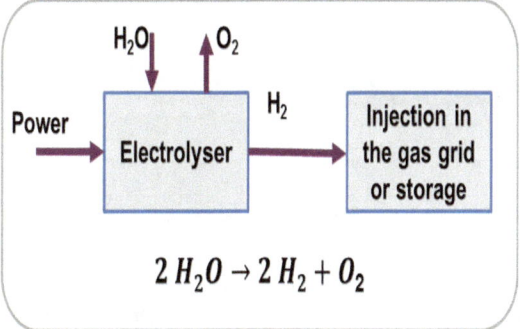

Figure 20.2: Power to hydrogen conversion chain.

The three main technologies of water electrolysis are either on the market or under development. As the electrolysis is a main component in a power-to-gas chain, a short description of the different electrolysis technologies is provided in the following section.

20.3.1 Low temperature electrolysis

Alkaline electrolysers and polymer electrolyte membrane (PEM) electrolysers operate at nearly 70 °C. The water splitting reaction (Eq. 20.1) is the following:

$$H_2O_{(l)} \rightarrow H_2 + \frac{1}{2}O_2 \ (\Delta_r H_{298} = +285 \text{kJ.mol}^{-1}) \tag{20.1}$$

Alkaline technology is the most mature with some units at several MW. The technology is based on an alkaline electrolyte, often potassium hydroxide. The anode and the cathode sides are separated by a diaphragm conducting the anion OH$^-$ and separating the oxygen and hydrogen produced. At the cathode the reduction occurs for the production of hydrogen (Eq. 20.2) and at the anode the oxidation results in oxygen production (Eq. 20.3).

$$\text{Cathode: } 2H_2O + 2e^- \rightarrow H_2 + 2OH^- \tag{20.2}$$

$$\text{Anode: } 2OH^- \rightarrow H_2O + \frac{1}{2}O_2 + 2e^- \tag{20.3}$$

The reported flexibility of this technology is between 15 and 100% of the nominal capacity and the typical consumption is between 49 and 58 kWhe/kgH$_2$ [3]. A new generation of electrolyser operating at 30 bars is emerging on the market with a higher flexibility from few percents to 110% [4].

The PEM technology is becoming mature with several offers on the market at the MW scale. The electrolyte is a polymer membrane, typically in Nafion, which transports protons from the anode (Eq. 20.4) to the cathode where protons are reduced into hydrogen (Eq. 20.5).

$$\text{Anode: } H_2O \rightarrow 2H^+ + \frac{1}{2}O_2 + 2e^- \tag{20.4}$$

$$\text{Cathode: } 2H^+ + 2e^- \rightarrow H_2 \tag{20.5}$$

The membrane is sandwiched between both electrodes, including a catalyst typically platinum at the cathode. The advantages of this technology is its flexibility from 0 to 200% of the nominal power and its ability to operate at relatively high pressure (60 bars and even higher). The typical electricity consumption is between 52 and 63 kWhe/kgH$_2$ at the system level [3], slightly higher than alkaline technology.

20.3.2 High temperature electrolysis

The solid oxide electrolysis cell technology is under development and the operation of the first demonstrator at a power of 150 kWe has started [5]. A ceramic electrolyte of ZrO$_2$ doped with Y$_2$O$_3$ is used. It operates at a temperature between 700 and 800 °C allowing the conduction of OH$^-$ anions through the electrolyte. Instead of water electrolysis, steam electrolysis occurs (Eq. 20.6) allowing a reduction of the electricity consumption by avoiding the energy consumption associated with water vaporisation (Eq. 20.7). In addition, the use of a high temperature allows the reduction of the Gibbs energy required for water splitting, thus reducing the electricity consumption at the price of an increase of the heat consumption and a lower internal cell resistance. The typical electricity consumption is from 40 kWhe/kgH$_2$ at the stack level for an efficiency of nearly 100% Higher Heating Value (HHV). However, to ensure this lower electricity consumption, a source of heat at a temperature above 200 °C for the vaporisation of water is required.

$$H_2O_{(g)} \rightarrow H_2 + \frac{1}{2}O_2 \; (\Delta_r H_{298} = +244 \text{kJ.mol}^{-1}) \tag{20.6}$$

$$H_2O_{(l)} \rightarrow H_2O_{(g)} \; (\Delta_r H_{298} = +41 \text{kJ.mol}^{-1}) \tag{20.7}$$

20.3.3 Hydrogen injection into the gas grid

This power-to-gas configuration is the most efficient chain but has some limitations. Indeed, the content in the gas grid is limited because of the hydrogen embrittlement [6] of some steel alloys composing the gas grid and welds. The LHV of hydrogen is 3.5 times lower than the LHV of methane (10.8 MJ/Nm^{3}H$_2$ against 36.1 MJ/Nm^{3}CH$_4$); a fluctuation of hydrogen injection will modify the LHV of the gas in the network. Technically, to inject pure hydrogen into the grid, natural gas is first extracted from the grid, mixed with hydrogen in a mixer unit to achieve the content specification of the grid and then reinjected. This procedure avoids all issues of mixing of pure hydrogen with natural gas in the transmission grid at 60–80 bars and sharp fluctuations of LHV in the network.

b)   Power to methane:
A complementary approach of power to hydrogen is the conversion of hydrogen into methane by reacting it with a source of CO$_2$ (Figure 20.3). The carbon in CO$_2$ acts as a means to store and transport hydrogen.

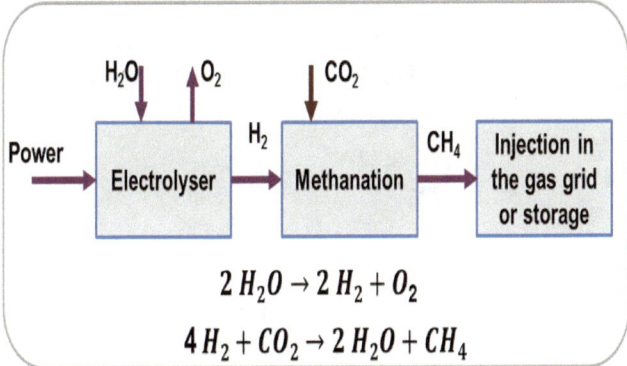

Figure 20.3: Power to methane conversion chain.

CO$_2$ can be captured with a solvent from industrial flue gas, whose content is typically between 10 and 20 vol.% of CO$_2$. For the CO$_2$ capture systems, we invite the reader to refer to Chapters 7 8 of this book for further information. We can notice that at the system scale, a heat integration with the exothermic methanation reaction (cf. following section) allows an optimisation of energy consumption of the CO$_2$ capture unit at the stripper level to release CO$_2$ from the solvent. Atmospheric CO$_2$ capture technology (cf. Chapter 11) is currently under development and offers an alternative solution but requires more energy as the CO$_2$ concentration in the atmosphere is around 400 ppmv.

Another interesting source of CO$_2$ is biogas, a typical composition of biogas is 60 vol.% of biomethane and 40 vol.% of CO$_2$. The coupling of a methanization unit and a power to methane unit offers several advantages. First, as the CO$_2$ has to be separated from the biomethane, there is no need for a specific CO$_2$ capture unit, second the gas upgrading (for example by membrane technology, cf. Chapter 10) can be mutualised, and finally, the injection unit to the grid can be mutualised (Figure 20.4). This synergy allows a cost reduction.

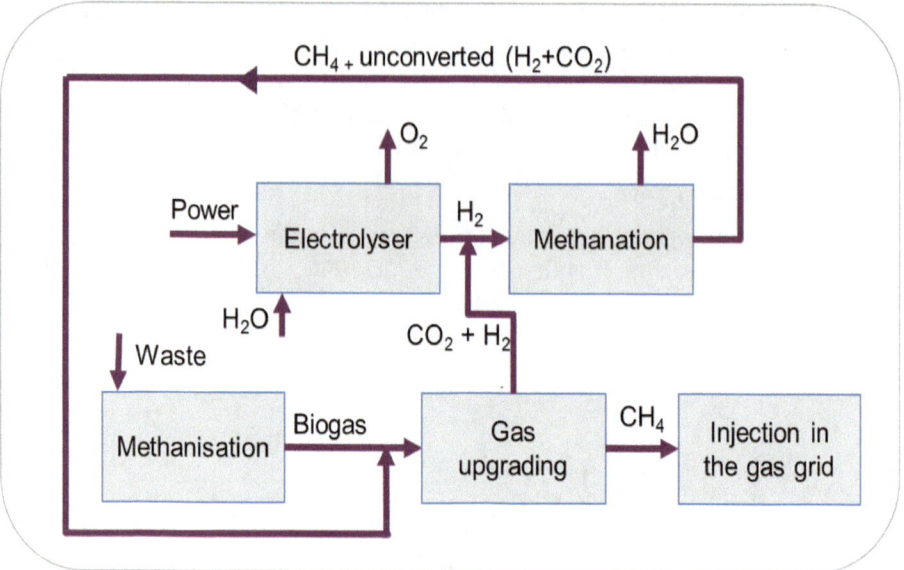

Figure 20.4: Configuration of a coupled methanization – power to methane unit.

The conversion yield from electricity to hydrogen or to methane depends on the efficiency of the electrolyser and of the methanation unit but is also impacted by the consumption of all auxiliaries. Considering only the efficiency of electrolysers given previously and a methanation efficiency of 78% (cf. following paragraph), the global conversion yield based on HHV varies from 50% to 62% in the case of low temperature electrolysis and up to 76% in the case of heat integration between methanation and high-temperature electrolysis.

20.4 The methanation reaction: Thermodynamics and kinetics

This section presents some of the theoretical background on the methanation reaction, a simple thermodynamic analysis is proposed, followed by a description of some commonly assumed mechanism and kinetic models. Then the main deactivation mechanisms observed with nickel catalysts are reported.

The principle of methane synthesis from hydrogen and carbon oxides or methanation reaction has been first described by Sabatier and Sanderens [7] in 1902.

The main industrial application of the methanation reaction was the removal of carbon monoxide traces from hydrogen feed gas in ammonia synthesis plants. In the 1970s, the synthesis of synthetic natural gas (SNG) from a syngas produced by coal gasification was developed as an industrial process [8]. Only recently, with the emerging power-to-gas concept did carbon dioxide methanation draw a real interest.

Carbon monoxide (Eq. 20.8) and carbon dioxide hydrogenation (Eq. 20.9) reactions are both reversible and strongly exothermic:

$$CO + 3H_2 \rightleftharpoons CH_4 + H_2O \left(\Delta_r H_{298} = -206 \text{ kJ.mol}^{-1}\right) \tag{20.8}$$

$$CO_2 + 4H_2 \rightleftharpoons CH_4 + 2H_2O \left(\Delta_r H_{298} = -165 \text{kJ.mol}^{-1}\right) \tag{20.9}$$

Both reactions are catalysed by metals such as nickel and are linked by the endothermic reverse water gas shift reaction (Eq. 20.10):

$$CO_2 + H_2 \rightleftharpoons CO + H_2O \left(\Delta_r H_{298} = +41 \text{kJ.mol}^{-1}\right) \tag{20.10}$$

Solid carbon can also be formed on the catalyst by the Boudouard reaction (Eq. 20.11):

$$2CO \rightleftharpoons C(s) + CO_2 \left(\Delta_r H_{298} = -172 \text{kJ.mol}^{-1}\right) \tag{20.11}$$

The thermodynamic equilibrium of the methanation reactions is strongly influenced by pressure and temperature. According to the Van't Hoff principle, methane production is favoured when the temperature is decreased, whereas it is reduced at high temperature. According to the Le Chatelier rule, due to the volume contraction during the CO and CO₂ methanation reactions, methane production is favoured when the pressure is increased. The gas composition at thermodynamic equilibrium can be evaluated by minimisation of the free Gibbs energy procedure. Figure 20.5 shows the equilibrium composition at 0.1 MPa using thermodynamic data from Gordon and McBride [9] and H₂, CO₂, CO, CH₄, H₂O as gases and solid graphite. Under the studied conditions of stoichiometric CO₂ methanation (H₂/CO₂ = 4/1), no graphite formation was predicted, whereas for stoichiometric conditions of CO methanation (H₂/CO = 3/1) the formation of solid graphite was predicted. These results are consistent with other studies [10]. For temperatures lower than 300 °C, the main products are methane and water, when the temperature is increased, these fractions decrease and hydrogen and carbon dioxide fractions increase. Because of the water gas shift reaction (Eq. 20.10), the CO fraction increases for temperatures above 400 °C at 1 bar, consequently, the CO₂ fraction increases with limited temperature, showing a maximum at 550 °C at 1 bar and then decreases as shown in Figure 20.5.

To evaluate the reaction's progress, the CO₂ conversion (Eq. 20.12), methane selectivity (Eq. 20.13), and methane yield (Eq. 20.14) can be defined as follows based on molar flow at the inlet and outlet of a reactor:

$$X_{CO_2} = \left(F_{CO_2,\,inlet} - F_{CO_2,\,outlet}\right) / F_{CO_2,\,inlet} \tag{20.12}$$

$$S_{CH_4} = F_{CH_4,\,outlet} / \left(F_{CO_2,\,inlet} - F_{CO_2,\,outlet}\right) \tag{20.13}$$

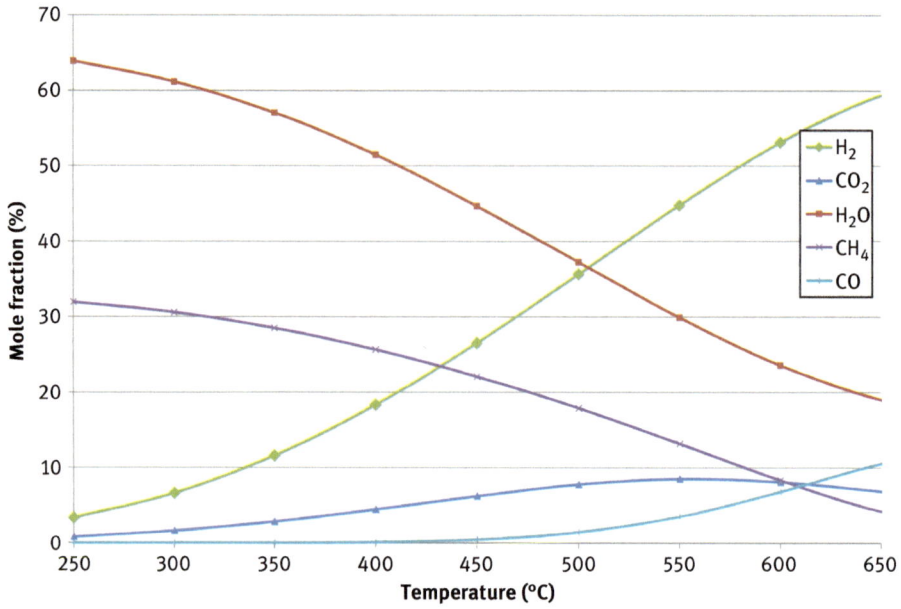

Figure 20.5: Equilibrium composition versus temperature at 0.1 MPa for $H_2/CO_2 = 4/1$.

$$Y_{CH_4} = X_{CO_2}.S_{CH_4} = F_{CH_4, outlet}/F_{CO_2, inlet} \qquad (20.14)$$

The evolution of methane yield with temperature and pressure at thermodynamic equilibrium is shown in Figure 20.6. From the thermodynamic equilibrium point of view, temperatures below 350 °C and pressures above 0.5 MPa allow a methane yield higher than 95%.

The catalysts used for the methanation reaction are generally composed of an active metallic phase (Ru, Rh, Pt, W, Ni, etc.) dispersed on an oxide support such as alumina (Al_2O_3), silica (SiO_2) or ceria (CeO_2). In their review article, Rönsch et al. [11] provide an analysis of the relative activity and selectivity of various metals. Ruthenium is the most active catalyst; however, its high cost prevents its use in industrial-scale methanation. Nickel is less active but exhibits a high methane selectivity and is relatively cheap and abundant making it a material of choice for industrial-scale methanation. The catalyst support strongly influences the catalyst performance because of its interactions with the active phase and its porosity, which are dependent on the nature of the support and on its preparation method. We can say that not only the active metal dispersion but also the nature of the support and its preparation method influence the activity, selectivity and stability of the catalyst.

Several mechanisms for CO_2 methanation can be found in the literature. The first family of mechanisms assumes adsorption and dissociation of CO_2 on the active phase of the catalyst and then hydrogenation of CO to methane, whereas

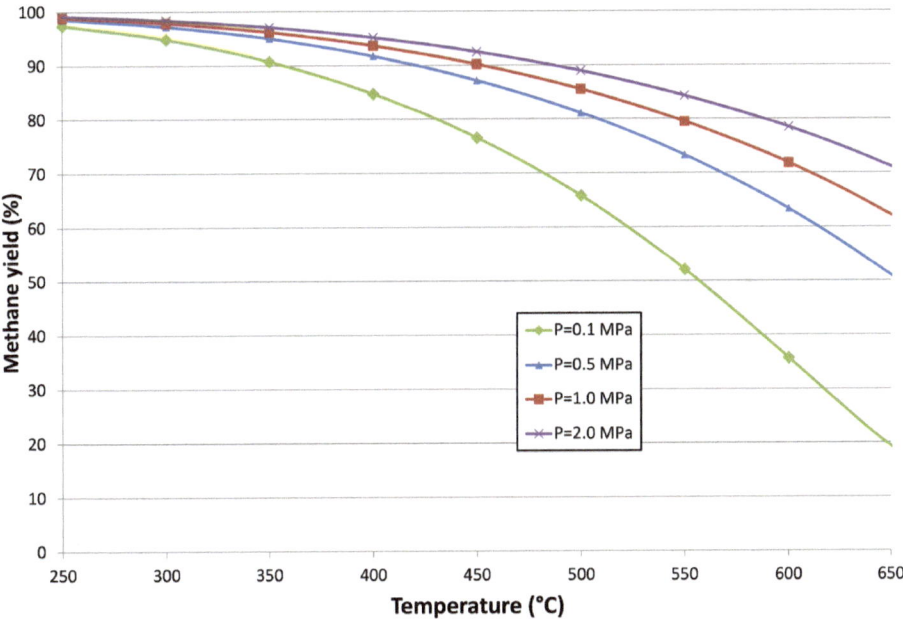

Figure 20.6: Equilibrium methane yield versus temperature for H$_2$/CO$_2$ = 4/1.

the second family assumes the direct hydrogenation of CO$_2$ to methane. The importance of this direct route and the role of hydrogen in the C–O bond cleavage is still considered an open issue for Ni/alumina catalysts [12]. For Ni/Ceria-Zirconia catalysts Aldana et al. [13] consider that CO$_2$ is directly activated on the ceria-zirconia support to form carbonates further hydrogenated into formate, methoxy and then methane.

From an engineering point of view, intrinsic kinetic models applicable in operating conditions encountered in methanation reactors are needed, such models must take into account temperature and reactant partial pressure. As high methane yield is desired, the reaction equilibrium constant and products partial pressure must also be taken into account. The reaction rate is generally expressed under the following form:

$$r_{CH_4} = \frac{(\text{kinetic term}, k).(\text{potential term})}{(\text{adsorption term})}$$

According to the Arrhenius law, the rate constant increases with the reaction temperature and has the following form:

$$k = k_0 e^{-E_a/RT} \tag{20.15}$$

k_0 is the pre-exponential factor (mol/kg$_{catalyst}$/s or mol/m$^3_{bed}$/s), E_a is the activation energy, R the gas constant and T the reaction temperature.

Koschany et al. [12] proposes a review of intrinsic kinetic models, either simple power laws or Langmuir–Hinshelwood–Hougen–Watson (LHHV) assuming various rate determining steps.
Among the kinetic models shown in Table 20.1 the last one is widely used and corresponds to the Xu and Froment mechanism [20] identified for steam methane reforming and methanation conditions over 15 wt.% Ni/MgAl$_2$O$_4$ catalyst at temperatures between 300 and 400 °C and pressures between 0.3 and 0.8 MPa.

However, when using one of these models on a given catalyst, the kinetic constants need to be identified by means of dedicated experiments. Such identification was performed for example by Ducamp et al. [21] for a commercial Ni/alumina catalyst. These experiments are particularly difficult to perform with active catalysts since the catalyst must operate in a chemical regime where no physical limitation (intraparticle heat and mass transport or external heat or mass transport) occurs.

Depending on the catalyst composition, preparation method, calcination and reduction temperature different initial catalytic activity can be obtained; however, the catalytic activity will change over time because of various deactivation mechanisms. For nickel-based catalysts, Bartholomew [22] proposed to group the deactivation mechanism depending on their type that can be chemical, mechanical or thermal as depicted in Table 20.2.

For a given catalyst, whose mechanism will be significant depends on parameters such as the e-duct gas quality, the reactor type and operating temperature. The understanding and the control of these mechanism is essential as a fast catalyst deactivation that could lead to an unwanted shut-down of the power-to-gas unit and an increase of the maintenance costs.

It should be mentioned that most of the data in the open literature on nickel catalyst deactivation has been collected in the case of CO methanation (mainly in the context of coal-to-gas processes). In the case of stœihiometric CO$_2$ methanation, deactivation mechanisms such as nickel carbonyl formation and migration and carbon deposition could be of less importance since the ratio of H$_2$ to CO is high.

It is well known that nickel catalysts are very sensitive to sulphur compounds as sulphur will adsorb strongly (almost irreversibly) on nickel and block the active sites. For example, nickel catalyst poisoning has been measured for H$_2$S concentration of the order of tens of ppb [22]. So it is essential that the feed gas is cleaned upstream the methanation reactor, unless special sulphur-resistant catalysts are used.

One important cause of deactivation is the thermal degradation of the catalyst. The growth of supported metal nanoparticles results in a loss of surface. Two mechanisms are proposed, the first is growth by interparticle transport, also called Ostwald ripening, the other is particle migration and coalescence and is referred to as sintering [23]. The catalyst thermal degradation is favoured by high temperatures

Table 20.1: Overview of intrinsic kinetic models presented in literature (adapted from Koschany et al. [12]).

Catalyst (Ni wt.%)	T (°C)	P_{max} (bar)	Rate equation	Ref.
Ni/SiO$_2$ (60)	280–400	30	$r_{CH_4} = \dfrac{k P_{CO_2} P_{H_2}^A}{\left(1 + K_{H_2} P_{H_2} + K_{CO_2} P_{CO_2}\right)^5}$	[14]
Ni/Al$_2$O$_3$ (28)	200–230	1	$r_{CH_4} = \dfrac{k P_{CO_2}}{\left(1 + A_{CO_2} P_{CO_2}\right)}$	[15]
Ni/SiO$_2$ (3)	227–327	1.4	$r_{CH_4} = \dfrac{k P_{CO_2}^{0.5} P_{H_2}^{0.5}}{\left(1 + K_1 P_{CO_2}^{0.5} P_{H_2}^{0.5} + K_2 P_{CO_2}^{0.5} / P_{H_2}^{0.5} + K_3 P_{CO}\right)^2}$	[16]
Ni/SiO$_2$ (58)	275–320	17	$r_{CH_4} = k P_{CO_2}^{0.66} P_{H_2}^{0.21}$ $r_{CH_4} = \dfrac{k P_{CO_2} P_{H_2}}{\left(1 + K_{H_2} P_{H_2} + K_{CO_2} P_{CO_2}\right)}$	[17]
Ni	250–350	–	$r_{CH_4} = \dfrac{k P_{H_2} P_{CO_2}^{1/3}}{1 + K_{CO_2} P_{CO_2} + K_{H_2} P_{H_2} + K_{H_2O} P_{H_2O}}$	[18]
Ni/La$_2$O$_3$/Al$_2$O$_3$ (17)	240–320	1	$r_{CH_4} = \dfrac{k P_{H_2}^{1/2} P_{CO_2}^{1/3}}{\left(1 + K_{H_2} P_{H_2}^{1/2} + K_{CO_2} P_{CO_2}^{1/2} + K_{H_2O} P_{H_2O}\right)^2}$	[19]

(continued)

Table 20.1 (continued)

Catalyst (Ni wt.%)	T (°C)	P_{max} (bar)	Rate equation	Ref.
Ni/MgAl$_2$O$_4$ (15)	300–400	10	$r_1 = \frac{k_1}{P_{H_2}^{2.5}}\left(P_{CH_4}P_{H_2O} - \frac{P_{H_2}^3 P_{CO}}{K_1}\right)/DEN^2$	[20]
			$r_2 = \frac{k_2}{P_{H_2}}\left(P_{CO}P_{H_2O} - \frac{P_{H_2}P_{CO_2}}{K_2}\right)/DEN^2$	
			$r_3 = \frac{k_3}{P_{H_2}^{3.5}}\left(P_{CH_4}P_{H_2O}^2 - \frac{P_{H_2}^4 P_{CO_2}}{K_3}\right)/DEN^2$	
			$DEN = 1 + K_{CO}P_{CO} + K_{H_2}P_{H_2} + K_{CH_4}P_{CH_4} + K_{H_2O}P_{H_2O}/P_{H_2}$	

Table 20.2: Deactivation mechanisms (adapted from Bartholomew [22]).

Mechanism	Type	Description
Poisoning	Chemical	Chemisorption of poisons on reactive sites (i.e., H$_2$S, SO$_2$)
Vapor formation	Chemical	Reaction of gas with active phase to produce volatile compound (i.e., nickel carbonyls formation)
Vapor–solid and solid–solid reactions	Chemical	Reaction of fluid, support, or promoter with catalytic phase to produce inactive phase
Fouling	Mechanical	Physical deposition of species on the catalyst surface or pores (i.e., carbon deposition and polymerisation)
Attrition	Mechanical	Size reduction of particles because of abrasion (and possible transport of fine particles through the reactor)
Crushing	Mechanical	Loss of catalytic material because of mechanical or thermally induced stresses
Thermal degradation	Thermal	Thermally induced loss of active phase surface area or support area (i.e., metallic particles sintering)

and high steam content, so the catalyst thermal degradation kinetics will depend on the type of reactor and the conversion rate.

Since the Sabatier reaction is highly exothermic ($\Delta_r H_{298} = -165 \text{kJ.mol}^{-1}$), the main challenge is the control of the reaction temperature. From the thermodynamics, it appears that the reaction temperature must be limited to ensure a high CO$_2$ conversion and methane selectivity. To prevent catalyst thermal deactivation, catalyst temperature must be limited too; however, the catalyst temperature must be high enough to provide a high catalyst productivity. The methanation reactor design and operation have to provide a trade-off between these two points of view.

20.5 Catalytic reactors technologies

The methanation unit, as part of the power-to-gas system, will influence its global capital and operating expenditures, energetic efficiency and life cycle in several ways:

- Local gas injection specifications (although not harmonised at the European level) require the injected gas being essentially methane, with a very low content of other gases (e.g., in France H$_2$ < 6 vol.%, CO$_2$ < 2.5 vol.%) [24]. To achieve these specifications, a high carbon dioxide conversion rate and methane selectivity can be reached in the methanation unit or a less efficient (possibly of smaller size and cost) methanation unit can be used followed by a gas upgrading unit for H$_2$ and CO$_2$ removal and possibly recycling.

- The intermittency of the renewable electric production leading to varying electricity cost and the objective of minimising the cost of SNG produced result in the need of dynamic operation of the power-to-gas system. This dynamic operation can be characterised by different states such as production, hot stand-by and cold stand-by and ramping between each of these states [25].
- During the methanation process, the energetic efficiency based on HHV for total conversion is 77.8% (if no conversion occurred, the energetic efficiency would be 100%!), the rest being released as reaction heat. The use of this reaction heat in other components of the power-to-gas system is essential to achieve high overall efficiency.

The catalytic methanation reactors are either state-of-the-art technologies adapted from large-scale, stationary coal-to-SNG systems, or new concepts designed to reach the specific requirements for smaller scale, more compact units and unsteady operation because of fluctuating availability of renewable hydrogen. A review of (CO) methanation reactors for coal-to-SNG as well as for biomass-to-SNG was written by Kopyscinski [8]. The two most common methanation reactor architectures are the cascade of adiabatic reactors and the multitubular reactor described below.

Adiabatic fixed-bed reactors are used in many methanation processes such as TREMP (Haldor Topsøe) [8] or Lurgi (currently Air Liquide) [26]. These reactors being used with no external or internal cooling, the temperature rises rapidly in the reactor until the thermodynamic equilibrium is reached. It is then necessary to use several reactors in cascade and to cool the flow at the outlet of each reactor. An example of a cascade of four adiabatic reactors at 25 bar and 250 °C is depicted in Figure 20.7. At the outlet of the first reactor, high temperatures (771 °C) and moderate methane yield (50%) are reached, then the gases are cooled down to 250 °C by an external heat exchanger before entering the second reactor. At the outlet of the second reactor, a lower temperature (630 °C) and higher methane yield (77%) are reached. Thanks to the cooling between each reactor, the methane yield obtained at the outlet of each reactor is increased, so by multiplying the number of reactors, high methane yield can be obtained. The high temperature at the outlet of the first reactor can be detrimental for the catalyst, so part of the outlet flux is often recycled to the inlet of the first reactor, leading to a dilution of the inlet flux by steam and methane and consequently a reduction of the adiabatic temperature rise (724 °C in this example).

This kind of reactor is attractive for its simple design, and ability to produce superheated steam but requires special high-temperature resistant catalysts. This technology has been deployed for large stationary coal-to-SNG units of over one billion standard cubic meter per year [11].

Considering the need to overcome the thermodynamic limitation and limit the catalyst deactivation, many options of cooled reactors have been developed for exothermic reactions and adapted to CO_2 methanation. The general idea is to limit the distance between the heat generation by the catalyst bed and the heat removal by

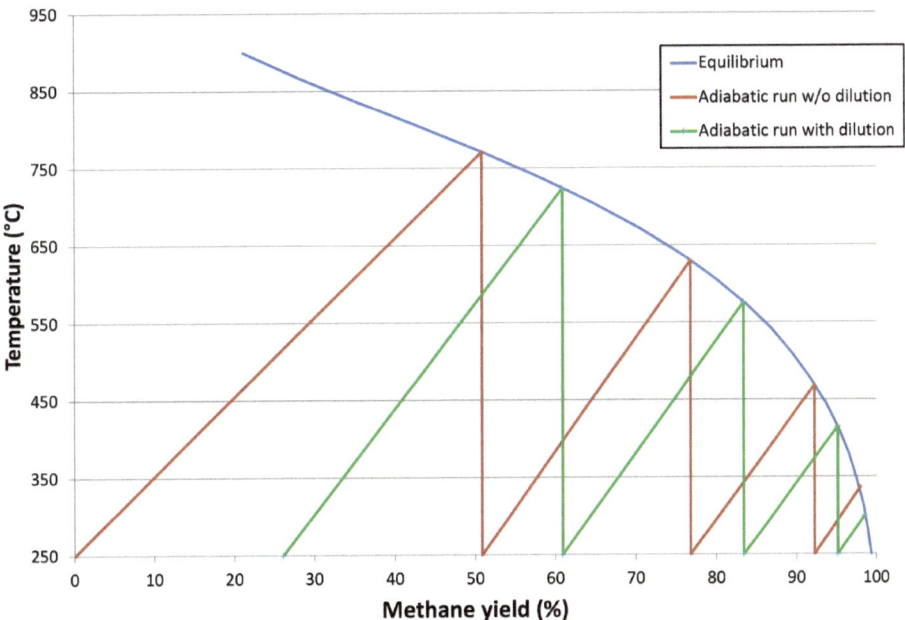

Figure 20.7: Example of a cascade of four adiabatic reactors with and without recycling part of the first reactor flow. Reaction conditions: 2.5 MPa; 250◦C; H$_2$/CO$_2$ = 4/1.

the cooling circuit. The technical solution can consist of a multitubular reactor where the catalyst is inserted inside the reactor tubes and the coolant circulates in the shell, a tube bundle or a plate reactor where the catalyst is inserted in the shell and coolant circulates within the plate.

Multitubular reactors are the most commonly used cooled reactors as their design is relatively simple and their manufacturing follows well-established standards. The catalysts in the form of pellets or spheres are inserted into centimetre-sized pipes with length of a few metres. This type of reactor commonly contains thousands to tens of thousands tubes. The cooling is performed by circulation of a thermal fluid on the shell side; typically thermal oil is used for temperatures up to 350 °C and molten salts are preferred for higher temperatures.

When using this kind of reactor for CO$_2$ methanation with active catalysts, a thermal runaway is very likely to occur close to the reactor inlet. This runaway can be limited by diluting the reactants. A number of experimental and numerical studies have been performed to better understand the behavior of such reactors, Schlereth et al. [27] performed an interesting numerical study of CO$_2$ methanation without dilution and showed how the coolant temperature at which the thermal runaway occurs can be shifted by changing the tube diameter.

Ducamp et al. [21] recently published experimental and modelling results of an annular fixed-bed reactor. The reactor ignition occurs between 200 and 225 °C, leading to a strong increase of CO_2 conversion rate, and then, the further increase from 250 to 275 °C leads to a slight maximum temperature increase, whereas the measured CO_2 conversion is constant (Figure 20.8).

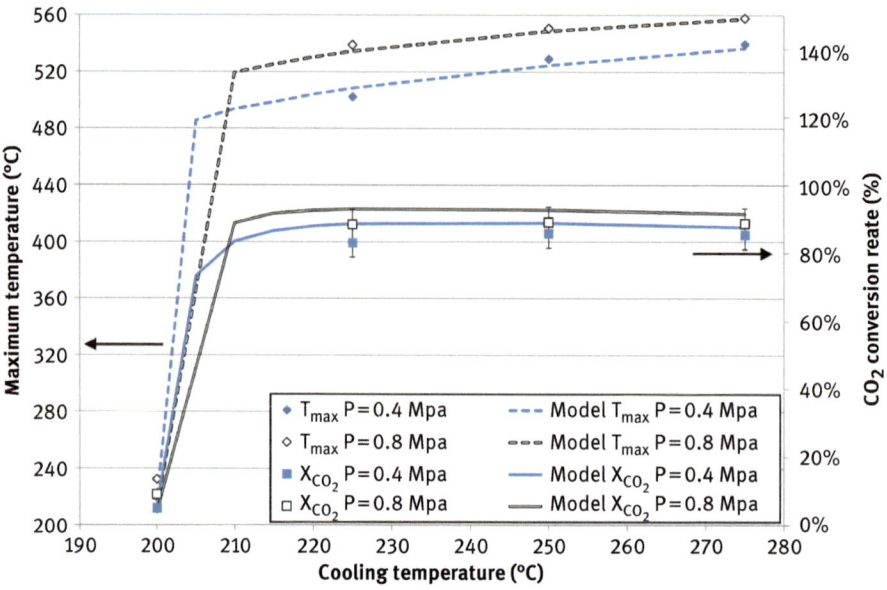

Figure 20.8: Comparison of measured and experimental maximum bed temperature and CO_2 conversion of an annular fixed bed reactor for different operating temperatures at 0.4 and 0.8 MPa [21].

The simulation showed that the thermodynamic equilibrium was reached locally at the peak temperature location (between 510 and 540 °C, Figure 20.9), then the net reaction rate being very low, the temperature decreases along the reactor because of the continuous wall cooling, leading to a shift of the thermodynamic equilibrium allowing higher CO_2 conversion rates. The same analysis showed that strong diffusion limitations occur because of the low CO_2 diffusion rate within the catalyst particles.

The experiments from Ducamp were performed with a rather small annular channel diameter (inner diameter 10 mm, outer diameter 25 mm) and a significant dilution (CO_2 flow rate 5 L (STP) min^{-1}, H_2 flow rate was 20 L (STP) min^{-1} and argon flowrate 50 L (STP) min^{-1}).

Limiting the temperature increase in the reactor is a common goal of many recent reactor developments as it can help reducing the catalyst deactivation (in situations where thermal sintering is the main deactivation mechanism) and overcoming the thermodynamic conversion limitation. Among the different reactor architectures to limit the temperature, we can mention the following:

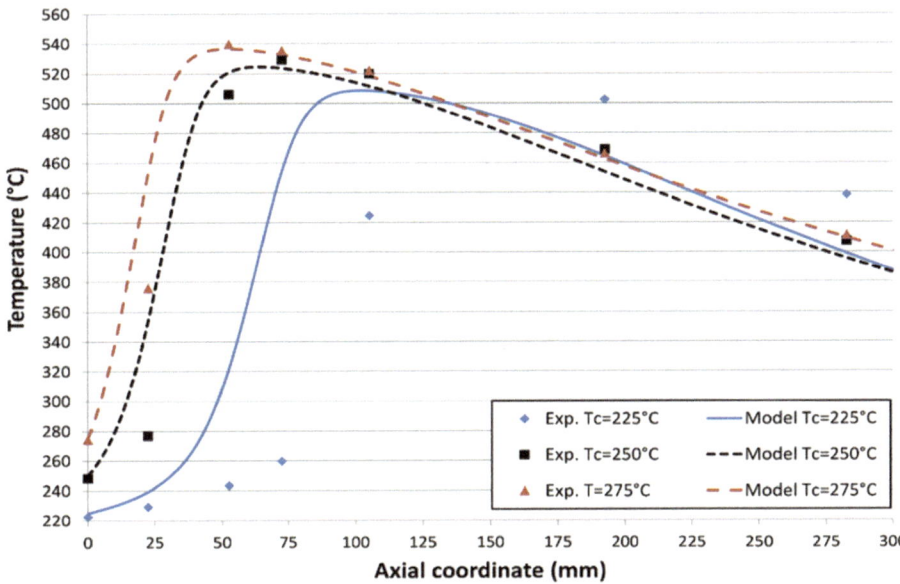

Figure 20.9: Comparison of measured and experimental bed temperature of an annular fixed bed reactor for different operating temperatures at 0.4 [21].

- Diluting the educt by an inert gas (generally steam); with this way the gas heat capacity is increased, the reactants partial pressure is reduced and if the diluant is a product, the thermodynamic equilibrium is shifted toward lesser conversion. The main drawback of this option is the large dilution needed and corresponding compression costs.
- Reducing the reactive channels size, by this way, the surface-to-volume ratio is increased to favour the heat exchange and the channels shape can be optimised to reduce the mass and heat diffusion lengths, this option leads to milli-structured reactors architectures (Figure 20.10) where layers of cooling and reactive channels are superposed [28]. In this case, manufacturing techniques such as diffusion bonding or brazing are needed and even with small-scale channels, the occurrence of a hot spot cannot always be avoided.
- If instead of using a fixed-bed of catalyst, a thin layer of catalyst can be coated [29] on a conductive structure such as an open cell foam [30], [31], a honeycomb [32], or a 3D mesh [33], then both the catalyst load and the diffusion lengths are reduced, leading to a very good heat management. If the catalyst can be deposited directly on the wall of a cooling channel, forming a so-called microreactor, then an even better heat management can be obtained. The catalyst coating has to ensure the long-term anchoring of the catalyst on the structure. In case of catalyst deactivation, the smaller catalyst inventory could lead to rapid performance reduction, so options to replace the catalysts are searched for.

Figure 20.10: Milli-structured fixed bed reactor [28].

- In the above-mentioned reactor architectures, the catalyst is fixed in the reactor. An alternative is the use of a fluidised bed reactor where the small-scale catalyst particles are fluidised by the reactant upward flow allowing a very good heat removal capacity, and thus quasi-isothermal reaction [34]. This reactor requires attrition-resistant catalyst.
- An alternative to the fluidised-bed reactor is the slurry bubble reactor where a heat transfer liquid is present in the reactor vessel and the reactants are injected from the bottom in gas phase [35]. The heat removal capacity of such a reactor can be very high; however, strong mass transfer limitation and thermal inertia can be encountered.

Rönsch [11] proposed a comparison of different methanation concepts based on the operation mode (adiabatic, polytropic or isothermal) characterised by the Semenov number used by Kiewidt and Thöming [36]. In this comparison (Table 20.3), state-of-the-art reactors are represented by the adiabatic fixed-bed reactors, the cooled fixed-bed and the fluidised-bed reactors with high technological readiness level (TRL) of 7 to 9. The other concepts described by Rönsch are micro-reactor (catalyst layer of a few tens of microns coated on cooled reactor walls) and three-phase reactor (catalyst suspended in a liquid phase), both concepts providing nearly isothermal reaction. Two additional concepts need to be described, the structured reactor (catalyst layer of a few tens of microns coated on a 3D structure like a honeycomb or open cell foam) and milli-structured reactor (packed catalyst of less than 1 mm in small cooled channels).

A performance evaluation of fixed-bed, milli-structured and metallic foam reactor channels for CO_2 methanation, performed at the laboratory scale with a commercial nickel/alumina catalyst, was published recently [28]. Well-established criteria such as CO_2 conversion (Eq. 20.12) and methane yield (Eq. 20.14) can reflect the gas quality

Table 20.3: Comparison of different methanation concepts (data from Rönsch [11]).

Reactor type	Operation mode	Reactor stages	Gas recycling	TRL	Temperature (°C)	Catalyst
Adiabatic	Adiabatic	2–7	Usually	250–700	Packed	9
Fixed-bed	Polytropic	1–2	Sometimes	250–500	Packed	7
Fluidised-bed	Isothermal	1–2	Sometimes	300–400	Fluidised	7
Micro-reactors	Polytropic	1–2	No	250–500	Coated	3–4
Three-phase	Isothermal	1–2	No	300–350	Fluidised	4–5
Structured	Polytropic	1–2	No	250–350	Coated	4–5
Milli-structured	Polytropic	1–2	No	250–500	Packed	4–5

at the outlet of one reactor, but additional criteria such as the space-time yield STY (h^{-1}) (Eq. 20.16), volumetric methane productivity VMP (mol.h^{-1}.L^{-1}) (Eq. 20.17), mass methane productivity MMP (mol.h^{-1}.kg^{-1}) (Eq. 20.18) and maximum temperature elevation are necessary.

$$STY = y_{CH_4}.GHSV \tag{20.16}$$

$$VMP = \frac{F_{CH_4, outlet}}{V_R} \tag{20.17}$$

$$MMP = \frac{F_{CH_4, outlet}}{m_c} \tag{20.18}$$

where GHSV is the gas hourly space velocity (h^{-1}), V_R the reactor volume (m) and m_c the catalyst mass (kg).

The main results of this evaluation are reported in Table 20.4 below. The annular fixed-bed reactor channel, despite a dilution ratio of 3, showed a large temperature elevation (277 °C) and a high methane yield, but a low space-time yield (2183 h^{-1}) and a low catalyst productivity (VMP 19 mol.h^{-1}.L^{-1}, MMP 28 mol.h^{-1}. kg^{-1}). The milli-structured reactor channel was used without dilution at a high GHSV (15800 h^{-1}). A thermal runaway was observed, leading to an important temperature excursion (206 °C) and a high methane yield (82 %). Both space-time yield

Table 20.4: Performance indicator of three different methanation reactor architectures (data from [28]).

Reactor type	y_{CH_4} (%)	STY (h^{-1})	VMP- (mol.h^{-1}.L^{-1})	MMP- (mol.h^{-1}.kg^{-1})	ΔT_{max} (°C)
Annular fixed-bed	83	2,183	19	28	227
Milli-structured	82	12,956	114	153	206
Metallic foam	65	1,755	21	482	15

(12956 h^{-1}) and productivity (*VMP* 114 mol.h^{-1}.L^{-1}, *MMP* 153 mol.h^{-1}.kg^{-1}) were high, because of the small reactor volume and catalyst mass. The metallic foam reactor channel showed a low (65 %) methane yield, despite a high pressure (1.5 MPa) and low GHSV (2700 h^{-1}). Because of the high equivalent thermal conductivity of the foam, combined with a low catalyst loading, the temperature elevation was very low (11 °C). The space-time yield was low (1755 h^{-1}) and the catalyst volume productivity was low (*VMP* 21 mol.h^{-1}.L^{-1}), whereas the catalyst mass productivity was very high (*MMP* 492 mol.h^{-1}.kg^{-1}).

Practically, the choice of a reactor architecture is not easy and criteria such as the technological readiness level, the availability of efficient and temperature- resistant catalysts, installation size and operating profile should be taken into account.

20.6 Reactor modelling, simulation and optimisation

Reactor modelling is a powerful tool for reactor design and optimisation; however, physical and chemical phenomena involved in catalytic methanation reactors are coupled and occur at different space and timescales making the modelling a challenging task. The space scales can be identified as the macroscopic scale of the reactor, the mesoscopic scale of the catalyst particle or bed porosity and the microscopic scale of the catalyst inner porosity and surface where the reactants diffuse, and react on the active sites to form the products that diffuse back to the catalyst boundary. Whereas the dynamics of start-up and transition from one operating point to another can be described at the minute or second scale, the catalyst performance variation with time because of deactivation has to be taken into consideration by specific modelling at a much longer timescale.

Even within the restriction of homogenised and steady-state conditions, many modelling options can be chosen regarding the geometric model, the phase model and the kinetic model:

– The geometric model of the reactor can be represented in one, two or three dimension of space. The one dimension of space allows the representation of axial gradient composition and temperature profiles but does not represent accurately the profiles in the other dimensions. This can be justified for adiabatic reactors but when an external cooling is applied, radial temperature gradients exists, which cannot be accurately described by such a model. For quantitative evaluations, a two-dimensional model should be preferred [27].When the two-dimensional hypothesis cannot be justified, a three-dimensional model can be needed.

- The gas and solid phases in the reactor can be treated separately or not. In the first case, diffusion phenomena leading to temperature and species concentration variation within the catalyst particle or between the catalyst outer surface and the gas are modelled and coupled to the heat and mass transport equations at the reactor scale, the resulting model is called a heterogeneous model. In the other case, the temperature and species concentration are supposed equal, leading to the pseudo-homogeneous model. In the case of CO$_2$ methanation, these diffusion limitations can result in changes in reaction rate because of higher temperature or lower reactant concentration but can also result in changes in the reaction itself, shifting for example from reverse water gas shift to water gas shift reaction as shown by Ducamp et al. [21].
- A detailed micro-kinetic model is seldom used for reactor modelling because of the computational burden; instead, simplified LHHW approaches are generally used. The model proposed by Xu and Froment [20] is often used in the literature; however, as the catalyst activity is dependent not only on the formulation, but also on elaboration, pre-treatment and ageing, such a model needs to be adapted to the catalyst used.

20.7 Demonstration projects

Pr Hashimoto in Japan proposed the concept of energy storage of renewable sources by power-to-gas in the 1990s. The first demonstrator was implemented in 1996 by the team of Pr Hashimoto on the roof of the Tohoku University [37]. In this unit, a PV solar panel fed a seawater electrolysis and the CO$_2$ was captured from methane combustion. The hydrogen and CO$_2$ were converted into SNG, which was then stored in a tank.

A new interest in power-to-gas has emerged since the middle of 2000s with the development of intermittent renewable energies in the context of the energy transition. The first significant demonstration unit of 250 KWe was implemented in Germany by ZSW in Stuttgart [38]. This unit allowed the tests of a catalytic methanation, which was then upscaled at the Audi Power-to-Gas unit of 6.3 MWe in Wertle [39]. Several demonstration projects have been launched since 2013 in Europe. We can notice a large Innovation Action project, STORE&GO, funded in the framework of the H2020 programme. This project plans the implementation of three different power-to-methane units in Europe. It includes a benchmark of three technologies of methanation: two catalytic [40], [41] and one biological [42].

The following table lists the main Power to Methane projects in Europe, including the implementation of demonstration units (TRL7) of a power of electrolyser above 100 kWe.

Table 20.5: Main power to methane projects in Europe.

Project	project start-up	Location of the demonstration unit	Electrolysis technology and power	CO$_2$ source	Methanation technology
Power-to-gas 250 [38]	2010	Stuttgart (DE)	Alkaline – 250 kW	Cylinder	Shell-and-tube catalytic reactors
Audi e-gas [39]	2011	Wertle (DE)	Alkaline – 6 MWe	Biogas	Shell-and-tube catalytic reactors
BioPower-to-gas [43]	2013	Allendorf (DE)	PEM – 400 kWe	Biogas	Biological methanation
CO2-SNG [44]	2014	Łaziska Górne (PL)	Alkaline – 100 kWe	Industrial emission from a coal power plant	Millistructured HEX catalytic reactor
Biocat [45]	2014	Avedøre (DK)	Alkaline – 1 MWe	Waste water treatment	Biological methanation
JUPITER1000 [46]	2016	Fos sur mer (FR)	Alkaline – 500 kWe + PEM – 500 kW	Industrial emission from an ASCOMETAL factory	Millistructured HEX catalytic reactor
STORE&GO [40]	2016	Falkenhagen (DE)	Alkaline – 1 MWe	Biogas	Honeycomb HEX catalytic reactor
STORE&GO [42]	2016	Solothurn (CH)	Alkaline – 700 kWe	Waste water treatment	Biological methanation
STORE&GO [41]	2016	Troia (IT)	Alkaline – 200 kWe	Atmospheric capture	Millistructured HEX catalytic reactor
RAG [47]	2018	Pilsbach (AT)	Alkaline – 600 kWe first and ultimately 13 MWe	Not specified	Underground microbiological methanation
METHYCENTRE [48]	2018	Céré La Ronde (FR)	PEM – 200 kWe	Biogas	Milli-structured HEX catalytic reactor

20.8 Techno-economic assessment and regulations

20.8.1 Production costs of hydrogen and methane

The natural gas was produced over millions and millions of years by nature. Its cost corresponds to its extraction and its transportation to end-users. In 2017, the wholesale day-ahead gas prices on gas hubs in the EU was nearly 15–25€/MWh$_{HHV}$ [49]. As the SNG has to be synthesised from hydrogen and carbon dioxide, its production cost will be higher than the cost of natural gas.

The production cost of hydrogen by power-to-gas should be from 70–90€/MWh $_{HHV}$ to 50–60€/MWh$_{HHV}$, respectively by 2020 and 2050 [50]. We notice that a cost of 50€/MWh$_{HHV}$ of hydrogen corresponds to nearly 2€/kgH$_2$. For SNG, the production cost is higher because of additional CAPEX and OPEX of the CO$_2$ capture unit and methanation unit. Its production cost is expected to be in the range of 100–160€/MWh$_{HHV}$ by 2020 and 70–110€/MWh$_{HHV}$ by 2050 according to the configuration of the system, operating time and the electricity cost [50].

The CAPEX and OPEX for a power-to-methane unit of 10 MWe in 2030 coupled to a biogas unit are described in the following table.

For an electricity cost at 25€/MWh and an operation time of 3,000 h per year, the production cost of methane is 126€/MWh. If heat and oxygen can be valorised, this cost drops to 106€/MWh. Sometimes large discrepancies appears on different studies as reported by Götz et al. [51] but the costs remain in the range given in Table 20.6. To be competitive with the fossil natural gas a carbon tax about 220–300€/t$_{CO2}$ should be implemented [50].

Table 20.6: CAPEX and OPEX in 2030 of a Power to methane unit coupled to a biogas unit (synthesis of data from [50]).

	Alkaline electrolyser	Catalytic methanation	Hydrogen storage and compression / CO$_2$ and CH$_4$ compression / connection to the gas grid	Auxiliaries / groundwork	Total
CAPEX €/kWe	400	310	200	240	1150€
OPEX (% of CAPEX)	5%	5%	1.8%		3.7%

20.8.2 Regulatory issues

In demonstration projects, the proposition of regulatory changes is often included for determining the conditions of the economic balance of power-to-gas as a solution for intermittent renewable electricity storage. Indeed, power-to-gas brings large energy storage capacity. It allows a reduction of investment to improve the resilience of the electricity grid, thanks to the available gas grid and underground storage. For an emergence of the power-to-gas market, regulatory changes should be proposed in the future:

- a specific revenue for the energy storage service proposed;
- a legal status of the renewable gas produced by power-to-gas should be defined;
- a certificate of origin of renewable gases should be implemented;
- implementation of significant CO_2 taxes for an incentive of CO_2 utilisation and a reduction in the consumption of fossil energies.

2.9 Conclusion

Power-to-gas is a solution for large-scale and long-term storage of intermittent renewable electricity. It consists of converting electricity into gas, either hydrogen or methane. To overcome the limitations related with hydrogen injection into the gas grid, hydrogen can be converted into methane by CO_2 hydrogenation. CO_2 can be captured from industrial flue gas, from the atmosphere, but CO_2 can also be taken from biogas allowing the complete carbon yield of the methanization unit and pooling of units such as the injection units

The CO_2 hydrogenation reaction known as Sabatier reaction is an exothermic and balanced reaction. The control of the temperature is one of the major factors to guarantee a high conversion rate, a high selectivity to methane and a sufficient lifetime of the nickel catalysts.

The catalytic methanation reactors are either state-of-the-art technologies adapted from large-scale, stationary coal-to-SNG systems, or new concepts designed to reach the specific requirements for smaller scale, more compact units and unsteady operation because of fluctuating availability of renewable hydrogen.

More than ten demonstration projects were launched in Europe since 2010 aiming at evaluating the efficiency and operating constraints of different technological options for electrolysis, CO_2 capture and methanation at different scales.

As SNG has to be synthesised from hydrogen and carbon dioxide, its production cost is higher than the extraction and transport cost of fossil natural gas. The emergence of a power-to-methane system at industrial scale requires a SNG production cost at medium and long terms below 100€/MWh combined with regulatory changes such as the implementation of significant CO_2 taxes and incentives for the production and use of renewable gases.

References

[1] D. Jones, A. Sakhel, M. Buck, and P. Graichen, "The European Power Sector in 2017." https://sandbag.org.uk/wp-content/uploads/2018/01/EU-power-sector-report-2017.pdf, 2018.

[2] *Directive 2009/28/Ec Of The European Parliament And Of The Council of 23 April 2009 on the promotion of the use of energy from renewable sources and amending and subsequently repealing Directives 2001/77/EC and 2003/30/EC*. EC, 2009.

[3] "Study on Early Business Cases for H2 in Energy Storage and More Broadly Power to Applications." https://www.fch.europa.eu/sites/default/files/P2H_Full_Study_FCHJU.pdf, 2017.

[4] "McPhy." https://mcphy.com/en/our-products-and-solutions/electrolyzers.

[5] "Sunfire Delivers World's Most Efficient Steam Electrolysis Module To Salzgitter Flachstahl Gmbh." https://www.sunfire.de/en/company/press/detail/sunfire-delivers-the-worlds-most-efficient-steam-electrolysis-module-to-salzgitter-flachstahl-gmbh.

[6] L. Briottet, I. Moro, and P. Lemoine, "Quantifying the hydrogen embrittlement of pipeline steels for safety considerations," *Int. J. Hydrogen Energ*, vol. 37, no. 22, pp. 17616–17623, Nov. 2012.

[7] P. Sabatier and J. B. Senderens, "Nouvelles synthèses du méthane," *C. R. Acad. Sci. Paris*, vol. 134 pp. 514, 1902.

[8] J. Kopyscinski, T. J. Schildhauer, and S. M. A. Biollaz, "Production of synthetic natural gas (SNG) from coal and dry biomass – A technology review from 1950 to 2009," *Fuel*, vol. 89, no. 8, pp. 1763–1783, Aug. 2010.

[9] B. J. McBride, S. Gordon, and M. A. Reno, "Coefficients for calculating thermodynamic and transport properties of individual species," NASA Technical Memorandum 4513 1993.

[10] J. Gao *et al.*, "A thermodynamic analysis of methanation reactions of carbon oxides for the production of synthetic natural gas," *RSC Adv.*, vol. 2, no. 6, pp. 2358, 2012.

[11] S. Rönsch *et al.*, "Review on methanation – From fundamentals to current projects," *Fuel*, vol. 166, pp. 276–296, Feb. 2016.

[12] F. Koschany, D. Schlereth, and O. Hinrichsen, "On the kinetics of the methanation of carbon dioxide on coprecipitated NiAl(O)x," *Appl. Catal. B: Environ.*, vol. 181, pp. 504–516, Feb. 2016.

[13] P. A. U. Aldana *et al.*, "Catalytic CO2 valorization into CH4 on Ni-based ceria-zirconia. Reaction mechanism by operando IR spectroscopy," *Catal. Today*, vol. 215, pp. 201–207, Oct. 2013.

[14] J. N. Dew, R. R. White, and C. M. Sliepcevich, "Hydrogenation of carbon dioxide on nickel-kieselguhr catalyst," *Ind. Eng. Chem.*, vol. 47, no. 1, pp. 140–146, 1955.

[15] T. Van Herwijnen, H. Van Doesburg, and W. a De Jong, "Kinetics of the methanation of CO and CO2 on a nickel catalyst," *J. Catal.*, vol. 28, no. 3, pp.391–402, 1973.

[16] G. D. Weatherbee and C. H. Bartholomew, "Hydrogenation of CO2 on Group VIII Metals," *J. Catal.*, vol. 77, pp. 460–472, 1982.

[17] J. H. Chiang and J. R. Hopper, "Kinetics of the hydrogenation of carbon dioxide over supported nickel," Ind. Eng. Chem. Prod. Res. Dev., vol. 22, no. 2, pp. 225–228, 1983.

[18] H. Inoue and M. Funakoshi, "Kinetics of methanation of carbon monoxide and carbon dioxide.," J. Chem. Eng. Jpn., vol. 17, no. 6, pp. 602–610, 1984.

[19] T. T. T. Kai, "Kinetics of the methanation of carbon dioxide over a supported ni-la2o3 catalyst," Can. J. Chem. Eng., vol. 66, no. 1, pp. 343–347, 1988.

[20] J. Xu and G. F. Froment, "Methane steam reforming, methanation and water-gas shift : 1. intrinsic kinetics," *AIChE. J.*, vol. 35, no. 1, pp. 88–96, 1989.

[21] J. Ducamp, A. Bengaouer, and P. Baurens, "Modelling and experimental validation of a CO_2 methanation annular cooled fixed-bed reactor exchanger," Can. J. Chem. Eng., vol. 95, no. 2, pp. 241–252, Feb. 2017.

[22] C. H. Bartholomew, "Mechanisms of catalyst deactivation," Appl. Catal. A: Gen., vol. 212, no. 1–2, pp. 17–60, Apr. 2001.

[23] T. W. Hansen, "Sintering and particle dynamics in supported metal catalysts," .

[24] GRTgaz, "Prescriptions Techniques applicables aux canalisations de transport de GRTgaz et aux installations de transport, de distribution et de stockage de gaz raccordées au réseau de GRTgaz," Prescriptions Techniques V3 du 01. 02.2007, 2007.

[25] E. Frank, J. Gorre, F. Ruoss, and M. J. Friedl, "Calculation and analysis of efficiencies and annual performances of Power-to-Gas systems," Appl. Energ., vol. 218, pp. 217–231, May 2018.

[26] C. Krier, M. Hackel, C. Hägele, H. Urtel, C. Querner, and A. Haas, "Improving the Methanation Process," Chem. Ing. Tech., vol. 85, no. 4, pp. 523–528, Apr. 2013.

[27] D. Schlereth and O. Hinrichsen, "A fixed-bed reactor modeling study on the methanation of CO2," Chem. Eng. Res. Des., vol. 92, no. 4, pp. 702–712, Apr. 2014.

[28] A. Bengaouer, J. Ducamp, I. Champon, and R. Try, "Performance evaluation of fixed-bed, millistructured, and metallic foam reactor channels for CO_2 methanation," Can. J. Chem. Eng., vol. 96, no. 9, pp. 1937–1945, Sep. 2018.

[29] V. Meille, "Review on methods to deposit catalysts on structured surfaces," Appl. Catal. A: Gen., vol. 315, pp. 1–17, Nov. 2006.

[30] E. Bianchi, T. Heidig, C. G. Visconti, G. Groppi, H. Freund, and E. Tronconi, "An appraisal of the heat transfer properties of metallic open-cell foams for strongly exo-/endo-thermic catalytic processes in tubular reactors," Chem. Eng. J., vol. 198–199, pp. 512–528, Aug. 2012.

[31] M. Frey, A. Bengaouer, G. Geffraye, D. Edouard, and A.-C. Roger, "Aluminum Open Cell Foams as Efficient Supports for Carbon Dioxide Methanation Catalysts: Pilot-Scale Reaction Results," Energy Technology, vol. 5, no. 11, pp. 2078–2085, Nov. 2017.

[32] D. Schollenberger, S. Bajohr, M. Gruber, R. Reimert, and T. Kolb, "Scale-Up of Innovative Honeycomb Reactors for Power-to-Gas Applications – The Project Store&Go," Chem.Ing. Tech., vol. 90, no. 5, pp. 696–702, May 2018.

[33] S. Danaci, L. Protasova, J. Lefevere, L. Bedel, R. Guilet, and P. Marty, "Efficient CO2 methanation over Ni/Al2O3 coated structured catalysts," Catal. Today, vol. 273, pp. 234–243, Sep. 2016.

[34] J. Kopyscinski, T. J. Schildhauer, and S. M. A. Biollaz, "Fluidized-Bed Methanation: Interaction between Kinetics and Mass Transfer," Ind. Eng. Chem. Res., vol. 50, no. 5, pp. 2781–2790, Mar. 2011.

[35] J. Lefebvre, M. Götz, S. Bajohr, R. Reimert, and T. Kolb, "Improvement of three-phase methanation reactor performance for steady-state and transient operation," Fuel. Process. Technol., vol. 132, pp. 83–90, Apr. 2015.

[36] L. Kiewidt and J. Thöming, "Predicting optimal temperature profiles in single-stage fixed-bed reactors for CO2-methanation," Chem. Eng. Sci., vol. 132, pp. 59–71, Aug. 2015.

[37] K. Hashimoto et al., "Global CO2 recycling – novel materials and prospect for prevention of global warming and abundant energy supply," Mater. Sci. Eng. A, vol. 267, no. 2, pp. 200–206, Jul. 1999.

[38] M. Specht, "250-kW P2G research plant of the ZSW." https://www.zsw-bw.de/en/research/renewable-fuels/topics/power-to-gas.html.

[39] S. Rieke, "Erste industrielle power-to-gas-anlage mit 6 megawatt," GWF – Gas/ Erdgas, pp. 660-664, Sept. 2013.

[40] "Store&Go Project, Falkenhagen site." https://www.storeandgo.info/demonstration-sites/germany.

[41] "Store&Go Project, Troia site." https://www.storeandgo.info/demonstration-sites/italy.

[42] "Store&Go Project, Solothurn site." https://www.storeandgo.info/demonstration-sites /switzerland.

[43] "BioPower2Gas Project." http://www.biopower2gas.de/mediathek/.

[44] "CO2-SNG Project." http://www.innoenergy.com/innovationproject/our-innovation-projects /co2-sng.

[45] "BioCat Project." http://biocat-project.com.

[46] "Jupiter 1000 Project," https://www.jupiter1000.eu, 2016.

[47] "Underground Sun Conversion Project." https://www.underground-sun-conversion.at/en. html.

[48] "Méthycentre Project." https://MethyCentre.eu.

[49] DG Energy, "Quarterly Report on European Gas Markets." https://ec.europa.eu/energy/sites/ ener/files/documents/201410_q3-4_quaterly_report_gas_market.pdf.

[50] GRTGaz, "Etude portant sur l'hydrogène et la méthanation comme procédé de valorisation de l'électricité excédentaire." http://www.grtgaz.com/fileadmin/engagements/documents/fr/ Power-to-Gas-etude-ADEME-GRTgaz-GrDF-complete.pdf, Sep-2014.

[51] M. Götz *et al.*, "Renewable Power-to-Gas: A technological and economic review," *Renew. Energ.*, vol. 85, pp. 1371–1390, Jan. 2016.

James McGregor

21 Fischer–Tropsch synthesis using CO_2

21.1 Introduction

Fischer–Tropsch synthesis (FTS) is a gas-to-liquid (GTL) technology principally employed for the production of hydrocarbon fuels. The traditional feedstock for this reaction is syngas, a mixture of carbon monoxide and hydrogen and frequently smaller amounts of carbon dioxide and/or methane. This can be derived from a variety of sources including the gasification of coal or steam reforming of natural gas. Increasing attention is, however, being paid to alternative and renewable sources of syngas such as the gasification of biomass. A sustainable alternative is the use of CO_2 in place of CO. This presents an opportunity to produce liquid hydrocarbons from CO_2 reclaimed as part of carbon capture schemes.

The increasing concentration of CO_2 in the atmosphere is well established, reaching 411 ppm in June 2018, with that value forecast to increase to 800 ppm by the end of the century without additional CO_2 reduction measures [1]. In addition to playing a role in reducing CO_2 emissions, the necessity to capture CO_2 from industrial and other sources means that it will become a readily available and likely low-cost feedstock, which can be utilised to reduce societal reliance on fossil resources such as natural gas and crude oil.

A leading driver of such change in Europe is the policy of Energiewende in Germany [2, 3]. Backed by legislation, this policy requires an 80–95% reduction in the emission of greenhouse gases (with respect to 1990 levels) by 2050 with a target that 80% of the country's gross power consumption should come from renewable sources by the same date. Such decarbonisation, both in Germany and globally, by necessity has a large focus on the transportation sector, where the dominant energy source is liquid hydrocarbon fuels. While this demand will likely reduce in the road passenger sector as increasing numbers of electric cars are produced, there is currently no viable alternative for aerospace, shipping or road freight transport. This is illustrated in Figure 21.1, where the fuel energy densities of hydrocarbon fuels such as diesel and gasoline are compared with other alternative sources and are shown to be notably greater than, for example, alcohols, hydrogen or batteries [4]. The production of liquid fuels from sustainable sources such as CO_2 will therefore play an increasingly important role.

This chapter introduces the use of CO_2 as a feedstock for FTS, providing a general introduction (Section 21.1), insights into the reaction mechanism (Section 21.2), details on catalysts and reactors (Sections 21.3 and 21.4) and finally a brief view of the future outlook in this area (Section 21.5).

https://doi.org/10.1515/9783110665147-021

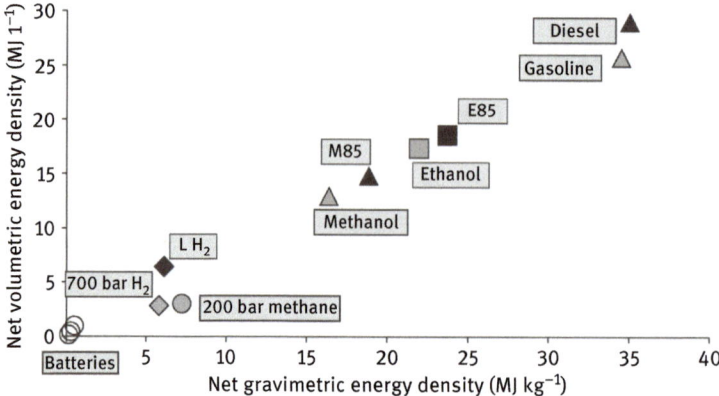

Figure 21.1: Net gravimetric and volumetric energy densities for various potential transportation power sources. Reproduced with permission from [4].

21.1.1 Background

The target of FTS employing CO_2 as the carbon source is typically to produce hydrocarbons for use either directly as a fuel or as a feedstock for further downstream processing into petrochemical products. The main reaction products are typically linear alkanes and alkenes ranging from gas-phase species such as methane to long-chained ($\geq C_{20}$) waxes. The target hydrocarbons are however generally diesel (C_{12}–C_{18}) or gasoline (C_5–C_{11}) range species. FTS products are often superior to crude oil-derived products for the same application. For example, they typically do not contain heteroatoms such as sulphur that have to be removed from conventional feeds. The use of FTS to generate feedstocks for the chemical industry is also crucial as, at present, the manufacture of products from plastics to pharmaceuticals relies on crude oil; FTS provides an alternative source of the requisite raw materials.

The distribution of hydrocarbon products from FTS is described by the Anderson–Schulz–Flory (ASF) distribution [5]:

$$\frac{W_n}{n} = (1-\alpha)^2 \alpha^{n-1}$$

where W_n is the weight fraction of linear products, with carbon number n, and α is the chain growth probability: this is shown graphically in Figure 21.2 [6]. The ASF distribution assumes that the probability of chain growth is constant and that alkanes are formed via a stepwise addition of monomer to the growing polymer chain.

The ASF distribution can also be considered by plotting $\log(W_n/n)$ versus n. This produces a linear plot, where α can be determined from the gradient (gradient = $\log \alpha$). As Figure 21.2 illustrates, the individual product formed in the greatest

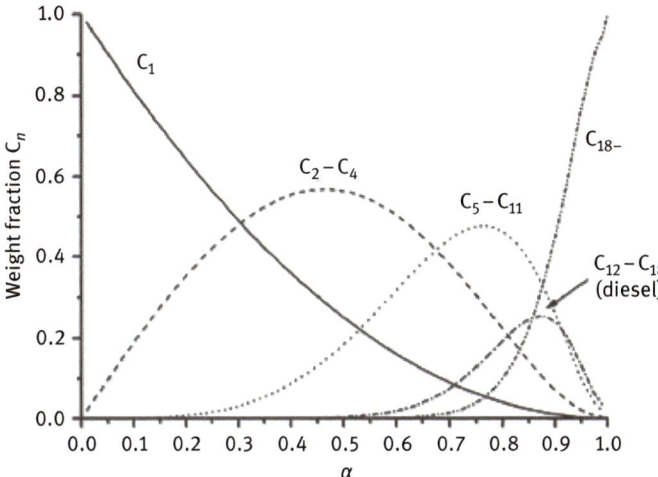

Figure 21.2: Anderson-Schulz-Flory distribution showing the relationship between chain growth probability (α) and weight fraction of hydrocarbons of chain length n. Reproduced with permission from [6].

concentration in the ASF distribution is most commonly methane. In order to minimise this and increase the formation of long-chained products, α should be as high as possible; this however is dependent on numerous factors including catalyst selection (Section 21.3).

21.1.2 Power-to-X

CO_2-FTS can play a crucial role in the development of so-called Power-to-X (P2X) technologies. P2X can be considered as a form of chemical energy storage, converting surplus electricity to an energy vector. This is particularly important in the case of renewables, where there is typically a mismatch between the times of maximal power supply and highest power demand. The "X" may be various products such as gas, liquids, heat and fuels. In the case of FTS, P2X may take the form of hydrogen generation via electrolysis [7–9]. This synthesised hydrogen can then be reacted with CO or CO_2 to produce fuels or other chemical products through FTS chemistry. While hydrogen can be used directly as a fuel, for example, in hydrogen fuel cells, its use a feedstock for FTS allows for the production of liquid hydrocarbon fuels such as diesel and kerosene which can be used directly with the existing infrastructure.

A further modification of this process is to simultaneously produce CO from CO_2 alongside H_2 production from H_2O via a co electrolysis process (Scheme 21.1) [10]. The resulting syngas can then be utilised directly in conventional Fischer–Tropsch reactors. An advantage of the co-electrolysis is the high purity of the syngas produced.

$$^{(+4)}CO_2 + H_2O \rightarrow {}^{(+2)}CO + H_2 + O_2 \qquad \text{Scheme 21.1}$$

An alternative P2X scenario is to focus on the conversion of CO_2 to methane, so-called power-to-gas [11, 12]. The CH_4 produced can be utilised directly in the existing natural gas infrastructure. In addition, methanation of CO_2 has also been proposed by NASA as a means to produce propellant fuel from CO_2 in the Martian atmosphere or to produce water from exhaled CO_2 in life support systems [6]. A schematic of the various P2X options, including FTS preceded by electrolysis, is shown in Figure 21.3 [13].

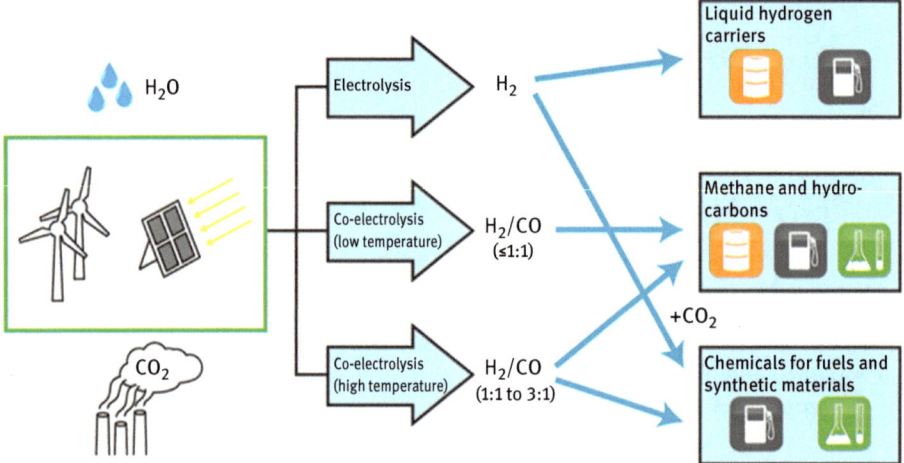

Figure 21.3: Power-to-X technologies to convert excess renewable energy and CO_2 to chemical energy vectors as a form of energy storage and a means of creating liquid and gaseous fuels to drop into the existing infrastructure. Reproduced with permission from [13].

22.2 Reaction mechanism

Superficially, the use of CO_2 in place of CO does not have a significant impact on the chemistry of FTS and does not, for instance, change the overall mass balance requirements. A number of factors, however, make CO_2-FTS more challenging than CO-FTS. Thermodynamically, the former process has a reaction enthalpy at 573 K of -128 kJ mol^{-1} which compares with -166 kJ mol^{-1} for conventional FTS (Scheme 21.2). Additionally, the typically lower heat of adsorption for CO_2 than CO results in reduced surface coverage and higher H/C ratios, resulting in a greater preference for methanation when CO_2 is used [11]. The effect of employing CO_2 as the reactant for the synthesis of hydrocarbons is described in Section 21.2.1, while the formation of selected other products – notably oxygenates – is discussed in Section 21.2.2.

21.2.1 FTS from CO_2 via RWGS

The indirect synthesis of hydrocarbons from carbon dioxide via the reverse water gas shift (RWGS) reaction represents the most viable route to widespread adoption of this technology at present. The general reaction scheme is shown in Scheme 21.2 [14–17].

1. $CO_2 + H_2 \rightleftharpoons CO + H_2O \quad \Delta H_{573\,K} = 38 \text{ kJ mol}^{-1}$

2. $CO + 2H_2 \rightarrow (CH_2)_n + H_2O \quad \Delta H_{573\,K} = -166 \text{ kJ mol}^{-1}$ Scheme 21.2

Steps 1 and 2 can be conducted sequentially in different reactors, but are more commonly conducted in a single reactor using an appropriate catalyst (Section 21.3). A detailed kinetic study of the two-step process has been conducted by Wiilauer et al. [18] Step 2, the Fischer–Tropsch reaction, was shown to be significantly slower than Step 1, RWGS; the difference being seven orders of magnitude at maximum rates. This difference in rate means that unconverted CO is present in the product stream when both reactions take place over a single catalyst.

There are two temperature regimes for FTS: low and high temperature (LTFTS and HTFTS, respectively). The former describes the reaction conditions of ~ 200–250 °C, and the latter 320–375 °C. The primary difference between these regimes is the chain length of the hydrocarbon product with high-temperature FTS yielding shorter chains. In the case where the produced hydrocarbon is methane, CO_2 hydrogenation is known as the Sabatier reaction. While nickel favours the formation of CH_4 from both CO and CO_2 [19–21], cobalt catalysts show differing selectivity depending on which carbon oxide is used: CO produces a typical FT product distribution, while the use of CO_2 results primarily in methanation [22–24]. In contrast iron catalysts yield FT hydrocarbons from both CO and CO_2 [25]. The behaviour of different catalysts is discussed in more detail in Section 21.3.

An alternative indirect route to produce liquid hydrocarbon fuels from CO_2 involves the initial conversion of CO_2 to methanol [18, 26, 27]. followed by the methanol-to-hydrocarbon (MTH, or MTG in the specific case of gasoline synthesis) reaction. MTH is a zeolite-catalysed process, which is commercialised on an industrial scale [28] and proceeds via the so-called hydrocarbon pool mechanism [29–31]. The use of hybrid catalysts may allow the conversion of CO_2 to hydrocarbons in a single reactor environment [32].

The use of bifunctional catalysts containing an acidic zeolite component has also been proposed as a means of upgrading the products from CO_2-FTS to longer chain hydrocarbons in the gasoline range [33]. This employs a sodium-promoted iron catalyst for the RWGS and FTS stages, while the zeolite component catalyses oligomerisation, isomerisation and aromatisation reactions. A schematic of the process is shown in Figure 21.4.

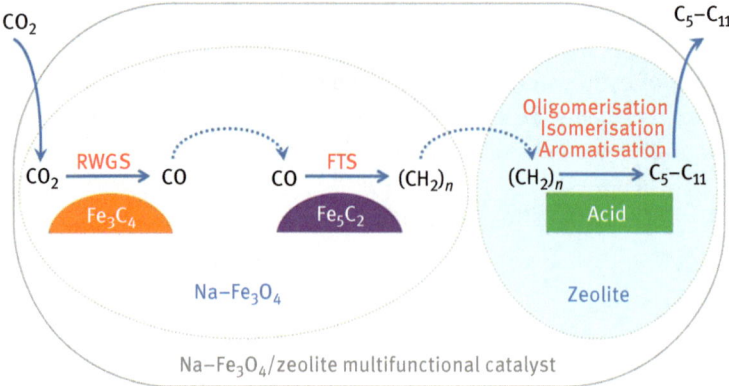

Figure 21.4: Schematic reaction scheme for the conversion of CO_2 to gasoline range hydrocarbons over a promoted iron catalyst [33].

Despite being a well-established process, the mechanism of the RWGS reaction remains unclear with both the formate mechanism and redox mechanism proposed and investigated. The formate mechanism has been advanced by Shido and co-workers [34] and has subsequently received further support through spectroscopic studies [35]. This mechanism involves the reaction of CO with hydroxyl groups on the catalyst support to form a bidentate formate (COOH-) intermediate which binds to the surface via both oxygen atoms. Steady-state isotopic techniques, however, have suggested that formation and reactivity of such formate species are highly dependent on reaction conditions and that they are likely to be present only as spectator species [36, 37]. The alternative mechanism typically proposed for RWGS is a redox mechanism, whereby CO_2 adsorption results in oxidation of the catalyst surface, reducing CO_2 to CO. Hydrogen then re-reduces the catalyst surface to its original state by reacting to form H_2O. While a number of different RWGS mechanisms are possible, steady-state isotopic techniques have suggested that, over Pt/CeO_2, the most likely is a carbonate route as shown in Figure 21.5 [36].

The second step in the two-step FTS process from CO_2 is the FT reaction involving CO and H_2. FTS is well established to be a step-wise chain growth polymerisation reaction [38–40]. A number of different reaction mechanisms have, however, been proposed. These include the enol mechanism, where condensation reactions between surface enol species formed from CO hydrogenation result in polymerisation [17, 41]; the formate mechanism, where gas-phase CO inserts into a surface hydroxyl or alkoxy group forming COOH/COOR [42, 43]; and the CO insertion mechanism where surface-bound CO inserts into a metal-alkyl surface species, leading to chain growth [44, 45]. The most commonly presented mechanism is, however, the carbide mechanism, whereby carbon, from the reactant CO (or CO_2), interacts with the metal catalyst forming a metal carbide phase which provides the catalytically active sites [44, 46, 47]. None of these mechanisms, however, fully explain all experimental observations over

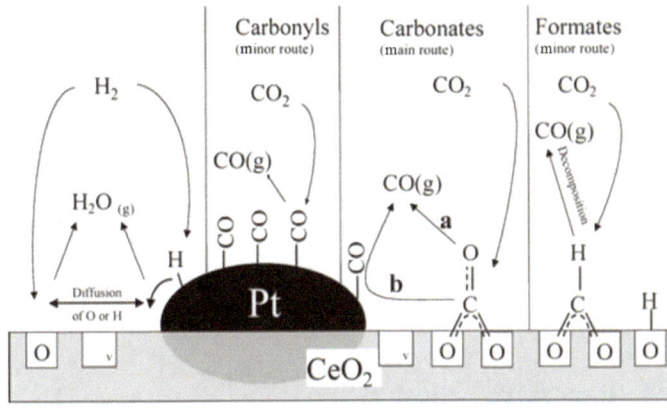

v = oxygen vacancy

Figure 21.5: Proposed RWGS mechanism over Pt/CeO$_2$. Reprinted with permission from [36] Copyright 2004, American Chemical Society.

all catalysts [39]. A discussion of the pros and cons of individual mechanisms over the most common catalysts for CO$_2$-FTS is presented in Section 21.3.

A challenge associated with CO$_2$-FTS is the additional water produced with respect to CO-FTS (Scheme 21.2). Water plays a direct role in catalyst deactivation. For instance, over Co/Al$_2$O$_3$ water promotes the formation of the inactive CoAl$_2$O$_4$ phase [48], while over iron catalysts water has been shown to impede reaction kinetics limiting the (per pass) conversion [49]. A detailed investigation into the influence of water on CO-FTS has been conducted by Satterfield and co-workers [50].

Lee and co-workers have proposed a detailed mechanism describing the direct conversion of CO$_2$ to higher hydrocarbons [18]. Specifically, CO$_2$ initially adsorbs onto the surface of the catalyst and is reduced at a Fe(II) site. The catalyst also adsorbs and dissociates H$_2$. The formed hydrogen radical can react with surface-bound carbon oxide yielding formic acid, and CO, which can subsequently convert to oxygenate products. Additionally, an Fe(III)CH$_2$ radical is formed which is the carbon–carbon propagation species, and polymerisation of this yields ≥C$_2$ hydrocarbons.

A further consideration is where a mixed feed of CO/CO$_2$/H$_2$ is employed. Traditional sources of syngas, such as coal gasification, produce both CO and CO$_2$ while future sources of syngas, in particular biomass gasification, also yield significant quantities of CO$_2$ (and often CH$_4$) alongside CO and H$_2$ [51, 52]. Some processes require the removal of CO$_2$ from the feed prior to syngas conversion; however, there are economic and resource utilisation benefits to converting both carbon oxides simultaneously. Indeed, CO$_2$ has been deliberately added to CO/H$_2$ mixtures when employing iron catalysts in order to limit the formation of CO$_2$ from syngas via undesirable water gas shift (WGS) reactions [53]. This is successful as WGS is an equilibrium-limited reaction; therefore, increasing the partial pressure of CO$_2$ limits the extent of the forward reaction. Due to its WGS activity, iron is the preferred catalyst for FTS where CO_2 is present as a minor component in the syngas feed, as is also the case where CO_2/H_2 feeds without CO are employed (Section 21.3). Cobalt has, however, also been investigated as a catalyst for mixed feeds [22, 54, 55]. These studies show that the product distribution is dependent on the CO/CO_2 ratio, with a higher partial pressure of CO_2 yielding shorter chained products and a greater percentage of alkanes rather than alkenes. As the partial pressure of CO increases, the product distribution increasingly resembles that of a typical FT processes with longer chained products and increasing quantities of alkenes, and more closely follows an ASF distribution (Section 21.1.1).

21.2.2 Synthesis of alternative products

While CO_2-FTS typically produces linear alkanes and alkenes, branched hydrocarbon products can be targeted through a judicious selection of catalyst [56, 57]. Of particular interest is the production of isobutene, which has applications as a gasoline additive [58]. Additionally, it is also possible to direct the synthesis towards oxygenate products such as alcohols. The production of methanol, which can be an intermediate in the production of gasoline-range hydrocarbons (Section 21.2.1), is well known; however, higher hydrocarbons can also be produced with the formation of ethanol, and higher hydrocarbons from a mixed $H_2/CO/CO_2/CH_4$ feed over a Fischer–Tropsch catalyst being successfully demonstrated [9]. Such products, in particular ethanol and butanol, can themselves be directly used as fuels [59]. The production of higher alcohols from CO_2 follows a similar two-step process as that for hydrocarbon synthesis described in Section 21.2.1 and is illustrated in Scheme 21.3:

1. $$CO_2 + H_2 \rightleftharpoons CO + H_2O$$

2. $$2CO + 4H_2 \rightarrow CH_3CH_2OH + H_2O \qquad \text{Scheme 21.3}$$

where $>C_2$ alcohols are formed via polymerisation as per hydrocarbon products in conventional FTS. Catalysts active for this reaction include composite FTS (Fe or Rh) and methanol synthesis (Cu) materials [60, 61]. Subramani and Gangwal have reviewed the production of ethanol from syngas [62], while alternative methods of producing long-chained alcohols from CO_2 are discussed in Section 21.4.3.

21.3 Catalysts

That the second step in the two-step mechanism (Scheme 21.2) for CO_2-FTS is the conventional hydrogenation and polymerisation of CO indicates that active sites of the same nature as those required in CO-FTS are also a pre-requisite for CO_2-FTS. There are, however, a number of key differences, which are discussed in the following sections.

The most commonly investigated active metals for CO-FTS are Ru, Co and Fe. Ruthenium is well established to be the most active catalyst for FTS but is not commercially relevant due to its prohibitive expense. Iron catalysts were originally the most common but are being supplanted by cobalt for most applications. Co catalysts yield more wax products than Fe; however, these can be converted through cracking or other reactions to shorter chained products.

21.3.1 Cobalt

Cobalt catalysts have become the most popular catalysts for conventional FTS from syngas due to the balance of good performance and low cost (relative, for example, to ruthenium). Their high performance derives in part from their low activity towards WGS relative to that of iron catalysts. They are generally operated in the LTFTS regime (Section 21.2.1) to limit excessive CH_4 formation at higher temperatures.

In contrast to conventional FTS, when CO_2 is used as the feed gas, cobalt catalysts favour methanation [22, 23]. A typical ASF distribution is not followed and this is indicative that a different mechanism operates when CO_2 is used as the reactant. Increasing the ratio of CO_2:H_2 does increase the yield of longer chained products but reduces the CO_2 conversion obtained [23]. The differences in product distribution between using CO and CO_2 as the reactant are indicated in Figure 21.6, which shows an ASF plot describing selectivity to hydrocarbons [63]. This clearly shows the preference towards the formation of low carbon number products, in particular CH_4, from CO_2.

While the carbide mechanism (Sections 21.2.1 and 21.3.2) is the most commonly proposed mechanism for FTS, recent theoretical studies have cast doubt on this, at least over Co catalysts [64, 65] Density functional theory (DFT) studies have suggested that over Co, FTS proceeds via an oxymethylidyne intermediate formed from hydrogenation of carbon monoxide. C–C bond formation then proceeds through a similar mechanism to homogeneously catalysed hydroformylation reactions. Complementary experimental studies have also suggested that the carbide route may not be the predominate mechanism over Co. For instance, the hydroformylation of alkenes with syngas under conditions comparable to FTS shows enhanced oxygenate production in the presence of solvents that hinder secondary hydrogenation reactions [66, 67]. This may imply that oxygenate species are the primary product of CO-FTS over cobalt with hydrocarbons formed via a subsequent hydrogenation step. Elsewhere, in situ X-ray diffraction (XRD) studies have provided further support that Co-carbides are inactive for FTS [68]. In order to confirm the reaction mechanism, it would necessary to conduct detailed and complex experimental investigations applying a range of advanced *operando* techniques.

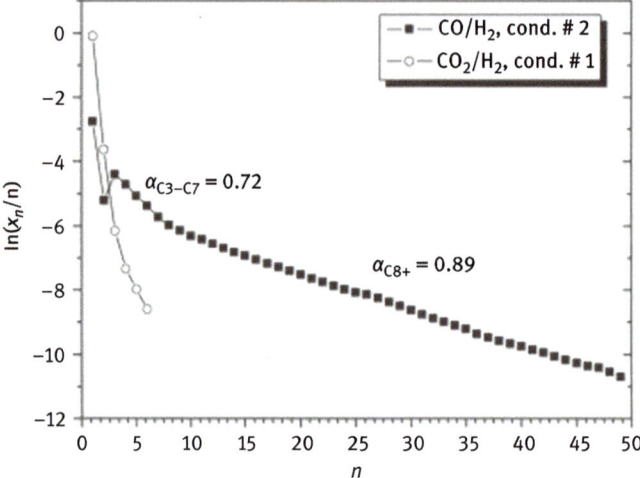

Figure 21.6: ASF plot of hydrocarbon selectivity over Co/Al_2O_3 using CO and CO_2 as the reactant. Reproduced with permission from [63].

21.3.2 Iron

Iron catalysts possess the ability to catalyse RWGS and WGS alongside FTS yielding olefinic products [33]. In contrast to cobalt catalysts, they can be employed for both HTFTS and LTFTS (Section 21.2.1). In conventional FTS, their WGS activity results in the production of a significant quantity of CO_2 as a co-product, making it a less desirable catalyst than cobalt. The exception to this is in the case of syngas derived from coal, where the WGS capacity of iron increases the ratio of H_2:CO from ~1 to nearer the 2:1 ratio required by FT stoichiometry [69].

The ability to catalyse RWGS works in the favour of iron as a catalyst for FTS from CO_2. FTS from CO_2 over iron catalysts follows the two-step mechanism described in Section 21.2.1 proceeding via RWGS followed by the conventional Fischer–Tropsch reaction from CO. Saeidi et al. have reviewed the efficacy of different iron catalysts under different operating conditions, identifying that the conversion of CO_2 typically falls within the range 19–68%, with C_2–C_{5+} alkene selectivities of approximately 80% [14].

The active phase of iron catalysts is the subject of some debate [39]; however, the carbide mechanism is perhaps the most widely proposed. The changing composition of iron catalysts during FTS from CO_2 has been described by Riedel (Figure 21.7) on the basis of Mössbauer spectroscopy studies [70]. Initially, the reactants adsorb on the catalyst surface and react, predominately, to form CO and H_2O alongside carbon deposition. Deposited carbon migrates into the subsurface region of the catalyst forming a carbide phase (proposed to be Fe_5C_2, however, the

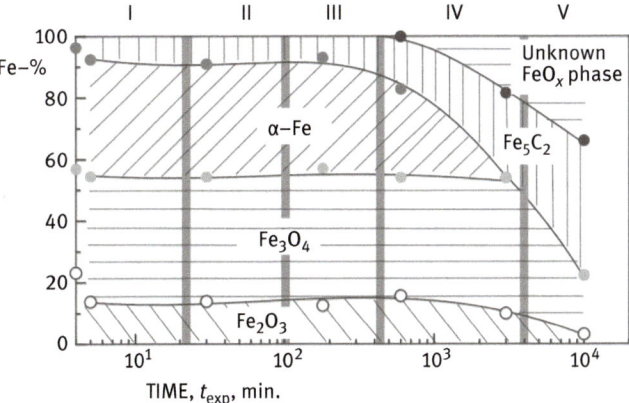

Figure 21.7: Evolution of iron phase composition as a function of time during Fischer–Tropsch synthesis determined by Mössbauer spectroscopy. Reproduced with permission from [70].

precise form of the carbide remains uncertain [71, 72]), which is correlated with FTS activity. While carbon deposition is necessary for activity, as it is in the range of diverse catalytic processes [73], it also plays a direct role in catalyst deactivation through the formation of carbonaceous overlayers, which prevent access to catalytically active sites.

Potassium is widely added to iron FTS catalysts as a promotor. The mechanism of promotion by potassium is debated; however, it most likely exerts an electronic effect influencing adsorption and reaction on the active metal, and provides new basic adsorption sites [6, 74]. The most effective concentration of potassium as a promotor is a key area of difference between CO_2-FTS and conventional FTS. In conventional FTS, large quantities of promotor act as a catalyst poison; however, this is not the case for CO_2-FTS. It is speculated that on alumina-supported catalysts an alanate phase ($KAlH_4$) is formed [75], and that this phase suppresses over-hydrogenation by acting as a sink for H_2.

21.4 Reactors

The reactor requirements for CO_2-FTS are broadly similar to those for conventional FTS; however, the differences in the chemistry of the process mean that innovative reaction engineering solutions are required in order to achieve comparable conversions. A key consideration in reactor design and selection for both CO- and CO_2-FTS is heat management. This is as a consequence of (i) the significant exotherm of the Fischer–Tropsch reaction (~150 kJ mol$^{-1}_{CO}$); and (ii) the sensitivity of the product distribution to temperature.

21.4.1 Gas-phase reactors

Fixed-bed reactors (FBRs) for FTS traditionally employ a multitubular approach, in the form of a shell and tube exchanger, whereby individual tubes are packed with catalyst and arranged within a larger tube containing a heat transfer fluid, often water. The first conventional FTS reactors commercialised, in Germany under the Nazi regime, were FBRs employing cobalt catalysts [76]. The initial tube-cooled design was rapidly superseded by a tube-in-tube approach, which provided more effective temperature control through provision of a larger heat exchange per catalyst volume. More recent gas-phase reactor designs have, however, predominately employed fluidised beds – either fixed or circulating. The choice of reactor type is intimately related to the nature of the catalyst; fluidised beds place greater demand on the structural integrity of the catalyst particles and hence support materials exhibiting high attrition resistance are required. Over time there has been increasing application of slurry-based reactors (Section 21.4.2); however, Shell's Pearl GTL facility – the largest in the world and operational since 2011 – employs multitubular FBRs [77]. A schematic of a multitubular FBR is shown in Figure 21.8 [78].

A key consideration when starting from CO_2 rather than CO that has a direct effect on reactor design is the presence and inhibiting effect of water (Section 21.2.1). In order to mitigate against these effects, more advanced reactor designs are required. Rohde and co-workers have proposed the use of a packed-bed reactor with an integrated silica membrane [79]. Using H_2 or H_2/CO_2 as a sweep gas they demonstrated an increase in CO_2 conversion of 50% as compared to an FBR without a membrane. Other membrane materials proposed in the literature include zeolites [49, 80]. Elsewhere, reactors incorporating a recycle loop have been developed for CO_2-FTS [18]. A CO_2 conversion of ~41% was achieved in a single FBR employing an iron catalyst; however, this was increased upon introduction of the recycle for H_2, CO_2, CO and light hydrocarbons. The final conversion achieved was dependent on the recycle ratio with higher ratios being more effective; for example, a recycle ratio of 6 resulted in increasing the conversion to 88%. Increasing conversion has significant benefits, not just in more efficient use of reactants but also in reducing the need for costly downstream separation steps. Where the choice is made to recycle liquid wax products alongside light gases, then the reactor must operate in trickle-flow with a high liquid content [81]. From these studies, it is clear that adequate consideration of the reaction engineering of the CO_2-FTS process is a key element in the drive towards commercialisation, and is as important as catalyst design.

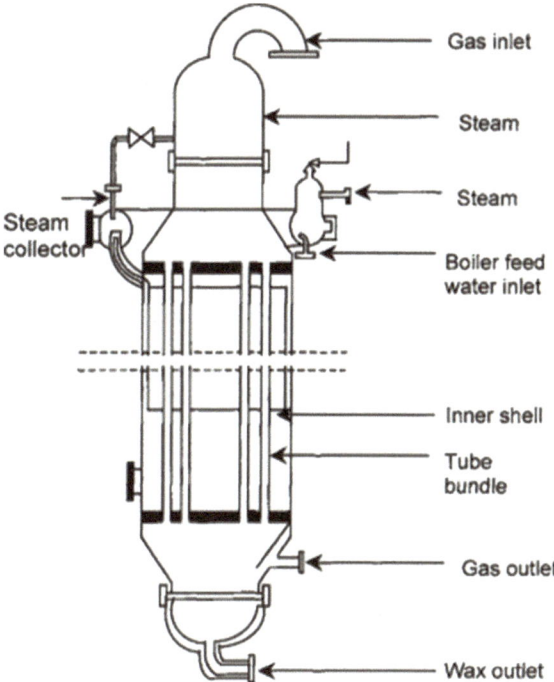

- Gas inlet
- Steam
- Steam
- Steam collector
- Boiler feed water inlet
- Inner shell
- Tube bundle
- Gas outlet
- Wax outlet

Figure 21.8: Schematic of a fixed-bed multitubular reactor designed for the Arge FT process. Reproduced with permission from [78].

21.4.2 Liquid-phase Fischer–Tropsch synthesis

In addition to two-phase (gas/solid) fixed-bed and fluidised bed reactors, three-phase (gas/solid/liquid) slurry reactors are also employed for CO-FTS [82, 83]. An advantage of slurry reactors over FBRs is that the risk of hot spot formation is removed, thereby allowing them to be operated at higher temperatures without unexpected coke deposition or catalyst degradation occurring. Additionally, mass transfer is improved as large catalyst pellets are no longer required. Such reactors have been operated commercially for CO-FTS by Sasol, Syntroleum and Rentech among others [83]. The choice of solvent can play a critical role in the performance of slurry reactors with water, alcohols, hydrocarbons and ionic liquids all having been the subject of investigation. Fan and co-workers achieved promising results using polyethylene glycol with a nanoparticulate iron catalyst exhibiting high activity at mild conditions of 150 °C and 3 MPa total pressure with a 2:1 H_2:CO ratio [84]. A significant technical challenge in their successful operation that does not exist for FBRs is the separation of wax products from the slurry phase. An additional limitation is the slow (as compared to gas-phase reactors) mass transfer of reactants to active sites on the surface of the catalyst.

Supercritical fluids have been proposed as reaction media in order to combine the benefits of gas-phase and slurry-phase reactors without the attendant disadvantages of either [85, 86]. Supercritical fluids are completely miscible with gases and have low surface tension and low viscosity giving them enhanced mass transfer properties. Disadvantages of supercritical processing include higher capital costs due to the need to operate at high pressure and to employ corrosion resistant materials. The most common solvents used in supercritical FTS are alkanes, in particular hexane and pentane [85, 87]. While still at a research scale, supercritical media present an interesting future opportunity in both conventional and CO_2-FTS, with the opportunity for reduced methane and carbon dioxide by-product selectivities and increased hydrocarbon chain length with respect to gas-phase processing.

The hydrogenation of CO_2 to form higher hydrocarbons and oxygenates can also be conducted in the liquid phase, using water as a solvent under subcritical conditions, also known as high-temperature water (HTW) [88, 89] While not necessarily following the same mechanism as FTS, the production of similar products from the same reactants means that it is worthy of further consideration. HTW describes water above 180 °C and below 374 °C (the supercritical point of water) using sufficient pressure to avoid the phase change to steam and maintain it in the liquid phase [90]. Many of the early studies of CO_2 hydrogenation in this field focused on investigations into the origin of life and the hypothesis that the complex molecules required for the evolution of life were formed from the reaction of CO_2 and water around high-temperature hydrothermal vents with metal-containing minerals acting as catalysts or reductants [91–93]. A key feature of these reactions, typically catalysed by zero-valent transition metals, is that gas-phase hydrogen is not required with water acting directly as the hydrogen source. Other studies have employed biomass-derived species as hydrogen-donating molecules [89]. While the majority of studies in this area focussed on the synthesis of C_1 and C_2 oxygenates, evidence for the production of longer chained species such as butenal, isobutene, phenol and diphenyl ether has been observed [91, 92, 94, 95]. By informed choice of catalyst and reaction conditions, it is possible to synthesise oxygenate species up to a chain length of C_9 [96]. For instance, Figure 21.9 shows the concentration of key products obtained over iron-based materials obtained from a mixture of CO_2 (25 bar) and water at 300 °C [96]. The formation of linear oxygenates such as 2-octanone is apparent. The mechanism of this reaction is unclear; however, the iron present likely acts as a catalyst, providing a lower activation energy surface-mediated reaction pathway, and as a reductant. The possibility of an alternative route to polymerise CO_2 to oxygenates or hydrocarbons presents an intriguing possibility albeit one some way from commercial development.

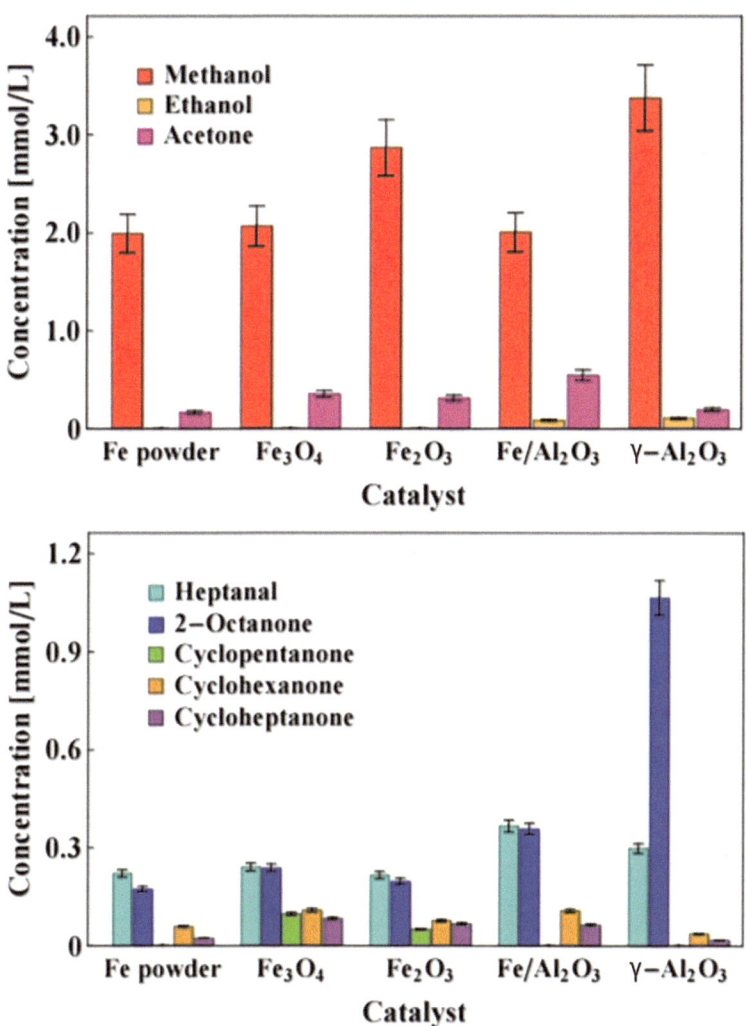

Figure 21.9: Composition of product mixture from the hydrothermal conversion of CO_2 in the presence of solid catalysts/reductants. Reaction conditions: 0.56 g catalyst, 7 mL H_2O, P_{CO_2}initial= 25 barg, 300 °C, reaction time 4 h. Adapted from [96].

21.5 Future perspectives

Sustainability must be at the centre of the hydrocarbon industry of the twenty-first century. For that reason, both the use of CO_2 and integration of FTS within P2X schemes are critical to future development. The use of CO_2, either from carbon capture schemes or as a component of syngas from renewable sources such as biomass gasification, can contribute to decarbonisation of the transportation fuel sector and reduce societal reliance on fossil sources. In parallel, the integration of FTS with P2X schemes makes maximal use of renewable energy, storing it when supply exceeds demand to produce a necessary energy vector to replace that currently derived from crude oil.

A number of challenges remain before CO_2-FTS is fully realised. Improved catalysts are necessary, overcoming problems with, for example, deactivation caused by produced water or through carbon deposition, and increasing selectivity for fuel range hydrocarbons. Additionally, novel reaction engineering approaches are needed to ensure that the process can be delivered at appropriate scale while remaining economically viable. Membrane reactors may be able to selectively remove water [49, 79, 80] while recycle loops allow for improved overall conversions [18, 81] The use of sub- or supercritical reaction media represents an intriguing avenue for research although this is further from commercial implementation than the gas-phase or slurry-phase reactors.

The integration of fundamental research and pilot-scale testing will be crucial in developing the new catalysts, processes and reactor environments required to make CO_2-FTS a commercial reality. Innovations such as the Diamond Pilot Plant, at the University of Sheffield, incorporating an FTS reactor as part of wider P2X concept, provide an essential resource in this regard, linking academic activities with industrial research and development [97]. Overall, FTS employing carbon dioxide as a feedstock has a bright future and the potential to play an important role in delivering on decarbonisation targets.

References

[1]   Earth's CO2 Home Page, https://www.co2.earth/, (accessed 26 October 2018).
[2]   BMWi – Federal Ministry for Economic Affairs and Energy – Energy Concept for an Environmentally Sound, Reliable and Affordable Energy Supply - 28 September 2010, http://www.bmwi.de/English/Redaktion/Pdf/energy-concept%2Cproperty%3Dpdf%2Cbereich%3Dbmwi%2Csprache%3Den%2Cwb%3Dtrue.pdf, (accessed 26 October 2018).
[3]   BMWi – Federal Ministry for Economic Affairs and Energy – Fourth Monitoring Report; The Energy of the Future; 2014 - Summary, https://www.bmwi.de/Redaktion/EN/Publikationen/vierter-monitoring-bericht-energie-der-zukunft-kurzfassung.pdf, (accessed26 October 2018).
[4]   Z. Jiang, T. Xiao, V. L. Kuznetsov and P. P. Edwards, Philos. Trans. A. Math. Phys. Eng. Sci., 2010, 368, 3343–64.
[5]   R. B. Anderson, R. A. Friedel and H. H. Storch, J. Chem. Phys., 1951, 19, 313–319.
[6]   R. W. Dorner, D. R. Hardy, F. W. Williams and H. D. Willauer, Energy Environ. Sci., 2010, 3, 884–890.
[7]   M. Becherif, H. S. Ramadan, K. Cabaret, F. Picard, N. Simoncini and O. Bethoux, Energy Procedia, 2015, 74, 371–380.
[8]   O. Ehret and K. Bonhoff, Int. J. Hydrogen Energ., 2015, 40, 5526–5533.
[9]   D. Mignard and C. Pritchard, Chem. Eng. Res. Des., 2006, 84, 828–836.

[10] S. R. Foit, I. C. Vinke, L. G. J. de Haart and R.-A. Eichel, Angew. Chemie Int. Ed., 2017, 56, 5402–5411.
[11] W. Wang, S. Wang, X. Ma and J. Gong, Chem. Soc. Rev., 2011, 40, 3703–3727.
[12] B. Mutz, H. W. P. Carvalho, S. Mangold, W. Kleist and J.-D. Grunwaldt, J. Catal., 2015, 327, 48–53.
[13] Germany pushes forward in energy conversion – DTU Energy, https://www.energy.dtu.dk/english/news/2016/04/germany-pushes-forward-in-energy-conversion?id=c29ce25f-36fe-4480-9ebd-7ae0df2a7246, (accessed 26 October 2018).
[14] S. Saeidi, N. A. S. Amin and M. R. Rahimpour, J. CO2 Util., 2014, 5, 66–81.
15   P. Kangvansura, L. M. Chew, W. Saengsui, P. Santawaja, Y. Poo-arporn, M. Muhler, H. Schulz and A. Worayingyong, Catal. Today, 2016, 275, 59–65.
[16] L. M. Chew, P. Kangvansura, H. Ruland, H. J. Schulte, C. Somsen, W. Xia, G. Eggeler, A. Worayingyong and M. Muhler, Appl. Catal. A Gen., 2014, 482, 163–170.
[17] T. Herranz, S. Rojas, F. J. Pérez-Alonso, M. Ojeda, P. Terreros and J. L. G. Fierro, Appl. Catal. A Gen., 2006, 311, 66–75.
[18] S. Lee, J. Kim, W. Lee, K. Lee, M. C.-S. in S. Science and U. 2004, Stud. Surf. Sci. Catal., 2004, 153, 73–78.
[19] J. Yang Lim, J. McGregor, A. J. Sederman and J. S. Dennis, Chem. Eng. Sci., 2016, 141, 28–45.
[20] S. Fujita, M. Nakamura, T. Doi and N. Takezawa, Appl. Catal. A Gen., 1993, 104, 87–100.
[21] J. L. Falconer and A. E. Zağli, J. Catal., 1980, 62, 280–285.
[22] Y. Zhang, G. Jacobs, D. E. Sparks, M. E. Dry and B. H. Davis, Catal. Today, 2002, 71, 411–418.
[23] R. W. Dorner, D. R. Hardy, F. W. Williams, B. H. Davis and H. D. Willauer, Energy & Fuels, 2009, 23, 4190–4195.
[24] J. Y. Lim, J. McGregor, A. J. Sederman and J. S. Dennis, Chem. Eng. Sci., 2016, 152, 754–766.
[25] T. Riedel, M. Claeys, H. Schulz, G. Schaub, S.-S. Nam, K.-W. Jun, M.-J. Choi, G. Kishan and K.-W. Lee, Appl. Catal. A Gen., 1999, 186, 201–213.
[26] M. Fujiwara, R. Kieffer, H. Ando and Y. Souma, Appl. Catal. A Gen., 1995, 121, 113–124.
[27] P. S. Sai Prasad, J. W. Bae, K.-W. Jun and K.-W. Lee, Catal. Surv. from Asia, 2008, 12, 170–183.
[28] F. J. Keil, Microporous Mesoporous Mater., 1999, 29, 49–66.
[29] A. Zachariou, A. Hawkins, D. Lennon, S. F. Parker, Suwardiyanto, S. K. Matam, C. R. A. Catlow, P. Collier, A. Hameed, J. McGregor and R. F. Howe, Appl. Catal. A Gen., 2019, 569, 1–7.
[30] S. Suwardiyanto, R. F. Howe, E. K. Gibson, C. R. A. Catlow, A. Hameed, J. McGregor, P. Collier, S. F. Parker and D. Lennon, Faraday Discuss., 2017, 197, 447–471.
[31] U. Olsbye, S. Svelle, K. P. Lillerud, Z. H. Wei, Y. Y. Chen, J. F. Li, J. G. Wang and W. B. Fan, Chem. Soc. Rev., 2015, 44, 7155–7176.
[32] K. Fujimoto and T. Shikada, Appl. Catal., 1987, 31, 13–23.
[33] J. Wei, Q. Ge, R. Yao, Z. Wen, C. Fang, L. Guo, H. Xu and J. Sun, Nat. Commun., 2017, 8, 15174.
[34] T. Shido and Y. Iwasawa, J. Catal., 1993, 141, 71–81.
[35] G. Jacobs, U. M. Graham, E. Chenu, P. M. Patterson, A. Dozier and B. H. Davis, J. Catal., 2005, 229, 499–512.
[36] Alexandre Goguet, Frederic C. Meunier, Daniele Tibiletti, A. John P. Breen and R. Burch, J.Phys. Chem. B 2004, 108, 20240–20246.
[37] D. Tibiletti, A. Goguet, F. C. Meunier, J. P. Breen and R. Burch, Chem. Commun., 2004, 1636–1637.
[38] O. O. James, B. Chowdhury, M. A. Mesubi and S. Maity, RSC Adv., 2012, 2, 7347.
[39] S. Saeidi, S. Najari, F. Fazlollahi, M. K. Nikoo, F. Sefidkon, J. J. Klemeš and L. L. Baxter, Renew. Sustain. Energy Rev., 2017, 80, 1292–1311.
[40] M. E. Dry, Appl. Catal. A Gen., 1996, 138, 319–344.
[41] B. H. Davis, Fuel Process. Technol., 2001, 71, 157–166.

[42] J. Schweicher, A. Bundhoo and N. Kruse, J. Am. Chem. Soc., 2012, 134, 16135–16138.

[43] K. Park, G. H. Gunasekar , N. Prakash, K.-D. Jung and S. Yoon, ChemSusChem, 2015, 8, 3410–3413.

[44] H. Schulz, Catal. Today, 2013, 214, 140–151.

[45] R. A. van Santen and A. J. Markvoort, ChemCatChem, 2013, 5, 3384–3397.

[46] F. Fischer and H. Tropsch, Brennstoff-Chemie, 1926, 7, 97–104.

[47] V. Ponec and W. A. van Barneveld, Ind. Eng. Chem. Prod. Res. Dev., 1979, 18, 268–271.

[48] W. Li, X. Nie, X. Jiang, A. Zhang, F. Ding, M. Liu, Z. Liu, X. Guo and C. Song, Appl. Catal. B Environ., 2018, 220, 397–408.

[49] T. Riedel, G. Schaub, Ki-Won Jun and K.-W. Lee, Ind. Eng. Chem. Res., 2001, 40, 1355–1363.

[50] C. N. Satterfield, R. T. Hanlon, S. E. Tung, Z. M. Zou and G. C. Papaefthymiou, Ind. Eng. Chem. Prod. Res. Dev., 1986, 25, 407–414.

[51] S. T. Chaudhari, A. K. Dalai and N. N. Bakhshi, Energy & Fuels, 2003, 17, 1062–1067.

[52] A. Demirbas, Energy Sources, 2004, 26, 715–730.

[53] S. L. Soled, E. Iglesia, S. Miseo, B. A. DeRites and R. A. Fiato, Top. Catal., 1995, 2, 193–205.

[54] T. Riedel, M. Claeys, H. Schulz, G. Schaub, S.-S. Nam, K.-W. Jun, M.-J. Choi, G. Kishan and K.-W. Lee, Appl. Catal. A Gen., 1999, 186, 201–213.

[55] Y. Yao, X. Liu, D. Hildebrandt and D. Glasser, Appl. Catal. A Gen., 2012, 433–434, 58–68.

[56] B. Rongxian, T. Yisheng and H. Yizhuo, Fuel Process. Technol., 2004, 86, 293–301.

[57] X. Ni, Y. Tan, Y. Han and N. Tsubaki, Catal. Commun., 2007, 8, 1711–1714.

[58] C. Erkey, J. Wang, W. Postula, Z. Feng, C. V. Phillip, A. Akgerman and R. G. Anthony, Ind. Eng. Chem. Res., 1995, 34, 1021–1026.

[59] J. Swana, Y. Yang, M. Behnam and R. Thompson, Bioresour. Technol., 2011, 102, 2112–2117.

[60] T. Inui, T. Yamamoto, M. Inoue, H. Hara, T. Takeguchi and J.-B. Kim, Appl. Catal. A Gen., 1999, 186, 395–406.

[61] T. Inui and T. Yamamoto, Catal. Today, 1998, 45, 209–214.

[62] V. Subramani and S. K. Gangwal, Energy & Fuels, 2008, 22, 814–839.

[63] C. G. Visconti, L. Lietti, E. Tronconi, P. Forzatti, R. Zennaro and E. Finocchio, Appl. Catal. A Gen., 2009, 355, 61–68.

[64] Oliver R. Inderwildi, A. Stephen J. Jenkins and D. A. King, J. Phys. Chem. C, 2008, 112, 1305–1307.

[65] O. R. Inderwildi, D. A. King and S. J. Jenkins, Phys. Chem. Chem. Phys., 2009, 11, 11110–11112.

[66] Y. Zhang, K. Nagasaka and X. Qiu, Catal. Today, 2005, 104, 48–54.

[67] X. Q. Qiu, N. Tsubaki, K. Fujimoto and Q. M. Zhu, Fuel Process. Technol., 2004, 85, 1193–1200.

[68] H. Karaca, O. V. Safonova, S. Chambrey, P. Fongarland, P. Roussel, A. Griboval-Constant, M. Lacroix and A. Y. Khodakov, J. Catal., 2011, 277, 14–26.

[69] Y. Cao, Z. Gao, J. Jin, H. Zhou, M. Cohron, H. Zhao, H. Liu and W. Pan, Energy & Fuels, 2008, 22, 1720–1730.

[70] T. Riedel, H. Schulz, G. Schaub, K.-W. Jun, J.-S. Hwang and K.-W. Lee, Top. Catal., 2003, 26, 41–54.

[71] Q. Zhang, J. Kang and Y. Wang, ChemCatChem, 2010, 2, 1030–1058.

[72] E. de Smit and B. M. Weckhuysen, Chem. Soc. Rev., 2008, 37, 2758–2781.

[73] C. H. Collett and J. McGregor, Catal. Sci. Technol., 2016, 6, 363–378.

[74] W. Li, H. Wang, X. Jiang, J. Zhu, Z. Liu, X. Guo and C. Song, RSC Adv., 2018, 8, 7651–7669.

[75] R. W. Dorner, D. R. Hardy, F. W. Williams and H. D. Willauer, Appl. Catal. A Gen., 2010, 373, 112–121.

[76] F. Asinger, Paraffins: Chemistry and Technology, Elsevier.

[77] Pearl GTL, Shell Qatar, https://www.shell.com.qa/en_qa/projects-and-sites/pearl-gtl.html, (accessed 25 October 2018).

[78] A. P. Steynberg, M. E. Dry, B. H. Davis and B. B. Breman, Stud. Surf. Sci. Catal., 2004, 152, 64–195.

[79] Martin P. Rohde, and Dominik Unruh and G. Schaub, Ind. Eng. Chem. Res., 2005, 44, 9653–9658.

[80] R. L. Espinoza, E. du Toit, J. Santamaria, M. Menendez, J. Coronas and S. Irusta, Stud. Surf. Sci. Catal., 2000, 130, 389–394.

[81] C. G. Visconti, Ind. Eng. Chem. Res., 2014, 53, 1727–1734.

[82] H. Itoh, S. Nagano and E. Kikuchi, Appl. Catal., 1990, 67, 215–221.

[83] B. H. Davis, Catal. Today, 2002, 71, 249–300.

[84] X.-B. Fan, Z.-Y. Tao, C.-X. Xiao, F. Liu and Y. Kou, Green Chem., 2010, 12, 795.

[85] R. M. Malek Abbaslou, J. S. Soltan Mohammadzadeh and A. K. Dalai, Fuel Process. Technol., 2009, 90, 849–856.

[86] X. Huang and C. B. Roberts, Fuel Process. Technol., 2003, 83, 81–99.

[87] K. Yokota, Y. Hanakata and K. Fujimoto, Fuel, 1991, 70, 989–994.

[88] D. Roman-Gonzalez, A. Moro, F. Burgoa, E. Pérez, A. Nieto-Márquez, Á. Martín and M. D. Bermejo, J. Supercrit. Fluids, 2018, 140, 320–328.

[89] M. Andérez-Fernández, E. Pérez, A. Martín and M. D. Bermejo, J. Supercrit. Fluids, 2018, 133, 658–664.

[90] N. Akiya and P. E. Savage, Chem. Rev., 2002, 102, 2725–2750.

[91] Ge Tian, Hongming Yuan, Ying Mu, A. Chao He and S. Feng, Org. Lett., 2007, 10, 2019–2021.

[92] G. Tian, C. He, Y. Chen, H.-M. Yuan, Z.-W. Liu, Z. Shi and S.-H. Feng, ChemSusChem, 2010, 3, 323–324.

[93] C. He, G. Tian, Z. Liu and S. Feng, Org. Lett., 2010, 12, 649–651.

[94] Q. Chen and Y. Qian, Chem. Commun., 2001, 0, 1402–1403.

[95] Q. W. Chen and D. W. Bahnemann, J. Am. Chem. Soc., 2000, 122, 970–971.

[96] L. Quintana Gomez, PhD thesis, University of Sheffield, 2017.

[97] Diamond Pilot Plant – The Diamond – The University of Sheffield, https://www.sheffield.ac.uk/diamond/pilotplant, (accessed 31 October 2018).

Part V: **Electrochemical reactions of CO₂**

Dongwei Du and Shanwen Tao

22 Electrochemical conversion of CO_2 into formate or formic acid

22.1 Introduction

22.1.1 Background

It is generally believed that the rising amount of CO_2 in the atmosphere is the key factor for the climate change because of its greenhouse effects [1]. The concentration of CO_2 in the atmosphere has increased from 278 ppm before the Industrial Revolution to over 400 ppm at present [2]. The massive consumption of fossil resources in global energy production and the chemical industry are the major contributors to CO_2 accumulation. It is predicted that the effects of climate change will last for up to 1,000 years even if emission of greenhouse gases halted [3]. Therefore, the reduction of CO_2 emission and conversion of CO_2 to useful materials have become a significant issue to the modern society.

In general, there are four methods to reduce the CO_2 in the atmosphere [4]: (1) improving the efficiency of current energy processes; (2) using non-carbon or low carbon sources for energy generation; (3) CO_2 capture and sequestration (CCS); and (4) CO_2 utilisation. The last two approaches seem to be the most direct and useful technologies to address the release of CO_2 in the atmosphere. However, since CCS involves the storage of the captured and sequestered CO_2 in geological subsurfaces or oceans, then there is still a potential that it will leak back to the atmosphere. Therefore, CO_2 utilisation seems a good approach to address the present challenge. Since 80–85% of the world energy consumption is provided by carbon-based fossil fuels [5], converting CO_2 into useful fuels or chemicals can not only lower the CO_2 emissions but also reduce the demand on fossil fuels.

Generally, CO_2 conversion can be grouped into four categories, namely, chemical methods, photocatalytic reduction, biotransformation and electrocatalytic reduction. The global annual CO_2 production was 41 Gt in 2017 with only a very small percentage of this recycled, which means that there is a huge potential for further development in this field [6].

22.1.2 Electrochemical reduction of CO_2

Energy generated from low-carbon sources such as renewable energy (e.g., solar, wind, geothermal, and wave) and nuclear energy is generally in the form of electricity; however, the places where there are sufficient amount of these resources are usually far away from the areas with a high-energy demand [7]. Therefore, storing

https://doi.org/10.1515/9783110665147-022

the redundant electric energy in the form of chemical energy would help solve this energy imbalance issue. In this respect, CO_2 would be an ideal feedstock to store the energy by converting it into useful fuels or chemicals. It is possible to transport the synthesised fuels to the users' sites. As a result, electrocatalysis of CO_2 has aroused great attention among various CO_2 conversion methods.

According to the number of electrons transferred per molecular of CO_2 during the reaction, the electrochemical reduction of CO_2 can be divided into one-, two-, four-, six- and eight-electron pathways in aqueous and nonaqueous electrolytes. The major products and their thermodynamic electrochemical half-reactions are shown in Table 22.1 [8]. It can be concluded that the main products are as follows: one-electron, oxalic acid ($H_2C_2O_4$) or oxalate ($C_2O_4^{2-}$); two-electron, carbon monoxide (CO) and formic acid (HCOOH) or formate ($HCOO^-$); four-electron, formaldehyde; six-electron, methanol (CH_3OH), ethylene (CH_2CH_2) and ethanol (CH_3CH_2OH); eight-electron, methane (CH_4). These reactions proceed very slowly even when catalysts are applied. In addition, the products of electroreduction are commonly a mixture rather than a single product. In terms of the type of product and the amount of each component, they are strongly dependent on large number of factors, including the kind of electro-catalysts, electrode support, electrolyte, pressure,

Table 22.1: Equilibrium potentials for various CO_2 electroreduction reactions (vs. SHE) in aqueous solution, at 298 K and 1.0 atm [8].

Thermodynamic electrochemical half-reactions	Electrode potentials (V vs. SHE)
$CO_2(g) + 4H^+ \rightarrow \circledR C(s) + 2H_2O(l)$	0.210
$CO_2(g) + 2H_2O(l) + 4e^- \rightarrow C(s) + 4OH^-$	−0.627
$CO_2(g) + 2H^+ + 2e^- \rightarrow HCOOH(l)$	−0.250
$CO_2(g) + 2H_2O(l) + 2e^- \rightarrow HCOOH(aq) + OH^-$	−1.078
$CO_2(g) + 2H^+ + 2e^- \rightarrow CO + H_2O(l)$	−0.106
$CO_2(g) + H_2O(l) + 2e^- \rightarrow CO + 2OH^-$	−0.934
$CO_2(g) + 4H^+ + 4e^- \rightarrow CH_2O(l) + H_2O(l)$	−0.070
$CO_2(g) + 3H_2O(l) + 4e^- \rightarrow CH_2O(l) + 4OH^-$	−0.898
$CO_2(g) + 6H^+ + 6e^- \rightarrow CH_3OH(l) + H_2O(l)$	0.016
$CO_2(g) + 5H_2O(l) + 6e^- \rightarrow CH_3OH(l) + 6OH^-$	−0.812
$CO_2(g) + 8H^+ + 8e^- \rightarrow CH_4(g) + 2H_2O(l)$	0.169
$CO_2(g) + 6H_2O(l) + 8e^- \rightarrow CH_4(g) + 8OH^-$	−0.659
$2CO_2(g) + 2H^+ + 2e^- \rightarrow H_2C_2O_4(aq)$	−0.500
$2CO_2(g) + 2e^- \rightarrow C_2O_4^{2-}(aq)$	−0.590
$2CO_2(g) + 12H^+ + 12e^- \rightarrow CH_2CH_2(g) + 4H_2O(l)$	0.064
$2CO_2(g) + 8H_2O(l) + 12e^- \rightarrow CH_2CH_2(g) + 8OH^-$	−0.764
$2CO_2(g) + 12H^+ + 12e^- \rightarrow CH_3CH_2OH(l) + 3H_2O(l)$	0.084
$2CO_2(g) + 9H_2O(l) + 12e^- \rightarrow CH_3CH_2OH(l) + 12OH^-$	−0.744

temperature, cell configuration and applied potential [9]. Overall, the reaction system is very complicated, and the end products are affected by many variables.

22.1.3 Current status of formic acid industry

Among the many products achieved from CO_2 reduction, formic acid is one of the most favoured for many reasons. Firstly, the economic possibility of large-scale electrochemical reduction of CO_2 to formic acid is considered to be feasible [4].

Secondly, formic acid is the simplest carboxylic acid existing naturally, and its demand is dramatically expanding year by year because of its various applications. The global formic acid market is estimated to reach 620 million dollars by 2019 [10]. A significant percentage (35%) of formic acid consumption is used in agriculture as a preservative because of its natural antibacterial properties [11]. The addition of formic acid to the livestock feed not only greatly reduces the growth of bacteria but also promotes the fermentation processes and lowers the fermentation temperature. In addition, more nutrients remain in the silage. In industry, formic acid is commonly used in the production of leather, manufacturing of rubber and dyeing of textiles since it evaporates without leaving any residue unlike mineral acids.

Thirdly, formic acid has been proposed as a fuel for fuel cells [12] and is also a promising hydrogen carrier with a hydrogen content of 4.4 wt% [13]. However, the energy density of formic acid is 2.1 kWh/L, which is less than half the value of methanol, thereby limiting its application for portable and transport applications.

Additionally, formate treatment has been proven to be a more effective and environmentally friendly treatment for slippery roads than traditional salts in some countries such as Switzerland and Austria.

So far, there are four commercial routes to produce formic acid: 1. Hydrolysis of methyl formate; 2. oxidation of hydrocarbons; 3. hydrolysis of formamide; 4. acidolysis of alkali formate. The traditional routes used to synthesise formic acid neither straightforward nor environment friendly [14]. Electrochemical conversion from CO_2 to formic acid is therefore a promising alternative synthesis method and more importantly, this sustainable process will reduce CO_2 emissions if the electricity is generated from renewable resources.

22.2 Electrochemical synthesis of formate and formic acid

As described previously, the overall reaction of formic acid production is a combination of an oxidation and reduction reaction at anode and cathode respectively. Figure 22.1 shows the basic configuration for electroreduction of CO_2 into formate/formic acid.

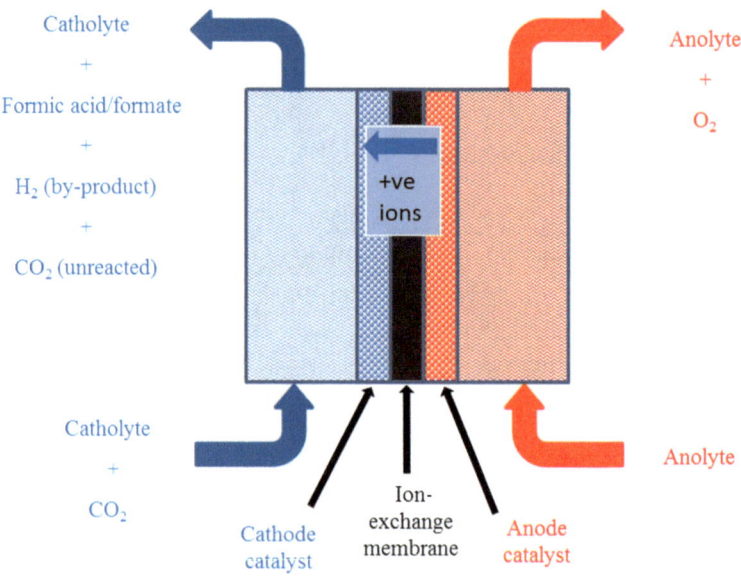

Figure 22.1: Schematic diagram of electro-conversion of CO_2 into formate/formic acid.

During the electrolysis, side reactions always compete with formate/formic acid production, leading to reduced Faradaic and energy efficiencies. The most common side reaction is water electrolysis, during which hydrogen, instead of formic acid, is formed at the cathode. CO_2 electroreduction to formate/formic acid can take place at different pH values. As the dissociation constant of formic acid is 1.6×10^{-4} at 298 K ($pk_a = 3.8$), formic acid is produced when the pH is below 3 while formate is formed at a higher pH [15]. Most of the experiments that have been carried out so far have used an alkaline electrolyte; therefore, the electrochemical reactions are presented as follows:

Anode:

$$2OH^- \rightarrow H_2O + \frac{1}{2}O_2 + 2e^- \tag{22.1}$$

(or $H_2O \rightarrow \frac{1}{2}O_2 + 2H^+ + 2e^-$ in acidic conditions)

Cathode:

$$CO_2(aq) + H_2O + 2e^- \rightarrow HCOO^-(aq) + OH^- \tag{22.2}$$

(or $CO_2(aq) + 2H^+ + 2e^- \rightarrow HCOOH(aq)$ in acidic conditions)

$$2H_2O + 2e^- \rightarrow 2OH^- + H_2(g) \tag{22.3}$$

(or $2H^+ + 2e^- \rightarrow H_2(g)$ in acidic conditions)

Equation 22.3 is more thermodynamically favourable than eq. 22.2. In order to achieve high selectivity for formic acid/formate, the development of proper catalysts is required to kinetically suppress the hydrogen evolution. Normally, the major catalysts for CO$_2$ electroreduction to formic acid/formate are related to the elements In, Pb, Sn, Hg and Zn.

22.2.1 Electrocatalysts

Many materials have been reported as electro-catalysts for electrochemical reduction of CO$_2$ into formates or formic acid. The properties of these catalysts are described in details below.

22.2.1.1 Lead (Pb)

In 1998, Kaneco et al. investigated a lead wire electrode in the electroreduction of CO$_2$ to formic acid at ambient temperature and pressure [16]. A KOH/methanol-based electrolyte was applied in their experiments because of the high CO$_2$ solubility in methanol. The results showed that the highest Faradaic efficiency of formic acid was 66% at –2.0 V versus Ag/AgCl (sat. KCl) with a current density of 4 mA cm^{-2}. The remaining products obtained were CO and H$_2$. Köleli and Balun prepared Pb-granule electrodes in a fixed-bed reactor to convert CO$_2$ to formic acid in 0.2 M aqueous K$_2$CO$_3$ solution [17]. The maximum Faradaic efficiency of 94% was obtained at –1.8 V (vs. SCE), 50 bar and 80 °C. They found that the Faradaic efficiency and current density significantly depended on the applied potential, pressure and temperature. An increase in temperature should lead to a decrease in Faradaic efficiency for formate because of the decreased CO$_2$ solubility at high temperature. Kwon and Lee synthesised a nanolayered Pb electrode on Pt quartz crystal by employing a stepwise potential deposition method [18]. Compared with the commercial Pb plate, the nanolayered Pb electrode showed a higher Faradaic efficiency towards formate production, which was mainly because of its good Pb crystallinity.

Lee and Kanan's study demonstrated that Pb film obtained through the reduction of PbO$_2$ exhibited up to 700-fold lower H$^+$ reduction activity than a Pb foil while not compromising its activity for CO$_2$ reduction to formate [19]. Moreover, the oxide-derived Pb could perform with a high selectivity for CO$_2$ conversion into formate, even at very low CO$_2$ concentrations in a N$_2$-saturated NaHCO$_3$ solution. It was supposed that the coverage of a thin layer, which can block the H$^+$ reduction and catalyse CO$_2$ reduction is much higher on the oxide-derived Pb electrode than that on Pb foil. The mechanisms of high selectivity for formate over CO and H$_2$ on Pb were investigated by Jung's group [20]. They suggested a proton-coupled electron transfer mechanism via a formate intermediate (*OCHO), which was consistent with the observations, specifically, the strong O affinity, weak C-species binding and *H of Pb catalyst jointly contributed to the high selectivity for formate production and suppression of H_2 production.

22.2.1.2 Tin (Sn)

Tin and its oxides have been widely reported as excellent cathode materials for the electrochemical reduction of CO_2 to formate/formic acid.

Lv et al. employed tin foil as the working electrode in a $KHCO_3$ solution with an undivided cell [21]. The Faradaic efficiency was up to 91% at −1.8 V versus Ag/AgCl while it gradually decreased with increasing electrolysis time because of oxidation of formate at the anode. Zhang et al. developed a practical method to solve this problem [22]. Specially, the anode was coated with a Nafion film on the surface, which blocks the diffusion of formate onto it (Figure 22.2).

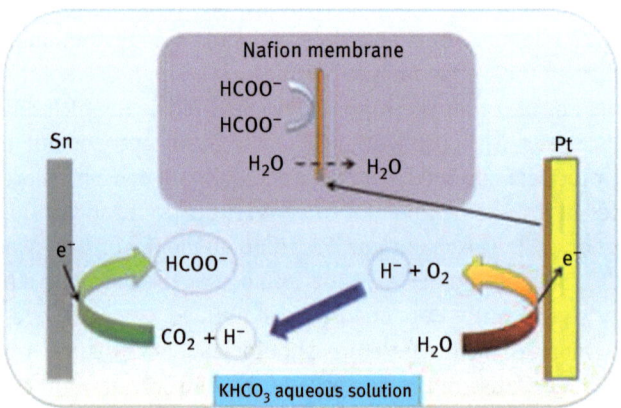

Figure 22.2: Schematic of electrochemical reduction of CO_2 on Sn cathode and Pt@Nafion anode [22].

Besides tin foil, other types of tin electrode have been employed in CO_2 reduction. Wu et al. studied the change of morphology and the corresponding Faradaic efficiency on a Sn particle gas diffusion electrode (GDE) during long-term electrolysis [23]. The Faradaic efficiency towards formate formation degraded from 90% to 56% after 60 h because of the reduced cathode potential and aggregation of fractured Sn particles. To overcome the high cost of Sn GDEs, a novel Sn GDE consisting of a roll-pressed GDL with PTFE binder and a sprayed Sn catalyst layer with Nafion binder was developed [24], which dramatically decreased the fabrication cost to 8% of its original cost. By adding an anion exchange membrane at the Sn GDE cathode side, Yang et al. achieved a current density of 140 mA cm^{-2} with formic acid Faradaic efficiencies up to 94% at a cell voltage of 3.5 V [25].

Combining tin with graphene is a good strategy to improve the CO$_2$ reduction performance. Lei et al. constructed tin quantum sheets confined in a graphene catalyst, which provides a nine times larger CO$_2$ adsorption capacity relative to bulk tin [26]. Catalyst composed of SnS$_2$ nanosheets supported on reduced graphene oxide also exhibits high activity and selectivity for reduction of CO$_2$ into formate, a Faradaic efficiency of 84.5% and a current density of 13.9 mA cm^{-2} were achieved at −1.4 V versus Ag/AgCl [27].

Electrodeposition of Sn onto different substrates such as Sn foil and Cu foil is also a good technique to prepare binder-free electrodes for CO$_2$ reduction [28, 29]. In addition, it was found that the deposited Sn–Pb alloy onto carbon paper with a surface composition of Sn$_{56.3}$Pb$_{43.7}$ could achieve a Faradaic efficiency of 79.8% with a partial current density of 45.7 mA cm^{-2} towards formate production [30]. However, formation of non-conductive PbO on the surface was a potential problem. On the contrary, tin oxides (SnO$_x$) can improve the performance on Sn electrodes for CO$_2$ reduction [31]. Wu et al. investigated the effect of thickness of the SnO$_x$ layer on CO$_2$ reduction [32], the current density exhibited a negligible dependence on the thickness of the oxide layer, while the selectivity of formate and CO demonstrated a strong relationship with oxide thickness (Figure 22.3). A comparative study has been carried out to explore the electrocatalytic properties of tin-oxide particles prepared from SnCl$_2$ or SnCl$_4$ precursors [33]. Results indicated that SnO$_x$ prepared from SnCl$_2$ exhibited a higher Faradaic efficiency (64%). The possible reason is that a SnCl$_2$ versus SnCl$_4$ precursor favours retention of the Sn(II) valence state in the surface layer, which benefits the formic acid production.

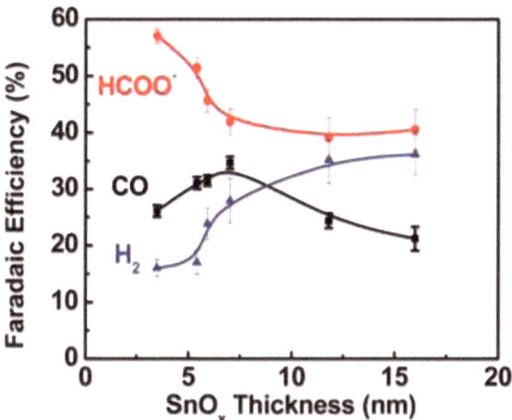

Figure 22.3: CO$_2$ reduction electrolysis data at a cell potential of −1.2 V for Sn GDEs made of the as-received and annealed Sn nanoparticles. Faradaic efficiencies of CO, H$_2$ and HCOO$^-$ versus thickness of SnO$_x$ layer [32].

22.2.1.3 Indium (In)

Indium has been proven to have good catalytic activity with a high selectivity for formic acid/formate in CO_2 electroreduction [34, 35]. As early as 1983, Kapusta and Hackerman found that the current efficiency for CO_2 reduction to formate reached 95% on an indium electrode; however, the overall energy efficiency was very low because of the high over-potential of the reaction [34]. Ikeda et al. investigated CO_2 reduction in a 0.1 mol L^{-1} tetraethylammonium perchlorate (TEAP) electrolyte with an indium electrode [35], they demonstrated that at −2.0 and −2.4 V versus Ag/AgCl, the Faradaic efficiency of formic acid was around 85%. To increase the electrochemical surface area, indium nanofoams with dendritic nanostructures were developed for CO_2 reduction [36]. They exhibited a Faradaic efficiency of 86% for formate at −0.86 V versus RHE and with a current density of 5.8 mA cm^{-2}. In addition to the aqueous electrolyte, ionic liquids are more attractive in CO_2 capture and conversion because of their higher CO_2 solubility and electrical conductivity [37, 38]. Watkins and Bocarsly studied the performance of indium, tin and lead disc working electrodes for CO_2 reduction in a room temperature ionic liquid, 1-ethyl-3-methylimidazolium trifluoroacetate [Emim][TFA] [39]. The Faradaic efficiency reached over 90% at −1.9 V versus Ag/AgCl and the production of formate was up to 3 mg h^{-1} cm^{-2}. The high current density and low electrode potential could be attributed to the stabilising effect of the ionic liquid on the carbon dioxide intermediate and the increased solubility of carbon dioxide in the matrix.

Compared to tin and lead, the cost of indium would be high while only exhibiting similar activities as Sn and Pb for the electrochemical reduction of CO_2 to formate/formic acid. From this point of view, it is not a favourable cathode catalyst.

22.2.1.4 Copper (Cu)

Normally, the primary product of CO_2 electrochemical reduction on copper is methane [40, 41], although some researchers found that a small percentage of formic acid could be produced during the electrolysis process [42, 43]% . Sen et al. prepared a copper nanofoam electrode, which exhibits a high Faradaic efficiency for formic acid at a lower potential [44]. The electrode was prepared through the electrodeposition method. The Faradaic efficiency for formate reached 29% at −1.1 V versus Ag/AgCl on 60 s electrodeposited copper foam (highest value at copper). XRD results indicated that the amount of (200) facet was 22% higher in the copper foam than in the normal copper foil, which may determine the high selectivity for formate. Interestingly, when an oxide-derived copper foam electrode was applied to CO_2 reduction, the preferable formation of (100) and (111) facets led to a high Faradaic efficiency of 23% towards formic acid [45]. These studies indicate that the products formed in the electrochemical reduction of CO_2 are greatly related to the microstructure, particularly, whose crystal faces are exposed to the reactants. Selective adsorption of certain species or intermediates may change the reaction process, thus the final products are also different.

Recently, Guo et al. reported Cu-CDots nanocorals as a highly efficient catalyst for CO_2 reduction to formate [46]. The over-potential for the reduction of CO_2 to formate on this catalyst was only 0.13 V. Furthermore, the total Faradaic efficiency of the CO_2 reduction products was 79% (including formate [68%] and methanol) at −0.7 V versus RHE in 0.5 M $KHCO_3$. This study provides a possibility for conversion of CO_2 into formate with high efficiency while using low-cost Cu as the electrocatalyst.

22.2.1.5 Bismuth (Bi)

Bismuth, a nontoxic and low-cost by-product of Pb, Cu and Sn refining, also shows catalytic activity for CO_2 conversion to formate [47]. Zhang et al. reported a microstructured Bi catalyst exhibiting a low over-potential of 600 mV and a high Faradaic efficiency of 90% at −1.45 V versus SCE in aqueous $KHCO_3$ electrolyte [48]. It is supposed that a nanosized electrode could provide more active sites for CO_2 reduction. Therefore, a variety of nanostructured Bi catalysts were developed, such as nanoparticles [49], nanosheets [50], nanodendrites [47] and nanoflakes [51]. Bi nanosheets with thickness of 10 nm were prepared through electrochemical reduction of Bi^{3+}, [50]. The nanosheets can efficiently reduce CO_2 to formate because of the numerous low-coordination sites and high electron state around the Fermi level, which was verified by DFT calculation. By using a similar electrochemical reduction method, Bi nanoparticles were fabricated in situ on BiOCl nanosheets [49]. A Faradaic efficiency of 92% was obtained at −1.5 V versus SCE, far higher than 55% achieved on the commercial Bi powders. Zhong et al. designed a Bi nanodendrite/carbon paper electrode for selective conversion of CO_2 to formate. This novel electrode demonstrated a maximum Faradaic efficiency of 96.4% with a current density of 15.2 mA cm^{-2}. Moreover, it exhibited negligible degradation over 50-h continuous electrolysis.

22.2.1.6 Zinc (Zn)

Zn was also reported as a cathode for the electrochemical reduction of CO_2 to formic acid. Pt was used in almost all the CO_2 electroreduction experiments as the counter electrode. Pt is considered to be insert in general; however, it was found that Pt anodes could be dissolved at high anodic potentials and that the Pt cations can be reduced at the cathode after diffusion [52]. This may affect the activity of the cathode catalysts. Yadav and Purkait firstly applied Co_3O_4, a cheap and easily available material as the anode, in which Zn was the electrocatalyst to generate formic acid [53]. Both Zn and Co_3O_4 were prepared through the electrodeposition method. In their two-electrode system, the maximum Faradaic efficiency was 78.54% in $KHCO_3$ solution at 1.5 V. The authors also compared the catalytic activity of the Zn cathode with Sn and it was found that the activity of Zn was slightly lower [54]. In this respect, Sn is still a better choice, although the cost of Zn is lower.

22.2.1.7 Cobalt

Cobalt is an uncommon catalyst for the electroredution of CO_2 into formate, but Gao et al.'s work has shown that the electrocatalytic activity of thin layers of Co_3O_4 can be greatly increased when the thickness is reduced to below 2 nm [55]. An activity 20 times higher was obtained on the 1.72 nm thick Co_3O_4 than that of the bulk, and the formate efficiency remained at over 60% for 20 h. They also fabricated a partially oxidised four-atom-thick cobalt layer in 2016 [56]; it was found that the atomically thin layer exhibited a higher intrinsic activity and selectivity towards formate production. Moreover, the partial oxidation of the atomic layers could further improve the intrinsic activity, performing current densities about 10 mA cm^{-2} over 40 h at −0.85 V versus SCE, with approximately 90% formate selectivity. Their recent studies revealed that the presence of oxygen(II) vacancies in Co_3O_4 single-unit-cell layers could favour the rate-limiting proton transfer step for CO_2 reduction to formate, thus accelerating the speed of CO_2 reduction [57].

22.2.1.8 Carbon

Carbon-based materials are commonly used for water oxidation [58] and oxygen reduction [59, 60] reactions because of their high surface area, low cost and significant electrocatalytic activity. Very recent studies have demonstrated the ability of carbon materials as electrocatalysts for CO_2 reduction, with CO being the major product [61, 62]. For example, Kumar et al. applied metal-free carbon nanotube fibres (CNF) for CO_2 reduction in an ionic liquid [62]. The CNF exhibited an exceptionally higher current density when compared to that of bulk Ag for the conversion of CO_2 to CO. Interestingly, it was found that the polyethylenimine (PEI) functionalised nitrogen-doped carbon nanotubes could reduce CO_2 to formate in aqueous media [63]. In 0.1 M $KHCO_3$ solution at −1.8 V versus SCE, a current density of 9.5 mA cm^{-2} was reached with 87% Faradaic efficiency towards formate. This excellent performance could be attributed to the synergistic effect of nitrogen-doping as well as a PEI overlayer. Besides the carbon nanotube, graphene also exhibits good performance for selective reduction CO_2 to formate. Wang et al. were first to report nitrogen-doped graphene for the reduction of CO_2 to formate in an aqueous electrolyte

[64]. The doping of nitrogen remarkably enhanced the current density, Faradaic efficiency and stability for conversion of CO$_2$ to formate. Other heteroatom-doped graphene such as boron-doped graphene (BG) was developed for the selective electroreduction of CO$_2$ to formate [65]. The boron-doping introduces asymmetric spin density in the graphene, favouring the CO$_2$ adsorption on BG, thus promoting the CO$_2$ reduction to formate. The results showed that under −1.4 V versus SCE in 0.1 M KHCO$_3$, a Faradaic efficiency of 66% was obtained.

22.2.1.9 Other catalysts

Recently, Kortlever et al. developed a reversible catalyst by electrodepositing palladium on a polycrystalline platinum substrate [66]. The catalyst could reduce bicarbonate to formic acid at a low over-potential while also being able to directly reduce CO$_2$ at more negative potentials. Moreover, the catalyst exhibited reversible formic acid oxidation properties. Palladium nanoparticles at sizes of 3.8–10.7 nm were synthesised for formate production from CO$_2$ [67]. The Faradaic efficiency of formate on 3.8 nm Pd was 86% while 98% was observed on 6.5 nm Pd. Isaacs et al. reported that CO$_2$ could be electrochemically reduced to formic acid by metal complex catalysts such as polymeric M-tetrakis aminophthalocyanines (M = Co, Ni) and hexa-aza-macrocycle complexes (Co, Ni, Cu) [68, 69]. The experiments indicated that the metallic centre affected the kinetics of polymerisation and the polymer morphology, therefore determining the final product [69]. It was also found that electrodes based on Fe supported on carbon black (Vulacan XC-72R) could convert CO$_2$ to formic acid [70, 71]. In 2008, Reda et al. found that a tungsten-containing formate dehydrogenase enzyme (FDH1) adsorbed to an electrode surface could electrochemically convert CO$_2$ to formate [72]. Formate was the only product with only a very small over-potential required with FDH1 (Table 22.2), making it a promising method for practical application. A novel synthesis of formic acid was developed in a room temperature ionic liquid via the reaction of electro-activated CO$_2$ and protons on pre-anodised platinum [37]. Strong acid bis(trifluoromethane)-sulfonimide (H[NTf$_2$]) was investigated on a platinum (Pt) microelectrode, 1-ethyl-3-methylimidazolium bis (trifluoromethylsulfonyl) imide ([C$_2$mim][NTf2]) was used to reduce CO$_2$, and formic acid was the main product with H$_2$ as the only side product.

In summary, the representative work on electrocatalysts for CO$_2$ reduction to formate/formic acid is shown in Table 22.2. The table includes the catalyst type, working potential, electrolyte, Faradaic efficiency and current density. Among the reported metal and oxide cathodes towards formic acid production, tin, tin oxides and Co$_3$O$_4$ seem to be the best choices in terms of cost and selectivity. Moreover, the heteroatom-doped carbon materials have shown promise as robust metal-free catalysts for the next generation of catalysts.

Table 22.2: Summary of electrocatalysts for CO_2 reduction to formate.

Catalysts	Potential	Electrolyte	Formate Faradaic efficiency (%)	Current density (mA cm^{-2})	Ref.
Pb wire	−2.0 V vs. Ag/AgCl	0.3 M KOH in methanol 15 °C	66	4	[16]
Pb granule	−1.8 V vs. SCE	0.2 M K$_2$CO$_3$ 50 bar 80 °C	94	0.72	[17]
Pb on stainless steel	−1.8 V vs. anode	Catholyte: 0.2 M K$_2$HPO$_4$ + H$_3$PO$_4$ pH 7 Anolyte: 0.2 M KOH	93	2	[14]
Electrodeposited porous Pb on Cu plate	−1.7 V vs. SCE	0.5 M KHCO$_3$ 5 °C	96.8	N/A	[89]
Nanolayered Pb		0.1 M KHCO$_3$ 5 °C	94.1	N/A	[18]
Pb foil	−1.83 V vs. SCE	0.5 M NaOH 4 °C	90	2.5	[90]
Pb plate	N/A	Catholyte: 0.45 M KHCO$_3$ + 0.5 M KCl Anolyte: 1 M KOH	57	10.5	[91]
Metallic Pb	−2.4 V vs. Ag/AgCl	0.1 M TEAP/H$_2$O 100 °C	78.9	N/A	[92]
Sn foil	−1.8 V vs. Ag/AgCl	0.1 M KHCO$_3$	91	2.5	[21]
Sn foil	−2.0 V vs. SCE	0.5 M KHCO3	63.49	28	[93]
Sn gas diffusion electrode (GDE)	−1.8 V vs. Ag/AgCl	0.5 M KHCO$_3$	72.99	13.45	[24]
Reduced Nano-SnO$_2$/graphene	−1.8 V vs. SCE	0.1 M NaHCO$_3$	93.6	10.2	[94]
SnS$_2$/graphene	−1.4 V vs. Ag/AgCl	0.5 M NaHCO$_3$	84.5	13.9	[27]
Sn quantum sheets confined in graphene	−1.8 V vs. SCE	0.1 M NaHCO$_3$	89	21.1	[26]
Nanoporous Sn foam	−2.0 V vs. Ag/AgCl	0.1 M NaHCO$_3$	90	23.5	[29]
Sn dendrite	−1.36 V vs. RHE (−2.0 V vs. SCE)	0.1 M KHCO$_3$	71.6	17.1	[95]
Sn/SnO$_x$ thin film	−0.7 V vs. RHE	0.5 M NaHCO$_3$	40	1.7	[31]
Sn nanoparticles of a 3.5 nm SnO$_x$ layer	−1.2 V vs. CE	0.1 M KHCO$_3$	64	3	[32]
Porous SnO$_2$/carbon cloth	−1.6 V vs. Ag/AgCl	0.5 M NaHCO$_3$	87	45	[96]

Table 22.2 (continued)

Catalysts	Potential	Electrolyte	Formate Faradaic efficiency (%)	Current density (mA cm^{-2})	Ref.
Sn powder coated with Nafion	−1.6 V vs. NHE	0.5 M NaHCO$_3$	70	27	[97]
In disc	−1.9 V vs. Ag/ AgCl	[Emim][TFA] + 33% H$_2$O	>90	3.5	[39]
In dendrite	−0.86 V vs. RHE	0.1 M KHCO$_3$	86	5.8	[36]
Bi	−1.45 V vs. SCE	0.5 M KHCO$_3$	90	~1.6	[48]
Bi nanoparticles	−1.5 V vs. SCE	0.5 M KHCO$_3$	92	N/A	[49]
Bi nanoparticles	−1.6 V vs. SCE	0.5 M KHCO$_3$	98.4	~9.7	[98]
Bi nanodendrites	−1.8 V vs. SCE	Catholyte: 0.5 M NaHCO$_3$ Anolyte: 0.1 M H$_2$SO$_4$	96.4	15.2	[47]
Zn	−1.5 V vs. anode	0.5 M KHCO$_3$	78.54	<2	[53]
Ultrathin Co$_3$O$_4$ (1.72 nm)	−0.88 V vs. SCE	0.1 M KHCO$_3$	64.3	0.68	[55]
Partially oxidised atomic Co	−0.85 V vs. SCE	0.1 M NaSO$_4$	90.1	~10	[56]
Cu nanofoam	−1.1 vs. Ag/AgCl	0.1 M KHCO$_3$	29	~10	[44]
Oxide-derived Cu foam	−0.45V vs. RHE	0.1 M KHCO$_3$	23	~9.4	[45]
Cu-CDots nanocorals	−0.7V vs. RHE	0.5 M KHCO$_3$	68	~4.2	[46]
Nitrogen-doped graphene	−0.84 vs. RHE	0.5 M KHCO$_3$	73	7.5	[64]
Boron-doped graphene	−1.4 V vs. SCE	0.1 M KHCO$_3$	66	~1.5	[65]
Tungsten-containing formate dehydrogenase enzyme (FDH1)	−0.41 ~ -0.81 V	0.02 M Na2CO3	~100	N/A	[99]
[Fe$_4$S$_4$(SR)$_4$]$^{2-}$	−1.7 V	DMF	59	N/A	[100]
[Ru(bpy)$_2$(CO)$_2$]$_2^+$	−1.3 V vs. SCE	Saturated H$_2$O/ DMF (9: 1, v/v) pH 9.5	84.3	3.3	[101]
[(bpy)$_2$Ru(dmbbbpy)] (PF$_6$)$_2$ (dmbbbpy = 2,2′-bis(1-methylbenzimidazol-2-yl)-4,4′-bipyridine)	−1.65 V vs. Ag/ AgCl	MeCN + 2.5% H$_2$O	89	N/A	[102]
[(bpy)$_2$Ru(dmbbbpy)Ru (bpy)$_2$](PF6)$_4$	−1.55 V vs. Ag/ AgCl	MeCN + 2.5% H$_2$O	90	N/A	[102]

Table 22.2 (continued)

Catalysts	Potential	Electrolyte	Formate Faradaic efficiency (%)	Current density (mA cm^{-2})	Ref.
Iridium(III) trihydride complex (POCOP)IrH$_2$ POCOP = C$_6$H$_3$-2,6-(OP^{t}Bu$_2$)$_2$	−1.73 V vs. Fc/Fc$^+$	MeCN + 12% H$_2$O	>99	N/A	[103]
Pd$_{70}$Pt$_{30}$/C	−0.4 V vs. RHE	0.1 M KH$_2$PO$_4$ / 0.1 M K$_2$HPO$_4$ pH 6.7	88	5	[104]

Note: SCE: saturated calomel electrode; RHE: reversible hydrogen electrode; NHE: normal hydrogen electrode; TEAP: tetraethylammonium perchlorate; [Emim][TFA]: 1-ethyl-3-methylimidazolium trifluoroacetate [Emim][TFA]; DMF: dimethyl formamide; MeCN: methyl cyanide

22.2.2 Electrolyte

Besides the electrocatalysts themselves, other factors such as the electrolyte, temperature and pressure are also crucial for the electrochemical reduction of CO_2 into formic acid or formate. The type of electrolyte frequently used in electroreduction of CO_2 can be roughly classed into aqueous and non-aqueous solutions. The non-aqueous electrolytes such as ionic liquids have been studied for the electroreduction of CO_2 because of their high CO_2 solubility, wide potential window and extremely low volatility. Martindale and Compton applied H[NTf$_2$] dissolved in [C$_2$mim][NTf$_2$] as an electrolyte for the electroreduction of CO_2 to formic acid with a Pt electrode [37]. In BMIMBF$_4$, CO_2 was reduced on a Cu electrode at −2.4 V versus Ag/AgCl at room temperature [73]. Snuffin et al. synthesised a novel ionic liquid EMIMBF$_3$Cl, which performed a high current density for CO_2 reduction at −1.8 V versus silver wire with a Pt electrode [74]. The use of non-aqueous electrolyte tetrabutylammonium perchlorate (TBAP) in methanol makes the electrochemical reduction of CO_2 to formic acid and acetic acid at low overpotentials possible, such as −0.3 V versus SCE (sat. KCl) when polyaniline (PANI)-Cu$_2$O nanocomposite was used as the cathode [75]. A large advantage of using a low over-potential is that the overall energy efficiency would be much higher [76]. A non-aqueous aprotic electrolyte was tried using a lead wire cathode for CO_2 reduction [77]. The propylene carbonate (PrC) containing TEAP was reported as the electrolyte. Cyclic voltammetry and infrared reflectance spectroscopy confirmed that the CO_2 reduction reaction was a mass transfer process with no CO formed as an intermediate or final product under these operating conditions. This indicates the use of non-aqueous solvent TEPA-PrC can suppress the formation of CO.

In the aqueous electrolytes, many studies demonstrated that the efficiency and selectivity for CO_2 electroreduction depended on the solvent, including cations and anions [78]. The Faradaic efficiency was enhanced in the presence of CO_3^{2-} and HCO_3^-, which could be because of the involvement of the anions during the electroreduction. The Faradaic efficiency of formic acid production increased with the anions in the order: $PO_4^{3-} < SO_4^{2-} < CO_3^{2-} < HCO_3^-$. Ogura et al. studied the differences of Cl^-, Br^- and I^- by conducting CO_2 electroreduction in 3 M KCl, KBr and KI solution using a copper mesh electrode [79]. It revealed that the electron transferred from the adsorbed halide anion to CO_2, accelerated the CO_2 reduction. A higher halide anion adsorption to the electrode lead to faster CO_2 conversion, forming a higher reduction current. Additionally, higher halide anion adsorption limited the proton adsorption, resulting in a more negative hydrogen evolution potential. Regarding the cations, since small cations such as Li^+ and Na^+ are strongly hydrated, they will not adsorb on the electrode, thus they can take a lot of water molecules to the cathode, which supplies the protons required during electroreduction. Contrary to this, large cations are not easily hydrated, so they tend to adsorb on the cathode. Bhugun et al. reported the effects of cations for formic acid formation finding that the formic acid production decreased with increasing concentration of cations with the ability of reactivity improvement in the order of $Mg^{2+} = Ca^{2+} > Li^+ > Na^+$ [80]. Thorson et al. investigated the relationship between the size of cations and CO selectivity on silver electrodes [81], specifically, larger cations ($Na^+ < K^+ < Rb^+ < Cs^+$) improved the CO production while suppressing H_2 formation.

It can be concluded that the type of anions influences the pH at the electrode, which then determines the proton content, thus it finally affects the reaction kinetics; cations of different sizes are adsorbed on the electrode surface to a different extent, changing the electrical double layer structure and affecting the reaction kinetics and energetics [9]. Furthermore, the solubility of CO_2 in solution is another key factor that affects the performance of the electrode, because of reduced mass transfer resistance at higher CO_2 solubilities.

22.2.3 Temperature and pressure

Temperature and pressure are important parameters for the electrochemical reduction of CO_2 into formic acid. It was reported that the current density was 20 times greater when the temperature changed from 275 to 333 K [82]. Mizuno et al. found that the Faradaic efficiency for formic acid on a Pb electrode firstly increased and then decreased from 20 to 100 °C [83]. When it comes to Tin and Indium electrodes, the Faradaic efficiency continued to decrease from 20 to 100 °C. It is expected that the increase in temperature promotes all the reaction rates, including those for the competing reactions, while also reducing the solubility of CO_2 and improving the

CO_2 diffusion rate. Generally, a higher efficiency of CO_2 electroreduction can be obtained at a lower temperature.

The solubility of CO_2 can be improved with higher pressure. From this point of view, higher pressures are better for CO_2 electroreduction with the current density expected to be higher. Electrochemical reduction of CO_2 in aqueous solutions (both inorganic and tetraalkylammonium salts) and non-aqueous electrolytes under high pressure were investigated by Ito et al. in 1980s [84, 85]. In 1995, Asano et al. studied the electrochemical reduction of carbon dioxide under high pressure on various electrodes in an aqueous $KHCO_3$ electrolyte [86]. It was found that the electrochemical reduction of CO_2 at large partial current densities was accomplished on many electrodes under a CO_2 pressure of 30 atm, for example, 397 mA cm^{-2} on Pd and 383 mA cm^{-2} on Ag; however, the Faradaic efficiency to formate on Pd and Ag was only 44 and 16% respectively. The Faradaic efficiency on Sn and Pb was 92% and 95% with high current densities of 163 and 156 mA cm^{-2} respectively [86], indicating that Pb and Sn are good cathode materials for the electrochemical reduction of CO_2 to formic acid at high pressure. Todoroki et al. explored the electroreduction of CO_2 with In, Pb and Hg under high pressure in an aqueous solution [87]. The Faradaic efficiency of formate grew with rising pressure, with the value increasing past 90% at 20 atm. As the pressure continuously increased, the rate-determining step of the reaction changed from CO_2 diffusion to electron transfer at the electrode surface. Recently, electrochemical reduction of CO_2 into formic acid under high pressure at a Sn cathode in an aqueous KCl or Na_2SO_4 electrolyte was investigated. It was found that the Faradaic efficiency decreased against time; this is possibly because of the anode oxidation of the formed formic acid [88].

22.3 Challenges and future opportunities

Large amounts of CO_2 will be available in the following decades as fossil fuels are still currently the most important energy resources on earth. With the possible application of CO_2 capture processes, it is desired to find an outlet for the captured CO_2. Also, the demand for formic acid will be booming with the development of a global economy. Electrochemical reduction of CO_2 to formic acid is therefore a useful technology to solve these problems. The combination of CO_2 capture and electrochemical reduction of CO_2 can be used for renewable electricity storage as well.

As discussed previously, many catalysts have been developed in order to achieve a high conversion. Among the reported metal and oxide cathodes towards formic acid production, tin and tin oxides have proven to be best choices in terms of cost and selectivity. However, there are still several challenges limiting the practical applications: (1) low catalyst activity, (2) low selectivity and (3) low catalyst stability. It is obvious that the over-potential for catalysts in CO_2 electroreduction is normally too high, meaning that their activities are poor and thereby reactions exhibit low-energy efficiencies. Although some of the catalysts exhibited high selectivity, their stability is not high enough for industrial-scale application. Normally, the stability test carried out in the literature is less than 100 h with longer-term stability tests not yet reported. The active sites on the electrode could be blocked by the intermediates or by-products produced during the CO_2 reduction, leading to the degradation of catalyst activity. For cells not divided by a separation membrane, anode oxidation of the produced formic acid will also reduce the efficiency.

In conclusion, electroreduction of CO_2 to formic acid or formate is still at an early stage with many issues that need to be solved. To overcome these challenges and reach the requirements for commercialisation, several directions should be focused on in the future: (1) Exploration of new cathode catalysts. Several metals and oxides such as tin/tin oxide are good cathode catalysts with high selectivity to formic acid. Some progress has been made but the technology is still not sufficient for large-scale use. Therefore, the next generation of novel catalysts might be obtained through a combination of the existing materials as well as the tailoring of the catalyst microstructure. The combined materials could demonstrate different properties from individual components because of the mutual effect of each substance. Catalysts with high activity, low over-potential, high selectivity and good stability may be obtained by optimising the compositions and other important parameters such as particle size. The desired number of active sites could be increased by tailoring the microstructure of these materials. Materials with different microstructures may have different selectivities and catalytic properties when prepared through different methods. For electrodeposited metal catalysts, the selectivity is greatly related to the current density and deposit time leading to various microstructures, thus various active faces may be exposed to CO_2 and H_2O resulting in different products. Simultaneously, the surface area and mass transport property would also be changed, which may form ideal catalysts for CO_2 reduction. Additionally, the electrolyte and working conditions such as temperature and pressure should be optimised according to the designed electrode. (2) Further understanding of the reaction mechanisms. The studies on the CO_2 conversion mechanisms are insufficient presently [105], although some of the literature has attempted to explore the fundamentals using both experimental and theoretical modelling methods [106, 107]. A better understanding of the fundamental mechanisms will significantly help researchers in optimising the operation conditions of the catalyst directly improving its performance. This will also help to identify better electrocatalysts for CO_2 reduction to formic acid. (3) To explore suitable anodes for two-electrode cells that can be used for real applications. Most of the research activities on electrochemical reduction of CO_2 to formate/formic acid focus on three-electrode cells; however, two-electrode cells are used for real commercial applications. Therefore, it is desired to explore suitable anodes for both acidic and alkaline environments, which can be used as a matched electrode in order to build two-electrode cells. Formic acid production rate, Faradaic and overall energy efficiencies, and long-term stability of the two-electrode cells are important parameters that need to be investigated to achieve real applications of this technology. With the efforts of researchers in the areas of catalysis, electrochemistry, materials science and other relevant areas, electrochemical synthesis will provide a low carbon and sustainable process for production of useful hydrocarbons, such as formic acid.

References

[1]   R.J. Lim, M. Xie, M.A. Sk, J.-M. Lee, A. Fisher, X. Wang, K.H. Lim, Catal. Today, 233 (2014) 169–180.
[2]   http://www.esrl.noaa.gov/gmd/ccgg/trends/ (accessed 30th November 2018).
[3]   S. Solomon, G.-K. Plattner, R. Knutti, P. Friedlingstein, Proceedings of the national academy of sciences, 106 (2009) 1704–1709.
[4]   A.S. Agarwal, Y. Zhai, D. Hill, N. Sridhar, ChemSusChem, 4 (2011) 1301–1310.
[5]   M. Aresta, Carbon dioxide: utilization options to reduce its accumulation in the atmosphere, in: Carbon Dioxide as Chemical Feedstock, Wiley-VCH Verlag GmbH & Co., 2010, pp. 1–13.
[6]   C. Le Quéré, R.M. Andrew, P. Friedlingstein, S. Sitch, J. Pongratz, A.C. Manning, J.I. Korsbakken, G.P. Peters, J.G. Canadell, R.B. Jackson, Earth System Science Data Discussions, (2017) 1–79.
[7]   C.S. Song, Catal. Today, 115 (2006) 2–32.
[8]   J. Qiao, Y. Liu, F. Hong, J. Zhang, Chem. Soc. Rev., 43 (2014) 631–675.
[9]   H.-R. Jhong, S. Ma, P.J. Kenis, Curr. Opin. Chem. Eng., 2 (2013) 191–199.
[10]   C. Le Quéré, R. Moriarty, R.M. Andrew, J.G. Canadell, S. Sitch, J.I. Korsbakken, P. Friedlingstein, G.P. Peters, R.J. Andres, T.A. Boden, Earth. Syst. Sci. Data., 7 (2015) 349–396.
[11]   T. Kahl, K. Schröder, F. Lawrence, W. Marshall, H. Höke, R. Jäckh, Wiley-VCH, Weinheim, Germany, (2002) 1.
[12]   X. Yu, P.G. Pickup, J. Power. Sources., 182 (2008) 124–132.
[13]   M. Grasemann, G. Laurenczy, Energy. Environ. Sci., 5 (2012) 8171–8181.
[14]   K. Subramanian, K. Asokan, D. Jeevarathinam, M. Chandrasekaran, J. Appl. Electrochem., 37 (2007) 255–260.
[15]   C. Oloman, H. Li, Chemsuschem, 1 (2008) 385–391.
[16]   S. Kaneco, R. Iwao, K. Iiba, K. Ohta, T. Mizuno, Energy, 23 (1998) 1107–1112.
[17]   F. Köleli, D. Balun, Appl. Catal. A: Gen., 274 (2004) 237–242.
[18]   Y. Kwon, J. Lee, Electrocatalysis, 1 (2010) 108–115.
[19]   C.H. Lee, M.W. Kanan, ACS Catal., 5 (2014) 465–469.
[20]   S. Back, J.-H. Kim, Y.-T. Kim, Y. Jung, Phys. Chem. Chem. Phys., 18 (2016) 9652–9657.
[21]   W. Lv, R. Zhang, P. Gao, L. Lei, J. Power. Sources., 253 (2014) 276–281.
[22]   R. Zhang, W. Lv, G. Li, M.A. Mezaal, X. Li, L. Lei, J. Power. Sources., 272 (2014) 303–310.
[23]   J. Wu, B. Harris, P.P. Sharma, X.-D. Zhou, ECS Trans., 58 (2013) 71–80.
[24]   Q. Wang, H. Dong, H. Yu, RSC Advances, 4 (2014) 59970–59976.
[25]   H. Yang, J.J. Kaczur, S.D. Sajjad, R.I. Masel, Journal of CO2 Utilization, 20 (2017) 208–217.
[26]   F. Lei, W. Liu, Y. Sun, J. Xu, K. Liu, L. Liang, T. Yao, B. Pan, S. Wei, Y. Xie, Nat. Ccmmun., 7 (2016) 12697.

[27] F. Li, L. Chen, M. Xue, T. Williams, Y. Zhang, D.R. MacFarlane, J. Zhang, Nano. Energy., 31 (2017) 270–277.
[28] C. Zhao, J. Wang, Chem. Eng. J., 293 (2016) 161–170.
[29] D. Du, R. Lan, J. Humphreys, S. Sengodan, K. Xie, H. Wang, S. Tao, ChemistrySelect, 1 (2016) 1711–1715.
[30] S.Y. Choi, S.K. Jeong, H.J. Kim, I.H. Baek, K.T. Park, ACS Sustain. Chem. Eng., 4 (2016) 1311–1318.
[31] Y. Chen, M.W. Kanan, J. Am. Chem. Soc., 134 (2012) 1986–1989.
[32] J. Wu, F.G. Risalvato, S. Ma, X.-D. Zhou, J. Mater. Chem. A. Mater., 2 (2014) 1647–1651.
[33] C. Zhao, J. Wang, J.B. Goodenough, Electrochem. commun., 65 (2016) 9–13.
[34] S. Kapusta, N. Hackerman, J. Electrochem. Soc., 130 (1983) 607–613.
[35] S. Ikeda, T. Takagi, K. Ito, Bull. Chem. Soc. Jpn., 60 (1987) 2517–2522.
[36] Z. Xia, M. Freeman, D. Zhang, B. Yang, L. Lei, Z. Li, Y. Hou, ChemElectroChem, 5 (2018) 253–259.
[37] B.C. Martindale, R.G. Compton, Chem. Commun., 48 (2012) 6487–6489.
[38] B.A. Rosen, W. Zhu, G. Kaul, A. Salehi-Khojin, R.I. Masel, J. Electrochem. Soc., 160 (2013) H138–H141.
[39] J.D. Watkins, A.B. Bocarsly, ChemSusChem, 7 (2014) 284–290.
[40] K. Manthiram, B.J. Beberwyck, A.P. Alivisatos, J. ACS., 136 (2014) 13319–13325.
[41] K.-J. Yim, D.-K. Song, C.-S. Kim, N.-G. Kim, T. Iwaki, T. Ogi, K. Okuyama, S.-E. Lee, T.-O. Kim, RSC Adv., 5 (2015) 9278–9282.
[42] S. Kaneco, K. Iiba, N.-H. Hiei, K. Ohta, T. Mizuno, T. Suzuki, Electrochimica acta., 44 (1999) 4701–4706.
[43] S. Kaneco, H. Katsumata, T. Suzuki, K. Ohta, Energy & Fuels, 20 (2006) 409–414.
[44] S. Sen, D. Liu, G.T.R. Palmore, ACS Catal., 4 (2014) 3091–3095.
[45] S. Min, X. Yang, A.-Y. Lu, C.-C. Tseng, M.N. Hedhili, L.-J. Li, K.-W. Huang, Nano. Energy., 27 (2016) 121–129.
[46] S. Guo, S. Zhao, J. Gao, C. Zhu, X. Wu, Y. Fu, H. Huang, Y. Liu, Z. Kang, Nanoscale, 9 (2017) 298–304.
[47] H. Zhong, Y. Qiu, T. Zhang, X. Li, H. Zhang, X. Chen, J. Mater. Chem. A. Mater., 4 (2016) 13746–13753.
[48] X. Zhang, T. Lei, Y. Liu, J. Qiao, Applied Catalysis B: Env., 218 (2017) 46–50.
[49] H. Zhang, Y. Ma, F. Quan, J. Huang, F. Jia, L. Zhang, Electrochem. Commun., 46 (2014) 63–66.
[50] S. Panpan, X. Wenbin, Q. Yanling, Z. Taotao, L. Xianfeng, Z. Huamin, ChemSusChem, 11 (2018) 848–853.
[51] S. Kim, W.J. Dong, S. Gim, W. Sohn, J.Y. Park, C.J. Yoo, H.W. Jang, J.L. Lee, Nano. Energy., 39 (2017) 44–52.
[52] S. Cherevko, A.R. Zeradjanin, G.P. Keeley, K.J. Mayrhofer, J. Electrochem. Soc., 161 (2014) H822–H830.
[53] V.S.K. Yadav, M.K. Purkait, New. J. Chem., 39 (2015) 7348–7354.
[54] V.S.K. Yadav, M.K. Purkait, Inorganica. Chim. Acta., 124 (2016) 177–183.
[55] G. Shan, J. Xingchen, S. Zhongti, Z. Wenhua, S. Yongfu, W. Chengming, H. Qitao, Z. Xiaolong, Y. Fan, Y. Shuyang, L. Liang, W. Ju, X. Yi, Angew. Chem. Int. Ed., 55 (2016) 698–702.
[56] S. Gao, Y. Lin, X. Jiao, Y. Sun, Q. Luo, W. Zhang, D. Li, J. Yang, Y. Xie, Nature, 529 (2016) 68–71.
[57] S. Gao, Z. Sun, W. Liu, X. Jiao, X. Zu, Q. Hu, Y. Sun, T. Yao, W. Zhang, S. Wei, Nat. Ccmmun., 8 (2017).
[58] Y. Zhao, R. Nakamura, K. Kamiya, S. Nakanishi, K. Hashimoto, Nat. Commun., 4 (2013) 1–7.
[59] K. Gong, F. Du, Z. Xia, M. Durstock, L. Dai, Science., 323 (2009) 760–764.

[60] R. Liu, D. Wu, X. Feng, K. Müllen, Angew. Chem., 122 (2010) 2619–2623.

[61] P.P. Sharma, J. Wu, R.M. Yadav, M. Liu, C.J. Wright, C.S. Tiwary, B.I. Yakobson, J. Lou, P.M. Ajayan, X.D. Zhou, Angew. Chem., 127 (2015) 13905–13909.

[62] B. Kumar, M. Asadi, D. Pisasale, S. Sinha-Ray, B.A. Rosen, R. Haasch, J. Abiade, A.L. Yarin, A. Salehi-Khojin, Nat. Commun., 4 (2013).

[63] S. Zhang, P. Kang, S. Ubnoske, M.K. Brennaman, N. Song, R.L. House, J.T. Glass, T.J. Meyer, J. Am. Chem. Soc., 136 (2014) 7845–7848.

[64] H. Wang, Y. Chen, X. Hou, C. Ma, T. Tan, Green Chem., 18 (2016) 3250–3256.

[65] N. Sreekanth, M.A. Nazrulla, T.V. Vineesh, K. Sailaja, K.L. Phani, Chem. Commun., 51 (2015) 16061–16064.

[66] R. Kortlever, C. Balemans, Y. Kwon, M.T. Koper, Catal. Today, 244 (2015) 58–62.

[67] M. Rahaman, A. Dutta, P. Broekmann, ChemSusChem, (2017).

[68] M. Isaacs, F. Armijo, G. Ramírez, E. Trollund, S. Biaggio, J. Costamagna, M.J. Aguirre, J. Mol. Catal. A. Chem., 229 (2005) 249–257.

[69] M. Isaacs, J. Canales, M. Aguirre, G. Estiú, F. Caruso, G. Ferraudi, J. Costamagna, Inorganica chimica acta, 339 (2002) 224–232.

[70] S. Pérez-Rodríguez, F. Barreras, E. Pastor, M. Lázaro, Int. J. Hydrogen. Energy., 41 (2016) 19756–19765.

[71] C. Ampelli, C. Genovese, B. Marepally, G. Papanikolaou, S. Perathoner, G. Centi, Faraday Discuss., 183 (2015) 125–145.

[72] T. Reda, C.M. Plugge, N.J. Abram, J. Hirst, Proceedings of the National Academy of Sciences, 105 (2008) 10654–10658.

[73] M. Feroci, M. Orsini, L. Rossi, G. Sotgiu, A. Inesi, J. Org. Chem., 72 (2007) 200–203.

[74] L.L. Snuffin, L.W. Whaley, L. Yu, J. Electrochem. Soc., 158 (2011) F155–F158.

[75] A.N. Grace, S.Y. Choi, M. Vinoba, M. Bhagiyalakshmi, D.H. Chu, Y. Yoon, S.C. Nam, S.K. Jeong, Appl. Energ., 120 (2014) 85–94.

[76] R. Chaplin, A. Wragg, J. Appl. Electrochem., 33 (2003) 1107–1123.

[77] B. Eneau-Innocent, D. Pasquier, F. Ropital, J.-M. Léger, K. Kokoh, Appl. Catal. B., 98 (2010) 65–71.

[78] M. Jitaru, D. Lowy, M. Toma, B. Toma, L. Oniciu, J. Appl. Electrochem., 27 (1997) 875–889.

[79] K. Ogura, J.R. Ferrell, III, A.V. Cugini, E.S. Smotkin, M.D. Salazar-Villalpando, Electrochimica acta, 56 (2010) 381–386.

[80] I. Bhugun, D. Lexa, J.M. Saveant, J. Phys. Chem., 100 (1996) 19981–19985.

[81] M.R. Thorson, K.I. Siil, P.J. Kenis, J. Electrochem. Soc., 160 (2013) F69–F74.

[82] J. Ryu, T. Andersen, H. Eyring, J. Phys. Chem., 76 (1972) 3278–3286.

[83] T. Mizuno, K. Ohta, A. Sasaki, T. Akai, M. Hirano, A. Kawabe, Energ. Source., 17 (1995) 503–508.

[84] K. Ito, S. Ikeda, M. Okabe, Denki Kagaku, 48 (1980) 247–252.

[85] K. Ito, S. Ikeda, T. Iida, A. Nomura, Denki Kagaku, 50 (1982) 463–469.

[86] K. Asano, T. Hibino, H. Iwahara, J. Electrochem. Soc., 142 (1995) 3241–3245.

[87] M. Todoroki, K. Hara, A. Kudo, T. Sakata, J. Electroanal. Chem., 394 (1995) 199–203.

[88] O. Scialdone, A. Galia, G.L. Nero, F. Proietto, S. Sabatino, B. Schiavo, Electrochimica acta., 199 (2016) 332–341.

[89] W. Jing, W. Hua, H. Zhenzhen, H. Jinyu, Front. Chem. Sci. Eng., 9 (2015) 57–63.

[90] B. Innocent, D. Liaigre, D. Pasquier, F. Ropital, J.-M. Léger, K. Kokoh, J. Appl. Electrochem., 39 (2009) 227–232.

[91] M. Alvarez-Guerra, S. Quintanilla, A. Irabien, Chem. Eng. J., 207 (2012) 278–284.

[92] S. Ikeda, T. Takagi, K. Ito, Bull. Chem. Soc. Jpn., 60 (1987) 2517–2522.

[93]   J. Wu, F.G. Risalvato, F.-S. Ke, P. Pellechia, X.-D. Zhou, J. Electrochem. Soc., 159 (2012) F353–F359.
[94]   S. Zhang, P. Kang, T.J. Meyer, J. Am. Chem. Soc., 136 (2014) 1734–1737.
[95]   D.H. Won, C.H. Choi, J. Chung, M.W. Chung, E.H. Kim, S.I. Woo, ChemSusChem, 8 (2015) 3092–3098.
[96]   F. Li, L. Chen, G.P. Knowles, D.R. MacFarlane, J. Zhang, Angew. Chem., 129 (2017) 520–524.
[97]   G.S. Prakash, F.A. Viva, G.A. Olah, J. Power. Sources., 223 (2013) 68–73.
[98]   Y. Qiu, J. Du, W. Dong, C. Dai, C. Tao, Journal of CO2 Utilization, 20 (2017) 328–335.
[99]   T. Reda, C.M. Plugge, N.J. Abram, J. Hirst, Proc. Natl. Acad. Sci. U. S. A., 105 (2008) 10654–10658.
[100] M. Tezuka, T. Yajima, A. Tsuchiya, Y. Matsumoto, Y. Uchida, M. Hidai, J. ACS., 104 (1982) 6834–6836.
[101] H. Ishida, H. Tanaka, K. Tanaka, T. Tanaka, J. Chem. Soc. Chem. Commun., (1987) 131–132.
[102] M.M. Ali, H. Sato, T. Mizukawa, K. Tsuge, M. Haga, K. Tanaka, Chem. Commun., (1998) 249–250.
[103] S.T. Ahn, E.A. Bielinski, E.M. Lane, Y. Chen, W.H. Bernskoetter, N. Hazari, G.T.R. Palmore, chem. Comm., 51 (2015) 5947–5950.
[104] R. Kortlever, I. Peters, S. Koper, M.T. Koper, ACS Catal., 5 (2015) 3916–3923.
[105] A.J. Martín, G.O. Larrazábal, J. Pérez-Ramírez, Green. Chem., 17 (2015) 5114–5130.
[106] A.A. Peterson, F. Abild-Pedersen, F. Studt, J. Rossmeisl, J.K. Nørskov, Energy. Environ. Sci., 3 (2010) 1311–1315.
[107] P. Hirunsit, J. Phys. Chem. C., 117 (2013) 8262–8268.

Stephen M. Lyth

23 Electrochemical conversion of CO_2 to carbon monoxide

23.1 Introduction

Carbon dioxide in the environment is one of the biggest problems mankind is facing. The repercussions of continuing to increase the atmospheric concentration of CO_2 unchecked could be cataclysmic [1]. To help reduce atmospheric CO_2 emissions in the future, there is an abundance of CO_2 capture and utilisation technologies under development, as outlined in this book. Electrochemical reduction of carbon dioxide is an option that has the potential to convert this greenhouse gas into useful products in an efficient, straightforward, low cost and scalable manner. Some advantages of electrochemical conversion of CO_2 include operation at ambient temperature and pressure, the possibility of using low-cost catalysts, fast response times and the potential for long-term continuous operation [2–5].

Recently, electrochemical CO_2 conversion has been having somewhat of a renaissance, and it is easy to think of this is a relatively young field. However, the first paper on the CO_2 reduction reaction (CO2RR) was published almost 150 years ago, long before anthropogenic global warming was around to provide a motivation. [6] Until climate change was known to be a threat, the main motivations for research into CO_2 conversion were the study of fundamental electrochemistry, and scrubbing CO_2 from the air in submarines or during manned space flight. [7] These are still also relevant today. A Web of Science search using several variations of the phrase "electrochemical reduction of carbon dioxide" reveals that before 1980 only 12 studies were published.[1] There was a flurry of interest in the 1990s, peaking at 32 publications in 1995, but falling again to seven manuscripts in 2005. Interest then began to accelerate again, with 19 papers published in 2010, 114 in 2015 and 217 published in 2017. This gives a total of 1,161 published reports to date.

Electroreduction of CO_2 can result in many different products through multiple different pathways. An emphasis is sometimes placed on the generation of higher hydrocarbons and C_2 products, since these can be used directly as fuels, or in the production of value-added chemicals. However, it is difficult to generate these high value products

---

[1] Full Web of Science search term: "electrochemical reduction of CO2" OR "electroreduction of CO2" OR "electroreduction of carbon dioxide" OR "electrochemical reduction of carbon dioxide", August 2018

Stephen M. Lyth, Platform of Inter/Transdisciplinary Energy Research (Q-PIT); International Institute for Carbon-Neutral Energy Research (I2CNER), Kyushu University; Department of Mechanical Engineering, University of Sheffield

https://doi.org/10.1515/9783110665147-023

electrochemically with high yield or selectivity. There will usually be several different compounds in the product feed, which will then have to be separated before use. This adds cost to the system and reduces the overall efficiency. Thus, in commercial applications, a major consideration when designing systems is high selectivity, [8] removing the need for added separation steps, simplifying the system. Carbon monoxide can be generated electrochemically with selectivity approaching 100%, on a variety of different cathode catalysts. [9–11] As such, this could be one of the most technologically relevant products of the CO2RR, and it is indeed a highly valued commodity. As a simple precursor, it can be converted into liquid fuels and other useful products such as plastics, with high efficiency, using the Fischer–Tropsch process. Some of these processes are outlined in more detail in the other sections of this book.

In this chapter, we will focus specifically on the generation of carbon monoxide as the major product of CO2RR on different types of cathode catalyst. Over the years, a host of different metals have been investigated for CO evolution under different conditions. Most research has been published on metal cathode catalysts. In aqueous electrolytes, gold (Au), silver (Ag), zinc (Zn), palladium (Pd) and gallium (Ga) are the main CO-producing metals. Electroreduction of CO_2 on copper (Cu) produces hydrocarbons such as methane, ethylene, alcohols and aldehydes, but CO plays a crucial role as an adsorbed intermediate species. Nickel (Ni) and platinum (Pt) also generate CO molecules during the CO2RR, but the CO binds strongly to the surface, preventing further reduction to hydrocarbons and promoting the hydrogen evolution reaction (HER). In non-aqueous electrolytes, cadmium (Cd), tin (Sn) and indium (In) switch from formate to CO generation due to a lack of proton availability. Other metals do not primarily form CO. The order of selectivity of metals for the CO2RR to CO has been estimated to be Au > Ag > Cu > Zn $\gg$ Cd > Sn > In > Pb > Tl $\approx$ Hg. [12–14] Here we will focus on Au and Ag as the most effective metal electrocatalysts for CO generation. Then we will move on to more recently discovered metal-free- and transition-metal-containing nitrogen-doped carbons as novel and exciting alternative catalysts for the CO2RR to CO.

One of the common problems faced when designing catalysts with high selectivity for the CO2RR is the competing hydrogen evolution reaction (HER), which occurs in the same potential range. Therefore, some of the best catalysts are the ones that suppress HER, allowing the CO2RR to dominate. This can be an issue in aqueous electrolytes, and especially so in acidic electrolytes with high proton availability. As such, much of the early work on the CO2RR was performed in non-aqueous organic electrolytes, [12, 15] and modern research is generally performed in basic-to-neutral aqueous electrolytes. However, another solution is to allow the HER to proceed in tandem with the CO2RR. Mixtures of H_2 and CO are known as syngas, which can be used directly in the Fischer–Tropsch reaction without the requirement of an additional source of hydrogen.

Several groups have performed technoeconomic analysis for electrochemical CO_2 conversion, usually focussing on selective reduction to CO in combination with the Fischer–Tropsch reaction. In 2013, Jhong et al. performed a cost analysis on the generation of CO via electrolysis. [16] Critically, they found that the cost would be relatively high and highly dependent on the operating current density. According to their estimates, at current densities below 250 mA/cm² the process would not be cost-effective (Figure 23.1a). At that time, this target was significantly higher than the state-of-the-art catalysts could achieve, but in recent years this value has been reached in several systems (notably by the same group). [17–19] In 2015, Dimitriou et al. performed an analysis of the economic viability of commercially proven carbon dioxide utilisation technologies for transport fuels from biogas (i.e. not including electrochemical methods). All of their scenarios included conversion from CO_2 to CO using the reverse water-gas shift reaction, followed by conversion to diesel and gasoline via Fischer–Tropsch synthesis. They concluded that the production of liquid hydrocarbon fuels using existing technologies at the time was not economically feasible, partly due to low conversion efficiencies. [20] It was suggested that future research should aim at developing novel CO_2 to CO conversion technologies. Electrochemical conversion could be a candidate. In 2016, Li et al. performed a gross-margin technoeconomic study focussing on electrochemical conversion of CO_2 to synthetic fuels. [21] In this case, the source of CO_2 was a 500 MW fossil fuel plant, and electroreduction of CO_2 to CO was assumed to be followed by conversion to diesel via the Fischer–Tropsch process. Electrochemically, this model considered the cell operating voltage, the current density, the Faradaic efficiency and catalyst durability. They concluded that CO and HCOOH are the most economically feasible products, and that liquid fuels could be produced for prices ranging from 3.80 to 9.20 USD per gallon, mainly depending on conversion efficiency. One of the key targets emerging from this study was the requirement of low-cost catalysts but with high electroreduction current densities above 140 mA/cm² (Figure 23.1b). The above studies suggest that electroreduction of CO_2 to

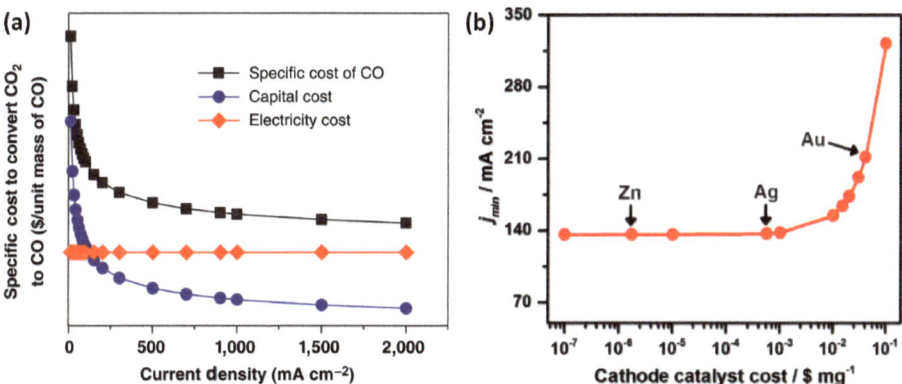

Figure 23.1: (a) Dependence of CO_2 to CO conversion cost on the partial CO evolution current density. Reprinted from [16] with permission from Elsevier. (b) Minimum operating current density required for CO generation as a function of cathode catalyst cost (at 100% Faradaic efficiency). Reprinted with permission from [21]. Copyright 2016 American Chemical Society.

CO could be economically feasible, provided cheap, durable and efficient catalysts can be found that can operate at high current density.

The concept of CO_2 electrolysis is similar to that of water electrolysis. The basic requirements are an electrolyte for ion transport, a cathode for the electroreduction of CO_2 and an anode for the counter-reaction (e.g. the oxygen evolution reaction). While water electrolysis is an established technology applied at industrial scale [22], the electrochemical conversion of CO_2 is not yet technologically mature. To catch up, new advances are needed. The thermodynamically determined reduction potential for the CO_2 to CO half-reaction is −0.52 V versus the standard hydrogen electrode (SHE), at neutral pH. However, much more negative potentials are required experimentally for the onset of CO2RR. The main reason for this is the intermediate species formed when CO_2 accepts an electron. The reduction potential for this electron transfer step is much more negative, at −1.9 V versus SHE. For the CO2RR to proceed at room temperature and at a reasonable cell voltage, a catalyst is therefore required to lower the energy barrier for the formation of intermediate species. The difference between the thermodynamic reduction potential and the experimentally observed onset potential is known as the overvoltage. The overvoltage is directly related to the cell efficiency, so minimising the overvoltage is crucial. The search for and optimisation of suitable catalysts to minimise the overpotential takes up the majority of studies in the literature.

The pH of the electrolyte is an important factor in the performance of electrochemical CO_2 conversion cells. The CO2RR potential does not vary significantly with pH, but the potential of the HER does. [12] As such, the HER is dominant in acidic solutions. Meanwhile, basic solutions can be problematic for CO_2. For example, CO_2 will react with KOH to form potassium bicarbonate, so studies that use KOH often use flowing electrolyte to avoid this problem. [18] More generally, the majority of studies are limited to pH neutral electrolytes such as $KHCO_3$. This is convenient for aqueous cells in the lab, but less so for industrially relevant membrane-based systems.

The efficiency of CO generation and the selectivity can also depend upon operational factors such as current density, applied potential, temperature, porosity, mass diffusion and unoptimised cell design. As such, electroreduction of CO_2 is a process that is extremely complex, and there is still much to discover. Multiphysics simulations have revealed the importance of improving CO_2 transport in the gas diffusion electrodes (GDEs) of electrochemical cells: the limited effects of CO_2 concentration in the feed, the CO_2 flow rate and the channel length. [23] As such there is a great deal of optimisation for this system to be performed before CO generation from the CO2RR can be industrially viable. The research community could learn, for example, from the more technologically mature fields of fuel cells and water electrolysis. Herein, recent progress on the CO2RR to carbon monoxide will be discussed in the context of different catalyst materials. This review will be limited to the materials with the highest current densities and Faradaic efficiencies, namely, gold-, silver- and nitrogen-doped carbons.

23.2 Gold

In early studies, gold emerged as one of the best cathode catalyst materials for conversion of CO$_2$ to CO with high selectivity and is considered to be the most active metal for this reaction. In 1987, Hori et al. investigated the CO2RR at a flat gold electrode in 0.5 M KHCO$_3$. [24] The onset of the reaction ranged from −0.8 to −1.2 V versus SHE, the maximum Faradaic efficiency for CO formation was 91% at −1.10 V versus SHE and the partial current density for CO evolution was 10 mA cm^{-2} (Figure 23.2a). These results were highly impressive at the time. The early development of gold cathodes for CO2RR has been described in several review articles. [12, 25] Herein we focus on more recent developments in the field.

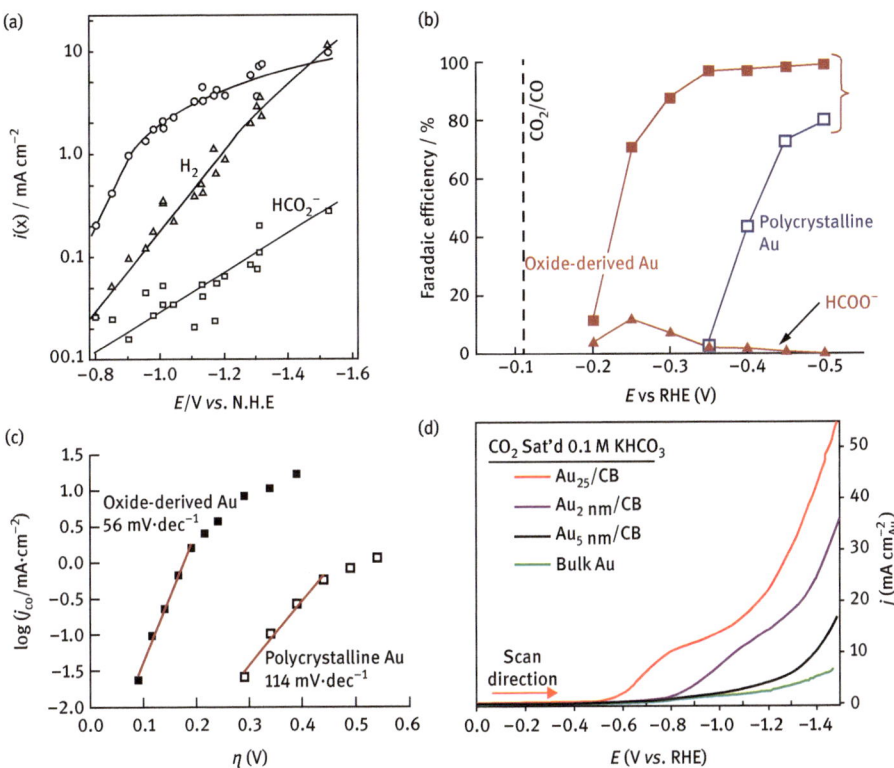

Figure 23.2: (a) Partial current densities of CO$_2$ electroreduction to CO, H$_2$ and formate on bulk gold electrodes. Republished with permission of the Royal Society of Chemistry from [24]. (b) Faradaic efficiencies and (c) partial current densities for the CO2RR on oxide-derived Au nanoparticles. Reprinted (adapted) with permission from [26]. Copyright 2012 American Chemical Society. (d) Total current densities for CO$_2$ reduction on 25 atom Au clusters (at 100% Faradaic efficiency). Reprinted (adapted) with permission from [27]. Copyright 2012 American Chemical Society.

In 2012, Chen et al. moved away from flat bulk electrodes and investigated the electroreduction of CO_2 on Au nanoparticles, in aqueous 0.5 M $NaHCO_3$. [26] These were obtained by electrochemically reducing thick Au oxide films. Selective reduction to CO was achieved compared with conventional flat gold electrodes, with an onset potential of −0.2 V versus RHE, and a maximum Faradaic efficiency for CO formation of 96% at −0.50 V versus RHE (Figure 23.2b-c). The partial current density for CO generation ranged from 2 to 4 mA/cm^2. Meanwhile, the activity was retained for at least 8 h. They attributed their performance to increased stabilisation of the intermediate on the roughened surface of the oxide-derived Au nanoparticles. However, the current densities are quite low, and some nanoparticle sintering was observed even after just 15 min of electroreduction.

The same year, Kauffman et al. performed an experimental and computational investigation of Au clusters containing 25 atoms for the CO2RR. [27] This system was found to promote CO formation with 100% Faradaic efficiency at −1.0 V versus RHE in 0.1 M $KHCO_3$. The onset potential was around −0.45 V versus RHE. The partial current density for CO was >50 mA/cm^2 at −1.4 V versus RHE, representing a significant leap compared to previous literature (Figure 23.2d). Density functional theory (DFT) indicated that this originated from a redistribution of charge within the cluster upon CO_2 adsorption, lowering the energy barrier for intermediate adsorption. In 2015, the same group performed an important technoeconomic study of the application of such gold clusters as a viable solution for large-scale conversion of CO_2 using an electrochemical reactor powered by photovoltaic energy. [10] They calculated that their model system would convert up to 800 L of CO_2 per gram of Au per hour, corresponding to conversion of 0.8–1.6 kg of CO_2 per gram of Au per hour, and made the important point that any electrochemical CO_2 conversion system will produce net CO_2 emissions if it does not integrate with renewable energy.

In 2013, Zhu et al. investigated the CO2RR on Au nanoparticles in 0.5 M $KHCO_3$. [28] They showed that the catalytic activity depended on the nanoparticle size, with 8 nm nanoparticles showing the best performance: an onset potential of −0.38 V versus RHE, and a maximum Faradaic efficiency of 90% for CO generation at −0.67 V versus RHE (Figure 23.3a-b). The maximum mass activity was 14 A/g, but neither the partial current density nor the electrode area was provided, making it impossible to convert to current density for comparison. DFT calculations suggested that the high activity was due to a greater proportion of edge sites, facilitating adsorption of the reaction intermediate and subsequent formation of CO. They further improved the activity by embedding the nanoparticles in butyl-3-methylimidazolium hexafluorophosphate, increasing the Faradaic efficiency to 97% at −0.52 V versus RHE.

Hall et al. took the innovative step of synthesising mesoporous Au-inverse opals for the CO2RR in 2015. [29] For optimised electrodes in 0.1 M $KHCO_3$ electrolyte, the partial current density for CO generation was rather low at 22 $\mu A\ cm^{-2}$ at −0.4 V versus RHE, dropping to 7 and 3 $\mu A\ cm^{-2}$ for 1.6 and 2.7 μm thick films, respectively. The highest Faradaic efficiency was 75% for the thickest films, which is also rather low.

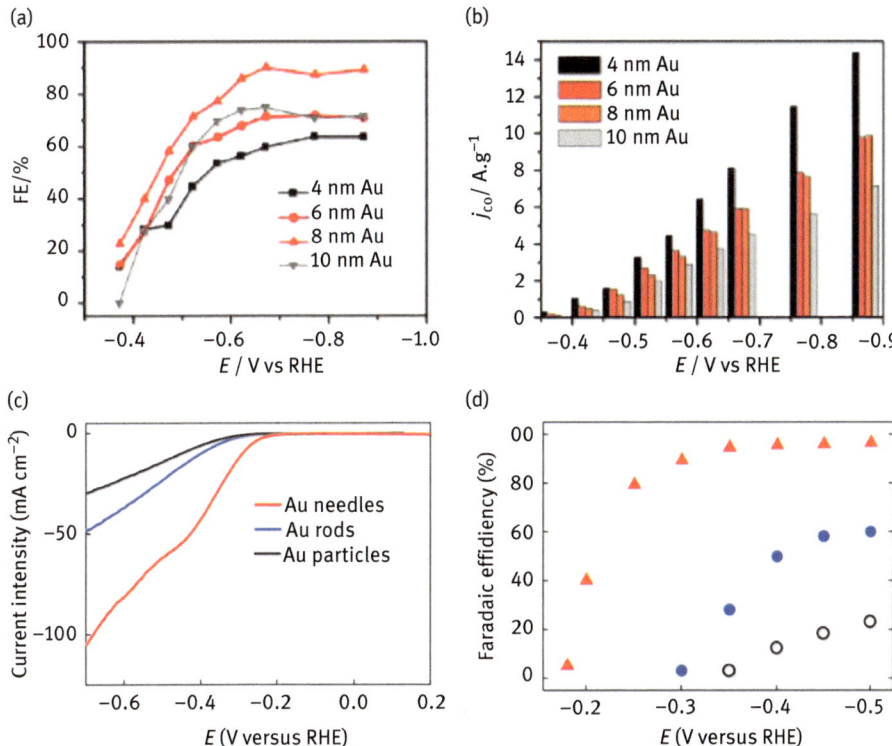

Figure 23.3: (a) Faradaic efficiency and (b) partial current density of CO generation on Au nanoparticles. Reprinted with permission from [28]. Copyright 2013 American Chemical Society. (c) Total current density and (d) Faradaic efficiency for the CO2RR on nanoscale Au needles compared with Au rods, and conventional Au nanoparticles. Reprinted with permission from Springer Nature [31], 2016.

Interestingly, they found that the HER was suppressed by up to 10-fold as the film thickness increased (while CO generation was largely unaffected). This led them to speculate that the HER is suppressed due to local changes in pH and the generation of diffusion gradients inside the Au mesopores.

It is well known that sharp metallic objects result in field enhancement at the tip under an applied electric field. [30] However, this concept has not been deeply explored in electrochemistry. In 2016, Liu et al. reported the novel use of electrical field enhancement to improve the CO2RR activity of Au. [31] Nanoscale gold needles were prepared via electrodeposition, and field enhancement concentrated K$^+$ ions at the active sites, enhancing the CO2RR activity (Figure 23.3c-d). This resulted in low onset potentials of around -0.2 V versus RHE, and high partial current density for CO generation (e.g. 22 mA cm^{-2} at -0.35 V vs RHE), with a Faradaic efficiency of >90%. They later expanded this work to grow fractal-like secondary and tertiary Au structures on the initial nanoscale needles, further enhancing the activity to 38 mA cm^{-2} with a maximum Faradaic efficiency of 95% for CO at an overpotential of 0.29 V versus RHE. [32] This may be due, for example, to the giant multistage field enhancement effect. [33]

In 2014, Kim et al. investigated Au–Cu bimetallic nanoparticles with different compositions for the electroreduction of CO_2 to CO. [34] Introducing Cu into the Au nanoparticles leads to slightly decreased CO Faradaic efficiency (~60%), increased HER and around 3% formate generation. However, the mass activity for CO generation was significantly enhanced (Figure 23.4a-c), which was simply attributed to decreasing the required Au content. Later, the same group investigated Au nanoparticles supported on TiC supports. [35] Here, the authors reported enhanced partial current density from 1 to 3 mA/cm^2 by using the TiC support (Figure 23.4d-e), which attributed to the strong metal–support interaction. This interaction was responsible for a shift in the electronic structure of the Au nanoparticles, in turn affecting the binding energy of the intermediate adsorbates.

In 2017, Jhong et al. reported a novel Au electrocatalyst for the CO2RR. [36] Multiwalled carbon nanotubes (MWNTs) were wrapped with a polymer (pyridine-polybenzimidazole, PyPBI) in order to provide anchoring points for the nucleation and attachment of Au nanoparticles. Their MWNT/PyPBI/Au catalyst exhibited high activity for CO generation in 1 M KCl, with impressive partial current density of 160 mA/cm^2, high Faradaic efficiency (>90%) and an onset potential of –1.1 V versus Ag/AgCl (Figure 23.5a-b). This high activity was attributed to the large electrochemically active surface area of the gold nanoparticles (23 m^2/g), as well as the highly dispersed, low catalyst loadings that were achieved (just 0.17 mg/cm^2). They performed stability measurements at constant potential of –1.6 V versus Ag/AgCl for 26 h. Rather than experiencing activity loss, the catalysts actually improved in performance by 12% during this test. In 2018, the same group investigated these MWNT/PyPBI/Au catalysts in an alkaline flow electrolysis cell with 2 M KOH electrolyte. [18] They obtained a very high partial current density of 200 mA/cm^2 at –0.7 V versus RHE, with an onset potential of –0.02 V versus RHE, stable for at least 8 h (Figure 23.5c). This work represented a significant step forward in the use of Au catalysts in the CO2RR, bringing technological applications a step closer.

Shi et al. investigated Au electrocatalysts in a range of different electrolytes, with a view to increasing the applicability of the CO2RR to CO at industrial scale. [37] Since CO_2 is a non-polar molecule, it has higher solubility in non-polar organic solvents. However, the CO2RR to CO produces water, and the effect of this should be considered. In 2017, they determined that tetrabutylammonium perchlorate (Bu$_4$NClO$_4$) mixed with propylene carbonate (PC) was the highest performing electrolyte, and that the presence of small amounts of water improved conductivity, decreased viscosity and enhanced CO_2 solubility. Meanwhile, the organic electrolyte was able to separate out the generated water at high concentrations (>6.8 wt%). They claimed Faradaic efficiencies of 84% at –2.6 V versus RHE, onset potentials of –1.2 V versus RHE (although these values seem unusual and may have mistakenly have converted to Ag/AgCl or SCE instead of RHE) and partial current density of 6.3 mA/cm^2, with no degradation after 4 h (Figure 23.6).

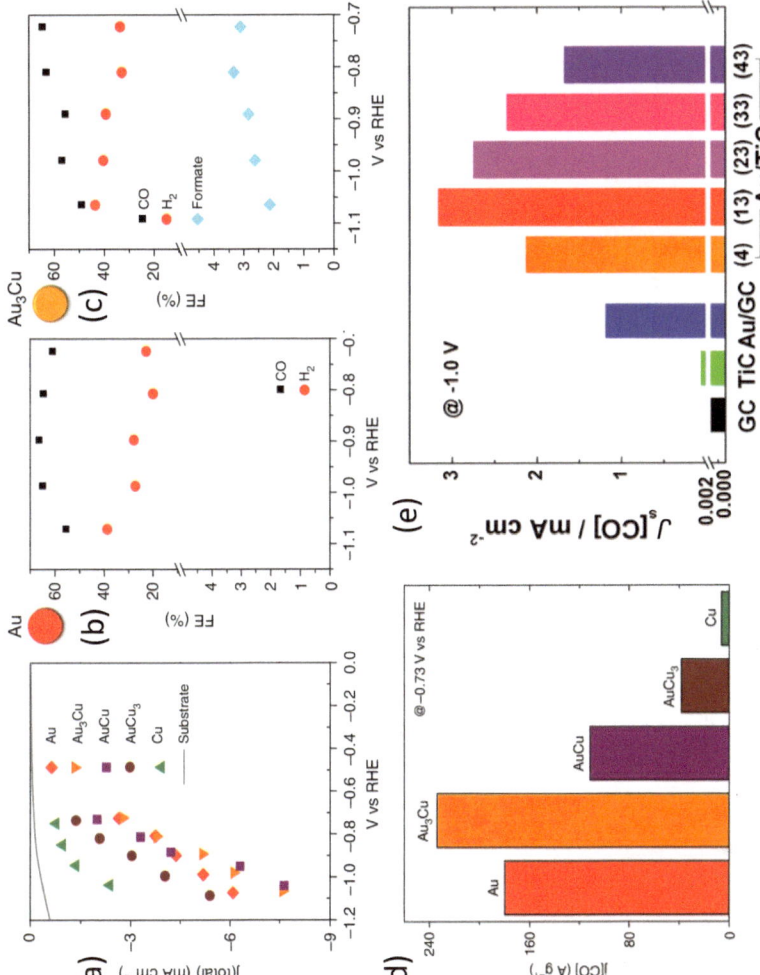

Figure 23.4: (a) Total current density, (b–c) Faradaic efficiencies, and (d) mass activity for the CO2RR on Au–Cu bimetallic nanoparticles, compared with Au nanoparticles. Reprinted with permission from Springer Nature [34], 2014. (e) Partial CO current density of Au nanoparticles supported on TiC. Reprinted with permission from [35]. Copyright 2017 American Chemical Society.

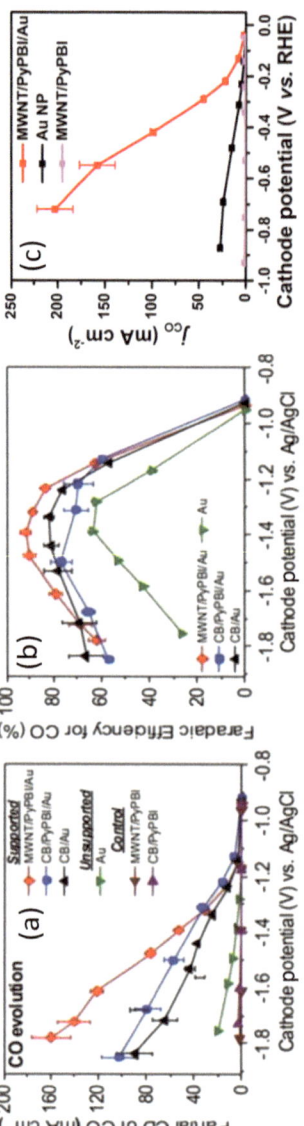

Figure 23.5: Performance of MWNT/PyPBI/Au electrocatalysts for the CO2RR: (a) partial CO current densities and (b) Faradaic efficiencies for CO formation in 1 M KCl. Published with permission from [36]. Copyright 2017 Wiley-VCH Verlag GmbH & Co. KGaA, Weinheim. (c) Partial current density of the same catalyst system in 2 M KOH. Reprinted with permission from [18].Copyright 2017 American Chemical Society.

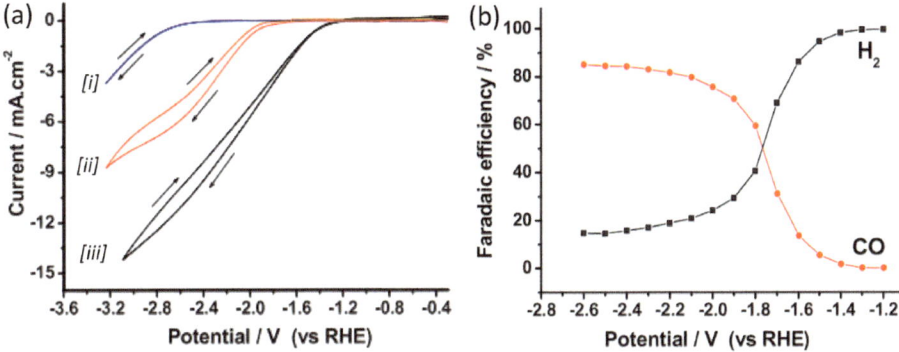

Figure 23.6: (a) Total CO2RR current density in [i] Ar-saturated Bu₄NClO₄/PC; [ii] CO₂-saturated Bu₄NClO₄/PC and [iii]: CO₂-saturated Bu₄NClO₄/PC + 6.8 wt% of H₂O. (b) Faradaic efficiency for CO and H₂ formation in CO₂-saturated Bu₄NClO₄/PC containing 6.8 wt% H₂O. Reprinted from [37] with permission from Elsevier.

Finally, Jin et al. investigated the CO2RR activity of ultrasmall Au nanoparticles on nitrogen-doped carbon in 2018. [38] The highest activity was achieved for 1.9 nm Au nanoparticles in 0.5 NaHCO₃ electrolyte, with a maximum Faradaic efficiency of 83% (at −0.73 V vs RHE) and an onset potential of −0.18 V versus RHE. This high activity was attributed to the large surface to mass ratio, and the fact that the nitrogen-doped carbon support acts as a Lewis base, increasing the surface charge density of the Au nanoparticles. The current density was not directly reported, but given the mass activity of 967 A/g and the catalyst loading of 0.2 mg/cm², it is estimated to be 193 mA/cm².

The above studies have made huge progress in improving the CO2RR activity of Au by making nanoparticles, by optimising nanoparticle size, by alloying, by decorating the nanoparticles on suitable support structures and by optimising the electrolyte and cell design. The current density of CO generation has increased from a few mA/cm² to over 200 mA/cm² at reasonable voltage, bringing about the possibility of real-world applications.

23.3 Silver

Silver has emerged as one of the most useful catalysts for electrochemical CO₂ conversion, balancing high selectivity for CO generation with lower cost than gold. Hori et al. showed in 1985 that electroreduction of CO₂ over Ag cathodes primarily results in CO in aqueous KHCO₃, with a Faradaic efficiency as high as 90% at −1.45 V versus SHE. [39] Later, the same group investigated the different interactions of CO₂ on Ag(111), Ag(100) and Ag(110) crystal faces in 0.1 M KHCO₃ (Figure 23.7a). [40] All of these surfaces produced CO as the main product, and Ag(110) was the most active with a partial current density of 12 mA/cm² at –1.6 V versus SHE. In one of the earliest attempts to fabricate an industrially relevant practical electrochemical CO_2 conversion cell (in 2003), the same group coated cation exchange membranes with silver, using electroless deposition. [41] These operated successfully for more than 2 h, but the reaction was mass diffusion limited due to slow transport of CO_2 through the silver electrode. They resolved this by increasing the porosity of the silver layer via ultrasonication. The partial current density for CO production achieved using this cell was 60 mA/cm², which was three times previously reported values. The Faradaic efficiency was up to 92% at –1.3 V versus SHE. This early important work by Hori et al. built the foundations for many later studies. Herein, we will focus on some more recent progress.

Delacourt et al. developed an electrochemical polymer electrolyte membrane cell for generating syngas from CO_2, using an Ag catalyst, in 2008. [42] To facilitate charge transfer between the Nafion membrane and the cathode, a buffer layer of $KHCO_3$ was used, significantly improving the performance (but increasing the complexity of cell design). The partial current density for CO generation was 30 mA/cm², at 1.7 V versus SCE, and the maximum Faradaic efficiency was 83% (Figure 23.7b). The onset potential was around –1.2 V versus SCE. One of the main reasons for this high performance was the use of GDEs, enhancing transport of CO_2 to the catalyst surface. Using this design, they also tailored the CO to H_2 ratio to match that required for methanol synthesis in the Fischer–Tropsch reaction (i.e. CO:H_2 = 1:2), at –2 V versus SCE. They investigated the differences in stability of the cell with various different catalyst configurations (up to 8 h), with an 80% unsupported Ag/20% acetylene black mix giving the best results.

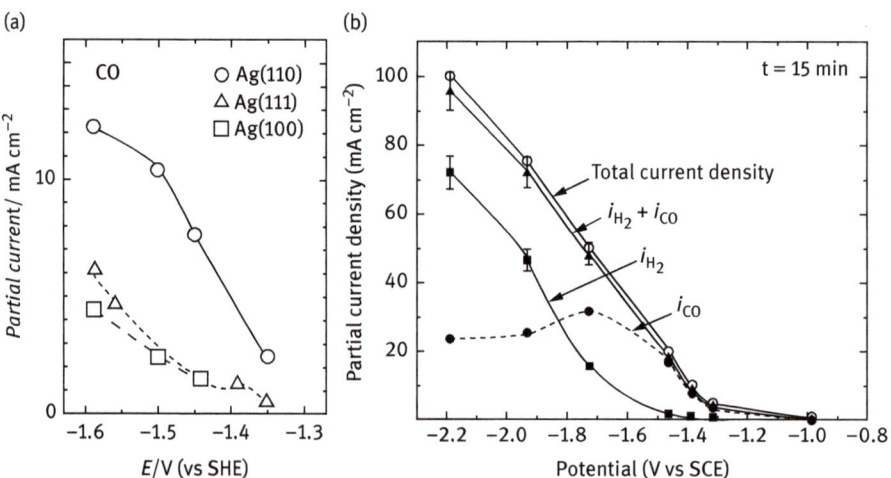

Figure 23.7: (a) Partial current density for CO formation on the different crystal facets of silver. Reprinted from [40] with permission from Elsevier. (b) Partial current density in a membrane reactor with different Ag catalyst configurations. Reprinted with permission from [42].

In 2011, Rosen et al. succeeded in achieving significantly enhanced CO_2 conversion activity on Ag electrodes by utilising an ionic liquid electrolyte. [43] The ionic liquid 1-ethyl-3-methylimidazolium tetrafluoroborate (EMIM-BF_4) was selected. The onset potential of CO generation was at a cell voltage of 1.5 V, compared to 2.1 V in the absence of ionic liquid, with a cell overpotential of just 0.17 V. The Faradaic efficiency was 96%, and the system operated for over 7 h. However, the cathode current density was not reported. The high activity was attributed to the EMIM-BF_4 ionic liquid which enhanced the ionic conductivity and CO_2 solubility, as well as forming a complex with CO_2, reducing the free energy barrier to CO formation, and essentially acting as a co-catalyst. The same group investigated the effect of electrolyte on the CO2RR in 2012. [44] The presence of large cations such as cesium and rubidium was reported to improve the partial CO current density and suppress the HER, resulting in high Faradaic efficiency. With Cs, a partial current density of 72 mA/cm² was obtained at −1.62 V versus Ag/AgCl. For Na, the partial current density was only 49 mA/cm² at the much more negative potential of −2.37 V versus Ag/AgCl. The effect of cation size on CO selectivity was explained by the interplay between cation hydration and the extent of cation adsorption on the Ag electrodes. The onset potential was −1.24 V versus Ag/AgCl and the maximum Faradaic efficiency was almost 100% (Figure 23.8c-d).

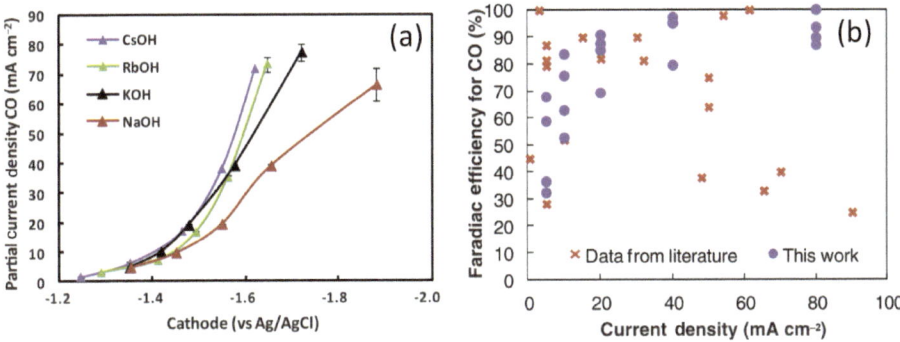

Figure 23.8: (a) Partial current density and (b) Faradaic efficiency of CO generation on Ag catalysts with different counterions in the electrolyte. Reprinted with permission from [44].

In 2014, Ma et al. developed titania-supported Ag nanoparticles for the CO2RR in 1 M KOH. [45] Compared with carbon black supports under the same conditions, the Ag/TiO_2 cathode displayed significantly enhanced activity. The partial current density was over 100 mA/cm² at −1.8 V versus Ag/AgCl, the onset potential was −1.3 V versus Ag/AgCl, and the Faradaic efficiency was over 90% (Figure 23.9a). This enhanced activity was attributed to (i) improved distribution and small size of Ag nanoparticles on the TiO_2 support; (ii) stabilisation of the reaction intermediate on

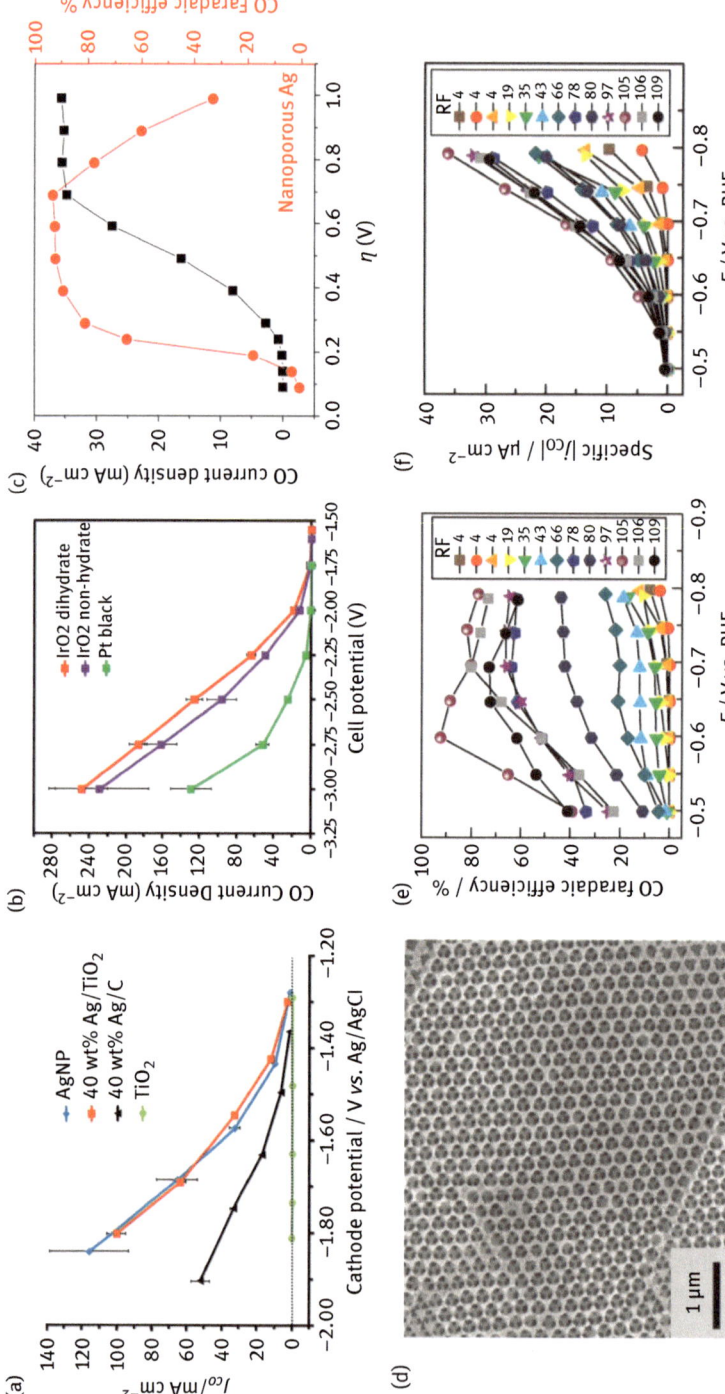

Figure 23.9: (a) Partial current density for CO generation on TiO$_2$-supported Ag nanoparticles. Published with permission from [45]. Copyright 2017 Wiley-VCH Verlag GmbH & Co. KGaA, Weinheim. (b) CO partial current density for different anode catalysts with an Ag cathode catalyst. Reprinted with permission from [46]. (c) Partial current density and CO Faradaic efficiency versus overpotential for mesoporous Ag nanoparticles. Reprinted with permission from Springer Nature [47], 2014. (d) Scanning electron microscopy image showing the porous structure of the inversion opal electrode. Graphs showing the variation in (e) CO Faradaic efficiency and (f) partial CO current density with roughness factor (RF). Published with permission from [48]. Copyright 2016 Wiley-VCH Verlag GmbH & Co. KGaA, Weinheim.

TiO$_2$ and (iii) TiO$_2$ serving as a redox electron carrier (effectively acting as a co-catalyst in a similar mechanism to the previously reported ionic liquid). In the same year, the same group optimised the anode catalyst of their Ag-based CO$_2$ conversion flow cell. By using the dihydrate form of IrO$_2$ in KOH, they achieved partial current density for CO generation of an astonishing 250 mA cm^{-2}, surpassing the target of their earlier technoeconomic analysis. [16] Even after performing 200 potential cycles, the cell retained 90% of its initial activity (Figure 23.9b). This work showed how important it is to optimise the whole system, rather than focussing on a single component such as the catalyst. [46]

Lu et al. reported a "nanoporous" Ag electrocatalyst able to electrochemically reduce CO$_2$ to CO in 0.5 M KHCO$_3$ electrolyte with 93.1% Faradaic efficiency and 34.7 mA/cm^2 partial current density, at a potential of -0.8 V versus RHE (Figure 23.9c). [47] Both the particle size and the pore size were around 100 nm. The high activity was attributed to enhanced electrochemical surface area (26.5 m^2/g) of the porous cathode, as well as an improvement in the intrinsic activity of their material by virtue of facile intermediate adsorption on the high negative curvature within the mesopores. In 2016, Yoon et al. investigated the CO2RR on mesoporous Ag-inverse opal electrodes, with a pore size of around 200 nm. [48] These were fabricated via an electroplating technique incorporating a polystyrene sacrificial template. They tuned the surface roughness by varying the deposition potential. The electrodes were measured in 0.1 M KHCO$_3$ electrolyte, but a very low partial current density of 35 µA/cm^2 was obtained. The variation in roughness resulted in a wide range in CO selectivity, from 5 to >80% at −0.8 V versus RHE. This trend was attributed to mesopore-induced transport limitations, resulting in local pH variations (Figure 23.9d-e).

In order to optimise the composition and structure of CO$_2$ conversion cells, Kim et al. investigated the effects of varying the hydrophobicity of GDEs in 1 M KOH (Figure 23.10a). [17] They concluded that electrolyte flooding must be minimised in order to improve gas permeability and accessibility to the Ag catalyst. The optimal hydrophobicity was achieved when 20% polytetrafluoroethylene (PTFE) was incorporated into the microporous layer of the cathode, and 10% PTFE was incorporated into the carbon fibre support. The partial current density achieved using these optimised electrodes reached an impressive 280 mA/cm^2 at a cathode potential of -2.2 V versus Ag/AgCl, with Faradaic efficiencies consistently over 90%, and onset potentials of −1.3 V versus Ag/AgCl. No deterioration in performance was observed after 4 h of operation.

In 2016, Ma et al. investigated the effect of incorporating MWNTs with Ag catalyst layers (Figure 23.10b). [19] In one case, they mixed Ag nanoparticles with MWNTs to form a homogeneously mixed cathode. In another case, they created a layer of MWNTs and then sprayed a layer of Ag nanoparticles on top. Both cases improved the performance compared to simple Ag nanoparticle cathodes, and the homogeneously mixed version was the best. In a cell operating at −3 V with 1 M KOH electrolyte, CO production was achieved with up to 350 mA cm^{-2} partial current density

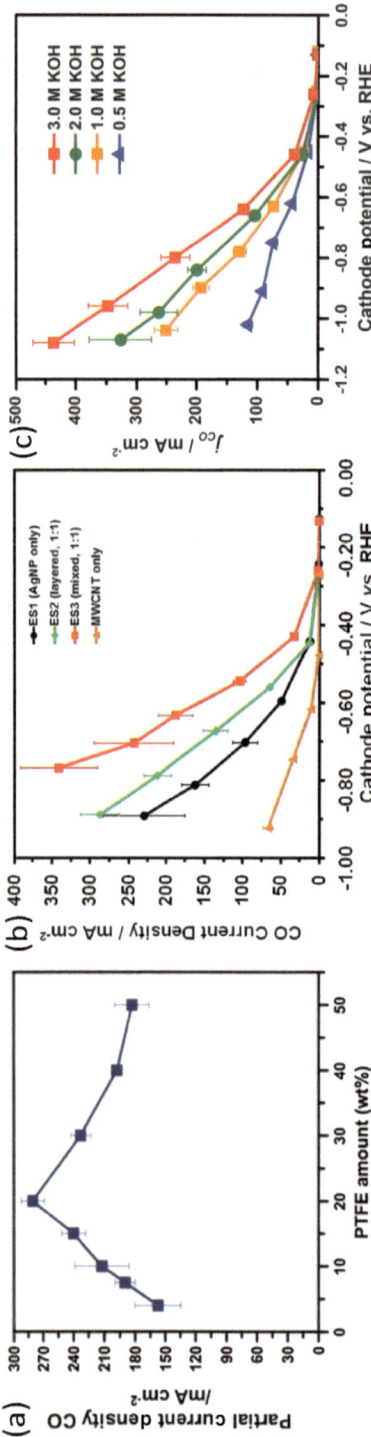

Figure 23.10: (a) Dependence of CO partial current density on the PTFE content in the microporous layer, at a reduction potential of −2.2 V versus Ag/AgCl. Reprinted from [17] with permission from Elsevier. (b) Partial CO current density for silver nanoparticles mixed with carbon nanotubes in different configurations. Reproduced from [19] with permission from The Royal Society of Chemistry. (c) High partial CO current density of Ag nanoparticles in different concentration KOH electrolytes. Reproduced from [11] with permission from The Royal Society of Chemistry.

with high Faradaic efficiency (>95%). Electrochemical impedance spectroscopy revealed that the enhancement could be attributed to lower charge transfer resistance, and this work showed how large improvements can be achieved simply by optimising the cathode catalyst layer.

The effect of electrolyte composition (KOH, KCl and KHCO$_3$) and concentration on CO evolution from Ag nanoparticles was investigated by Verma et al. in 2016. [11] By increasing the molarity from 0.5 to 3 M electrolyte solutions, large improvements in activity were achieved (Figure 23.10c). Records of partial current densities for an unprecedented 440 mA/cm^2 were achieved using 3 M KOH, which attributed to improved ion conductivity. The Faradaic efficiencies were close to 100%. The onset potential was found to vary depending on the electrolyte anion, from −0.13, to −0.46, to −0.6 V versus RHE for OH$^-$, HCO$_3^-$ and Cl$^-$, respectively.

We synthesised carbon foam as a novel catalyst support for Ag nanoparticles in 2017. This carbon foam was derived from sodium ethoxide pyrolysed at 600 °C and had large surface area of 900 m^2/g. [49, 50] Silver nanoparticles were decorated onto this scaffold by chemical reduction of silver nitrate. The resulting electrocatalyst was tested for the CO2RR in 1 M KOH, and compared with carbon black and graphene supports. [51] The resulting electrocatalyst displayed a partial CO current density of 120 mA/cm^2 at −2.06 V versus Ag/AgCl, compared with 58 and 30 mA/cm^2 for graphene and carbon black supports, respectively. The onset potential was − 0.145 V versus Ag/AgCl. The maximum CO Faradaic efficiency was 84% at −1.82 V versus Ag/Cl, but this dropped to around 60% at more negative potentials. Meanwhile, the maximum CO Faradaic efficiency for the carbon black support was just 25%, with the HER being dominant. The maximum Faradaic efficiency on graphene was also quite low, at 60%. These results showed that the nature of the support can have a significant impact on the selectivity. The high performance of the carbon foam support was attributed to the small size and uniformly distributed nanoparticles resulting from the amorphous and defect-rich nature of the surface. Much larger and poorly distributed Ag particles were observed on the other supports (Figure 23.11). The high proportion of HER on the carbon black and graphene catalyst supports is attributed to the relatively exposed carbon surface.

Very recently, Jeanty et al. demonstrated large-scale conversion of CO$_2$ to CO with Ag-based electrocatalysts, using large area 100 cm^2 GDEs in 0.4 M K$_2$SO$_4$ electrolyte. Using this system, they obtained current densities of 150 mA/cm^2 and Faradaic efficiencies of 60%. Their cell was stable for several hundred hours. [52]

The above studies represent impressive progress in the electrochemical conversion of CO$_2$ to CO. From promising selectivity but with low partial current densities (e.g. 10 mA/cm^2) in early studies, cells are now being produced with CO outputs as high as 440 mA/cm^2 at 100% Faradaic efficiency, well in the range of industrially relevant rates. Key milestones have been the development of suitable nanostructure in the Ag catalyst to maximise the electrochemically active surface area; the use of catalyst supports to improve electronic conductivity and gas diffusion; and optimisation

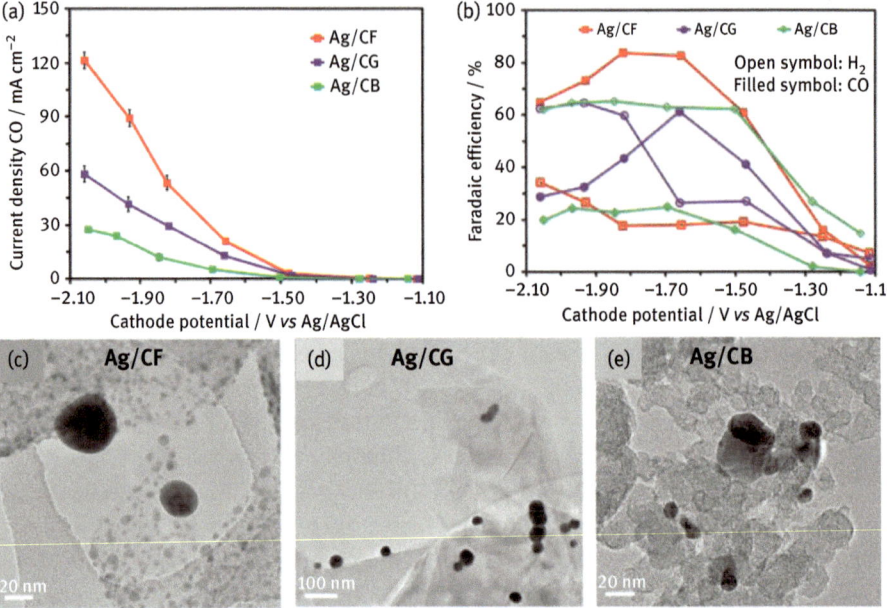

Figure 23.11: (a) Partial current density and (b) Faradaic efficiency of CO generation on carbon foam, graphene and carbon black supports. (c–e) SEM images of Ag nanoparticles on the three different supports. Reproduced with permission from [51].

of the electrochemical cell components and structure. Huge increases in current density have been made through quite simple engineering and optimisation, and hopefully similar improvements can be seen in the near future through further innovations.

23.4 Metal-free and transition metal-doped nitrogen-doped carbons

Until recently, bulk metals and metal nanoparticles have dominated the research landscape of heterogeneous electrocatalysts for the CO2RR. However, moving away from metal nanoparticles, electrocatalysts could have several advantages. In some cases, the metal catalysts used for CO production are relatively expensive (e.g. Ag and Au), limiting their usefulness at an industrial scale. Another issue is durability and stability – metal nanoparticles can undergo electrochemical degradation processes such as support detachment, ripening, dissolution, sintering and agglomeration. [53] Few long-term durability tests beyond 8 h have been reported for the CO2RR on Au or Ag catalysts. It is thus desirable to search for alternative catalysts. Nitrogen-doped carbons and porphyrin-like transition metal-containing nitrogen-doped carbons have been investigated as Pt-free catalysts for the oxygen reduction reaction (ORR) in fuel cells for decades. Their activity is competitive with conventional platinum catalysts, and they are now beginning to apply commercially. [54–57] Recently, it has become apparent that such electrocatalysts are also relevant to the CO2RR.

The first paper to successfully report the electroreduction of CO$_2$ to CO over nitrogen-doped carbon- catalysts was published in 2013 (Figure 23.12a). [58] Kumar et al. synthesised nominally metal-free nitrogen-doped carbon nanofibers by pyrolysis of electrospun polyacrylonitrile mats at 1050 °C. These were tested in EMIM-BF$_4$ ionic liquid electrolyte to boost the activity. They observed a high CO Faradaic efficiency (98%), and the maximum partial CO current density (4 mA/cm^2) was similar to Ag metal catalysts reported at the time. The high activity was partly attributed to the microstructure of the nanofibers, leading to large surface area and porosity (although these were not quantified in the paper). It was also attributed to enhanced adsorption of reaction intermediates onto the nitrogen-doped carbon surface. The ionic liquid was reported to act as a homogeneous co-catalyst for the reaction, reacting with CO$_2$ to form an EMIM-CO$_2$ complex and lowering the energy barrier to subsequent electron transfer at the nitrogen-doped carbon cathode (Figure 23.12b). As such this is not strictly simply CO$_2$ reduction on nitrogen-doped carbon, but a hybrid heterogeneous/homogeneous system. It should be noted again that the use of ionic liquids is expensive, they are moisture sensitive and they would be difficult to implement in large-scale electrochemical cells.

In 2015, Wu et al. reported electrochemical CO$_2$ conversion to CO on nitrogen-doped carbon nanotubes grown from acetonitrile and dicyandiamide with ferrocene at 850C. [59] These were tested in aqueous 0.1 M KHCO$_3$ electrolyte, and the Faradaic efficiency for CO was 80% at a cell voltage of −0.78 V. The partial current density reached −2.25 mA/cm^2 at −1.7 V (Figure 23.12c-d). No degradation was observed during 10 h of testing. Meanwhile, nitrogen-free pristine carbon nanotubes did not show any activity for CO$_2$ electroreduction. The excellent activity was attributed to the high conductivity of carbon nanotubes, a low free energy barrier for CO$_2$ activation on nitrogen-doped carbon compared to a higher barrier for the HER and facile desorption of the produced CO. They used metal-free DFT calculations to conclude that the site with highest selectivity towards CO production is pyridinic nitrogen, but neglected to consider the possibility of metal-containing active sites arising due to the use of the ferrocene precursor.

In the same year, Sharma et al. synthesised nitrogen-doped carbon nanotube arrays from acetonitrile and dicyandiamide with ferrocene at up to 950 °C, and investigated their use as electrodes in the electroreduction of CO$_2$. [60] Catalysts synthesised at 850 °C displayed the highest activity of 80% Faradaic efficiency at a cell voltage of −1.1 V, and a partial CO current density of 4.5 mA/cm^2 at −1.5 V, in 0.1 M KHCO$_3$ (Figure 23.13a-b). They concluded that the incorporation of graphitic and pyridinic nitrogen significantly decreases the overpotential compared with pristine

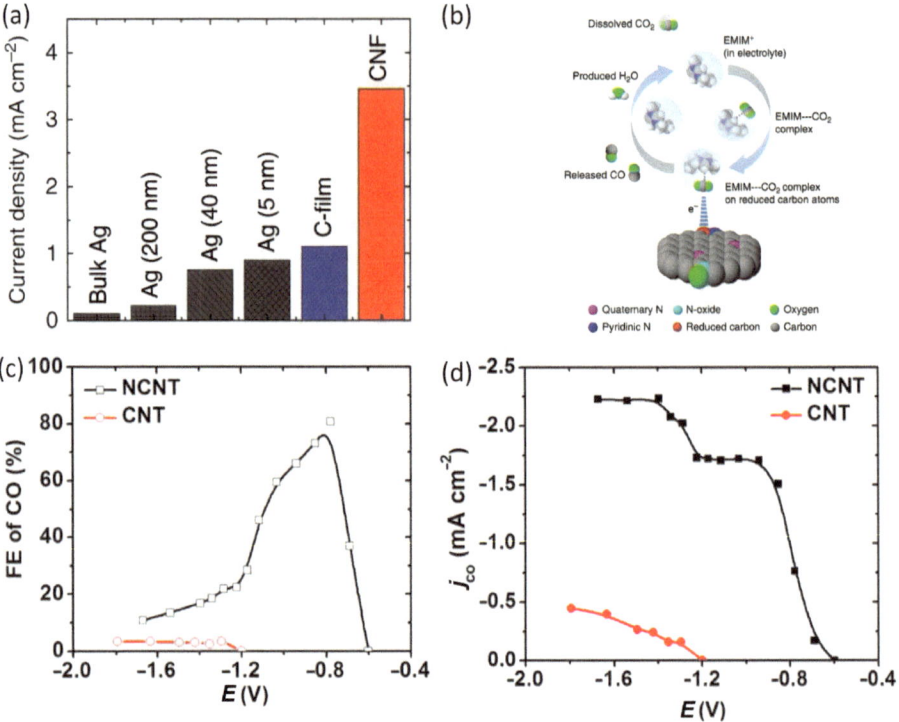

Figure 23.12: (a) Partial current density for CO formation on nitrogen-doped carbon fibres in the presence of EMIM-BF$_4$, and (b) proposed mechanism. Reprinted by permission from Springer Nature [58], 2013. (c) Faradaic CO efficiency and (d) partial current density on nitrogen-doped carbon nanotubes. Reprinted with permission from [59].Copyright 2015 American Chemical Society.

graphene, as well as increasing selectivity. However again, they did not seem to consider possible contamination of active metal sites from their ferrocene precursor.

In 2015, Varela et al. synthesised nitrogen-doped carbons decorated with atomically dispersed metals (Figure 23.13c-d). [61] These were derived from mixtures of metal chloride salts, polyaniline and Ketjen black, pyrolysed at 900 °C. The three samples investigated were Fe–N–C, Mn–N–C, and bimetallic FeMn–N–C (as well as metal-free N–C). The metal loadings varied between 4 and 6 wt%, and the nitrogen content between 6 and 7 wt%. The total current densities were similar for all metal-containing catalysts, reaching a maximum of around 35 mA/cm^2 at −0.95 V versus RHE, with an onset potential of -0.5 V versus RHE. In contrast, the metal-free catalyst produced less than 3 mA/cm^2. The CO selectivity for all the catalysts was similar, starting at around 80% at −0.5 V versus RHE in the low current density region. Unusually, there was a linear decrease in selectivity from −0.5 to −1.0 V versus RHE, resulting in a trade-off between CO selectivity and current density.

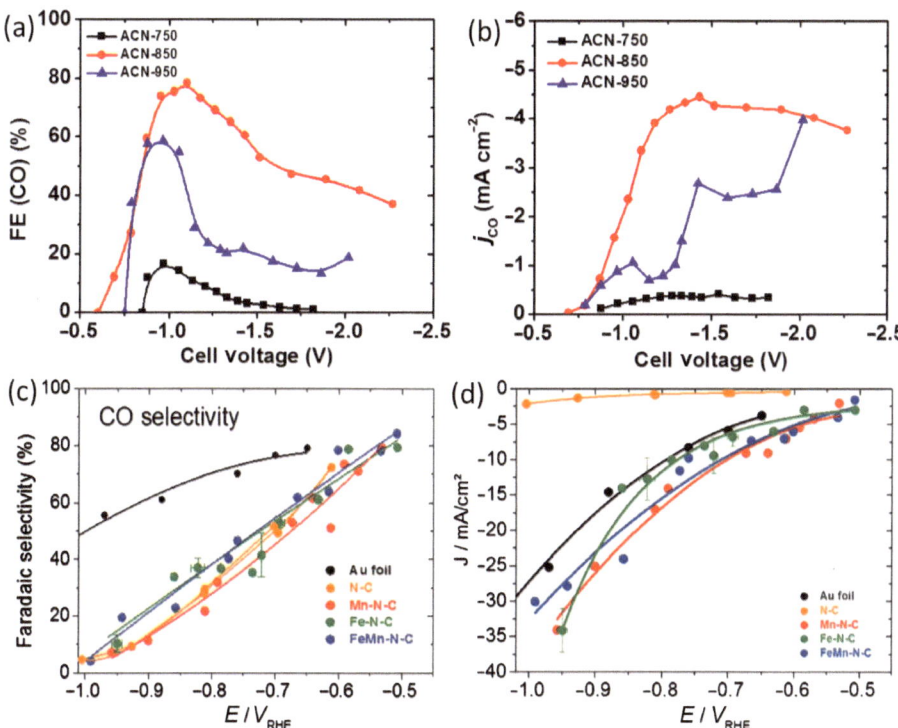

Figure 23.13: Faradaic efficiency and partial CO current densities for (a–b) nitrogen-doped carbon nanotube arrays. Published with permission from [60]. Copyright 2016 Wiley-VCH Verlag GmbH & Co. KGaA, Weinheim. (c-d) Transition-metal-doped nitrogen-doped carbons. Published with permission from [61]. Copyright 2015 Wiley-VCH Verlag GmbH & Co. KGaA, Weinheim.

In 2016, Li et al. reported nominally metal-free sulphur- and sulphur/nitrogen-doped carbon for the electroreduction of CO_2 to CO. [62] Their carbons were derived from pyrolysis of poly(4-styrenesulfonic acid-*co*-maleic acid) sodium salt at 800°C. The surface area was large, at around 1500 m^2/g. However, the maximum Faradaic efficiency achieved for CO generation was just 11.3% and the CO partial current density was just a few mA/cm². This group concluded that pyridinic nitrogen groups actively participate in binding of CO_2, while quaternary nitrogen and thiophenic groups were also assumed to be involved in the reduction process. Positively charged carbon atoms adjacent to pyridinic nitrogen were proposed to stabilise the intermediate, promoting the formation of CO. Surface basicity was also mooted as being important in supressing the HER.

In 2017, we reported highly selective electrochemical CO_2 reduction to CO using MWNTs-coated carbon nitride and pyrolysed at 1,000 °C, as a composite electrocatalyst for the CO2RR (Figure 23.14). [9] This work was carried out in an electrochemical

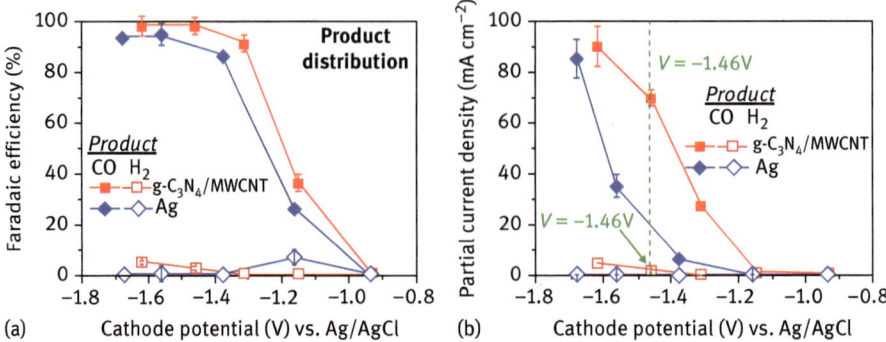

Figure 23.14: Faradaic efficiency and partial current density for CO generation on carbon-nitride-coated carbon nanotubes. [9].

flow cell, with 1 M KOH flowing electrolyte. The selectivity for CO formation was remarkably high at 98% at −1.45 to 1.60 V versus Ag/AgCl (no other products apart from H_2 were detected). Meanwhile, the partial current density for CO formation reached 90 mA cm^{-2} at −1.6 V versus Ag/AgCl, and the onset potential was around −1.2 V versus Ag/AgCl. The activity was attributed to the high conductivity of the underlying MWNTs and the reasonably high surface area of 250 m^2/g. This work represented a major leap in the performance of nitrogen-doped carbons for the CO2RR, and represents the highest performance reported to date. However, it was not clear if the performance was due to metal-free nitrogen-doped carbon sites, or due to metal contamination from the synthesis of the carbon nanotubes.

In 2017, Huan et al. synthesised Fe–N–C catalysts derived from blends of a Zn-based zeolitic imidazolate framework (ZIF-8), ferrous acetate and phenanthroline, and flash pyrolysed at 1,050 °C. [63] They concluded that the selectivity of CO formation over H_2 is determined by the ratio of isolated FeN_4 sites versus Fe-based nanoparticles. Meanwhile, Fe nanoparticles mainly catalysed the HER. Materials containing only FeN_4 sites reduced CO_2 to CO in 1 M KHCO$_3$, with an onset potential of −0.3 V versus RHE, and Faradaic efficiency over 90% at −0.6 V versus RHE (Figure 23.15a-b). They also found that the selectivity for CO can be improved by decreasing the electrolyte concentration. However, quite low current densities of 7.5 mA/cm^2 were reported.

In 2017, Ye et al. functionalised ZIF-8 with ammonium ferric citrate and pyrolysed the resulting material at up to 1,000 °C to obtain Fe–N–C catalysts with atomically dispersed porphyrin-like Fe sites, and surface areas up to 724 m^2/g. [64] They obtained Faradaic efficiencies for CO of 93% at −0.43 V versus RHE in 1 M KHCO$_3$, and partial current densities of around 10 mA/cm^2 at around −0.64 V versus RHE (Figure 23.15c). The partial CO current density was found to increase with increasing Fe content, suggesting that the active site is an Fe–N centre. In 2018, the same group used a similar ZIF-derived carbon doped with

Fe–N sites but with an extra ammonia treatment step. [65] This increased the maximum partial current density to 17.8 mA/cm^2, but at the expense of the CO Faradaic efficiency (85%), at −0.8 V versus RHE in 1 M KHCO$_3$. Their catalyst was stable for 10 h at −0.55 V versus RHE. They attributed the activity to higher loading of Fe–N active sites, larger surface area around 1200 m^2/g and optimised the pore structure due to ammonia treatment.

In 2018, Roy et al. synthesised a series of highly porous CO2RR electrocatalysts using a sacrificial support method (Figure 23.15d). [66] The nitrogen-doped carbon structure was formed by pyrolysis of aminoantipyrine, while metal salts provided the active centres. They found that the metal-free reference sample was most active for the CO$_2$ conversion to CO with partial CO current density of 0.26 mA/cm^2. No Faradaic efficiency was reported, but it is estimated to be 93% from the relative proportions of H$_2$ and CO. The HER activity tended to increase when metals were added to the carbon structure, and the partial CO current density decreased.

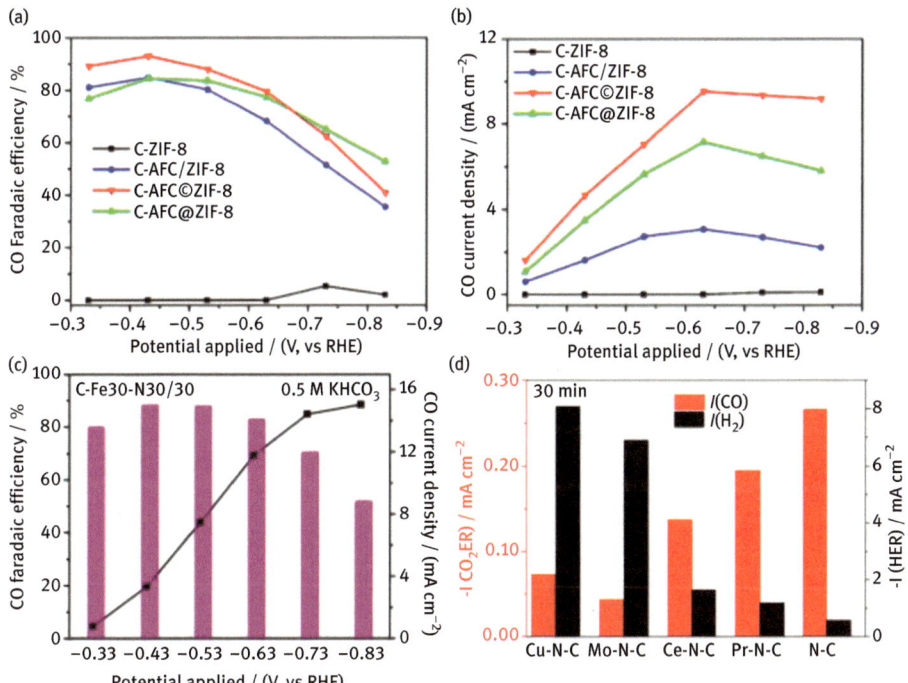

Figure 23.15: CO Faradaic efficiency and partial current density of the CO2RR on (a–b) Fe–N–C electrocatalysts derived from ZIF-8 and ammonium ferrous citrate. Reprinted from [64] with permission from Elsevier. (c) Similar catalysts with additional ammonia treatment. Reprinted from [65] with permission from Elsevier. Partial CO and HER current densities of different Me–N–C carbons prepared by the sacrificial support method. Reprinted from [66] with permission from Elsevier.

The above studies have confirmed experimentally that nitrogen-doped carbons with or without metal incorporation are interesting catalysts for the CO2RR. Several different groups have attempted to shed light on the mechanism for this reaction using computational studies. For example, a metal-free nitrogen-doped carbon nanotube system was studied from first principle calculations by Chai et al. in 2016. [67] They used DFT and ab initio molecular dynamics to show that the synergy between nitrogen doping and carbon nanotube curvature can effectively tune the catalytic activity and selectivity. They found that the activation barrier for CO_2 can be decreased from 1.3 eV for undoped carbon, to 0.58 eV at graphitic-type nitrogen-doped edges (Figure 23.16a). Flat nitrogen-doped graphene was shown to have strong selectivity for CO and HCOOH generation, while carbon nanotubes with high curvature were shown to be more selective for methanol production. This type of product selectivity has yet to be observed experimentally. The curvature was also found to be important in tuning the overpotential for given products, for example, from −1.6 to −0.12 V versus RHE for CO generation.

In the same year, Liu et al. performed DFT calculations on metal-free nitrogen-doped graphene. [68] Their calculations suggested that nitrogen doping reduces the energy barrier for the rate-limiting step, that is, formation of the COOH intermediate. They found that this intermediate would be strongly adsorbed, while CO and HCOOH products can easily desorb. They concluded that in all cases, formate is the major product, and especially at pyrrolic sites, where pure formate is produced (Figure 23.16b). However, this is in contrast to experimental studies, in which CO was the major product.

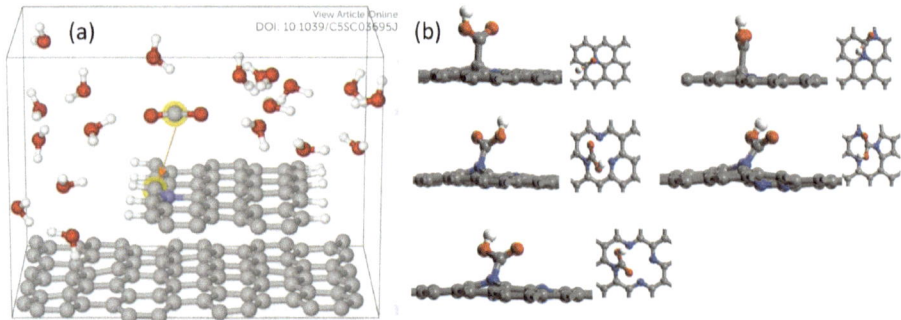

Figure 23.16: (a) Model of tertiary nitrogen edge sites for CO_2 binding. [67] – Published by the Royal Society of Chemistry. (b) CO_2 binding on various types of nitrogen-doped graphene. Republished with permission of the Royal Society of Chemistry, from [68].

Similarly, in 2017, Siahrostami et al. performed DFT calculations on metal-free nitrogen-doped graphene, investigating a wide range of defect sites. [69] They found that the two-electron reduction of CO_2 to CO was favoured, agreeing with most experimental results. They determined that the formation of the COOH inter-mediate was difficult at pyrrolic- and tertiary nitrogen-doped graphene, preventing the reaction on those sites. In contrast, the formation energy was relatively low on the pyridinic nitrogen site, with facile desorption of the formed CO (Figure 23.17a). Interestingly, they also found that some nitrogen-free defect sites are also active for the electroreduction of CO_2.

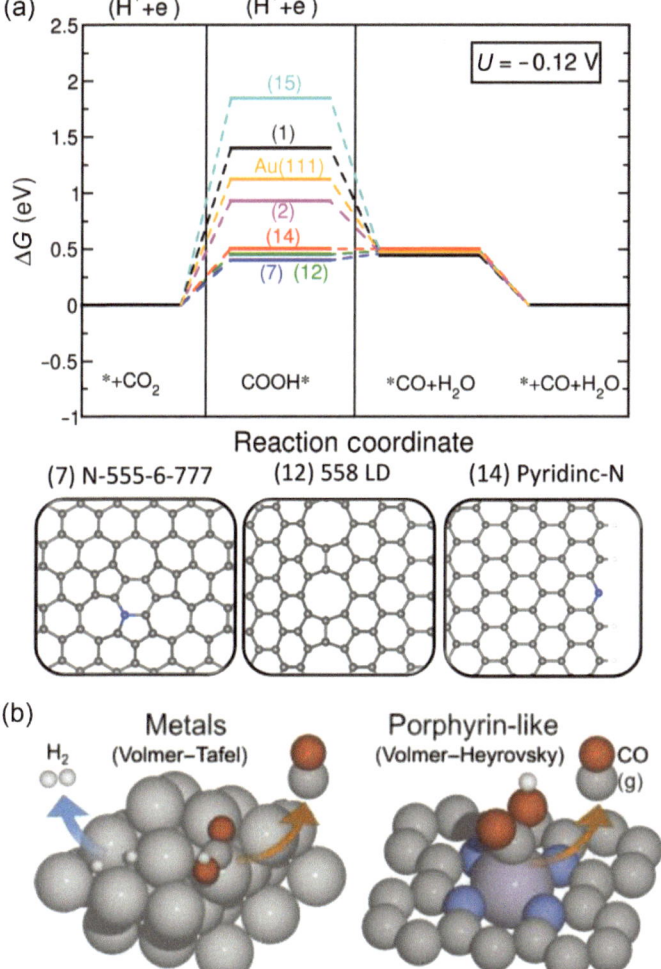

Figure 23.17: (a) Calculated reaction coordinates of CO_2 reduction on different nitrogen moieties and defects in graphene with the structures of the best three sites. Reprinted with permission from [69]. Copyright 2017 American Chemical Society. (b) CO_2 reduction on bulk metallic electrodes and at porphyrin-type structures in graphene, showing differences in selectivity for CO and H_2. Reprinted from [70] with permission from Elsevier.

Finally, Bagger et al. recently performed DFT calculations on carbon supports with atomically dispersed porphyrin-like metal sites. By analysing reaction intermediates, they found that this type of active site has the advantage of suppressing the HER, leading to enhanced CO_2 reduction (Figure 23.17b). [70]

The above studies describe both nominally metal-free nitrogen-doped carbon electrocatalysts, and metal-incorporated nitrogen-doped carbon electrocatalysts for the electroreduction of CO_2 to CO. In the metal-free cases, quite high partial current density and Faradaic efficiency have been achieved, However, in experimental studies, metal contamination is often an issue, and without sensitive elemental analysis techniques such as inductively coupled plasma atomic emission spectroscopy, it cannot be ruled out. Despite this, many DFT studies have been performed, suggesting that the metal-free system can indeed catalyse this reaction, and that pyridinic nitrogen sites are probably responsible. Therefore, metal-free catalysts could be quite promising. Other research groups have purposefully introduced metals into their nitrogen-doped carbon catalysts for the CO2RR. Experimentally, there has not been much difference in the activity of nominally "metal-free", or metal-containing catalysts, but one DFT study suggests that the metal centres can suppress the HER, leading to enhancement of the CO2RR. However, further DFT studies are probably required, especially on metal-containing systems. The highest activity achieved so far for this class of catalyst is 90 mA/cm^2, and 98% Faradaic efficiency for CO generation, but it is likely that these values will soon be surpassed.

23.5 Summary and prospects

In summary, there are many different types of catalysts that are highly active and selective for the electrochemical conversion of CO_2 to CO. Gold is a potential electrocatalyst, especially when in the form of small nanoparticles and decorated on a suitable support such as carbon nanotubes. Partial current densities up to 200 mA/cm^2 and Faradaic efficiency of up to 100% have been achieved. However, the price of Au is relatively high. Silver has been used to great effect as a CO2RR electrocatalyst when suitable nanostructure is introduced, a conductive carbon support is utilised and the electrochemical cell system is properly optimised. Current densities up to 440 mA/cm^2 and Faradaic efficiencies up to 100% have been reported. Nitrogen-doped carbons are a highly promising newcomer to the field of CO_2 electroreduction. Despite being simple to synthesise at relatively low cost, such catalysts have been reported to have Faradaic efficiency up to 98%, and partial current density up to 90 mA/cm^2. This is a research area where great strides forward could potentially be made. The activity of the above reactions can be enhanced by the use of the ionic liquid EMIM-BF$_4$, due to enhanced CO_2 solubility, and a possible synergistic homogeneous-type co-catalytic effect. This is interesting, but ionic liquids can be sensitive to moisture and result in added costs to the system.

There are several areas where the field of study of the CO2RR could be improved. Standardisation of experimental conditions would be useful and make it much easier to compare results from different research groups. For example, at present there is no commonly used standard reference catalyst (e.g. Ag nanoparticles purchased from a particular supplier). Different groups use different types of electrolyte, or the same electrolyte with different concentrations. Several different types of reference electrode are used, and potentials are quoted versus different reference electrodes. The partial current density is not always reported, and sometimes the molar production rate or turnover frequency is used instead. Some consistency or reporting both values would avoid confusion. The counter-electrode is often different in different experiments, and this can also have a major impact on the outcome. The cathode electrodes are prepared completely differently in different groups, for example spray painting (air brushing) or hand painting the electrocatalyst layers, and with completely different catalyst loadings. This will critically affect, for example, the current density by altering the microstructure of the catalyst layer. As such, standardisation could be of great benefit to the field. Another important point is durability, which is not often reported beyond 8 h. If this technology takes off, devices may need to operate at a high current density and constant potential for much longer periods of time. Durability tests should also be standardised by specifying a particular constant potential or constant current density, a particular electrolyte and a particular catalyst loading.

New catalysts for the electrochemical CO2RR are being reported every week as the topic gains popularity, and records are frequently being broken. As such, it is an interesting and exciting time to be part of this burgeoning research field.

References

[1] Blunden J, Arndt DS, Hartfield G, et al AmericAn meteorologicAl Society STATE OF THE CLIMATE IN 2017. doi: 10.1175/2018BAMSStateoftheClimate.1

[2] Scibioh MA, Viswanathan B, Scibioh MA, Viswanathan B (2018) Electrochemical Reduction of CO2. Carbon Dioxide to Chem. Fuels 307–371. doi: 10.1016/B978-0-444-63996-7.00007-9

[3] Kumar B, Brian JP, Atla V, et al (2016) New trends in the development of heterogeneous catalysts for electrochemical CO2 reduction. Catal. Today 270:19–30. doi: 10.1016/J. CATTOD.2016.02.006

[4] Ganesh I (2016) Electrochemical conversion of carbon dioxide into renewable fuel chemicals – The role of nanomaterials and the commercialization. Renew Sustain. Energ. Rev. 59:1269–1297. doi: 10.1016/J.RSER.2016.01.026

[5] Kenis PJA, Dibenedetto A, Zhang T (2017) Carbon Dioxide Utilization Coming of Age. ChemPhysChem 18:3091–3093. doi: 10.1002/cphc.201701204

[6] Royer ME (1870) Reduction de l'acide carbonique en acide formique. Comptes rendus l'Académie des Sci 70:731

[7] Fischer F (1912) Elektrothermische Methoden. In: Praktikum der Elektrochemie. Springer Berlin Heidelberg, Berlin, Heidelberg, pp 107–113

[8] Wang Y, Liu J, Wang Y, et al (2017) Tuning of co_2 reduction selectivity on metal electrocatalysts. Small 13:1701809. doi: 10.1002/smll.201701809

[9] Jhong H-RM, Tornow CE, Smid B, et al (2017) A nitrogen-doped carbon catalyst for electrochemical co_2 conversion to co with high selectivity and current density. ChemSusChem 10:1094–1099. doi: 10.1002/cssc.201600843

[10] Kauffman DR, Thakkar J, Siva R, et al (2015) Efficient electrochemical $CO2$ conversion powered by renewable energy. ACS Appl. Mater. Interfaces. 7:15626–32. doi: 10.1021/acsami.5b04393

[11] Verma S, Lu X, Ma S, et al (2016) The effect of electrolyte composition on the electroreduction of $CO2$ to CO on Ag based gas diffusion electrodes. Phys. Chem. Chem. Phys. 18:7075–7084. doi: 10.1039/C5CP05665A

[12] Hori Y (2016) $CO2$ Reduction Using Electrochemical Approach. Springer, Cham, pp 191–211

[13] Ikeda S, Takagi T, Ito K (1987) Selective formation of formic acid, oxalic acid, and carbon monoxide by electrochemical reduction of carbon dioxide. Bull. Chem. Soc. Jpn. 60: 2517–2522. doi: 10.1246/bcsj.60.2517

[14] Hori Y, Wakebe H, Tsukamoto T, Koga O (1994) Electrocatalytic process of CO selectivity in electrochemical reduction of $CO2$ at metal electrodes in aqueous media. Electrochim. Acta. 39:1833–1839. doi: 10.1016/0013-4686(94)85172-7

[15] JITARU M, LOWY DA, TOMA M, et al (1997) Electrochemical reduction of carbon dioxide on flat metallic cathodes. J. Appl .Electrochem. 27:875–889. doi: 10.1023/A:1018441316386

[16] Jhong HR, Ma S, Kenis PJA (2013) Electrochemical conversion of $CO2$ to useful chemicals: currentstatus, remaining challenges, and future opportunities. Curr. Opin. Chem. Eng. 2:191–199

[17] Kim B, Hillman F, Ariyoshi M, et al (2016) Effects of composition of the micro porous layer and the substrate on performance in the electrochemical reduction of $CO2$ to CO. J. Power. Sources. 312:192–198. doi: 10.1016/J.JPOWSOUR.2016.02.043

[18] Verma S, Hamasaki Y, Kim C, et al (2018) Insights into the low overpotential electroreduction of co_2 to co on a supported gold catalyst in an alkaline flow electrolyzer. ACS Energ. Lett. 3:193–198. doi: 10.1021/acsenergylett.7b01096

[19] Ma S, Luo R, Gold JI, et al (2016) Carbon nanotube containing Ag catalyst layers for efficient and selective reduction of carbon dioxide. J. Mater. Chem. A. 4:8573–8578. doi: 10.1039/C6TA00427J

[20] Dimitriou I, García-Gutiérrez P, Elder RH, et al (2015) Carbon dioxide utilisation for production of transport fuels: process and economic analysis. Energ. Environ. Sci. 8:1775–1789. doi: 10.1039/C4EE04117H

[21] Li X, Anderson P, Jhong H-RM, et al (2016) Greenhouse gas emissions, energy efficiency, and cost of synthetic fuel production using electrochemical co_2 conversion and the fischer–tropsch process. Energy & Fuels 30:5980–5989. doi: 10.1021/acs.energyfuels.6b00665

[22] Nishihara M, Terayama Y, Haji T, et al Proton-conductive nano zeolite-PVA composite film as a new water-absorbing electrolyte for water electrolysis. doi: 10.3144/expresspolymlett.2018.23

[23] Wu K, Birgersson E, Kim B, et al (2014) Modeling and experimental validation of electrochemical reduction of $co2$ to co in a microfluidic cell. J. Electrochem. Soc. 162: F23–F32. doi: 10.1149/2.1021414jes

[24] Hori Y, Murata A, Kikuchi K, Suzuki S (1987) Electrochemical reduction of carbon dioxides to carbon monoxide at a gold electrode in aqueous potassium hydrogen carbonate. J. Chem. Soc. Chem. Commun. 0:728. doi: 10.1039/c39870000728

[25] Lim RJ, Xie M, Sk MA, et al (2014) A review on the electrochemical reduction of $CO2$ in fuel cells, metal electrodes and molecular catalysts. Catal. Today 233:169–180. doi: 10.1016/J.CATTOD.2013.11.037

[26] Chen YH, Li CW, Kanan MW (2012) Aqueous co2 reduction at very low overpotential on oxide-derived au nanoparticles. J. Am. Chem. Soc. 134: 19969–19972. doi: Doi 10.1021/Ja309317u

[27] Kauffman DR, Alfonso D, Matranga C, et al (2012) Experimental and computational investigation of au $_{25}$ clusters and co$_2$: a unique interaction and enhanced electrocatalytic activity. J. Am. Chem. Soc. 134:10237–10243. doi: 10.1021/ja303259q

[28] Zhu WL, Michalsky R, Metin O, et al (2013) Monodisperse au nanoparticles for selective electrocatalytic reduction of co2 to co. J. Am. Chem. Soc. 135: 16833–16836. doi: Doi 10.1021/Ja409445p

[29] Hall AS, Yoon Y, Wuttig A, Surendranath Y (2015) Mesostructure-induced selectivity in co$_2$ reduction catalysis. J Am Chem Soc 137:14834–14837. doi: 10.1021/jacs.5b08259

[30] Lyth SM, Silva SRP (2015) Electron field emission from water-based carbon nanotube inks. ECS J. Solid. State. Sci. Technol. 4:. doi: 10.1149/2.0051504jss

[31] Liu M, Pang Y, Zhang B, et al (2016) Enhanced electrocatalytic CO2 reduction via field-induced reagent concentration. Nature 537:382–386. doi: 10.1038/nature19060

[32] Saberi Safaei T, Mepham A, Zheng X, et al (2016) High-density nanosharp microstructures enable efficient co$_2$ electroreduction. Nano. Lett. 16:7224–7228. doi: 10.1021/acs.nanolett.6b03615

[33] Watts PCP, Lyth SM, Henley SJ, Silva SRP (2008) Secondary nanotube growth on aligned carbon nanofibre arrays for superior field emission. J. Nanosci. Nanotechnol. 8:2147–2150

[34] Kim D, Resasco J, Yu Y, et al (2014) Synergistic geometric and electronic effects for electrochemical reduction of carbon dioxide using gold–copper bimetallic nanoparticles. Nat. Commun. 5:4948. doi: 10.1038/ncomms5948

[35] Kim J-H, Woo H, Choi J, et al (2017) Co2 electroreduction on au/tic: enhanced activity due to metal–support interaction. ACS Catal. 7:2101–2106. doi: 10.1021/acscatal.6b03706

[36] Jhong H-R "Molly," Tornow CE, Kim C, et al (2017) Gold nanoparticles on polymer-wrapped carbon nanotubes: an efficient and selective catalyst for the electroreduction of co$_2$. ChemPhysChem 18: 3274–3279. doi: 10.1002/cphc.201700815

[37] Shi J, Shen F, Shi F, et al (2017) Electrochemical reduction of CO2 into CO in tetrabutylammonium perchlorate/propylene carbonate: Water effects and mechanism. Electrochim Acta. 240:114–121. doi: 10.1016/J.ELECTACTA.2017.04.065

[38] Jin L, Liu B, Wang P, et al (2018) Ultrasmall Au nanocatalysts supported on nitrided carbon for electrocatalytic CO2 reduction: the role of carbon support in high selectivity. Nanoscale. doi: 10.1039/C8NR04322A

[39] Hori Y, Kikuchi K, Suzuki S (1985) Production of co and ch 4 in electrochemical reduction of co 2 at metal electrodes in aqueous hydrogencarbonate solution. Chem. Lett. 14:1695–1698. doi: 10.1246/cl.1985.1695

[40] Hoshi N, Kato M, Hori Y (1997) Electrochemical reduction of CO2 on single crystal electrodes of silver Ag(111), Ag(100) and Ag(110). J. Electroanal. Chem. 440:283–286. doi: 10.1016/S0022-0728(97)00447-6

[41] Hori Y, Ito H, Okano K, et al (2003) Silver-coated ion exchange membrane electrode applied to electrochemical reduction of carbon dioxide. Electrochim Acta. 48:2651–2657. doi: 10.1016/S0013-4686(03)00311-6

[42] Delacourt C, Ridgway PL, Kerr JB, Newman J (2008) Design of an electrochemical cell making syngas (co+h[sub 2]) from co[sub 2] and h[sub 2]o reduction at room temperature. J. Electrochem. Soc. 155:B42. doi: 10.1149/1.2801871

[43] Rosen BA, Salehi-Khojin A, Thorson MR, et al (2011) Ionic liquid-mediated selective conversion of CO_2 to CO at low overpotentials. Science 334:643–4. doi: 10.1126/science.1209786

[44] Thorson MR, Siil KI, Kenis PJA (2012) Effect of cations on the electrochemical conversion of co2 to co. J. Electrochem. Soc. 160:F69–F74. doi: 10.1149/2.052301jes

[45] Ma S, Lan Y, Perez GMJ, et al (2014) Silver supported on titania as an active catalyst for electrochemical carbon dioxide reduction. ChemSusChem 7:866–874. doi: 10.1002/cssc.201300934

[46] Ma S, Luo R, Moniri S, et al (2014) Efficient electrochemical flow system with improved anode for the conversion of co2 to co. J. Electrochem. Soc. 161:F1124–F1131. doi: 10.1149/2.1201410jes

[47] Lu Q, Rosen J, Zhou Y, et al (2014) A selective and efficient electrocatalyst for carbon dioxide reduction. Nat. Commun. 5:3242. doi: 10.1038/ncomms4242

[48] Yoon Y, Hall AS, Surendranath Y (2016) Tuning of silver catalyst mesostructure promotes selective carbon dioxide conversion into fuels. Angew. Chemie. 128:15508–15512. doi: 10.1002/ange.201607942

[49] Lyth SM, Shao H, Liu J, et al (2014) Hydrogen adsorption on graphene foam synthesized by combustion of sodium ethoxide. Int. J. Hydrogen. Energy. 39:376–380. doi: 10.1016/j.ijhydene.2013.10.044

[50] Liu J, Takeshi D, Sasaki K, Lyth SM (2014) Defective graphene foam: a platinum catalyst support for pemfcs. J. Electrochem. Soc. 161:F838–F844. doi: 10.1149/2.0231409jes

[51] Ma S, Liu J, Sasaki K, et al (2017) Carbon foam decorated with silver nanoparticles for electrochemical co_2 conversion. Energy. Technol.. doi: 10.1002/ente.201600576

[52] Jeanty P, Scherer C, Magori E, et al (2018) Upscaling and continuous operation of electrochemical CO2 to CO conversion in aqueous solutions on silver gas diffusion electrodes. J. CO2 Util. 24:454–462. doi: 10.1016/J.JCOU.2018.01.011

[53] Liu J, Takeshi D, Sasaki K, Lyth SM (2014) Platinum-decorated nitrogen-doped graphene foam electrocatalysts. Fuel. Cells. 14:728–734. doi: 10.1002/fuce.201300258

[54] Thompson ST, Wilson AR, Zelenay P, et al (2018) ElectroCat: DOE's approach to PGM-free catalyst and electrode R&D. Solid. State. Ionics. 319:68–76. doi: 10.1016/j.ssi.2018.01.030

[55] Mufundirwa A, Harrington GF, Smid B, et al (2017) Durability of template-free Fe-N-C foams for electrochemical oxygen reduction in alkaline solution. J. Power. Sources.. doi: 10.1016/j.jpowsour.2017.07.025

[56] Zitolo A, Ranjbar-Sahraie N, Mineva T, et al (2017) Identification of catalytic sites in cobalt-nitrogen-carbon materials for the oxygen reduction reaction. Nat. Commun. 8:. doi: 10.1038/s41467-017-01100-7

[57] Gokhale R, Thapa S, Artyushkova K, et al (2018) Fully synthetic approach towards transition metal-nitrogen-carbon oxygen reduction electrocatalysts. ACS Appl. Energy. Mater. acsaem.8b00537. doi: 10.1021/acsaem.8b00537

[58] Kumar B, Asadi M, Pisasale D, et al (2013) Renewable and metal-free carbon nanofibre catalysts for carbon dioxide reduction. Nat. Commun. 4:2819. doi: 10.1038/ncomms3819

[59] Wu J, Yadav RM, Liu M, et al (2015) Achieving highly efficient, selective, and stable co2 reduction on nitrogen-doped carbon nanotubes. ACS Nano. 9:5364–71. doi: 10.1021/acsnano.5b01079

[60] Sharma PP, Wu J, Yadav RM, et al (2015) Nitrogen-doped carbon nanotube arrays for high-efficiency electrochemical reduction of co_2 : on the understanding of defects, defect density, and selectivity. Angew. Chemie. Int. Ed. 54:13701–13705. doi: 10.1002/anie.201506062

[61] Varela AS, Ranjbar Sahraie N, Steinberg J, et al (2015) Metal-doped nitrogenated carbon as an efficient catalyst for direct co 2 electroreduction to co and hydrocarbons. Angew. Chemie. Int. Ed. 54:10758–10762. doi: 10.1002/anie.201502099

[62] Li W, Seredych M, Rodríguez-Castellón E, Bandosz TJ (2016) Metal-free nanoporous carbon as a catalyst for electrochemical reduction of co 2 to co and ch 4. ChemSusChem 9:606–616. doi: 10.1002/cssc.201501575

[63] Huan TN, Ranjbar N, Rousse G, et al (2017) Electrochemical reduction of co$_2$ catalyzed by fe-n-c materials: a structure–selectivity study. ACS Catal. 7:1520–1525. doi: 10.1021/acscatal.6b03353

[64] Ye Y, Cai F, Li H, et al (2017) Surface functionalization of ZIF-8 with ammonium ferric citrate toward high exposure of Fe-N active sites for efficient oxygen and carbon dioxide electroreduction. Nano. Energ. 38:281–289. doi: 10.1016/J.NANOEN.2017.05.042

[65] Yan C, Ye Y, Lin L, et al (2018) Improving CO2 electroreduction over ZIF-derived carbon doped with Fe-N sites by an additional ammonia treatment. Catal. Today. doi: 10.1016/J.CATTOD.2018.03.062

[66] Roy A, Hursán D, Artyushkova K, et al (2018) Nanostructured metal-N-C electrocatalysts for CO2 reduction and hydrogen evolution reactions. Appl. Catal. B. Environ. 232:512–520. doi: 10.1016/J.APCATB.2018.03.093

[67] Chai G-L, Guo Z-X (2016) Highly effective sites and selectivity of nitrogen-doped graphene/CNT catalysts for CO2 electrochemical reduction. Chem Sci 7:1268–1275. doi: 10.1039/c5sc03695j

[68] Liu Y, Zhao J, Cai Q (2016) Pyrrolic-nitrogen doped graphene: a metal-free electrocatalyst with high efficiency and selectivity for the reduction of carbon dioxide to formic acid: a computational study. Phys. Chem. Chem. Phys. doi: 10.1039/C5CP07458D

[69] Siahrostami S, Jiang K, Karamad M, et al (2017) Theoretical investigations into defected graphene for electrochemical reduction of co$_2$. ACS Sustain. Chem. Eng. acssuschemeng.7b03031. doi: 10.1021/acssuschemeng.7b03031

[70] Bagger A, Ju W, Varela AS, et al (2017) Single site porphyrine-like structures advantages over metals for selective electrochemical CO 2 reduction. Catal. Today. doi: 10.1016/j.cattod.2017.02.028

Esperanza Ruiz Martínez and Josemaria Sanchez Hervas

24 Electrochemical conversion of CO$_2$ into alcohols

24.1 Introduction

Electrochemical conversion of CO$_2$ offers a simple, selectively controlled and environmental friendly process to transform waste CO$_2$ into alcohols [1–4], such as methanol (MeOH), ethanol (EtOH) and n- and i-propanol (PrOH), which have high-energy density (15.6 MJ·dm^{-3}, 21 MJ dm^{-3} and 27 MJ dm^{-3} for methanol, ethanol and n-propanol, respectively), octane rating, stable storage properties, ease of transport and established use [3, 5–7]. It also allows CO$_2$ recycling and chemical storage of intermittent renewable energy in the form of sustainable fuels and platform chemicals [5, 6, 8, 9].

The electrochemical conversion of CO$_2$ into alcohols requires the contribution of electricity to establish a potential between two or three electrodes (cathode or catalyst-working-electrode and anode or counter or reference-electrode/s). Electrolytes can be liquids (aqueous, non-aqueous or ionic liquids [ILs]) or solids (H$^+$, O^{2-}, K$^+$, etc. ion-conducting polymers or oxides). The process can occur in gaseous and liquid phase at different cell and electrode configurations and under different operation conditions [5, 9].

Unfortunately, the production of alcohols via CO$_2$ electrochemical conversion remains below that required for economic viability because of limited selectivity, low efficiency and lack of stability of present-day catalysts, implying the use of high over-potentials, resulting also in high-energy requirements [1, 5, 7, 10–12].

Many catalysts, such as metals, metal alloys, metal complexes, metal oxides/chalcogenides and so on have been used for generating alcohols [10, 11, 13–15]. Among the available metals, the most active for the electrochemical conversion of CO$_2$ to alcohols have been identified to be Mo, Ru, Ti, Fe, Ga, In, Pd, Pt and, in particular, Cu, which has been used in the form of metal foils, electrochemical deposits on carbon-based materials, chemically deposited Cu films on solid oxide electrolytes (SOE) and solid polymer electrolytes (SPE), single crystal Cu specimens, Cu alloys and Cu containing perovskite electrocatalysts [14–16]. Their use is discussed in this chapter.

This chapter reviews significant work performed in the field of electrochemical conversion of CO$_2$ to alcohols, taking into consideration the effect of the applied catalyst, electrolyte and electrode and electrochemical cell configuration, as well as working conditions. This review aims also to present the progress of diverse recent

Esperanza Ruiz Martínez, Josemaria Sanchez Hervas, Unit for Sustainable Thermochemical Valorization, CIEMAT, Madrid

https://doi.org/10.1515/9783110665147-024

process alternatives, such as the application of solid polymer or solid oxide electrochemical membrane reactors (SOEMRs) for the electrochemical conversion of CO_2 to alcohols, as well as the electrochemical promotion of catalytic CO_2 hydrogenation in solid oxide electrolyte membrane reactors (SOEMRs). Therefore, this chapter presents the current status, remaining challenges and future opportunities for electrochemical conversion of CO_2 into alcohols. Because of the enormous amount of work published on CO_2 electroreduction to alcohols, many of the results are presented in tabular form. Different tables along the chapter summarise the best performance data reported for Faradaic efficiencies (FE) and alcohol production yields or current densities (CD). The tables also include the reaction medium, operating conditions and catalyst-electrode and cell configuration in which the electrochemical conversion of CO_2 takes place.

24.2 Electrocatalysts

24.2.1 Cu-based

The formation of alcohols has been reported at Cu, electrodeposited Cu and Cu-coated electrodes, at ambient temperature in aqueous solutions [5, 16]. However, the traditional Cu catalysts exhibit high over-potential and low selectivity to alcohols [2, 14]. Considerable efforts have been made to enhance energy efficiency for CO_2 electroreduction by modifying surface structures, morphologies and compositions, resulting in different types of Cu-based catalysts:

Cu and Cu oxides. The electrochemical reduction of CO_2 to alcohols has been performed at various copper electrodes: Cu foil, Cu foil thermally oxidised in air and air-oxidised Cu electrodeposited on anodised or air-oxidised Ti foils [17]; Cu_2O-catalysed carbon clothes [5, 6, 18, 19]; electrodeposited Cu_2O thin film electrodes [20]; and Cu(core)/CuO(shell) catalysts [21], as listed in Table 24.1. It was suggested that Cu (I) species play a critical role in catalyst electrode activity and selectivity to methanol. However, these catalysts suffer from a lack of stability because of the partial reduction of the Cu oxide [5]. The selective electroreduction of CO_2 to ethanol has also been studied on copper (I) oxide films [22]. The same process was studied over Cu_2O coated on a graphite plate [23]; the maximum Faradaic efficiency for ethanol was about 16 and 96%, respectively. The performance of Cu_2O and Cu_2O/ZnO deposited on carbon papers for the electroreduction of CO_2 to methanol [6], ethanol [24] and n-propanol [19] was also studied. The presence of ZnO improves stability of the system [6, 19].

The performance of multi walled carbon nanotubes (MWCNTs) impregnated with Cu_2O for CO_2 electro-reduction to CH_3OH [15] was studied. The role of graphene as supporting catalyst for the electrochemical reduction of CO_2 to ethanol [2] and n-propanol [3] has also been investigated. Both carbon supported systems showed higher activity, selectivity and stability than unsupported Cu_2O. Best performance data obtained are shown in Table 24.1.

Table 24.1: Best performance data for aqueous electroreduction of CO_2 to alcohols on Cu and Cu oxides.

Electrode	Electrolyte/Cell	Overpotential (V)	CD (mA·cm⁻²)	Maximum FE (%)	Ref.
Graphene/Cu_2O–Cu foil	0.5 M $NaHCO_3$/ Two compartments	−0.9/−1.3	0.53/2.75	9.93/6.75 (EtOH)	[2]
Graphene/ZnO/Cu_2O–Cu foil	(glass frit)-Three electrodes	−0.9/−1.2	–	30/22 (n-PrOH)	[3]
$Cu_2O(1)$–carbon paper	0.5 M $KHCO_3$/ Two compartments	−1.30	6.93	45.7 (MeOH)	[6]
Cu_2O/ZnO(1:1)–carbon paper	(Nafion)	−1.30	10.64	17.7 (MeOH)	[6]
MWCNTs impregnated with Cu_2O (30 %)–Cu foil	0.5 M $NaHCO_3$/ Two compartments (Nafion)-Three electrodes	−0.8	7.5	38 (MeOH),	[15]
Pre-oxidised Cu foil	0.5 M $KHCO_3$/ Undivided	−0.05	0.069	240 (MeOH)	[17]
Cu_2O–carbon paper	0.1 M $KHCO_3$	−1.39	10	45 (MeOH), 10 (EtOH), 2.4 (n-PrOH)	[19]
Cu_2O/ZnO–carbon paper	0.1 M $KHCO_3$	−1.16	10	26 (MeOH), 5 (EtOH)	[19]
Cu_2O–stainless steel	$KHCO_3$/Undivided -Three electrodes	−1.10	5	38 (MeOH)	[20]
Cu_2O films (1.7/3.6 μm)	0.1 M $KHCO_3$	−0.99	–	9/16 (EtOH)	[22]
Cu_2O–graphite plate	0.5 M $KHCO_3$/ Undivided -Two electrodes	−2	4.5	96.2 (EtOH)	[23]

Cu alloys. Alloying Cu with other metals (i.e. Ni, Sn, Pb, Zn, Ag and Cd) may enhance electrocatalytic activity and CH_3OH selectivity [11, 25–28]. Electroreduction of CO_2 at nanostructured Cu–Au alloys showed that methanol and ethanol FE depend on their nanostructures and compositions [1]. Electroreduction of CO_2 to different alcohols has also been studied over the highly stable and recyclable catalyst, [PYD] @Cu–Pt composite ([4-(3-Phenoxy-2,2-bis(phenoxymethyl)propoxy)pyridine]@Cu–Pt) [29]. Best performance results obtained are listed in Table 24.2.

    Cu nanostructures and organic frameworks. Modified electrodes with different surface structures and morphologies were used to get more effective, selective and stable Cu-based catalysts for the CO_2 electroreduction to alcohols in aqueous solutions. Electrochemical CO_2 reduction to alcohols was realised at Cu-transition-metal-based perovskite oxides [11, 16, 30, 31]; nanostructured Cu/ZnO [32]; stepped Cu (110) [33]; oxide-derived Cu [34–39]; nanostructured Cu [12, 40–48]; Cu/CuO nanoparticles (NPs) supported on graphene [49, 50], N-doped graphene (NG) [49], carbon [51], carbon nanotubes (CNTs) [52] and TiO_2 [14, 53, 54]; Core (Cu sulphide)–shell (Cu) NPs (CSVE–Cu) [7] and Cu-based metal organics porous materials (MOPMs) [4]. Results for the best performing systems are shown in Table 24.3.

24.2.2 Other transition metals

The formation of CH_3OH by CO_2 electroreduction in aqueous solution was observed on Mo [56] or MoO_2 [57]–based catalysts, Mo–Bi alloy chalcogenide [58], MoS_2-rods/TiO_2 nanotubes [59] and on Ru-based electrodes, such as Teflon-supported Ru [60], conductive oxide mixtures (RuO_2, TiO_2, MoO_2, Co_3O_4 and Rh_2O_3) deposited on Ti foil [57], mixed Ru/Ti covered with Cu [61], Ru, Cu–Cd-modified Ru, Cu–Cd-modified RuO_x +IrO_x [62], RuO_2-coated diamond [63] and Pt modified with RuO_2–TiO_2 nanotubes/ nanoparticles [64]. The same process has also been studied at metal (Fe/Co/Ni/Cr)-based complexes of porphyrin [65], ES (Everitt's Salt, $K_2Fe(II)[Fe(II)(CN)_6]$) [66–68] and PB (Prussian blue, $KFe(III)[Fe(II)(CN)_6]$) [69]. TiO_2 with other metal or oxides deposited on Ti was also used to enhance CO_2 reduction to CH_3OH [57, 59, 61, 63, 65, 70, 71]. Results for the best performing systems are shown in Table 24.4.

24.2.3 Noble metals

CO_2 reduction in aqueous solutions over various Pt [67, 72, 73] and Pd [74–77]–based catalyst electrodes resulted in the production of methanol [5]. Results for the best performing systems are listed in Table 24.5.

Table 24.2: Best performance data for aqueous electroreduction of CO$_2$ to alcohols on Cu alloys.

Electrode	Electrolyte/Cell	Overpotential (V)	CD (mA·cm^{-2})	Maximum FE (%)	Ref.
Cu$_{63.9}$Au$_{36.1}$ on Cu films	0.5 M KHCO$_3$/ Two compartments H-type	−1	0.85	15.9 (MeOH), 12 (EtOH)	[1]
Cu–Cd alloy (Cu/Cd=38/62)	0.5 M KHCO$_3$/	−1.75	–	5 (MeOH)	[25]
Cu–Ni alloy (Cu/Ni=60/40)	Two compartments (Nafion 117)	−0.9	–	10 (MeOH)	[26]
Cu$_{88}$Sn$_6$Pb$_6$ alloy foil	1.5 M HCl−0.08 MLaCl$_3$/ Two compartments (Nafion 117)	−0.7	0.68	35.7 (MeOH)	[27]
[PYD]@Cu–Pt	0.5 M KCl/ Two compartments	−0.6/−1.2	22/7	37(MeOH)/24 (EtOH), 1 (n-PrOH)	[29]
Cu–Pt alloy	(Nafion 117)	−1.2	7	21 (EtOH), 1 (n-PrOH)	[29]

Table 24.3: Best performance data for aqueous electroreduction of CO_2 to alcohols on Cu nanostructures and organic frameworks.

Electrode	Electrolyte/Cell	Overpotential (V)	CD (mA·cm^{-2})	Maximum FE (%)	Ref.
MOA with DTA and Cu and Zn oxides (CuZnDTA)	0.5 M KHCO$_3$/ GDE filter-press cell (Nafion 117)	−1.41	10	46.5 (EtOH), 3.4 (MeOH)	[4]
CSVE–Cu	1 M KOH/ H-cell/Flow cell	−0.95/−0.92	7.3/126	15.1/ 24.7 (EtOH), 8/6.9 (n-PrOH)	[7]
CuO NPs–TiO$_2$–carbon paper	0.5 M KHCO$_3$/ GDE H type cell (Nafion 117)	−0.85	8.3	37.5 (EtOH)/5.6 (n-PrOH)	[14]
Perovskite (La$_{1.8}$Sr$_{0.2}$CuO$_4$) based carbon GDE	0.5 M KOH/ Two compartments (Nafion 117)	−2.30/−2.60	180	2 (MeOH), 30.7 (EtOH), 10 (n-PrOH)	[16]
Cu nanocluster-ZnO	0.1 M KHCO$_3$/ Two compartments (fritted glass)	−1.45	12	2.8 (MeOH), 10.2 (EtOH)	[32]
Cu(110)/Cu(S)-[4 (100)x(111)]	0.1 M KHCO$_3$	–	−5/−0.23	9.7/7.4 (EtOH), 1.5/4.6 (n-PrOH)	[33]
Cu$_2$O-derived Cu	0.1 M KHCO$_3$	−0.83	–	4.7 (EtOH), 8.2(n-PrOH)	[34]
Oxide-derived Cu4Zn film	0.1 M KHCO$_3$	−1.05	37.3	29.1 (EtOH), 4.4 (n-PrOH)	[38]
CuO-derived Cu NPs	0.2 M KI	−1.7	20	34.3 (EtOH)/1.8 (n-PrOH)	[39]
CuOxCly cubes-Cu foil	0.1 M KHCO$_3$	−1.05	49.7	20 (EtOH), 4.4 (n-PrOH)	[44]
Annealed denditric Cu	0.5 M KHCO$_3$	−1/−0.9	2.81/–	13(EtOH)/13.1 (n-PrOH)	[47]
Cu NPs-NG	0.1 M KHCO$_3$	−1.2	2.75	63 (EtOH)	[49]

Cu NPs-reduced graphene oxide (rGO)-Cu	0.1M NaHCO₃/ Undivided, Three electrodes	-0.4		64.2 (Liquid products)	[50]
Cu NPs-carbon GDL	1 M KOH/ GDE flow cell (Fumatech)	0.7	53	17.5 (EtOH)	[51]
20 % Cu/CNT	—	-1.7	—	38.4 (MeOH)	[52]
Cu NPs–TiO₂–carbon paper	0.2 M KI/ GDE H- type (Nafion 117)	-1.45	8.7	27.4 (EtOH)/6.2 (n-PrOH)	[53]
Cu/TiO₂ NPs-NG	0.1 M KHCO₃	-0.2/-0.75	0.1/0.7	19.5 (MeOH)/43.6 (EtOH)	[54]
Cu nanocrystal	0.1 M KHCO₃/ Two compartments (Selemion)	-0.85/-0.95	1.5/1.74	17.7/12.75 (EtOH),10.6/ 8.8 (n-PrOH)	[55]

Table 24.4: Best performance data for aqueous electroreduction of CO_2 to CH_3OH on other transition metals.

Electrode	Electrolyte/cell	Over-potential (V)	CD (mA·cm^{-2})	Maximum FE (%)	Ref.
Mo foil-Cu	0.2 M Na_2SO_4/undivided	−0.80	0.12	84	[56]
RuO_2+TiO_2 (35/65)-Ti	0.2 M Na_2SO_4/ Two compartment	−1.48	0.06	76	[57]
Mo−Bi alloy chalcogenide-carbon paper	0.5 M [bmin]BF4MeCN (1-butyl-3-methylimidazolium tetrafluoroborate in MeCN)	−0.7	12.1	71.2	[58]
MoS_2-rods/TiO_2 nanotubes	0.1 M $KHCO_3$/ Undivided, three electrodes	−1.30	0.75	44.9	[59]
Electroplated Ru on Cu foil	Na_2SO_4/ Two compartments (agar bridge)	−0.54	0.08	42	[60]
TiO_2/RuO_2 (75/25)-Ti	0.5 M $KHCO_3$/ Undivided-Rotating-disk electrode	−1.00	5	29.8	[61]
RuOx/Cu−Ti	0.5 M $NaHCO_3$/ Undivided-Three electrodes	−0.80	2	18.2–41.3	[62]
RuO_2-diamond (BDD)	0.4 M Britton-Robinson solution/ Two compartments	−0.80	5	8.12	[63]
RuO_2/TiO_2 nanotubes-Pt	0.5 M $NaHCO_3$/ Two compartments H-type cell (Nafion 117)	−0.80	1.2	60.5	[64]
Co(II) TPP−Pt plate	Aquopentacyanoferrate (II) (Na_3[Fe(CN)$_5$(H_2O)-0.1M KCl-CH_3OH/ Undivided, three electrodes	−0.5	0.07	32.9	[65]
ES supported-Pt	0.1 M KCl-10mM Na_3[Fe (CN)$_5$(H_2O)])-15 mM CH_3OH/ Two compartment (glass frit)	−0.7	−	56.3–45.3	[67]
ES-coated Pt	Na_3[Fe(CN)$_5$(H_2O)]-CH_3OH/ Two compartment (glass frit)	−0.9	−	15.5	[68]
Fe−C/Pan/PB-Pt	0.5 M KCl/ Two compartments H-type	−0.8	−	12.2	[69]

Table 24.5: Best performance data for aqueous electroreduction of CO$_2$ to methanol on noble metals.

Electrode	Electrolyte/cell	Over-potential (V)	CD (mA·cm^{-2})	Maximum FE (%)	Ref.
Indigo (C$_{16}$H$_{10}$N$_2$O$_2$)/ graphite-Pt	0.1 M KCl-10 mM Na$_3$[Fe(CN)$_5$(H$_2$O)])-	−0.7	–	70.23–37.2	[67]
Alizarin (C$_{14}$H$_8$O$_4$)/ graphite-Pt	15 mM CH$_3$OH/ Two compartment (glass frit)	−0.5--0.7	–	44.3–69.4	[67]
Pd disk (HNO$_3$/NaOH treated)	0.5 M NaClO$_4$+10 mM Pyridine/ Two compartment (glass frit)	–	0.04	30	[74]

24.2.4 Post-transition metals

The electrochemical reduction of CO$_2$ to methanol in aqueous medium was carried out over n- and p-type Ga and In containing semiconductors, that is, p-gallium phosphide (p-GaP) [78–80], p-gallium arsenide (p-GaAs) [81], n-gallium arsenide (n-GaAs) [79] and p-indium phosphide (p-InP) [81]. High selectivity for CH$_3$OH formation was observed but only at high over-potentials. Results for the best performing systems are listed in Table 24.6.

Table 24.6: Best performance data for aqueous electroreduction of CO$_2$ to methanol on post transition metals.

Electrode	Electrolyte/Cell	Overpotential (V)	CD (mA·cm^{-2})	Maximum FE (%)	Ref.
nGaP/Pyridine	10 mM Pyridinium pH 5.2	−0.06	0.27	90	[80]
n-GaAs-crystal-(111) As	0.2 M Na$_2$SO$_4$/two compartments (agar bridge)	−1.2--1.4	0.16–0.2	100	[81]
n-GaAs-crystal-(111) Ga		−1.2--1.4	0.34	30–80	[81]

24.3 Electrolytes

Even for the same electrocatalyst, the electrolyte choice has a severe influence on current density, product selectivity and energy efficiency of electrochemical conversion of CO$_2$ to alcohols [5, 82].

24.3.1 Liquid electrolytes

The use of non-aqueous solutions or ILs instead of water solutions may improve CO_2 electroreduction performance by enhancing the solubility of CO_2 and decreasing mass transfer limitations [82].

24.3.1.1 Aqueous

Aqueous electrolytes commonly utilised in electrochemical reduction of CO_2 to alcohols consisted of aqueous solutions of alkali cations (e.g. Na^+, K^+) and various anions (e.g. Cl^-, HCO_3^- or OH^-) [40, 83–85], which exhibit high conductivities in water and can provide the required protons [82]. Faradaic efficiency for CH_3OH formation, at similar potential, depends on the electrolyte solution applied [86]. As an example, Table 24.7 shows the effect of electrolyte composition on the CO_2 electroreduction to CH_3OH over a $Cu_{88}Sn_6Pb_6$ alloy foil [27]. The CO_2 reduction rate can be increased and product distribution can be controlled by varying the utilised cation and the pH of the aqueous solution.

However, the application of aqueous media in CO_2 electroreduction processes brings several drawbacks, such as low selectivity for CO_2 reduction, because of the competing hydrogen evolution reaction (HER); slow reaction kinetics, which results in enlarged over-potentials and costs; complicated and costly products separation and recovery; low CO_2 solubility and electrode surface poisoning by electrolyte impurities, which results in lesser yields [5].

24.3.1.2 Non-aqueous

Different non-aqueous electrolytes have been used to increase the solubility of CO_2 in the reaction medium [81, 82]. However, the leading product in non-aqueous organic solvents, such as methanol, propylene carbonate and dimethylformamide, seems to be CO regardless of the electrode materials [87]. In addition, none of these studies deal with the possible application of non-aqueous electrolytes in the electroreduction of CO_2 to alcohols.

24.3.1.3 Ionic liquids

Replacing the conventional aqueous electrolytes with ILs brings the following advantages [80, 87]: suppression of the competing HER [88]; enhancement of electrolyte conductivity [89] and higher CO_2 solubility at a wider range of temperatures [5], thereby reducing mass transport limitations and facilitating CO_2 electro-catalytic conversion at

Table 24.7: Effect of electrolyte composition on aqueous electroreduction of CO$_2$ to CH$_3$OH.

Electrode	Electrolyte/cell	Over-potential (V)	CD (mA·cm^{-2})	Maximum FE (%)	Ref.
Cu$_{88}$Sn$_6$Pb$_6$ alloy	2 M HCl/Two compartments (Nafion)	−0.65	0.24	34.3	[27]
Cu$_{88}$Sn$_6$Pb$_6$ alloy	1.5 M HCl−0.5 M NaCl/Two compartments (Nafion)	−0.7	0.36	28.2	[27]
Cu$_{88}$Sn$_6$Pb$_6$ alloy	1.5 M HCl−0.17 M MgCl$_2$/Two compartments (Nafion)	−0.7	0.38	34.1	[27]
Cu$_{88}$Sn$_6$Pb$_6$ alloy	1.5 M HCl−0.17 M CaCl$_2$/Two compartments (Nafion)	−0.7	0.39	29.6	[27]
Cu$_{88}$Sn$_6$Pb$_6$ alloy	1.5 M HCl−0.17 M BaCl$_2$/Two compartments (Nafion)	−0.7	0.41	36.3	[27]
Cu$_{88}$Sn$_6$Pb$_6$ alloy	1.5 M HCl−0.08 M AlCl$_3$/Two compartments (Nafion)	−0.7	0.43	17.8	[27]
Cu$_{88}$Sn$_6$Pb$_6$ alloy	1.5 M HCl−0.08 M NdCl$_3$/Two compartments (Nafion)	−0.7	0.61	34.6	[27]
Cu$_{88}$Sn$_6$Pb$_6$ alloy	1.5 M HCl−0.08 M LaCl$_3$/Two compartments (Nafion)	−0.7	0.68	35.7	[27]
Cu$_{88}$Sn$_6$Pb$_6$ alloy	1.5 M HCl−0.33 M ZrCl$_4$/Two compartments (Nafion)	−0.7	0.45	23.7	[27]

lower over-potentials. Moreover, ILs show high thermal and chemical stability, high viscosity, acutely low vapour pressure and can be recycled or reused continuously in a flow system [80]. On the contrary, they are much more expensive than water [80]. Some recent studies indicate the potential application of ILs in CO$_2$ electroreduction [88, 90], but few of them deal with the conversion of CO$_2$ to alcohols [58].

24.3.2 Solid electrolyte

Solid electrolytes are dense membranes where the transfer of ions is driven by electrical potential difference. Depending on the nature of the electrolyte, two main types of solid electrolytes can be distinguished: SPE and ceramic SOEs.

24.3.2.1 Solid polymer

The advantage of the electrochemical reduction of CO$_2$ is that water can be used as proton source. Nevertheless, aqueous electrolytes have many disadvantages such as high ohmic loss, possibility of leakage from the reactor, increased HER and mass transfer limitation of the solubilised CO$_2$ from bulk electrolyte to electrode surface. To tackle these problems, the concept of SPE may be used. In addition, SPEs may have many advantages in comparison with liquid electrolytes, such as good tensile strength; ease of handling, which allows the fabrication of thin films with low resistance; low convection, which reduces electrode erosion increasing its lifetime; and easy separation of the reaction products [92]. Depending on their conductive nature, two types of SPEs can be distinguished: cationic, mainly H^+ conductors such as CMI-7000, Nafion, SPEEK and so on, and anionic, mainly OH^- conductors such as CMI-7001, Alkali doped PVA and 1% Amberlyst/SPEEK [92–93]. The use of most of the SPEs is restricted to low operation temperatures (< 393 K). It was concluded that the use of SPEs alleviated the mass transfer limitations of CO_2 in the electrochemical cell [93]. Moreover, the use of anion exchange membranes may help supressing proton reduction [61].

The properties of the polymeric membrane, such as chemical structure, thickness and wetting properties mark the performance of the cell, making it possible to tailor the physical and chemical characteristics of the membrane towards CO_2 electroreduction (or even towards a specific product formation). Moreover, durability and pressure manipulation of the membrane should be also considered [87].

Many studies reported the use of SPE to separate anolyte from the aqueous catholyte in divided H-type cells. A couple of papers reported the use of SPE in flow type cells, in which CO_2 was sent to the reactor as a gas or in the form of CO_2 saturated solutions. The SPE can also be used for gas-phase electrocatalytic reduction of CO_2 [92–93]. Both the current density and product distribution varied as a function of the chemical properties of the membrane, mainly because of the different ionic conductivity and ion-exchange capacity [92–93].

24.3.2.2 Solid oxide

SOEs can be classified based on their conductive nature [91]:
- Alkaline (Na^+, K^+, Li^+, etc.) conductors, of which the most usual ones are β-alumina and NASICON.
- Anionic (O^{2-}, Cl^-, F^-, etc.) conductors, of which the most common ones are O^{2-} ion conductors, represented by two main SOEs: fluorite-type and perovskite-structured oxides.
- Protonic (H^+) conductors, different types of SOEs can be distinguished depending on the operation temperature: intermediate temperature electrolytes, represented by two main SOEs: solid acid (phosphates, arsenates, sulphates, seleniates, etc.) electrolytes (473–573 K) and perovskite-based electrolytes (723–873 K) and high temperature protonic conductors (> 923 K), such as $La_xSr_{1-x}CoO_{3-\delta}$ (LSC), $La_xSr_{1-x}MnO_{3-\delta}$ (LSM), $La_xSr_{1-x}Co_yFe_{1-y}O_{3-\delta}$ (LSCF) and so on.
- Mixed ionic-electronic conductors (MIEC), which exhibit both ionic and electronic conductivity. There are two leading MIEC conductors: Ceria-based and perovskite-based electrolytes.

Anionic (mainly yttria-stabilised-zirconia, YSZ) and alkaline (β-alumina and NASICON) SOEs are well known for acting as effective materials for electrochemical promotion (via EPOC) of catalytic CO$_2$ hydrogenation applications, whereas in the case of protonic conductors, promoter species (H$^+$) takes part not only in increasing the catalytic rate via EPOC but also in the electrochemical reaction itself, oxidising (H$_2$O) and reducing (CO$_2$) reactant gases involved in the process [91].

Oxygen and proton conducting SOEs have been also used for electrochemical reduction of CO$_2$ in solid oxide electrolysis cells (SOECs). In SOECs, electrolytes should have high oxygen ion or proton conductivity and negligible electron conductivity and should be stable under variable redox conditions and CO$_2$/steam atmospheres. Meanwhile the electrolyte should be easily shaped into a dense, thin and strong film to avoid mixing of the gases at the anode and cathode chambers. There are many types of materials available as the electrolytes for SOECs. YSZ is the most usual electrolyte material due to its high ionic conductivity, low electronic conductivity, high stability and mechanical strength, low cost and easy availability. However, the low ionic conductivity of YSZ limits its application at intermediate-low temperatures [94]. Intermediate temperature proton conductors offer some benefits over oxygen ion conductors: higher ionic conductivity and proper chemical compatibility with electrodes (Ni). The most deeply studied materials during the last years are mainly ABO3 (A = Ba, Sr; B = Ce, Zr) perovskite-type oxides, since they have the best proton conductivity. Among them, BaZr$_{0.1}$Ce$_{0.7}$Y$_{0.2}$O$_{3-\delta}$ (BZCY) is shown as the best material that combines both large bulk conductivity and proper chemical stability [95].

24.4 Electrochemical cells

The main limitations of electrochemical conversion of CO$_2$ are related to the slow kinetics, high energy consumption and the low energy efficiency of the process. Different reactor configurations have been developed in order to improve process performance [82].

24.4.1 Liquid-phase CO$_2$ conversion

24.4.1.1 Cell configuration

Different research groups have used a variety of cell configurations for CO$_2$ electroreduction in the liquid phase:

Undivided three-electrode cell. In traditional undivided three-electrode electrochemical cells [17, 18, 20, 56, 59, 61, 62], dense-plate-type electrodes are immersed in a liquid that acts as both anolyte and catholyte. In these cells, product recovery is difficult and requires an additional separation step, increasing process costs [82]. There are some studies in literature on electrochemical conversion of CO_2 to alcohols in undivided systems [16, 17, 20, 23, 50, 56, 59, 61, 62, 67]. Some results for the best performing systems are displayed in Table 24.8.

Table 24.8: Electroreduction of CO_2 to alcohols in undivided three-electrode cells.

Electrode	Electrolyte	Over-potential (V)	CD (mA·cm^{-2})	Maximum FE (%)	Ref.
Perovskite (La$_{1.8}$Sr$_{0.2}$CuO$_4$) based carbon GDE	0.5 M KOH	Up to −3	180	6.9 (EtOH), 1.7 (n-PrOH)	[16]
Cu-based carbon GDE		Up to −3	180	5 (EtOH), 1 (n-PrOH)	[16]
Perovskite (Pr$_{1.8}$Sr$_{0.2}$CuO$_4$)-based carbon GDE		Up to −3	180	7.4 (EtOH), 1.1 (n-PrOH)	[16]
Perovskite (Gd$_{1.8}$Sr$_{0.2}$CuO$_4$)-based carbon GDE		Up to −3	180	6.4 (EtOH), 1.4 (n-PrOH)	[16]
Perovskite (La$_{1.8}$Thr$_{0.2}$CuO$_4$)-based carbon GDE		Up to −3	180	4.9 (EtOH), 0.61 (n-PrOH)	[16]
Perovskite (YBa$_2$Cu$_3$O$_{6-7}$)-based carbon GDE		Up to −3	180	3.2 (EtOH), 0.9 (n-PrOH)	[16]
Pre-oxidised Cu foil (17 h, 403 K)	0.5 M KHCO$_3$	−1.55	7.1	–	[17]
Anodised Cu foil		−1.25	1.4	120 (MeOH)	[17]
Pre-oxidised Cu-TiO$_x$ (50 min., 573 K)		−1.06	0.74	180 (MeOH)	[17]
Pre-oxidised Cu-TiO$_x$ (45 min., 773 K)		−0.45	0.30	30 (MeOH)	[17]
Air-furnace oxidised Cu	0.5 M KHCO$_3$	−1.50	10	2 (MeOH)	[20]
Mo foil on Cu wire	0.2 M Na$_2$SO$_4$	−0.80	0.05	55 (MeOH)	[56]
Mo foil on Cu wire	0.05 M H$_2$SO$_4$	−0.69	0.31	46 (MeOH)	[56]
RuO$_x$ thermally deposited-Ti	0.5 M KHCO$_3$	−0.80	2	30.5–17.2 (MeOH)	[62]
RuO$_x$/Cd thermally deposited-Ti		−0.80	2	20.4–38.2 (MeOH)	[62]

Two compartment cell. Most of the studies on electrochemical conversion of CO_2 mainly to methanol have utilised a two-compartment electrochemical cell configuration. In these systems, the cathode and anode chambers are separated by an ion conducting membrane (H-type electrochemical cell) [1, 6, 15, 16, 19, 21, 25–27, 29, 32, 42, 55, 57, 63–66, 69, 96], an agar bridge [60, 81] or a glass frit [2, 3, 66–68, 77, 82], resulting in an improved separation of products and in the avoidance of reoxidation reactions. Some results for the best performing systems are shown in Table 24.9.

Membrane electrode assembly (MEA) flow cell. Continuous flow reactors have some advantages compared with their batch counterparts: increased mass transfer and improved mixing of different phases and better residence time, temperature and heat transfer control in the reactor [87, 88]. In MEA devices the two electrodes are separated by an ion-exchange membrane, pressed together on both sides of the membrane with no flow channels between them, enhancing contact and transport of species between electrodes [97]. The electrolyte/gas is fed to the electrodes, mostly using gas diffusion electrodes (GDEs) formed by immobilising the catalyst on a gas diffusion layer (GDL). Some studies reported the use of SPE in flow type cells, in which CO_2 was sent to the reactor in the liquid phase (CO_2 saturated electrolyte solution) [4, 6, 14, 19, 51, 53, 92, 98, 99]. For instance, the performance of Cu_2O and Cu_2O/ZnO has been studied in a filter-press electrochemical cell for the continuous electroreduction of CO_2 to methanol [6], ethanol and propanol [19]. Obtained results are summarised in Table 24.10. The electrodes including ZnO were stable, in contrast with Cu_2O-deposited on carbon papers, which showed a strong deactivation with time [6, 19].

24.4.1.2 Electrode structure

Optimising electrode performance, and consequently reactor performance, requires optimising all of the transport processes, which strongly depend on the electrode structure. Metallic electrodes applied in CO_2 electroreduction can be classified as follows: bulk metal (plate, foil, etc.) electrodes, metals electrodeposited on a metallic/glassy carbon support (GCE) or, metals supported on GDLs.

Bulk metal electrodes. The CO_2 electroreduction to alcohols was studied at various Cu foil electrodes (i.e. anodised Cu foil, Cu foil thermally oxidised and air-oxidised Cu foil) [17, 20, 42, 83]. However, in general, the small surface area, together with the low surface concentration of CO_2 at the electrode, resulting from the reduced CO_2 solubility in aqueous solutions, gave rise to a limited performance of bulk metal planar electrodes [83].

Metal/Glassy carbon (GCE) supported electrodes. Electrochemical conversion of CO_2 into alcohols has been carried out at different metal [2, 15, 20, 56, 57, 60–62, 64–69, 77, 81] and glassy carbon [23, 100, 101] supported electrodes. As an example, the CO_2 electrochemical conversion to methanol was studied over a Cu_2O thin film electrodeposited on stainless steel. FE (38%) were remarkably higher than those obtained for air-oxidised or anodised Cu electrodes [20].

Gas diffusion electrodes (GDEs). In a GDE the catalytic material (polymer bonded catalyst particles) is dispersed by different methods onto a porous structure

Table 24.9: Electroreduction of CO_2 to methanol in two compartment cells.

Electrode	Electrolyte/cell	Over-potential (V)	CD (mA·cm⁻²)	Maximum FE (%)	Ref.
Cu/CuO nanopowder-carbon GDE	1 M $KHCO_3$/ Two compartments (Nafion)	-1.40	17.3	2.5	[21]
RuO_2+TiO_2 (35/65)-Ti	0.05 M H_2SO_4/ Two compartment	-0.55	5	24	[57]
RuO_2+MoO_2+TiO_2 (25/30/45)-Ti		-0.55	5	12	[57]
RuO_2+Co_3O_4+SnO_2+TiO_2 (20/10/8/62)-Ti		-0.55	5	7	[57]
RuO_2+TiO_2 (35/65)-Ti	Phosphate buffer 0.2 M/ Two compartment	-1.44	0.08	35	[57]
RuO_2+Co_3O_4+SnO_2+TiO_2 (20/10/8/62)-Ti	0.2 M Na_2SO_4/ Two compartment	-1.49	0.05	53	[57]
RuO_2/TiO_2 nanoparticles-Pt	0.5 M $NaHCO_3$/ Two compartments H-type cell (Nafion)	-0.80	1.2	40.2	[64]
Fe(II) TPP (tetraphenylporphyrin)-Pt	2-hydroxyl-1-nitrosonaphthalene-3,6-disulphonatocobal(II) l in 0.1 M KCl-CH_3OH / Two compartment (H^+ membrane)	-0.5	0.36	12.2	[65]
Co(II) TPP-Pt		-0.5	0.39	15.1	[65]
Ni(II) TPP-Pt		-0.5	0.15	14.4	[65]
Cr(III) TPPCl-Pt		-0.5	0.1	11.6	[65]
Fe(III) TPPCl-Pt		-0.5	0.37	11.1	[65]
Indigo/graphite-Fe	0.1 M KCl-10 mM Na_3[Fe(CN)$_5$(H_2O)])-15 mM CH_3OH/ Two compartment (glass frit)	-0.7		42.8	[67]
2-aminoanthraquinone/graphite-Fe		-0.7		31	[67]
Alizarin ($C_{14}H_8O_4$)/ graphite/Fe		-0.7		30.8	[67]
p-benzoquinone/graphite-Fe		-0.7		36	[67]
Everitt's salt coated-Pt	[Fe(C_6H_2(OH)$_2$(SO_3)$_2$)$_2$]-CH_3OH/ Two compartment (glass frit)	-0.90	–	14.5	[68]
Everitt's salt coated-Pt	K[Cr(C_2O_4)$_2$(H_2O)$_2$)]-CH_3OH/ Two compartment (glass frit)	-0.9	–	14.5	[68]
Fe-S/Pan/PB-Pt (PAn: polyaniline)	0.5 M KCl/Two compartments H-type	-0.8	–	6.8	[69]
Fe-T/Pan/PB-Pt		-0.8	–	8.3	[69]
Fe-C/Pan/PB-Pt-self assembled	0.2 M KCl/ Two compartments H-type	-0.8	–	10.1	[69]
n-GaAs-crystal-(110)Ga	0.2 M Na_2SO_4/ Two compartments (agar bridge)	-1.2--1.4	0.13	14	[81]

Table 24.10: Electroreduction of CO$_2$ to alcohols in a membrane electrode assembly (MEA) flow cell.

Electrode	Electrolyte/Cell	Overpotential (V)	CD (mA·cm^{-2})	Maximum FE (%)	Ref.
Benchmark MOF HKUST-1	0.5 M KHCO$_3$/GDE filter-press cell	−0.9	10	10.3 (EtOH), 5.6 (MeOH)	[4]
[Cu$_2$(m3-adeninate)$_2$(m2-OOC(CH$_3$)$_2$)]n (CuAdeAce)	(Nafion 117)	−1.75	10	0.5 (EtOH), 0.7 (MeOH)	[4]
CuMOA with bis-bidentate dithiooxamidate (DTA) (CuDTA)		−1.41	10	4.1 (EtOH), 1.9 (MeOH)	[4]

(usually a carbon-based material) support [97]. The use of GDEs led to a significant enhancement in the performance of CO$_2$ electroreduction to alcohols in liquid phase by improving the mass transfer of CO$_2$ [4, 6, 18, 19, 21, 24, 36, 51, 53, 102–108].

24.4.2 Gas-phase CO$_2$ conversion

Solution-phase electrochemical cells are suitable for comparing inherent activities of different catalysts but impractical for the electrochemical conversion of CO$_2$ to alcohols on a preparative scale. Moreover, the rate of CO$_2$ reduction is limited by the relatively low solubility of CO$_2$ in aqueous solution [5].

24.4.2.1 Polymer electrolyte membrane cell

A polymer electrolyte membrane (PEM) cell is basically a MEA flow cell where CO$_2$ is directly fed as gas into the cathode compartment and a SPE is responsible for the conduction of protons or anions, separation of products gases and electrical insulation of the electrodes. This configuration may overcome mass transfer limitations and facilitate the separation of products [97]. The available literature in this regard is scarce [92, 93, 109–118]. The gas-phase electrocatalytic conversion of CO$_2$ to CH$_3$OH (with modest current efficiencies) was investigated at GDEs, which consisted of Cu electroplated on carbon paper over different cationic and anionic SPEs [92–93]. The same process was also carried out (with relatively high FE) at carbon-supported Pt [109] and Pt–Ru [110] GDEs. It seemed that Ru addition efficiently promotes CO$_2$ reduction to CH$_3$OH. Nanostructured carbon-based electrodes, such as CNTs are reported to promote electron/ion transport and inhibit active catalyst from mechanical and chemical degradation. Recently, Jimenez et al. [111] investigated the electrocatalytic conversion of CO$_2$ to methanol in the gas phase using CNT-supported Pt catalysts. Sebastian et al. [112] studied the co-electrolysis of CO_2 and water in a similar SPE system but using $IrRuO_x$ as anode. The main products were methanol on PtRu/C and a mixture of alcohols on Ru/C catalyst. More recently, Cu supported on different carbon-based materials, graphite, activated carbon and carbon nanofibers (CNFs), was employed to selectively catalyse CO_2 conversion to alcohols. IrO_2 was used as anode [113]. Other CNT-supported metal (Fe, Pt, Co) catalysts were also explored for electrocatalytic conversion of CO_2 to alcohols (mainly to isopropanol) [114–117]. Fe/CNTs showed better activity than Pt/CNTs but were unstable. Fe–Co/CNTs catalysts were more stable. Fe nanoparticles supported on nitrogen doped CNTs (NCNTs) yielded enhanced electrocatalytic activity and selectivity for CO_2 reduction to isopropanol. The main results obtained are summarised in Table 24.11.

24.4.2.2 Solid oxide electrolyte cell

A SOE cell consists of a dense SOE membrane and two or three porous electrodes. In closed circuit operation, the ions travel from one electrode to the other where they react with the gaseous content of that chamber. SOE cells have been used as electrochemical membrane reactors (SOEMRs) in which heterogeneous catalytic reactions were carried out. The typical configuration of a SOE cell is the double chamber cell, in which the anode and cathode are separated so that each of them can be exposed to different gas mixtures. The second cell configuration is the single chamber cell, where both electrodes are exposed to the same reaction mixture with one/two of them (counter/reference electrode) considering inert for the reaction of interest. In this case, the solid electrolyte simply acts as an ion conducting catalyst support. Solid electrolyte cells can be used to electrochemically enhance catalytic reaction rates by pumping promoting ions (O^{2-}, K^+, etc.) to or away from the catalyst electrode surface during the process, causing a drastic change in catalytic activity and/or product selectivity and thus in product yield [119].

Solid oxide electrolysers (double chamber SOE cells). Solid oxide electrolysers can exploit disposable exhaust heat streams to maximise energy efficiency and work at higher temperatures (typically > 923 K), offering both thermodynamic and kinetic benefits over low temperature co-electrolysis, resulting in a higher electrolysis efficiency [120–122]. At high temperatures, activation of CO_2 becomes easy and mass transfer is not the limiting step for the electrochemical reduction of CO_2, resulting also in higher current density [94]. Moreover, they enable the use of cheap oxide-derived electrodes, instead of the noble metals typically used in solution-based electrolysis cells [120].

Ideally, in a solid oxide electrolysis cell using oxygen-ion-conducting electrolytes, CO_2 and H_2O at the cathode receive electrons and decompose to form oxygen ions, which pass through the electrolyte membrane to reach the anode where they

Table 24.11: Gas phase electrocatalytic conversion of CO$_2$ to alcohols in PEM cells.

Electrode	Electrolyte	Over-potential (V)	CD (mA·cm^{-2})	Maximum FE (%)	Ref.
Cu$_2$O–carbon paper	SPE (CMI-7000)	−2	2.4	20 (MeOH)	[92]
Cu$_2$O–carbon paper	SPE (AMI-7001)	−2	3.7	5 (MeOH)	[92]
Cu–carbon paper	SPE (Nafion)	–	11.1	0.54 (MeOH)	[93]
Cu–carbon paper	SPE (SPEEK)	–	8.9	0.13 (MeOH)	[93]
Pt/C–carbon paper	SPE (Nafion 117)	−0.35	–	40 (MeOH)	[109]
Pt–Ru/C–carbon paper		−0.4	20	35 (MeOH)	[110]
Pt–CNTs-carbon paper		−0.45	15	75 (MeOH)	[110]
Pt–CNTs-carbon cloth	SPE (Nafion)	–	16	1.9 (MeOH)	[111]
Pt–Ru/C-carbon cloth	SPE (Nafion 115)	1.25	1	0.33 (MeOH)	[112]
Ru/C–carbon cloth		1.25	0.5	0.91 (MeOH+ EtOH + i-PrOH)	[112]
Cu–Graphite GDE	SPE (Sterion)	−1.7	30	80 (MeOH selectivity)	[113]
Cu-Activated carbon GDE		−3	30	45 (MeOH selectivity)	[113]
Cu–CNFs GDE		−2.5	30	5/10/12 (MeOH/EtOH/n-PrOH selectivity)	[113]
Pt–CNTs–carbon cloth	SPE (Nafion 117)	–	–	0.0005 (MeOH), 0.007 (EtOH), 0.016 (i-PrOH) μmoles/h cm^2	[114–117]
Fe–CNTs–carbon cloth		–	–	0.0005 (MeOH), 0.01 (EtOH), 0.022 (i-PrOH) μmoles/h cm^2	[114–117]
Fe- NCNTs-carbon cloth		–	–	0.001 (MeOH), 0.0005 (EtOH), 0.06 (i-PrOH) μmoles/h cm^2	[114–117]

lose electrons and are reoxidised to oxygen gas [94]. At the same time the reduced absorbed carbon species can react with the activated hydrogen to yield different products. For a solid oxide electrolysis cell using protonic electrolytes, H_2O at the anode is oxidised to oxygen accompanied with the production of protons, which then pass through the electrolyte membrane to arrive at the cathode where they react with adsorbed CO_2 to produce different value-added products [121]. Most of the work in this area has focusing on attempting to co-electrolyse CO_2 and steam to produce both CO and H_2 simultaneously for downstream processing in a Fischer–Tropsch (F–T) reactor [80, 86] to produce higher hydrocarbon products or alcohols [123]. However, this design has an inherent limitation: long gas transport networks connecting two separated processes lead to extra cost and large energy dissipation. Therefore, the integration of co-electrolysis of H_2O and CO_2 (favoured at high temperature) and F–T, or other synthesis processes (favoured at low temperature), in a single unit can potentially decrease the investment (compact equipment) and increase the energy efficiency (by proper heat integration) of combined system. In the future, to obtain more valuable products, the electrolyte, cathode and anode materials should be developed to enable SOECs to be operated at a temperature range of 473–873 K. At those temperatures, SOECs may combine the benefits of the low and the high temperature system to produce various products with high CD [94]. This integration can be realised by using intermediate temperature SOEs operating at temperatures compatible with methane (e. g. perovskites) [124], or with alcohol (e. g. solid acids) synthesis or by establishing a temperature gradient in the cell (e.g. in a tubular cell), which enables one part to be controlled at high temperature for co-electrolysis whereas the other is at reduced temperature for the synthesis reactions, resulting in a significant improvement in hydrocarbon yield [124–125].

Electrochemically promoted CO_2 hydrogenation (single-chamber SOE cells). Electrochemically promoted CO_2 hydrogenation to, among other compounds, alcohols (methanol and ethanol) and DME has been carried out in single-chamber SOEMRs, which consisted of catalyst films of Pt [126, 127], Pd [127], Cu [128] or Fe–TiO_2 [129], simultaneously acting as catalyst electrodes, deposited on ion conducting solid electrolyte supports, such as YSZ (an O^{2-} conductor) [127, 129] or K-βAl_2O_3 (a K^+ conductor) [126, 128]. In these systems, the application of small potentials between the catalyst- and the counter-electrode (Au) results in movement of promoting ions to (or from) the catalyst surface, providing an electrochemical modification of relative chemisorption between reactants that allows increasing the catalytic activity to higher reaction rate, and thus, the operation of the catalyst under milder conditions (with subsequent energy savings), tuning the selectivity to the desired product (increasing energy efficiency in product formation), enhancing catalyst tolerance to feed poisons (electrochemically retarding its chemisorption) and extending catalyst life time by in situ electrochemical regeneration, while monitoring and controlling the process (by an attached electrochemical sensor). These works studied the effect of metal based catalyst (Pt, Pd, Cu or Fe), solid oxide electrolyte (YSZ or K-βAl$_2$O$_3$), film preparation procedure ("paint-brushing", "dip-coating" or "electroless") and operating parameters (potential, temperature, H_2/CO_2 ratio, etc.) on the extent of the CO_2 hydrogenation reaction, on the selectivity to different fuels and on the level of electropromotion of the catalyst. Electrochemically promoted CO_2 hydrogenation was studied in an available bench-scale plant, using a tubular electrochemical catalyst configuration, concentrated CO_2 streams, representative of CO_2 capture exiting gases, and changing H_2/CO_2 ratios (from 1 to 4), to simulate a discontinuous renewable H_2 flow. Results (see Table. 24.12) revealed that catalytic CO_2 hydrogenation reactions can be electrochemically enhanced and the selectivity to the different target products can be modulated by modifying the applied potential [126–129].

Table 24.12: Electrochemically promoted catalytic CO_2 hydrogenation to alcohols in single-chamber SOEMRs.

Electrode	Electrolyte/cell	Over-potential (V vs. Au)	CO_2 conversion (%)/ ΔCO_2 conversion	Selectivity (%)/ΔSelectivity	Ref.
Dip-coated Pt	K-β-Al$_2$O$_3$/ tubular-single chamber, H_2/ CO_2=3	1	≈1/-	1.5/27 (MeOH), 8.5/16 (EtOH)	[126]
Paint-brushed Pt	YSZ/ tubular-single chamber, H_2/CO_2=2	0.5	24/3.2	8/800 (MeOH)	[127]
Electroless-coated Pd	YSZ/ tubular-single chamber, H_2/CO_2=2	≈O.C.	15/1.3	96/1.4 (DME)	[127]
Electroless-coated Cu	K-β-Al$_2$O$_3$/ tubular-single chamber, H_2/ CO_2=2/4*	2.5/1**	25/4.3 (*,**)	55.4/34 (MeOH),19.7/22 (EtOH), 87.1/3.4 (DME)**	[128]
Dip-coated Fe–TiO$_2$	YSZ/ tubular-single chamber, H_2/CO_2=3/4*	−1.5/-0.5**/1***	15/3.7	50/50 (MeOH) **,15/150 (EtOH)*, 47/1.7 (DME)***	[129]

24.5 Operating conditions

Despite the fact that operating conditions may highly influence product yields and selectivities, only a few reports to date have focused on evaluating the effect of pH, temperature and pressure on the liquid-phase CO_2 electroreduction performance [30, 130]. In the case of CO_2 electrocatalytic conversion to alcohols in gas phase, this evaluation is practically unexplored yet.

24.5.1 Liquid-phase CO_2 conversion

24.5.1.1 pH

Several reports can be found in the literature for the effect of pH in the electroreduction of CO_2 to CH_3OH [27, 30, 40, 60, 63, 84, 86, 130, 131]. The obtained results indicated that, in general, on decreasing the pH of the reaction medium, the rate of CO_2 reduction to alcohols significantly increases, with subsequent enhancement in both current density and Faradaic efficiency.

24.5.1.2 Temperature

The product distribution in the CO_2 electroreduction to alcohols is strongly affected by reaction temperature [60, 86, 132, 133]. Competitive hydrogen formation was significantly depressed, whereas solubility of CO_2 in the reaction medium was considerably increased with decreasing temperature.

24.5.1.3 Pressure

The study of the effect of pressure on alcohol formation from CO_2 electroreduction is almost unexplored yet. The electroreduction of CO_2 in aqueous solutions can be limited by the low solubility of CO_2 in the reaction medium. To speed up the reaction process, pressurised CO_2 is usually applied, which often causes a certain degree of change in product selectivity, current density and FE [103, 104, 134].

24.5.2 Gas-phase CO_2 conversion

24.5.2.1 Potential

In electrochemically promoted CO_2 hydrogenation to alcohols, applied potential considerably affects CO_2 conversion and product distribution, as a result of its effect on the chemisorptive bond strength of reactants (H_2 and CO_2) and intermediate surface species.

Formation of CH_3OH and C_2H_5OH are reported to be promoted under conditions where surface coverage of both reactants (CO_2 and H_2) are expected to be very similar, under low potassium or oxygen ion surface coverages, which correspond to positive potentials [126, 128] and to slightly positive or negative potentials [127, 129] for K-βAl_2O_3 and YSZ, respectively. In general, CO_2 does not adsorb on clean metal (Pt [126, 127], Cu [128], Fe [129]) surfaces, that is, in the absence of promoter ions (at highly positive potentials for K^+ or at nearly open circuit for O^{2-}), and, at the same time metals (Pt > Cu) are somewhat catalytic for hydrogen evolution (at very negative potentials). Therefore, a strong competition between CO_2 and H_2 could be expected, and thus according to this mechanism, it is expected that the formation of CH_3OH and C_2H_5OH (and DME) will be limited by the adsorption of both H_2 and CO_2. Thus, there is a given value of potential or promoter coverage, which optimises catalyst activity and selectivity to the desired product, which depends on temperature and gas composition.

24.5.2.2 Temperature

The effect of temperature on electrochemically promoted CO_2 hydrogenation to alcohols was studied over Cu–K–βAl_2O_3 [128] and Fe–TiO_2/YSZ [129]. For H_2/CO_2 ratios of three, which matched with the theoretical stoichiometry in methanol and ethanol synthesis reactions from CO_2 hydrogenation, so that they are thermodynamically favoured under these conditions, both CO_2 conversion and selectivity to CH_3OH exhibit a maximum at a certain temperature for whatever the applied potential [128, 129], whereas selectivity to C_2H_5OH [128] decreases as temperature increases. This agrees with the fact that equilibrium conversion of CO_2 to methanol and ethanol decreases as temperature increases. Moreover, C_2H_5OH can also be formed by hydrogenation or hydration of methanol. In addition, methanol converted to dimethyl ether at 523–573 K and both are transformed into hydrocarbons at temperatures of about 673 K, where reverse water gas shift and methanation reactions are also favoured. Decreasing the applied potential (around open circuit conditions) leads to a significant increase in the maximum CH_3OH selectivity, which is obtained at lower temperature (about 548 K) [128, 129].

24.5.2.3 H₂/CO₂ ratio

In general, in electrochemically promoted CO_2 hydrogenation to alcohols on Pt–K-βAl_2O_3 [126], Pt–YSZ [127] and Cu–K-βAl_2O_3 [128], there is an increase in the maximum CO_2 conversion and a slight decrease in promotion level on increasing the H_2/CO_2 ratio, supposedly, as a result of the increase in hydrogen availability [126–128] with respect to that stoichiometrically required (H_2/CO_2=3) for the synthesis reaction. Thus, selectivity to CH_3OH and C_2H_5OH slightly decreases [126, 127] on increasing the H_2/CO_2 ratio from 3 to 4. CH_3OH and C_2H_5OH selectivities exhibited a maximum for a H_2/CO_2 ratio of 2 (under optimum conditions of low promoter surface coverage), which is that required for stoichiometric synthesis of methanol and ethanol by the hydrogenation of the CO resulting from CO_2 dissociative adsorption, and therefore, this ratio thermodynamically favours the formation of CH_3OH and C_2H_5OH as the expenses of other CO_2 hydrogenation products [127, 128]. In contrast, over $Fe-TiO_2/YSZ$ [129], CO_2 conversion and selectivity to CH_3OH showed a maximum for a stoichiometric H_2/CO_2 ratio of 3.

24.5.2.4 Flow rate

The effect of gas flow rate on the electropromoted CO_2 hydrogenation to alcohols (CH_3OH and C_2H_5OH) over $Cu-K-\beta Al_2O_3$ [128] and $Fe-TiO_2-YSZ$ [129] has been analysed. Increasing gas flow rate for a H_2/CO_2 ratio of 3 at a constant temperature of 598 K, resulted in a decrease in the maximum CH_3OH and C_2H_5OH selectivity, which, in the case of Cu [128], also shifted to more negative potentials.

24.6 Conclusions

The electrochemical conversion of CO_2 to alcohols (mainly methanol, ethanol and n-/i-propanol) is reported to occur at various types of electrocatalysts, such as metals, metal alloys, metal complexes, metal oxides/chalcogenides and so on.

Among the available metals, the most active for the process have been identified to be Mo, Ru, Ti, Fe, Ga, In, Pd, Pt and, in particular, Cu which has been used as metal foils, single crystal Cu specimens, Cu alloys, Cu perovskites and Cu deposits on carbon-based materials or Nafion-based SPE.

Substantial efforts have been pursued to enhance the performance of Cu-based electrocatalysts through altering configuration, surface structures, morphologies and compositions. Alloying Cu with other metals improved the reversibility, selectivity and reaction rate for the CO_2 reduction to alcohols at lower over-potentials. Nanostructured copper has also been identified as a promising electrocatalyst for the electrochemical reduction of CO_2 into alcohols with improved performance in terms of Faradaic efficiency, selectivity, current density, stability and onset potential. $Cu_2S-Cu-V$ core–shell nanoparticles were also found to enhance CO_2 reduction to alcohols. An excellent alternative to improve electrode performance for CO_2 reduction is reported to be the use of GDEs with both, protonic and anionic, SPEs. In this way, Cu containing metal organic porous materials (MOPMs) supported in GDEs promoted CO_2 conversion to alcohols.

Among the studied materials, copper (Cu), molybdenum (Mo) and ruthenium (Ru), as well as their mixtures and oxidised forms have been reported to be the most active materials for the process. Particularly, Cu oxide electrodes (Cu (I) species in Cu$_2$O) seem to be the most promising materials in terms of combined product selectivity and current density, with FE greater than 100%. However, alcohol formation rates tend to diminish at longer reaction times. Cu$_2$O/ZnO-based electrodes presented better stability than Cu$_2$O itself. In addition, graphene/Cu$_2$O exhibited higher activity and selectivity for alcohol production compared to Cu$_2$O.

Different reactor configurations have been developed in order to improve process performance and bring the technology closer to industrial scale, such as the application of electrochemical membrane reactors, based on solid polymer (SPE) or solid oxide (SOE) electrolytes. Liquid–liquid solid polymer cells are the most studied. Little work has been done on continuous-flow MEA (membrane electrode assembly) cells, even though this might be the only scalable configuration. Works based on use of a gas phase at the cathode are emerging with the aim to overcome mass transfer limitations. Both SPE and SOE electrochemical membrane reactors can provide gas-phase co-electrolysis of CO$_2$ and H$_2$O to alcohols, but the efficiencies are still low and further developments are needed. In addition, electrochemical promotion of CO$_2$ conversion to alcohols in SOE cells is reported to be a promising alternative to improve the efficiency and selectivity of the process.

24.7 Prospects and challenges

Although the electrochemical conversion of CO$_2$ to alcohols shows great potential, significant technological advances are still necessary for the process to become viable and commercially applicable. The main challenge for the advancement of this technology is to increase the energy efficiency (or reduce the energy cost) of the process and the tendency is to maximise alcohol production with a minimum energy input. High-energy efficiency is achieved through a high selectivity (high Faradaic efficiency) and low over-potentials. On the other hand, reaction rate (measured as current density) determines the reactor size and the capital cost (CAPEX), whereas reaction conditions dictate the operating costs (OPEX) of the process.

Despite the many advances made in the electrochemical reduction of CO$_2$ in aqueous medium there are still a number of challenges to overcome such as the high over-potential required, the low solubility of CO$_2$ in water at room temperature and pressure, the formation of a mixture of products whose separation implies a cost and the plugging and deactivation of the electrodes by impurities. Therefore, there are still possibilities for improvement: optimisation of reaction conditions to enhance reaction rates, which allow increasing the concentration of CO$_2$ in the aqueous electrolyte, that is, by lowering temperature and increasing pressure; use of novel reaction media (non-aqueous solutions, ILs, etc.) that improve the solubility of CO_2 and avoid the formation of hydrogen; enhancement of the activity, selectivity and stability of the electrocatalysts by exploring new composite and nanostructured materials; optimisation of electrode/reactor and cell designs, such as the use of GDEs and electrolyte membrane-based electrochemical cells, which allow reduction of internal resistance and improved mass transfer, and thus enhance reaction rate or current density; development and testing in a continuous operation mode of easily scalable flow cells and so on.

One way to improve CO_2 reduction, to lessen over-potentials and increase current density is by increasing the operation temperature. SOE cells that conduct ions (O^{2-}, K^+, H^+, etc.) allow the conversion of CO_2 at higher temperatures (> 673 K). Although the electrochemical promotion of catalytic CO_2 hydrogenation to different products (CO, CH_4, CH_3OH, etc.) and the co-electrolysis of CO_2 and H_2O (to synthesis gas and CH_4) in SOE cells have undergone important advances, additional research is still necessary to improve the efficiency, selectivity, stability and durability of the electrocatalysts, to lower the operating temperature of the system (increase in energy efficiency and catalyst lifetime), lower the cost of the process (material cost minimisation, compact reactor design, etc.) and to obtain greater simplicity and scalability in the preparation of materials (electrocatalysts, solid electrolytes, etc.) and in reactor design. In the future, the electrolyte, cathode and anode materials should be developed to allow solid oxide electrolysers to be operated at a temperature around 573 K, in order to also produce alcohols (methanol, ethanol and propanol) with high CD, and in a continuous mode of operation.

References

[1] Jia F, Yu X, Zhang L. Enhanced selectivity for the electrochemical reduction of CO_2 to alcohols in aqueous solution with nanostructured Cu-Au alloy as catalyst. J. Power. Sources. 2014, 252, 85–89.

[2] Geioushy RA, Khaled MM, Hakeem AS, Alhooshani K, Basheer C. High efficiency graphene/Cu_2O electrode for the electrochemical reduction of carbon dioxide to ethanol. J. Electroanal. Chem. 2017, 785, 138–43.

[3] Geioushy RA, Khaled MM, Alhooshani K, Hakeem AS, Rinaldi A. Graphene/ZnO/Cu_2O electrocatalyst for selective conversion of CO_2 into n-propanol. Electrochim. Acta. 2017, 245, 456–62.

[4] Albo J, Vallejo D, Beobide G, Castillo O, Castaño P, Irabien A. Copper-based metal-organic porous Materials for CO_2 electrocatalytic reduction to alcohols. ChemSusChem 2017, 10, 1100–9.

[5] Albo J, Alvarez-Guerra M, Castaño P, Irabien A. Towards the electrochemical conversion of carbon dioxide into methanol. Green. Chem. 2015, 17, 2304–24.

[6] Albo J, Sáez A, Solla-Gullón J, Montiel V, Irabien A. Production of methanol from CO_2 electroreduction at Cu_2O and Cu_2O/ZnO-based electrodes in aqueous solution. Appl. Catal. B. 2015, 176–177, 709–17.

[7]   Zhuang TT, Liang ZQ, Seifitokaldani A, et al. Steering post-C-C coupling selectivity enables high efficiency electroreduction of carbon dioxide to multi-carbon alcohols. Nat Catal 2018, 1, 421–8.

[8]   Durst J, Rudnev A, Dutta A, et al. Electrochemical CO$_2$ reduction-A critical view on fundamentals, materials and applications. Chimia. 2015, 69, 769–76.

[9]   Whipple DT, Kenis PJA. Prospects of CO$_2$ utilization via direct heterogeneous electrochemical reduction. J Phys Chem Lett 2010, 1, 3451–58.

[10] Hirunsit P, Soodsawang W, Limtrakul J. CO$_2$ electrochemical reduction to methane and methanol on copper-based alloys: Theorical insight. J. Phys. Chem. 2015, 119, 8238–8249.

[11] Qiao J, Liu Y, Hong F, Zhang J. A review of catalysts for the electroreduction of carbon dioxide to produce low-carbon fuels. Chem. Soc. Rev. 2014, 43, 631–75.

[12] Kuhl KP, Hatsukade T, Cave ER, Abram DN, Kibsgaard J, Jaramillo TF. Electrocatalytic conversion of carbon dioxide to methane and methanol on transition metal surfaces. J. Am. Chem. Soc. 2014, 136, 14107–13.

[13] Zhang W, Hu Y, Ma L, et al. Progress and perspective of electrocatalytic CO$_2$ reduction for renewable carbonaceous fuels and chemicals. Adv. Sci. 2018, 5, 1700275, 1–24.

[14] Yuan J, Zhang J-J, Yang M-P, Meng W-J, Wang H, Lu J-X. CuO nanoparticles supported on TiO$_2$ with high efficiency for CO$_2$ electrochemical reduction to ethanol. Catal. 2018, 8, 171, 1–11.

[15] Malik MI, Malaibari ZO, Atieh M, Abussaud B. Electrochemical reduction of CO$_2$ to methanol over MWCNTs impregnated with Cu$_2$O. Chem. Eng. Sci. 2016, 152, 468–77.

[16] Schwartz M, Cook RL, Kehoe VM, MacDuff RC, Patel J, Sammells AF. Carbon dioxide reduction to alcohols using perovskite-type electrocatalysts. J. Electrochem. Soc. 1993, 140, 3, 614-18.

[17] Frese KW. Electrochemical reduction of CO$_2$ at intentionally oxidized copper electrodes. J. Electrochem. Soc. 1991, 138, 3338–44.

[18] Chang TY, Liang RM, Wu PW, Chen JY, Hsieh YC. Electrochemical reduction of CO$_2$ by Cu$_2$O-catalyzed carbon clothes. Mater. Lett. 2009, 63, 1001–3.

[19] Albo J, Irabien A. Cu$_2$O-loaded gas diffusion electrodes for the continuous electrochemical reduction of CO$_2$ to methanol. J. Catal. 2016, 343, 232–9.

[20] Le M, Ren M, Zhang Z, Sprunger PT, Kurtz RL, Flake JC. Electrochemical reduction of CO$_2$ to CH$_3$OH at copper oxide surfaces. J. Electrochem. Soc. 2011, 158, E45–E49.

[21] Lan Y, Ma S, Lu J, Kenis PJA. Investigation of a Cu(core)/CuO(shell) catalyst for electrochemical reduction of CO$_2$ in aqueous Solution. Int. J. Electrochem. Sci. 2014, 9, 7300–8.

[22] Ren D, Deng Y, Handoko AD, Chen CS, Malkhandi S, Yeo BS. Selective electrochemical reduction of carbon dioxide to ethylene and ethanol on copper (I) oxide catalysts. ACS Catal. 2015, 5, 2814–21.

[23] Yadav YSK, Purkait MK. Electrochemical studies for CO$_2$ reduction using synthesized Co$_3$O$_4$ (anode) and Cu$_2$O (cathode) as electrocatalysts. Energy Fuels 2015, 29, 6670–77.

[24] Ikeda S, Ito K, Noda H. 2009. Electrochemical reduction of carbon dioxide using gas diffusion electrodes loaded with fine catalysts.: AIP Conference Proceedings 1136, 108 (Accessed August 8, 2018, at https://doi.org/10.1063/1.3160110).

[25] Watanabe M, Shibata M, Kato A, Azuma M, Sakata T. Design of alloy electrocatalysts for CO$_2$ reduction III. The selective and reversible reduction of CO$_2$ on Cu alloy electrodes. J. Electrochem. Soc. 1991, 138, 3382–89.

[26] Watanabe M, Shibata M, Kato A, Sakata T, Azuma M. Design of alloy electrocatalysts for CO$_2$ reduction. Improved energy efficiency, selectivity, and reaction rate for the CO$_2$ electroreduction on Cu alloy electrodes. J. Electroanal. Chem. 1991, 305, 319–28.

[27] Schizodimou A, Kyriacou G. Acceleration of the reduction of carbon dioxide in the presence of multivalent cations. Electrochim. Acta. 2012, 78, 171–76.

[28] Lim RJ, Xie M, Sk MA, et al. A review on the electrochemical reduction of CO_2 in fuel cells, metal electrodes and molecular catalysts. Catal Today 2014, 233, 169–80.

[29] Yang H-P, Yue Y-NA, Qin S, Wang H, Lu J-X. Selective electrochemical reduction of CO_2 to different alcohol products by an organically doped alloy catalyst. Green. Chem. 2016, 18, 3216–20.

[30] Hori Y. Electrochemical CO_2 reduction on metal electrodes. In: Vayenas CG, White RE, Gamboa-Aldeco ME, eds. Modern Aspects of Electrochemistry, New York, NY, USA, Springer, 2008, 89–189.

[31] Brown Bourzutschky JA, Homs N, Bell AT. Hydrogenation of CO_2 and CO_2/CO mixtures over copper-containing catalysts. J. Catal. 1990, 124, 73–85.

[32] Andrews E, Ren M, Wang F, et al. Electrochemical Reduction of CO_2 at Cu Nanocluster / (1010) ZnO Electrodes. J. Electrochem. Soc. 2013, 160, H841–46.

[33] Hori Y, Takahashi I, Koga O, Hoshi N. Electrochemical reduction of carbon dioxide at various series of copper single crystal electrodes. J. Mol. Catal. A: Chem. 2003, 199, 39–47.

[34] Huang Y, Handoko AD, Hirunsit P, Yeo BS. Electrochemical reduction of CO_2 using copper single-crystal surfaces: effects of CO* coverage on the selective formation of ethylene. ACS Catal. 2017, 7, 1749–56.

[35] Handoko AD, Ong CW, Huang Y, et al. Mechanistic Insights into the selective electroreduction of carbon dioxide to ethylene on Cu_2O-derived copper catalysts. J. Phys. Chem. C. 2016, 120, 20058–67.

[36] Chen CS, Wan JH, Yeo BS. Electrochemical reduction of carbon dioxide to ethane using nanostructured Cu_2O-derived copper catalyst and palladium(II) chloride. J. Phys. Chem. C. 2015, 119, 26875–82.

[37] Lee S, Kim D, Lee J. Electrocatalytic production of C_3-C_4 compounds by conversion of CO_2 on a chloride-induced bi-phasic Cu_2O-Cu catalyst. Angew. Chemie. Int. Ed. 2015, 54, 14701–05.

[38] Ren D, Ang BS-H, Yeo BS. Tuning the selectivity of carbon dioxide electroreduction toward ethanol on oxide-derived CuxZn catalysts. ACS Catal. 2016, 6, 8239–47.

[39] Chi D, Yang H, Du Y, et al. Morphology-controlled CuO nanoparticles for electroreduction of CO_2 to ethanol. RSC adv. 2014, 4, 37329–32.

[40] Hori Y, Murata A, Takahashi R. Formation of hydrocarbons in the electrochemical reduction of carbon dioxide at a copper electrode in aqueous solution. J. Chem. Soc. Faradaic. Trans. 1 1989, 85, 2309–26.

[41] Loiudice A, Lobaccaro P, Kamali EA, et al. Tailoring copper nanocrystals towards C_2 products in electrochemical CO_2 reduction. Angew. Chemie. Int. Ed. 2016, 55, 5789–92.

[42] Kuhl KP, Cave ER, Abram DN, Jaramillo TF. New insights into the electrochemical reduction of carbon dioxide on metallic copper surfaces. Energ. Environ. Sci. 2012, 5, 7050–59.

[43] Kwon Y, Lum Y, Clark EL, Ager JW, Bell AT. CO_2 electroreduction with enhanced ethylene and ethanol selectivity by nanostructuring polycrystalline copper. ChemElectroChem 2016, 3, 1012–19.

[44] Gao D, Zegkinoglou I, Divins NJ, et al. Plasma-activated copper nanocube catalysts for efficient carbon dioxide electroreduction to hydrocarbons and alcohols. ACS Nano. 2017, 11, 4825–31.

[45] Raciti D, Livi KJ, Wang C. Highly dense Cu nanowires for low overpotential CO_2 reduction. Nano. Lett. 2015, 15, 6829–35.

[46] Kim D, Kley CS, Li Y, Yang P. Copper nanoparticle ensembles for selective electroreduction of CO_2 to C_2-C_3 products. Proc. Natl. Acad. Sci. USA 2017, 114, 10560–65.

[47] Rahaman M, Dutta A, Zanetti A, Broekmann P. Electrochemical reduction of CO_2 into multicarbon alcohol on activated Cu mesh catalysts: an identical location (IL) study. ACS Catal. 2017, 7, 7946–56.

[48]   Hoang TTH, Ma S, Gold JI, Kenis PJA, Gewirth AA. Nanoporous Copper Films by Additive-Controlled Electrodeposition: CO$_2$ Reduction Catalysis. ACS Catal. 2017, 7, 3313–21.

[49]   Song Y, Peng R, Hensley DK, et al. High-selectivity electrochemical conversion of CO$_2$ to ethanol using a copper nanoparticle/N-doped graphene electrode. Chem. 2016, 1, 1–8.

[50]   Hossain NM, Wen J, Chen A. Unique copper and reduced graphene oxide nanocomposite toward efficient electrochemical reduction of carbon dioxide. Sci. Rep. 2017, 7, 3184.

[51]   Ma S, Sadakiyo M, Luo R, Heima M, Yamauchi M, Kenis PJA. One-step electrosynthesis of ethylene and ethanol from CO$_2$ in an alkaline electrolyzer. J. Power. Sources. 2016, 301, 219–28.

[52]   Hossain SS, Rahman S, Ahmed S. Electrochemical reduction of carbon dioxide over CNT-supported nanoscale copper electrocatalysts. J. Nanomater. 2014, 374318, 10pp.

[53]   Yuan J, Liu L, Guo RR, Zeng S, Wang H, Lu JX. Electroreduction of CO$_2$ into ethanol over an active catalyst: copper supported on titania. Catal. 2017, 7, 220, 11pp.

[54]   Yuan J, Yang MP, Hu QL, Li SM, Wang H, Lu JX. Cu/TiO$_2$ nanoparticles modified nitrogen-doped graphene as highly efficient catalyst for the selective electroreduction of CO$_2$ to different alcohols. J. CO2 Util. 2018, 24, 334–40.

[55]   Ren D, Wong NT, Handoko AD, Huang Y, Yeo BS. Mechanistic insights into the enhanced activity and stability of agglomerated Cu nanocrystals for the electrochemical reduction of carbon dioxide to n-propanol. J. Phys. Chem. Lett. 2016, 7, 20–4.

[56]   Summers DP, Leach S, Frese KW. The electrochemical reduction of aqueous carbon-dioxide to methanol at molybdenum electrodes with low overpotentials. J. Electroanal. Chem. 1986, 205, 219–32.

[57]   Bandi A. Electrochemical reduction of carbon dioxide on conductive metallic oxides. J. Electrochem. Soc. 1990, 137, 2157–60.

[58]   Sun XF, Zhu QG, Kang XC, et al. Molybdenum-bismuth bimetallic chalcogenide nanosheets for highly efficient electrocatalytic reduction of carbon dioxide to methanol. Angew. Chem. Int. Ed. 2016, 55, 6771–5.

[59]   Li P, Hu H, Xu J, et al. New insights into the photo-enhanced electrocatalytic reduction of carbon dioxide on MoS$_2$-rods/TiO$_2$ NTs with unmatched energy band. App.l Catal. B. 2014, 147, 912–9.

[60]   Frese KW, Leach S. Electrochemical reduction of carbon dioxide to methane, methanol, and CO on Ru electrodes. J Electrochem Soc 1985, 132, 259–60.

[61]   Bandi A, Kuhne HM. Electrochemical reduction of carbon-dioxide in water-Analysis of reaction-mechanism on ruthenium-titanium-oxide. J. Electrochem. Soc. 1992, 139, 1605–10.

[62]   Popic JP, Avramovlvic ML, Vukovic NB. Reduction of carbon dioxide on ruthenium oxide and modified ruthenium oxide electrodes in 0.5 M NaHCO$_3$. J. Electroanal. Chem. 1997, 421, 105–10.

[63]   Spataru N, Tokuhiro K, Terashima C, Rao TN, Fujishima A. Electrochemical reduction of carbon dioxide at ruthenium dioxide deposited on boron-doped diamond. J. Appl. Electrochem. 2003, 33, 1205–10.

[64]   Qu JP, Zhang XG, Wang YG, Xie CX. Electrochemical reduction of CO$_2$ on RuO$_2$/TiO$_2$ nanotubes composite modified Pt electrode. Electrochim Acta. 2005, 50, 3576–80.

[65]   Ogura K, Yoshida I. Electrocatalytic reduction of CO$_2$ to methanol: Part 9: Mediation with metal porphyrins. J. Mol. Catal. 1988, 47, 51–7.

[66]   Ogura K, Yoshida I. Electrocatalytic reduction of carbon dioxide to methanol in the presence of 1,2-dihydroxybenzene-3,5-disulphonate ferrate (III) and ethanol. J. Mol. Catal. 1986, 34, 67–72.

[67] Ogura K, Fujita M. Electrocatalytic reduction of carbon dioxide to methanol: Part 7. With quinone derivatives immobilized on platinum and stainless steel. J. Mol. Catal. 1987, 41, 303–11.

[68] Ogura K, Takamagari K. Electrocatalytic reduction of carbon dioxide to methanol. 2. Effects of metal complex and primary alcohol. J. Chem. Soc. Dalton. Trans. 1986, 8, 1519–23.

[69] Ogura K, Endo N, Nakayama M, Ootsuka H. Mediated activation and electroreduction of CO_2 on modified electrodes with conducting polymer and inorganic conductor films. J. Electrochem. Soc. 1995,142, 4026–32.

[70] Monnier A, Augustynski J, Stalder C. On the electrolytic reduction of carbon dioxide at TiO_2 and TiO_2-Ru cathodes. J. Electroanal. Chem. 1980, 112, 383–85.

[71] Koudelka M, Monnier A, Augustynski J. Electrocatalysis of the cathodic reduction of carbon dioxide on platinized titanium dioxide film electrodes. J. Electrochem. Soc. 1984, 131, 745–750.

[72] Capello C, Fischer U, Hungerbühler K. What is a green solvent? A comprehensive framework for the environmental assessment of solvents. Green Chem. 2007, 9, 927–34.

[73] Eggins BR, McNeill J. Voltammetry of carbon dioxide: Part I. A general survey of voltammetry at different electrode materials in different solvents. J. Electroanal. Chem. 1983, 148, 17–24.

[74] Ohkawa K, Noguchi Y, Nakayama S, Hashimoto K, Fujishima A. Electrochemical reduction of carbon dioxide on hydrogen-storing materials: Part II Copper-modified palladium electrode.. J. Electroanal. Chem. 1993, 348, 459–64.

[75] Ohkawa K, Noguchi Y, Nakayama S, Hashimoto K, Fujishima A. Electrochemical reduction of carbon dioxide on hydrogen-storing materials: Part 3. The effect of the absorption of hydrogen on the palladium electrodes modified with copper. J. Electroanal. Chem. 1994, 369, 165–73.

[76] Podlovchenko BI, Kolyadko EA, Lu S. Electroreduction of carbon dioxide on palladium electrodes at potentials higher than the reversible hydrogen potential. J. Electroanal. Chem. 1994, 373, 185–7.

[77] Seshadri G, Lin C, Bocarsly AB. A new homogeneous electrocatalyst for the reduction of carbon dioxide to methanol at low overpotential. J. Electroanal. Chem. 1994, 372, 145–50.

[78] Halmann M. Photoelectrochemical reduction of aqueous carbon dioxide on P-type gallium phosphide in liquid junction solar cells. Nature 1978, 275, 115–6.

[79] Aurian-blajeni B, Halmann M, Manassen J. Electrochemical measurement on the photoelectrochemical reduction of aqueous carbon dioxide on p-Gallium phosphide and p-Gallium arsenide semiconductor electrodes. Sol. Energ. Mater. 1983, 8, 425–40.

[80] Jones JP, Surya Prakash GK, Olah GA. Electrochemical CO_2 reduction: Recent advances and current trends. Isr. J. Chem. 2014, 54, 1451–66.

[81] Canfield D, Frese KW. Reduction of carbon dioxide to methanol on n- and p-GaAs and p-lnP. Effect of crystal face, electrolyte and current Density. J. Electrochem. Soc. 1983, 130, 1772–3.

[82] Jhong HRM, Ma S, Kenis PJA. Electrochemical conversion of CO_2 to useful chemicals: current status, remaining challenges and future opportunities. Curr. Opin. Chem. Eng. 2013, 2, 191–99.

[83] Hori Y, Wakebe H, Tsukamoto T, Koga O. Electrocatalytic process of CO selectivity in electrochemical reduction of CO_2 at metal electrodes in aqueous media. Electrochim. Acta. 1994, 39, 1833–9.

[84] Murata A, Hori Y. Product selectivity affected by cationic species in electrochemical reduction of CO_2 and CO at a Cu electrode. Bull. Chem Soc. Jpn. 1991, 64, 123–7.

[85] Wu JJ, Risalvato FG, Ke FS, Pellechia PJ, Zhou XD. Electrochemical reduction of carbon dioxide I. Effects of the electrolyte on the selectivity and activity with Sn electrode. J Electrochem Soc 2012, 159, F353–59.

[86] Frese KW, Leach SC, Summers DP. Electrochemical reduction of aqueous carbon dioxide to methanol. US patent no. 4, 609, ; 2 September 4411986.

[87] Endródi B, Bencsik G, Darvas F, Jones R, Rajeshwar K, Janáky C. Continuous-flow electroreduction of carbon dioxide. Progr. Energ. Combust. Sci. 2017, 62, 133–154.

[88] Rosen BA, Salehi-Khojin A, Thorson MR, et al. Ionic liquid-mediated selective conversion of CO_2 to CO at low overpotentials. Science 2011, 334, 643–4.

[89] Welton T. Room-temperature ionic liquids. Solvents for synthesis and catalysis. Chem. Rev. 1999, 99, 2071–84.

[90] Barrose-Antle LE, Compton RG. Reduction of carbon dioxide in 1butyl-3-methylimidazolium acetate. Chem. Commun. 2009, 0, 3744–46.

[91] Vernoux P, Lizarraga L, Tsampas MN, et al. Ionically conducting ceramics as active catalyst supports. Chem. Rev. 2013, 113, 8192–260.

[92] Aeshala LM, Uppaluri RG, Verma A. Effect of cationic and anionic solid polymer electrolyte on direct electrochemical reduction of gaseous CO_2 to fuel. J. CO2 Util. 2013, 3–4, 49–55.

[93] Aeshala LM, Rahman SU, Verma A. Effect of solid polymer electrolyte on electrochemical reduction of CO_2. Sep. Purif. Technol. 2012, 94, 131–7.

[94] Zhang L, Hu S, Zhu X, Yang W. Electrochemical reduction of CO_2 in solid oxide electrolysis cells. J. Energ. Chem. 2017, 26, 593–601.

[95] Nguyen VN, Blum L. Syngas and synfuels from H_2O and CO_2: Current status. Chem. Ing. Tech. 2015, 87, 354–375.

[96] Peterson AA, Abild-Pedersen F, Studt F, Rossmeis J, Nørskov JK. How copper catalyzes the electroreduction of carbon dioxide into hydrocarbon fuels. Energ. Environ. Sci. 2010, 3, 1311–5.

[97] García-Merino I, Alvarez-Guerra E, Albo J, Irabien A. Electrochemical membrane reactors for the utilization of carbon dioxide. Chem. Eng. J. 2016, 305, 104–120.

[98] Li H, Oloman C. Development of a continuous reactor for the electro-reduction of carbon dioxide to formate – Part 2: Scale-up. J. Appl. Electrochem. 2007, 37, 1107–17.

[99] Surya Prakash CK, Viva FA, Olah GA. Electrochemical reduction of CO_2 over Sn-nafion coated electrode for a fuel-cell-like device. J. Power. Sources. 2013, 223, 68–73.

[100] Hernández RM, Márquez J, Márquez OP, et al. Reduction of carbon dioxide on modified glassy carbon electrodes. J. Electrochem. Soc. 1999, 146, 4131–36.

[101] Rail MD, Berben LA. Directing the reactivity of [HFe4N(CO)12]- toward H+ or CO_2 reduction by understanding the electrocatalytic mechanism. J. Am. Chem. Soc. 2011, 133, 18577–9.

[102] Cook RL, Mac Duff RC, Sammells AF. On the electrochemical reduction of carbon dioxide at in situ electrodeposited copper. J. Electrochem. Soc. 1988, 135, 1320–6.

[103] Hara K, Kudo A, Sakata T. Electrochemical CO_2 reduction on a glassy carbon electrode under high pressure. J. Electroanal. Chem. 1997,421, 1–4.

[104] Hara K, Kudo A, Sakata T. Electrochemical reduction of carbon dioxide under high pressure on various electrodes in an aqueous electrolyte. J. Electroanal. Chem. 1995, 391, 141–7.

[105] Furuya N, Yamazaki T, Shibata M. High performance Ru Pd catalysts for CO_2 reduction at gas-diffusion electrodes. J. Electroanal. Chem. 1997, 431, 39–41.

[106] Salehi-Khojin A, Jhong HRM, Rosen BA, et al. Nanoparticle silver catalysts that show enhanced activity for carbon dioxide electrolysis. J. Phys. Chem. C. 2013, 117, 1627–32.

[107] Mahmood MN, Masheder D, Harty CJU. Use of gas-diffusion electrodes for high-rate electrochemical reduction of carbon dioxide. I. Reduction at lead, indium- and tin-impregnated electrodes. J. Appl. Electrochem. 1987, 17, 1159–70.

[108] Hara K, Tsuneto A, Kudo A, Sakata T. Change in the product selectivity for the electrochemical CO_2 reduction by adsorption of sulfide ion on metal electrodes. J. Electroanal. Chem. 1997, 434, 239–43.

[109] Shironita S, Karasuda K, Sato M, Umeda M. Feasibility investigation of methanol generation by CO_2 reduction using Pt/C-based membrane electrode assembly for a reversible fuel cell. J. Power. Sources. 2013, 228, 68–74.

[110] Shironita S, Karasuda K, Sato K, Umeda M. Methanol generation by CO_2 reduction at a Pt-Ru/C electrocatalyst using a membrane electrode assembly. J. Power. Sources. 2013, 240, 404–10.

[111] Jiménez Z, García R, Camarillo R, Martínez F, Rincón J. Electrochemical CO_2 reduction to fuels using Pt/CNT Catalysts synthesized in supercritical medium. Energy Fuels 2017, 31, 3038–46.

[112] Sebastian D, Palella A, Baglio V, et al. CO_2 reduction to alcohols in a polymer electrolyte membrane co-electrolysis cell operating at low potentials. Electrochim Acta 2017, 241, 28–40.

[113] Gutiérrez-Guerra N, Moreno-López L, Serrano-Ruiz JC, Valverde JL, de Lucas-Consuegra A. Gas phase electrocatalytic conversión of CO_2 to syn-fuels on Cu based catalysts-electrodes. Appl. Catal. B 2016, 188, 272–82.

[114] Gangeri S, Perathoner S, Caudo G, et al. Fe and Pt carbon nanotubes for the electrocatalytic conversion of carbon dioxide to oxygenates. Catal. Today 2009, 143, 57–63.

[115] Centi G, Perathoner S. Problems and perspectives in nanostructured carbon-based electrodes for clean and sustainable energy. Catal. Today 2010, 150, 151–62.

[116] Genovese C, Ampelli C, Perathoner S, Centi G. Electrocatalytic conversion of CO_2 to liquid fuels using nanocarbon-based electrodes. J. Energy. Chem. 2013, 22, 202–213.

[117] Centi G, Perathoner S. Opportunities and prospects in the chemical recycling of carbon dioxide to fuels. Catal. Today 2009, 148, 191–205.

[118] Merino-García I, Albo J, Irabien A. Tailoring gas-phase CO_2 electroreduction selectivity to hydrocarbons at Cu nanoparticles. Nanotechnology 2018, 29014001 (9).

[119] Marnellos G, Stoukides M. Catalytic studies in electrochemical membrane reactors. Solid State Ion. 2004, 175, 597–603.

[120] Yan J, Chen H, Dogdibegovic E, Stevenson JW, Cheng M, Zhou XD. High-efficiency Intermediate temperature solid oxide electrolyzer cells for the conversion of carbon dioxide to fuels. J. Power. Sources. 2014, 252, 79–84.

[121] Wu G, Xie K, Wu Y, Yao W, Zhou J. Electrochemical conversion of H_2O/CO_2 to fuel in a proton-conducting solid oxide electrolyser. J. Power. Sources. 2013, 232, 187–92.

[122] Uhm S, Kim YD. Electrochemical conversion of carbon dioxide in a solid oxide electrolysis cell. Curr. Appl. Phy. 2014, 14, 672–9.

[123] Lei L, Liu T, Fang S, Lemmon JP, Chen F. The co-electrolysis of CO_2-H_2O to methane via a novel micro-tubular electrochemical reactor. J. Mater. Chem. A. 2017, 5, 2904–10.

[124] Xie K, Zhang Y, Meng G, Irvine JTS. Electrochemical reduction of CO_2 in a proton conducting solid oxide electrolyser. J. Mater. Chem. 2011, 21, 195–8.

[125] Xie K, Zhang Y, Meng G, Irvine JTS. Direct synthesis of methane from CO_2/H_2O in an oxygen-ion conducting solid oxide electrolyser. Energy. Environ. Sci. 2011, 4, 2218–22.

[126] Ruiz E, Cillero D, Martínez PJ, et al. Bench scale study of electrochemically promoted catalytic CO_2 hydrogenation to renewable fuels. Catal. Today 2013, 210, 55–66.

[127] Ruiz E, Cillero D, Martínez PJ, et al. Bench-scale study of electrochemically assisted catalytic CO_2 hydrogenation to hydrocarbon fuels on Pt, Ni and Pd films deposited on YSZ. J. CO2 Util. 2014, 8, 1–20.

[128] Ruiz E, Cillero D, Martínez PJ, et Al. Electrochemical synthesis of fuels by CO_2 hydrogenation on Cu in a potassium-ion conducting membrane reactor at bench scale. Catal. Today 2014, 236, 108–20.

[129] Ruiz E, Martínez PJ, Morales A, San Vicente G, de Diego G, Sánchez JM. Electrochemically assisted synthesis of fuels by CO_2 hydrogenation over Fe in a bench scale solid electrolyte membrane reactor. Catal. Today 2016, 268, 46–59.

[130] Gattrell M, Gupta N, Co A. A review of the aqueous electrochemical reduction of CO_2 to hydrocarbons at copper. J. Electroanal. Chem. 2006, 594, 1–19.

[131] Riyanto, Afiti TA. Electrochemical synthesis of ethanol from carbon dioxide using copper and carbon-polyvinyl chloride (PVC) electrode. Mater. Sci. Eng. 2018, 288, 012128.

[132] Azuma M, Hashimoto K, Hiramoto M, Watanbe M, Sakata T. Electrochemical reduction of carbon dioxide on various metal electrodes in low-temperature aqueous $KHCO_3$ Media. J. Electrochem. Soc. 1990, 137, 1772–8.

[133] Brisard GM, Camargo APM, Nart FC, Iwasita T. On-line mass spectrometry investigation of the reduction of carbon dioxide in acidic media on polycrystalline Pt. Electrochem. Commun. 2001, 3, 603–7.

[134] Hara K, Tsuneto A, Kudo A, Sakata T. Electrochemical reduction of CO_2 on a Cu electrode under high pressure. Factors that determine the product selectivity. J. Electrochem. Soc. 1994, 141, 2097–103.

Keerthiga Gopalram and Raghuram Chetty

25 Electrochemical conversion of CO_2 into hydrocarbons

25.1 Introduction

The demand to combat atmospheric concentrations of CO_2 has rapidly emerged as a priority for the survival of the planet and its inhabitants. The need is to devise a strategy for mitigation and utilisation of atmospheric carbon content and to prevent or reduce its emission from the fossil fuel-based combustion sources. Although there is no good technology available yet to achieve the desired objectives, policy-based decisions have been implemented to monitor, reduce CO_2 emission and to gain carbon credit within organisations.

CO_2 is the most oxidised form of carbon, and a strong indication of a CO_2 molecule's stability can be inferred by its large, negative standard Gibbs free energy of formation value (ΔG_{fo} = −349.67 kJ mol^{-1}) [1]. The CO_2 structure can change from linear to bent (CO_2 to $CO_2^{\cdot-}$) on an electron transfer to the molecule, which results in the activation of CO_2 for its utilisation.

This chapter confines itself to electrochemical reduction (ECR) of CO_2 and hence the emphasis is on ECR with the following merits: (i) it directs the intermediates towards reduction with less byproduct formation; (ii) potential for rapid industrial up-scaling, avoiding the use of expensive high-pressure/temperature reactors; (iii) operation at ambient conditions avoiding complex start-up processes; (iv) it enables the process production rate to be rapidly varied with easy switch on/off [2]; and (v) the mixture of hydrocarbons obtained from the ECR of CO_2 can be used as a blended fuel named hythane, thus avoiding separation [3].

Figure 25.1 compares the reactivity of CO_2 reduction by various processes such as petrochemical, biochemical, catalytic and electrocatalytic reactions. It can be inferred that the reactivity of electrocatalytic process of CO_2 reduction is near to industrial readiness [2], and it can be external energy independent by combining with a photovoltaic panel, to outperform photoelectrochemical reduction. However, major progress in its large-scale commercialisation has been limited to the two-electron reduction products of CO and formate. The formation of multi-carbon products involving multiple proton and electron transfers remains as one of the biggest scientific challenges to be addressed.

Keerthiga Gopalram, Department of Chemical Engineering, Indian Institute of Technology Madras, India and Department of Chemical Engineering, SRM Institute of Science and Technology Katankulathur, India
Raghuram Chetty, Department of Chemical Engineering, Indian Institute of Technology Madras, India

https://doi.org/10.1515/9783110665147-025

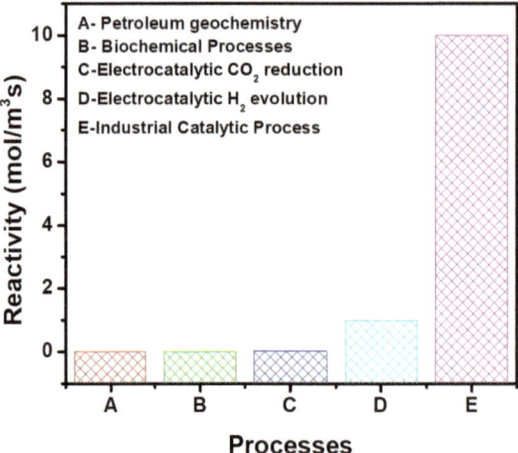

Figure 25.1: Comparison of reactivity of various CO_2 reduction methods [2].

In the ECR, CO_2 is made to adsorb on metal surfaces and activated by supplying electrons. Moreover, when employing an aqueous system for CO_2 reduction, the selection of metal is important as the applied energy can be wasted on hydrogen evolution instead of being used for CO_2 reduction [4]. CO_2 needs a lot of energy as the input with optimised conditions and an active catalyst for its reduction. A longstanding classification based on the hydrogen overvoltage characteristics has been structured as high hydrogen overvoltage, medium hydrogen overvoltage and low hydrogen overvoltage for ECR of CO_2.

Metals such as Pb, Hg, In, Sn, Cd and Bi have negligible CO adsorption yielding formate and hence are graded as poor catalysts towards CO_2 reduction, under the category of high hydrogen overvoltage characteristics. Metals like Cu, Au, Ag, Pd, Ga and Zn with medium hydrogen evolution reaction (HER) properties promote the formation of CO and hydrocarbons. Metals such as Ni, Fe, Pt, Ti, Co, Rh and Ir have low hydrogen overvoltage characteristics and exhibit strong CO adsorption [1]. Most literature cites the above classification for the ECR of CO_2; however, due to numerous contributions from researches a better understanding has been gained for electrode–electrolyte system [5–7]. This frames the objective of this chapter to discuss the classification of metals and their modification based on the specific hydrocarbons formed for ECR of CO_2.

Hydrocarbons (especially methane) have always been the choice of fuels from an environmental point of view. It is preferred for its high energy density of 56 MJ kg^{-1} [8] and low CO_2 emissions as compared to other fossil fuels or biofuels (Figure 25.2).

Methane, as well as other hydrocarbon formation using ECRs, has been studied by several authors, and most of them reported the use of Cu as an electrocatalyst [5, 9]. Kinetically sluggish ECR of CO_2 follows heterogeneous catalysis, which

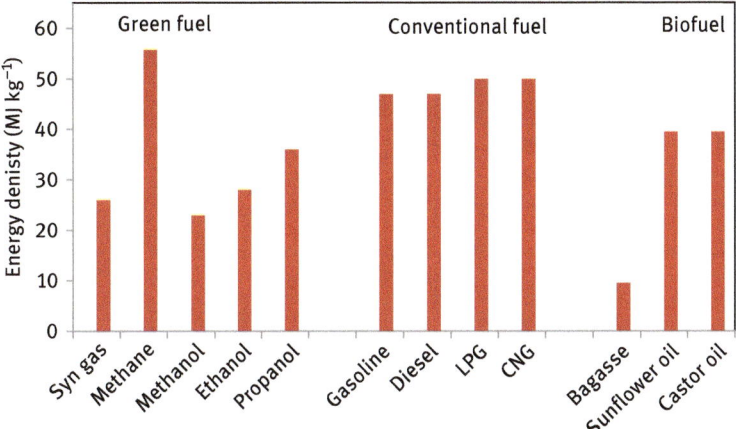

Figure 25.2: Comparison of energy density of conventional fuel with green and biofuels modified from Mallick et al., 2017 [8].

involves three major steps: (i) chemical adsorption of CO$_2$ on an electrocatalyst; (ii) electron transfer and/or proton migration to cleave C–O bonds and/or form C–H bonds; and (iii) desorption of the products back into the electrolyte [6].

25.1.1 Copper – an electrode for CO$_2$ reduction

Copper (Cu) is unique for the reduction of CO$_2$ as it can catalyse the breaking of both C–O bonds in CO$_2$ and hence further form hydrocarbons. It was basically derived from the medium hydrogen overvoltage group due to its excellent ability to adsorb and desorb metal to carbon bonds in the electrochemical CO$_2$ reduction [1]. The major products of CO$_2$ reduction are hydrocarbons (CH$_4$) and alcohols. Understanding copper surfaces and hence their modification has helped to lower the overpotential and to increase the selectivity for hydrocarbons [9].

However, the problem of deactivation of the copper electrode coupled to the formation of mixed products containing various gaseous and liquid species with competitive hydrogen evolution limits its commercialisation [6]. Hori et al. identified that the copper deactivation could be due to the presence of ions such as Fe^{2+} and Zn^{2+} in the electrolyte, which can be removed by pre-electrolysis of the electrolyte or adopting pulse mode electrolysis [10]. Lee et al. suggested that the cause of copper poisoning may be due to the adsorption of some species, such as graphite, originating from adsorbed CO (CO$_{ad}$) on the Cu surface, which can be prevented by pulse mode reduction [11].

Investigation of the possible modes of adsorption of CO$_2$ on Cu metal surface as described by Li et al. [12] is presented in Figure 25.3. Coordination can be either

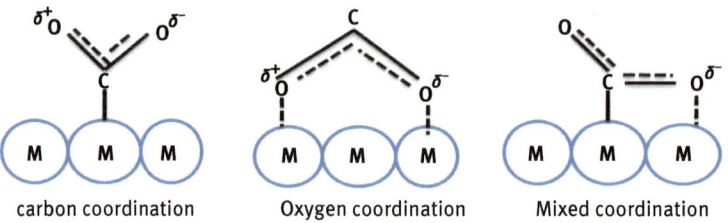

Figure 25.3: Possible structure of adsorbed CO_2 on metals; redrawn from Li et al. [12].

through (a) carbon coordination – with the carbon bonded to the electrode surface, (b) oxygen coordination – with either single oxygen or two oxygen bonded to the electrode surface or (c) mixed coordination – with a carbon and oxygen bonded to the electrode surface.

The choice for carbon and oxygen adsorption on the metal surface depends on the particle size, morphology, electrolyte and its applied potential. In addition to the design of catalysts, the electrolyte concentration and applied CO_2 pressure could also affect the hydrocarbon selectivity [9]. Thus, understanding Cu for ECR of CO_2 is a vast field by itself and here we summarise the advances in selectivity of methane for ECR of CO_2.

25.2 Classification of metals for methane formation

25.2.1 Effect of process conditions for methane formation

The chemical properties of the electrolyte medium affect the electrochemical reaction mechanism with sodium-supporting salts favouring high methane efficiency [11]. Methane was observed with various sodium-based supporting salts of Cl^-, I^-, Br^-, CN^- at a low temperature (243 K) on Cu electrodes [13]. Kaneco et al. reported a maximum Faradaic efficiency of 70.5% for a $NaClO_4$/methanol (243 K)-based system at −3 V (vs Ag/AgCl). The use of inorganic salts of halide anions (e.g. Na^+, K^+ and Cl^-), bicarbonate (HCO_3^-) or hydroxide (OH^-) is encouraged due to their high conductivities in water, altered buffer capacities and hence influence on the local pH at the Cu electrode, thus controlling the nature and the amount of products formed for ECR of CO_2 [14].

Varying the concentration of $KHCO_3$ for tuning the product selectivity of Cu electrodes was studied by Mul et al. [15]. The selectivity of CH_4 increased with the increase in $KHCO_3$ concentration. Electrolytes such as LiOH in methanol and CH_3COOK in methanol provide 63% and 23%, respectively, of methane. The comparison of Faradaic efficiency of Au_3Cu nanoparticles (NPs) when investigated in

0.1 M KH$_2$PO$_4$ and 0.5 M KHCO$_3$ electrolyte at −1.6 V versus Saturated Calomel Electrode (SCE) yielded 35% and 22% of methane, respectively [16].

A small pH shift selectively alters the buffer capacity of the electrolyte and tunes the CO$_2$ reduction over H$_2$ evolution [17]. As shown in Figure 25.4, if protonation of *CO were to form methane, then by alteration of pH during reduction, the electrodes would be prone to poisoning by its graphitic intermediates. Meanwhile, a pH-independent process was initiated by *CO coupling and C–C coupling, whereby a high local pH arising from low electrolyte concentration should contribute to the formation of ethylene [16]. Similar result was observed by Keerthiga et al., where a higher pH value leads to C$_2$ formation rather than C$_1$ when studied at 0.5 M KHCO$_3$ [18].

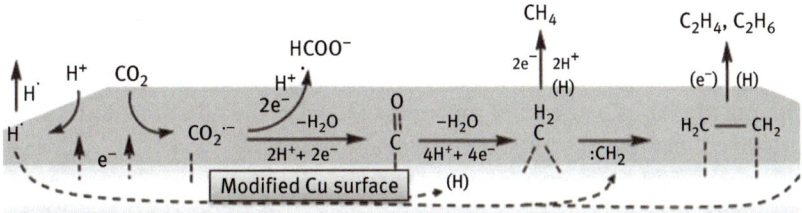

Figure 25.4: Simplified reaction network at the electrode surface for ECR of CO$_2$.

Literature reports the influence of temperature and pressure for ECR of CO$_2$. In 1989, Hori et al. found an inverse relationship between Faradaic efficiency and temperature for methane formation over a Cu electrode, with Faradaic efficiency as high as 65% at 0°C, while approaching 0% at 40 °C [10]. In contrast, the formation of ethylene and CO was found to increase on raising the temperature. The effect of pressure on the activity of copper was also studied, and it was found that an increase in pressure opposes hydrocarbon formation [8]. Thus, the wise choice of process conditions will aid in selective methane formation for ECR of CO$_2$.

25.2.2 Effect of nanostructures for methane formation

Highly active surfaces for ECR of CO$_2$ reduction were observed on Cu electrode nanofoams, which are selective for methane formation. Faradaic efficiencies of CH$_4$ up to 80% are reported by dispersing copper nanocatalysts on glassy carbon electrodes. Compared to high purity Cu foil electrodes, four times higher methane formation was reported on dispersed metal NPs. The efficiency remained constant at 71–90% over 1 h at a lower overpotential of −1.25 V [6]. In a similar study, methanation was greatly enhanced on well-dispersed Cu NPs (0.7–0.4 nm) supported on glassy carbon (n-Cu/C). The Tafel slope of 60 mV/dec indicates that equilibration of CO_2 to CO_2^- occurs before the rate determining step to methane formation [19]. Thus, isolated Cu NPs have been reported to give more active catalytic sites for methanation than those of dense aggregated NPs.

The association of Cu metal overlayers and lattice strain aids the design of advanced electrocatalysts for CO_2 reduction. The strain effects that stemmed from lattice mismatch between Cu and substrate metal could alter the binding energies of the reaction intermediates and hence influence the product formation. The thickness of the Cu overlayers greatly influenced the reaction activity and product selectivity for CO_2 electrolysis. Though high current densities were achieved on the Cu overlayers/Pt substrate (Cu of 5 nm on Pt and Cu of 15 nm on Pt), the Faradaic Efficiency (FEs) for hydrocarbon formation were significantly lower than those on polycrystalline Cu [20].

Attempts have been made to understand the size dependency of Cu towards CO_2 reduction in a CO_2–$KHCO_3$ system [7]. Bulk Cu produced mostly CH_4 in addition to smaller amounts of C_2H_6, H_2 and trace amounts of CO. On the other hand, 2–15 nm Cu NPs produced mainly H_2 (60–70% Faradaic efficiency) with CH_4 (10–15% Faradaic efficiency), and the remaining was CO and C_2H_6. Larger nanostructures appear to favour hydrocarbon formation, but there was no clear understanding of the specific surface structures that could impact product selectivity. Moreover, the Cu NP was reported to grow (7–25 nm) during CO_2 electrolysis and it has been reported to show higher partial current density (10 mA/cm^2 at −1.45 V vs Reversible Hydrogen Electrode (RHE)) than that produced by polycrystalline Cu for methane formation [7].

Tang et al. [21] found that roughened polycrystalline Cu electrodes exhibited better electrocatalytic performance for CO_2 reduction than their smooth counterpart. The roughened electrode provides numerous surface-active sites for CO_2 reduction and offers high specific surface area with higher current density than a smooth electrode. Roughened Cu is reported to be characterised by the presence of edges, steps and defects versus a smooth surface of low-index crystal facets.

The selectivity for ECR of CO_2 was influenced by the atomic configuration of the electrode. Cu (111) favours C_1 formation as the *CO intermediates were protonated to yield methane. On the other hand, Cu(100) facets were considered to be more selective for C_2 formation [8, 9]. Modification of the surface by electrodeposition influenced the formation of specific planes. Keerthiga et al. [22] reported chronoamperometric deposition of Cu on Cu (Cu/Cu) at two different $CuSO_4$ bath concentrations, 0.25 (high) and 0.025 M (low). Crystallographic studies revealed pure Cu and Cu/Cu were aligned towards (111) and (220) planes with a texture coefficient of 1.2 and 1.7, respectively. On ECR of CO_2, maximum methane formation was 13 mmol cm^{-2} at −1.6 V on a pure Cu electrode with (111) and (200) orientation. Similarly, Cu/Cu low (deposition from 0.025 M bath concentration) showed maximum ethane production of 6.7 mmol cm^{-2} at −1.6 V with dominating (220) orientation compared to its counterpart of Cu/Cu high [22, 9].

Thus, the understanding gained from different nanostructures such as isolated NPs based on their shape/size, crystalline facets/planes, lattice strains and metal over layers can be used to alter the Faradaic efficiency for the formation of methane.

25.2.3 Effect of bimetals and alloys for methane formation

In order to design new electrocatalysts, bimetallic or core–shell electrocatalyst are being investigated, where the electronic perturbations caused by one metal to another can favour the conversion of CO$_2$ into hydrocarbons. Recently, Guo et al. fabricated monodisperse Cu–Pt nanocrystals (NCs) with different atomic ratios for ECR of CO$_2$ [23]. In general, the incorporation of Pt facilitated protonation of adsorbed *CO, resulting in an enhanced Faradaic efficiency for CH$_4$. However, a search for the optimum atomic bimetallic ratio resulted in Cu$_3$Pt NCs being chosen for the highest current density and best selectivity towards CH$_4$ [23].

Xu et al. [24] studied oxidation of AuCu NPs as a function of composition and temperature. Alloying Au with Cu lowers the reduction overpotential for ECR of CO$_2$, where voltammetric measurements showed ~100 mV reduction in the onset potential for Au$_x$Cu$_y$ alloy when compared to bare Cu and Au electrodes. Kim et al. [25] also studied ordered monolayers of Au–Cu bimetallic NPs with varying compositions for ECR of CO$_2$. CO was selectively formed from the oxygen end of COOH attached to a Cu atom adjacent to the Au-CO$_2$ primary bond. On increasing the Cu content of the bimetallic NPs, the number of obtained hydrocarbon products increased with increasingly selective towards CO. Similar understanding should also be possible for methane formation.

Recently, alloying has drawn much attention, since it can enhance the performance of metal catalysts by tuning the stabilisation of key intermediates. In 2014 Zhao et al. devised a new strategy for the synthesis of phase stabilised Au$_3$Cu alloy for ECR of CO$_2$. The incorporation of Au helps lower the overpotential of ECR of CO$_2$, whereas Cu NPs help improve the stability of products [26].

Zhang et al. reported on product distribution of CuPd alloys, where the composition was varied to control the product distribution [27]. The synergistic approach of nanoalloy (CuPd) resulted in a maximum Faradaic efficiency for CH$_4$ production (51%). Methane formation on Cu$_2$Pd nanoalloy (36%) was higher than the yield obtained on a nano-Cu catalyst (13%) [27].

An alloy of Cu with transition metals (W, Pt, Rh, Pd) was investigated for ECR of CO$_2$. Methanol was the more favourable product on Cu$_3$Pd, Cu$_3$Pt and W-Au alloys, whereas Cu$_3$Rh resulted in methane formation. W–Au alloy has been successful in lowering the overpotential for methane formation compared to bare Cu electrode due to the free energy difference in formation of methane intermediates [28]. On the other hand, Cu$_3$X can form *CO, *COH and *HCO intermediates favouring methane. Thus, alloys and bimetallic species can be chosen to control product formation in ECR of CO_2, provided the binding energy change for the particular product distribution is known.

25.2.4 Effect of metal oxides for methane formation

Various methods are reported in literature for the synthesis of oxide films on Cu electrodes such as air oxidation, thermal oxidation, anodisation and electrodeposition [9, 30, 29]. Among these, thermal oxidation of Cu metal is widely used for the fabrication of oxide films, but this method suffers from the disadvantage of forming a detachable black oxide film on handling. Moreover, the temperature and time have to be optimised stringently to obtain either cuprous or cupric oxide.

ECR of CO_2 has been investigated on oxide-based electrodes to overcome stability issues and to enhance the selectivity. Modification of Cu metal by oxidation [29,30], deposition [31] and pulse mode reduction [32] is reported in the literature. Goncalve et al. [31] reported that Cu electrodes modified with Cu electrodeposits yield C_2 products rather than C_1 products. Similarly, Cu particles impregnated with Pb and Zn electrodes resulted in methane formation [13]. Kyriacou et al. [33] worked on modifying Cu thermally and non-thermally to give methane, ethylene and CO. On reviewing the work done on oxidised Cu electrodes, in 2011 Le et al. obtained methanol by electrodepositing Cu_2O on Cu [29]. Though the surface had mixed oxides, the authors have pinpointed that Cu species direct the formation of methanol in a bicarbonate electrolyte. In 2012, Li and Kanan went a step further, producing formic acid in aqueous electrolytes at a lower onset potential using conventionally oxidised Cu by varying the oxidation time (15 min to 13 h) and temperature (130–700°C) [30].

Keerthiga et al. [9] have compared native (furnace oxidation) and non-native (Bunsen burner oxidised) way of obtaining oxidised electrodes for ECR of CO_2. Both methods differ in terms of duration of oxidation and the atmosphere in which it is oxidised. Comparable Faradaic efficiency of ethane was obtained at −1.2 V versus NHE for flame-annealed Cu (34%) and conventionally oxidised Cu (40%) while bare Cu yielded methane (10%). When commercial Cu and copper oxide in powder form were evaluated for CO_2 reduction, methane was obtained as the sole product with low Faradaic efficiency, indicating the need of a metal–metal oxide interface for enhanced CO_2 conversion and C_2 product formation. Kanan and his group also support this fact as thin native SnO_x films seem to stabilise the CO_2^- intermediate and in situ-generated metal facets were responsible for product formation [34]. Similarly, as indicated by Xiao et al., the binding strength of CO_2 molecules can be altered by varying the subsurface oxygen impurities on modified Cu, leading to lower overpotential of the products formed [35]. Thus, as understood from the literature, metal oxides and their interfaces show promising results when studied for

ECR of CO$_2$ where temperature, time and stability of oxide are the variables to be controlled for selective methane formation.

25.2.5 Metals other than Cu for methane formation (Ru, Pt, Sn, Pb)

A Ru electrode in acidic media using 0.1 N H$_2$SO$_4$ as electrolyte was reported to yield methane in a reaction, which was affected by pH, temperature and electrolyte purity [36]. Kudo et al. [37] reported hydrocarbon formation over a Ni electrode using 0.1 M KHCO$_3$ as electrolyte at ambient as well as high pressure (60 atm) conditions. The authors reported that the high pressure enhanced the formation of hydrocarbons with decreased H$_2$ evolution. Hara et al. [38] reported Pt-loaded GDEs at 30 atm. pressure, where Faradaic efficiency for CH$_4$ formation reached up to 35% at a current density of 900 mA cm^{-2}. A small amount of ethane and ethylene formation was also found. Over Pd and Cd electrodes with 0.1 M KHCO$_3$ electrolyte, methane Faradaic efficiency was found to be 2.9% and 1.3%, respectively. Sn and Pb electrodes also produced small amounts of methane with Faradaic efficiencies of 0.2% each [5].

Non-metal-doped catalysts such as cobalt protoporphyrin immobilised on a pyrolytic graphite electrode reduced carbon dioxide in an aqueous acidic solution at relatively low overpotential (0.5 V) yielding CO and methane [39], whereas similar modification of polyacrylonitrile-based heteroatomic carbon nanofibres as renewable and metal-free electrocatalysts for CO$_2$ reduction did not significantly yielded hydrocarbons [5].

Thus apart from Cu, a group of transition metals like Ru, Pt, Ni and Pd were reported to favour methane formation under certain process conditions. However, understanding the synergistic effect of the metal with Cu can aid researchers search for selective methane formation.

25.2.6 Mechanism of methane formation

CH$_4$ formation and its structure–sensitivity relationships on Cu(111), Cu(100) and Cu (211) surfaces have been analysed for ECR of CO$_2$. Density functional theory (DFT) studies on (211) surfaces revealed enhanced intermediate stabilisation in addition to lower onset potentials for CH$_4$ evolution. The observation was confirmed for similar experimental conditions with a partial current density of 5 A/cm^2 for hydrocarbon formation at stepped surfaces [40]. Janick and co-workers concluded that the reduction of adsorbed *CO intermediates was a key step in hydrocarbon formation on Cu(111) [41].

Pioneering research work on Cu electrodes carried out by Hori in the 1980s [10, 42] reported that the ECR of CO$_2$ on a Cu electrode could effectively yield

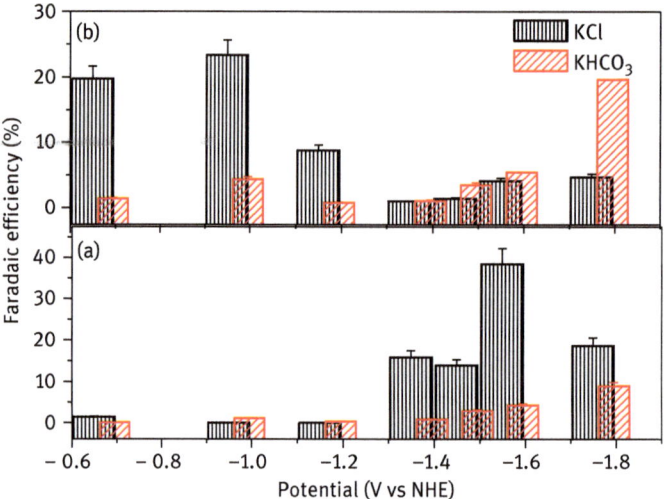

Figure 25.5: Faradaic efficiency versus potential for the products (a) ethane and (b) methane obtained on Cu in 0.5 M KCl and 0.5 M KHCO$_3$ electrolyte.

hydrocarbons and alcohols under ambient conditions. As shown in Figure 25.5, the product distribution varies significantly with applied potential when a Cu electrode was studied in bicarbonate and halide electrolytes.

There seems to be a major influence of the applied potential on ECR of CO$_2$: less negative potentials yield CO and HCOOH, whereas more negative potentials result in hydrocarbons and alcohols [21]. On the other hand, with the help of a highly sensitive reaction cell, a crystalline Cu electrode and advanced characterisation techniques for ECR of CO$_2$ offered 16 different products among which ethylene glycol, glycolaldehyde, hydroxyacetone, acetone and glyoxal were reported for the first time.

Figure 25.6 demonstrates the influence of applied potential on the products of CO$_2$ reduction. It can be observed that four major products namely CO, formate, CH$_4$ and C$_2$H$_4$ were observed at less negative potentials (−0.65 to −0.8 V vs RHE) and products like acetone, n-propanal, glycoaldehdye and hydroxyacetone are observed in trace amounts as shown in Figure 25.6 [43].

In order to identify possible intermediates for CO$_2$ reduction, in 2011 Schouten et al. [44] investigated the reduction of several small organic molecules using online mass spectrometry. A new mechanism for CO$_2$ reduction was proposed from the knowledge gained from the intermediates. Electrochemical dimerisation of CO lead to C$_2$ reaction pathway, whereas *CHO was regarded as the key intermediate for CH$_4$ formation. The reaction pathway suggested by Schouten et al. for CH$_4$ formation on Cu metal surface was predicted to be *CO →*CHO →*CH$_2$ →*CH$_3$ →CH$_4$ [44].

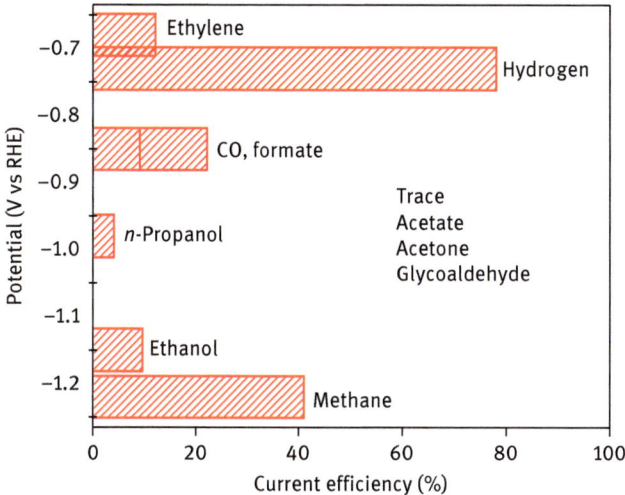

Figure 25.6: Electrochemical reduction of CO$_2$ performed at Cu electrode in 0.1 M KHCO$_3$. Reproduced from Jarmillo et al. [43].

25.3 Classification of metals based on ethylene formation

The energy density of ethylene (568 kg/m^3) is higher than methane (424 kg/m^3) with high energy density per unit volume [45]. Ethylene has greater ignition time than hydrogen facilitating easier storage and handling. Hence, targeting ethylene as product from CO$_2$ reduction has mandated the attention of current research. This section will summarise the metals and their modification favouring C$_2$ (ethylene and ethane) formation from CO$_2$ reduction.

25.3.1 Effect of process conditions for ethylene formation

Metal coverage by CO$_2$ is influenced by the process conditions such as pH, temperature and pressure, which can attenuate the metal's activation. In general, at high pressure, high *CO surface coverage and the release of OH$^-$ ions results in C–C coupling for ethylene formation with high local pH and low electrolyte concentration as reported by Kas et al. [16]. Similar results were observed for C$_2$ formation on a Cu electrode in 0.5 M KHCO$_3$ and 0.5 M KCl, where KCl at higher potential favoured C–C coupling with a Faradaic efficiency of 22.5%, whereas KHCO$_3$ preferably yielded methane (19 %) [46]. These studies suggest the use of high pressure and high local surface pH for C$_2$ formation.

25.3.2 Effect of nanostructures for ethylene formation

Planes and facets of metal surfaces greatly influence the product formation from ECR of CO_2. Cu (100) and Cu (111) facets were found to be selective for C_2 formation (ethylene and ethanol) and CH_4 formation, respectively [1]. In addition to having a unique facet for C_2 formation, low proton availability and high pH at the electrode surface favours C–C coupling [47]. Though there is a hunt for metals favouring C_2 products, Kim et al. have demonstrated a densely packed copper NP for C_2–C_3 products and reported Faradaic efficiency of 50% at lower overpotential [48]. This has gained attraction for solar-to-fuel applications, as stable operation has been proved for up to 10 h [48]. The Bell group has illustrated the opportunities for C_2H_4 formation on Cu grain boundaries [49], and the Kannan group reported a potentially dependent mixture of ethanol, acetate and traces of C_2H_4, C_2H_6 and propanol [50] for ECR of CO_2 on Cu grain boundary interfaces.

Smooth and thick electrodeposited Cu films (350 nm) showed selective ethylene formation, whereas aggregated high surface density defect sites, corners and edges were found to favour C_2 products from CO_2 [47]. The morphology of the catalyst showed an altered distribution of products and their FEs. C_2 hydrocarbons (C_2H_4 and C_2H_6) are formed on the dendritic, honeycomb and 3D structured Cu electrodes, while methane was suppressed on dendritic surfaces as reported by Gonçalves et al. [31]. Sen et al. claimed that porous Cu nanofoams would reduce CO_2 to C_2 products at 200 mV lower overpotential than smooth Cu electrodes [51]. As understood from the literature, Cu nanofoams could "trap" key intermediates in the pores and facilitate hydrocarbon formation as they possess a large number of C–C coupling surface sites. The enhanced residence time of adsorbed intermediates could yield 55% of C_2 products in 0.5 M $NaHCO_3$ [52].

Roughness and purity alter the adsorption characteristics towards C_2 formation. Mallik et al. have reported the performance of smooth and rough surfaces for ECR of CO_2 [8]. By preliminary pre-treatment a rough surface electrode was obtained, which remains stable for a longer time. The product formation from hydrocarbons to CO can be compared for electrolytically treated Cu and untreated Cu respectively [8]. The air-oxidised Cu electrode preferentially produces H_2 with very little CH_4 production [29]. Keerthiga et al. attempted to oxidise Cu using conventional furnace oxidation and compared the results with flame oxidation, which takes a shorter oxidation time. A comparable Faradaic efficiency for ethane and ethylene was observed at a shorter electrode preparation time than the conventional furnace-oxidised electrode [9].

The effect of nanostructures for ethylene formation has been demonstrated for planes of Cu, Cu nanoforms, NPs, dendrites and Cu grain boundary interfaces. The extensive product distribution of small Cu NPs was better understood for any nanostructures based on size, density of under coordinated sites, particle surface density and interparticle separation [21]. Kim et al. reported the rate determining step for C_2 products to be reductive CO coupling during reduction of CO_2. Thus, catalytically active cube-like structures will provide a framework for C$_2$ product (ethylene and ethanol) formation at low overpotential and in neutral pH aqueous media [48].

25.3.3 Effect of bimetals and alloys for ethylene formation

Appending one metal on another for the formation of a bimetallic catalyst will aid in obtaining synergistic properties for CO$_2$ reduction. As different metals have different affinities for the adsorption of carbon and oxygen, product formation is triggered by changes in binding energy based on the combination of the chosen metals. This hypothesis will aid in the production of selective and stable intermediates on the catalyst and thus enhance the selectivity of electrocatalysts [15].

Nanostructured Au core–Cu shell catalyst was studied for the ECR of CO$_2$, which was reported to give H$_2$, CO, formate, CH$_4$ and C$_2$H$_4$. Unique steps and kinks on the surfaces are achieved by thin Cu overlayers on the underneath Au (100) structure [51]. Similarly, biphasic NPs containing Cu$_2$O and Cu are reported to yield C$_1$–C$_4$ products. In situ X-ray absorbance spectroscopy has identified the retention of Cu$^+$ species during CO$_2$ reduction. Metals with stronger CO binding at Cu$^+$ species facilitated hydrocarbon formation, whereas the presence of Cu^{n+} surface sites may improve C^{2+} selectivity [11].

The CO$_2$ reduction process on Cu bimetals was monitored by introducing the isotope of ^{13}CO$_2$ along with ^{12}CO$_2$, and a substantial increase in ethylene and ethane along with ethanol and propanol was observed. The difference in the consumption of the product provides a clue as to the mechanism of product formation [52].

Verma and his group tested non-noble bimetallic electrocatalysts based on Ag–Co, whereby the change in the binding energy alters the intermediate formation and hence the product formation [15]. A synergetic effect arises from the combination of Ag and Cu with KHCO$_3$ electrolyte for C$_2$H$_4$ selectivity [53]. Ag would produce predominantly CO, which would be transferred to the neighbouring Cu atoms and subsequently get reduced to CH$_2^-$, a precursor in the formation of C$_2$H$_4$.

In an attempt to modify the Cu foil electrodes to nanopores, alloying and dealloying have been employed to suppress methane formation (<1%) with higher ethylene formation [54]. Thus, bimetallic combination of Cu with Au and Ag has been proved to be effective for ECR of CO$_2$.

25.3.4 Effect of metal oxides for ethylene formation

Co-catalysts can help tune the product distribution for ECR of CO$_2$. Cu$_2$O-derived Cu^0 nanostructures were observed to yield ethane (30%) on the addition of PdCl$_3$ salt compared to ethylene (3%) with complete suppression of methane in 0.1 M KHCO$_3$. The authors propose the formation of ethylene at Cu^0, which is then hydrogenated into ethane at the PdCl$_2$ NPs [55]. The use of electrodeposited copper oxide–copper bromide (Cu$_2$O–CuBr) composite films showed selective conversion of CO$_2$ to ethylene [56]. The composite electrodes were checked for durability, stability and electroactivity. The stability was maintained by the application of anodic pulses to the cathode surface and it remained stable up to 10 h for CO$_2$ electrolysis [56]. Moreover, the formation of CO intermediate was favoured at low temperatures (3–5 °C). Patents have reported copper-modified electrodeposits for demonstrating selective ethylene formation (to the detriment of ethane) rather than methane [57]. This novel ex situ copper electrodeposit leads to the steadiness of ECR of CO$_2$ without decomposition. Thus, the oxides of Cu mixed with CuBr and other combinations have been proved to be effective for C$_2$ formation.

25.3.5 Computational aspect for enhancement of ethylene formation

The mechanism of methane formation does not apply to the formation of ethylene for ECR of CO$_2$. The coupling of two *CO intermediates was the potential determining step for C$_2$H$_4$ formation [21]. Later work by Bell and his group partially supported this hypothesis by predicting a low-potential ethylene formation mechanism involving the coupling of bound *CO intermediates [52]. DFT calculations of Goa et al. suggest that hydrocarbon selectivity depends on the hydrogenation of the *CO intermediates to hydroxymethylidyne (*COH) or formyl (*CHO) [58]. Further, two *CHO species combine together to form ethylene by C–C coupling under relatively low overpotential [58]. Cu (100) facet is preferred for *CHO production, which is supported by online electrochemical mass spectrometry studies [59]. Thus, a new mechanistic pathway for C$_2$H$_4$ formation needs to be understood and practiced for selective formation of ethylene.

25.4 Conclusion

Classification of metals based on hydrocarbon formation has been attempted in this chapter. Grouping metals based on a particular product: methane or ethylene has been discussed pertaining to its process conditions, choice of metals and its nanostructures, combination of bimetals, alloys and metal oxides. Theoretical understanding for methane and ethylene formation is also discussed in this chapter. The understanding gained from this chapter will guide future researchers to choose their objective for ECR of CO$_2$, either to update on the existing modification or to choose a new electrode combination based on the lessons from the literature. Thus, development of ECR of CO$_2$ demands a separate classification of metals based on hydrocarbon formation and exhaustive experimental and theoretical results supporting its Faradaic efficiency.

References

[1] Gattrell, M., N. Gupta and A. Co. 2006. A review of the aqueous electrochemical reduction of CO$_2$ to hydrocarbons on Cu. Journal of Electroanalytical Chemistry, 594: 1–19.

[2] Shibata, H., 2009, Electrocatalytic CO$_2$ reduction catalysis engineering and reaction mechanism, Ph.D thesis, Technical University of Delft.

[3] Ortenzi, F., M. Chiesa, R. Scarcelli and G. Pede. 2008. Experimental tests of blends of hydrogen and natural gas in light-duty vehicles. International Journal of Hydrogen Energy, 33: 3225–3229.

[4] Song, C. S. 2006. Global challenges and strategies for control, conversion and utilization of CO$_2$ for sustainable development involving energy, catalysis, adsorption and chemical processing. Catalysis Today, 115: 2–32.

[5] Zhu D. D., J. L. Liu, and S. Z. Qiao. 2016. Recent Advances in Inorganic Heterogeneous Electrocatalysts for Reduction of Carbon Dioxide, Advance Matter., 28: 3423–3452.

[6] Zhang L., Z. J. Zahao, and J. Gong. 2017. Nanostructured Materials for Heterogeneous Electrocatalytic CO$_2$ Reduction and their Related Reaction Mechanisms, Angewandte Chemie International edition, 56: 11326–11353.

[7] Vickers J. W., D. Alfonso, and D. R. Kauffman. 2017. Electrochemical Carbon Dioxide Reduction at Nanostructured Gold, Copper, and Alloy Materials, Energy Technology, 5:1–22.

[8] Malik, K., S. Singh, S. Basu and A. Verma, 2017. Electrochemical reduction of CO$_2$ for synthesis of green fuel, WIREs Energy Environment, 244. doi: 10.1002/wene.244.

[9] Keerthiga, G. 2015. Electrochemical reduction of copper and zinc based electrodes for aqueous electrolytes, Ph.D thesis, Indian Institute of Technology, Madras.

[10] Hori, Y., A. Murata and R. Takahashi. 1989. Formation of hydrocarbons in the electrochemical reduction of carbon dioxide at a copper electrode in aqueous solution. Journal of Chemical Society, Faraday Transaction, 1: 2309–2326.

[11] Lee S., D. Kim and J. Lee. 2015. Electrocatalytic production of C$_3$-C$_4$ compounds by conversion of CO$_2$ on a chloride-induced bi-phasic Cu$_2$O-Cu Catalyst, Angewandte Chemie International edition, 54:14701–14705.

[12] Li, W. 2010.Electro catalytic reduction of CO$_2$ to small organic molecule fuels on metal catalysts pp.55–76 in: Y. H. Hu (Eds), Advances in CO$_2$ Conversion and Utilization, ACS symposium series, Oxford University Press, USA.

[13] Kaneco, S., H. Katsumata, T. Suzuki and K. Ohta. 2006. Electrochemical reduction of CO$_2$ to methane at the Cu electrode in methanol with sodium supporting salts and its comparison with other alkaline salts. Energy & Fuels, 20: 409–414.

[14] Monz J., Y. Malewski, R. Kortlever, F. J. Vidal-Iglesias, J. Solla- Gulln, M. T. M. Koper and P. Rodriguez. 2015. Enhanced electrocatalytic activity of Au@Cu core@shell nanoparticles towards CO$_2$ reduction, Journal of Materials Chemistry A, 3: 23690–23698.

[15] Surya S., Rajeev K. Gautam, Karan Malik and Anil Verma. 2017. Ag-Co bimetallic catalyst for electrochemical reduction of CO$_2$ to value added products, *Journal of CO$_2$ utlization,* 18: 139–146.

[16] Kas. R, R. Kortlever, H. Yilmaz, M.T.M Koper and G. Mul. 2015. Manipulating the Hydrocarbon Selectivity of Copper Nanoparticles in CO$_2$ Electroreduction by Process Conditions, Chem Electrochem, 2: 354–358.

[17] Noda, H., S. Ikeda, Y. Oda, K. Imai, M. Maeda and K. Ito. 1990. Electrochemical reduction of CO_2 at various metal electrodes in aqueous potassium hydrogen carbonate solution. Bulletin of Chemical Society Japan, 63: 2459–2462.

[18] Keerthiga. G and Raghuram Chetty. 2017. Electrochemical Reduction of Carbon Dioxide on Zinc-Modified Copper Electrodes, Journal of the Electrochemical Society, 164 (4): H1-H6

[19] Manthiram K., B. J. Beberwyck and A. P. Alivisatos. 2014. Enhanced electrochemical methanation of carbon dioxide with a dispersible nanoscale copper catalyst, Journal of American Chemical Society, 136:13319–13325.

[20] Reske, R., M. Duca, M. Oezaslan, K. J. P. Schouten, M. T. M. Koper and P. Strasser. 2013. Controlling Catalytic Selectivities during CO_2 Electroreduction on Thin Cu Metal Overlayers, The Journal of Physical Chemistry Letters, 4: 2410–2413.

[21] Tang, W., A. A Peterson, A. S. Varela, Z. P. Jovanov, L. Bech, W. J. Durand, S. Dahl, J. K. Norskov and I. B. Chorkendorff. 2012. The importance of surface morphology in controlling the selectivity of polycrystalline copper for CO_2 electroreduction, Physical Chemistry Chemical Physics, 14: 76–81.

[22] Keerthiga G., B. Viswanathan and Raghuram Chetty. 2015. Electrochemical Reduction of CO_2 on Electrodeposited Cu Electrodes: Crystalline Phase Sensitivity on Selectivity, Catalysis Today, 245: 68–73.

[23] Guo, X., Y. Znang, C. Deng, X. Li, Y. Xue, Y. M. Yan and K. Sun. 2015. Composition dependent activity of Cu–Pt nanocrystals for electrochemical reduction of CO_2, Chemical Communication, 51:1345–13.

[24] Xu Z., E. Lai, Y. Shao-Horn and K. Hamad-Schifferli. 2012. Compositional dependence of the stability of AuCu alloy nanoparticles, Chemical Communication, 48: 5626–5628.

[25] Kim C., H. S. Jeon, T. Eom, M.S. Jee, H. Kim, C. M. Friend, B. K. Min, and Y. J. Hwang. 2015. Achieving Selective and Efficient Electrocatalytic Activity for CO_2 Reduction Using Immobilized Silver Nanoparticles, Journal of American Chemical Society, 137 (43): 13844–13850.

[26] Zhao W., L. Yang, Y. Yin and M. Jin. 2014. Thermodynamic controlled synthesis of intermetallic Au_3Cu alloy nanocrystals from Cu microparticles, Journal of Materials Chemistry A, 2: 902–906.

[27] Zhang S., P. Kang, M. Bakir, A. M. Lapides, C. J. Dares and T. J. Meyer. 2015. Polymer-supported CuPd nanoalloy as a synergistic catalyst for electrocatalytic reduction of carbon dioxide to methane, Proceedings of National Academy of Sciences of the United States of America, 29, 112: 15809–15814.

[28] Hirunsit, P. and J. Limtrakul. 2015. CO_2 Electrochemical Reduction to Methane and Methanol on Copper-Based Alloys: Theoretical Insight, Journal of Physical Chemistry C, 119: 8238–8249.

[29] Le, M., M. Ren, Z. Zhang, P. T. Sprunger, R. L. Kurtz, and J. C. Flake. 2011. Electrochemical reduction of CO_2 to CH_3OH at copper oxide surfaces. Journal of Electrochemical Society, 158: E45-E49.

[30] Li, C. W. and M. W. Kanan. 2012. CO_2 reduction at lower overpotential on Copper electrodes resulting from the reduction of thick Cu_2O films. Journal of American Chemical Society, 134: 7231–7234.

[31] Goncalves, M. R., A. Gomesa, J. Condeco, T. R. C. Fernandes, T. Pardala, C. A. C. Sequeira and J.B. Branco. 2013. Electrochemical conversion of CO_2 to C_2 hydrocarbons using different ex situ copper electrodeposits. Electrochimica Acta, 102: 388–392.

[32] Yano, J., T. Morita, K. Shimano, Y. Nagami and S. Yamasaki. 2007. Selective ethylene formation by pulse-mode electrochemical reduction of carbon dioxide using copper and copper oxide electrodes. Journal of Solid State Electrochemistry, 11:554–557.

[33] Kyriacou, G. and A. Anagnostopoulos. 1992. Electroreduction of CO$_2$ on differently prepared copper electrodes: The influence of electrode treatment on the current efficiencies. Journal of Electro analytical Chemistry, 322: 233–246.

[34] Chen, Y. and M. W. Kanan. 2012. Tin oxide dependence of the CO$_2$ reduction efficiency on tin electrodes and enhanced activity for tin/tin oxide thin-film catalysts. Journal of American Chemical Society, 134: 1986–1989.

[35] Xiao H., W. A. Goddard, T. Cheng, and Y. Liu. 2017. Cu metal embedded in oxidized matrix catalyst to promote CO$_2$ activation and CO dimerization for electrochemical reduction of CO$_2$, Proceedings of National Academy of Science 114, 26: 6685–688.

[36] Frese KW Jr and Leach S. 1985. Electrochemical reduction of carbon dioxide to methane, methanol and CO on Ru electrodes. Journal of Electrochemical Society, 132:259–260.

[37] Kudo, A., S. Nakagawa, A. Tsuneta and T. Sakata. 1993. Electrochemical reduction of high pressure CO$_2$ on Ni electrodes. Journal of Electrochemical Society, 140: 1541–1546.

[38] Hara. K and T. Sakata. 1997. Electrocatalytic formation of methane from CO$_2$ on Pt gas diffusion electrode. Journal of Electrochemical Society, 144:539–545.

[39] Shen, J., R. Kortlever, Recep Kas, Yuvraj Y. Birdja, Oscar Diaz-Morales, Youngkook Kwon, Isis Ledezma-Yanez, Klaas Jan P. Schouten, Guido Mul and Marc T.M. Koper. 2015. Electrocatalytic reduction of carbon dioxide to carbon monoxide and methane at an immobilized cobalt protoporphyrin. Nature Communications. 6:8177. DOI: 10.1038/ncomms9177

[40] Peterson A., F. Abild-Pedersen, F. Studt, J. Rossmeisl and J. K. Norskov, 2010, How copper catalyzes the electroreduction of carbon dioxide into hydrocarbon fuels, Energy Environment and Science, 3: 1311–1315.

[41] Nie X., M. R. Esopi, M. J. Janik and A. Asthagiri. 2013. Selectivity of CO$_2$ reduction on copper electrodes: the role of the kinetics of elementary steps, Angewandte Chemie International edition 52: 2459–2462; Angewandte Chemie International edition, 2013. 125: 2519–2522.

[42] Hori Y., I. Takahashi, O. Koga and N. Hoshi. 2002. Selective Formation of C$_2$ Compounds from Electrochemical Reduction of CO$_2$ at a Series of Copper Single Crystal Electrodes, The Journal of Physical Chemistry B, 106:15–17.

[43] Kuhl K. P., E. R. Cave, D. N. Abram and T. F. Jaramillo. 2012. New insights into the electrochemical reduction of carbon dioxide on metallic copper surfaces Energy Environment and Science, 5: 7050–7059.

[44] Schouten K. J. P., E. Perez Gallent and M. T. M. Koper. 2013. Structure Sensitivity of the Electrochemical Reduction of Carbon Monoxide on Copper Single Crystals, ACS Catalysis, 3:1292–12995.

[45] Michael K. Smart, Vincent Wheatley and Anand Veeraragavan, Comparison between hydrogen, methane and ethylene fuels in a 3-D Scramjet at Mach 8, Final report of Air Force Research Laboratory, Arlington, Virginia 22203, published on 26.4.2016. http://www.dtic.mil/dtic/tr/fulltext/u2/1022771.pdf

[46] Keerthiga G., B. Viswanathan and Raghuram Chetty. 2018. Electrochemical Reduction of Carbon Dioxide on Cu Electrode: A Comparative Study in Bicarbonate and Chloride Electrolytes, Catalysis in Green Chemistry and Engineering, 2: 179–188.

[47] Mistry H., A. S. Varela, C. S. Bonifacio, I. Zegkinoglou, I. Sinev, Y.- W. Choi, K. Kisslinger, E. A. Stach, J. C. Yang, P. Strasser and B. R. Cuenya. 2016. Highly selective plasma-activated copper catalysts for carbon dioxide reduction to ethylene, Nature Communication, 7: 12123.

[48] Kim D., C. S. Kley, Y. Li, and P.Yang. 2017. Copper nanoparticle ensembles for selective electroreduction of CO$_2$ to C$_2$–C$_3$ products, PNAS Early Edition, (40) 114: 10560–10565.

[49] Kwon Y., Y. Lum, E. L. Clark, J. W. Ager and A. T. Bell. 2016. CO$_2$ Electroreduction with Enhanced Ethylene and Ethanol Selectivity by Nanostructuring Polycrystalline Copper, ChemElectro-Chem, 3: 1012–1019.

[50] Feng, X., K. Jiang, S. Fan and M. W. Kanan. 2016. A Direct Grain-Boundary-Activity Correlation for CO Electroreduction on Cu Nanoparticles, ACS Central Science, 2:169–174.

[51] Birdja Y.Y., Shen J. and Koper M.T.M. 2017. Influence of the metal center of metalloprotoporphyrins on the electrocatalytic CO_2 reduction to formic acid, Catalysis Today, 288: 37–47 and Ooka H., Costa, M.C. Figueiredo and M.T.M Koper. 2017. Competition between Hydrogen Evolution and Carbon Dioxide Reduction on Copper Electrodes in Mildly Acidic Media, Langmuir, 37: 9307–9313.

[52] Goodpaster J. D., A. T. Bell, and M. Head-Gordon. 2016. Identification of Possible Pathways for C–C Bond Formation during Electrochemical Reduction of CO_2: New Theoretical Insights from an Improved Electrochemical Model, The Journal of Physical Chemistry letters, 7:1471–1477.

[53] Nakato Y., S. Yano, T. Yamaguchi and H. Tsubomura.1991. Reactions and mechanism of the electrochemical reduction of carbon-dioxide on alloyed copper-silver electrodes, Denki Kagaku, 59: 491–498.

[54] Peng Y., T Wu, L Sun, J M V Nsanzimana, AC Fisher and X Wang. 2017. Selective Electrochemical Reduction of CO_2 to Ethylene on Nanopores-Modified Copper Electrodes in Aqueous Solution, ACS Appl Mater Interfaces. 9(38):32782–32789.

[55] Chen C. S., J. H. Wan and B. S. Yeo. 2015. Electrochemical Reduction of Carbon Dioxide to Ethane Using Nanostructured Cu_2O-Derived Copper Catalyst and Palladium(II) Chloride, The Journal of Physical Chemistry C, 119: 26875–26882.

[56] Norma R. de Tacconi, W. Chanmanee, B.H. Dennis and K. Rajeshwar. 2017. Composite copper oxide–copper bromide films for the selective electroreduction of carbon dioxide, Journal of Materials Research, 32 9:1727–1734.

[57] Patent Process for the selective electrochemical conversion of CO_2 into C_2 hydrocarbons, EP2686464A2, WO2012125053A3, WO2012125053A4.

[58] Gao S., X. Jiao, Z. Sun, W. Zhang, Y. Sun, C. Wang, Q. Hu, X. Zu, F. Yang, S. Yang, L. Liang, J. Wu and Y. Xie. 2016. Ultrathin Co_3O_4 Layers Realizing Optimized CO_2 Electroreduction to Formate, Angewandte Chemie International edition, 128: 708–712.

[59] Roberts F.S, Kuhl K.P and Nilsson A., High selectivity for ethylene from carbon dioxide reduction over copper nanocube electrocatalysts. 2015. Angew Chem Int Ed Engl, 54(17): 5179–82.

Volodymyr Tabas and Benjamin R. Buckley

26 Non-reductive CO$_2$ electrochemistry

26.1 Introduction

Carbon dioxide utilisation (CDU) technologies have recently gained increased interest as they offer an attractive alternative to the well-established carbon dioxide storage processes [1]. This is due to the fact that carbon dioxide is abundant, cheap and non-toxic; and waste carbon dioxide under CDU protocols can be converted into useful chemicals or fuels. Carbon dioxide has been used in the manufacture of salicylic acid, urea and cyclic carbonates for 50–100 years; however, due to carbon dioxide's relative inertness ($\Delta_f H°$ of –394 kJ mol^{-1}) these processes are significantly energy demanding with reactions taking place at high temperatures and pressures. Efficient chemical incorporation of CO$_2$ is restricted to reactive substrates, such as epoxides [2] to produce cyclic carbonates and amines to produce carbamates [3]. Alternative technologies are therefore sought to enable carbon dioxide insertion into less energetic substrates.

Electrosynthesis has been widely studied within the electrochemical community, but represents an underused tool, with great potential for the construction of chemical bonds [4]. As a direct result, there are currently only a few processes at an industrial or pilot plant scale for the electrosynthesis of organic molecules. This presents electrosynthesis as a potential step changing technology, particularly since the use of electrons can be directly linked to many aspects associated with green chemistry and the recent developments in the supply of renewable energy. The formation of a thermodynamically and kinetically stable C–C bond is arguably the most desirable application for CO$_2$ utilisation, since many feedstocks and even fuels could be produced. In general, these electrochemical carboxylation reactions could occur at room temperature, in part due to the fact that the energy of the electrons is related to the voltage applied to the system. Coupled with the underuse of electrosynthetic protocols, attractive and straightforward procedures for the carboxylation of carbon nucleophiles with CO$_2$ as the electrophile also remain largely underdeveloped [5].

This chapter concentrates on the formation of carboxylic acids from carbon dioxide under electrochemical conditions. The direct formation of carboxylic acids from carbon dioxide and alkenes, alkynes and halides under "green" conditions has been described as one of the most challenging CDU processes and the hydrocarboxylation of ethene and CO$_2$ has been described as a "dream process". Traditionally carboxylic acids are prepared through the reaction of carbon dioxide with organometallic nucleophiles, such as alkyl/aryl lithium or Grignard reagents; however, these processes

Volodymyr Tabas, Benjamin R. Buckley, Department of Chemistry, Loughborough University, Ashby Road, Loughborough Leicestershire, LE11 3TU United Kingdom

https://doi.org/10.1515/9783110665147-026

are moisture/air sensitive and generate a large inorganic waste stream [6]. Thus, activation of CO_2 in an alternative process, such as electrochemistry, should be of significant interest as the high thermodynamic stability of CO_2 is negated by single-electron reduction.

26.2 Cell parameters

Electrochemical reduction of either carbon dioxide or the substrate takes place at the cathode surface, so a corresponding oxidation must take place at the anode. This can be either a metal that is usually present as a sacrificial electrode (e.g. Mg, Al, Zn) or an organic reducing agent with inert electrodes (e.g. Pt or carbon). Typically, carbon dioxide incorporation reactions tend to be carried out in undivided cells, although there are some examples that employ a divided cell system (Figure 26.1) [7].

A major challenge is in the development of electrochemical systems that do not require a sacrificial agent, as the production of stoichiometric quantities of inorganic or organic waste will inhibit the "green" credentials of the processes. For example, Figure 26.2 illustrates three types of typical cell set-ups shown in the context of alkene decarboxylation. In the undivided cell, a solid metal anode is

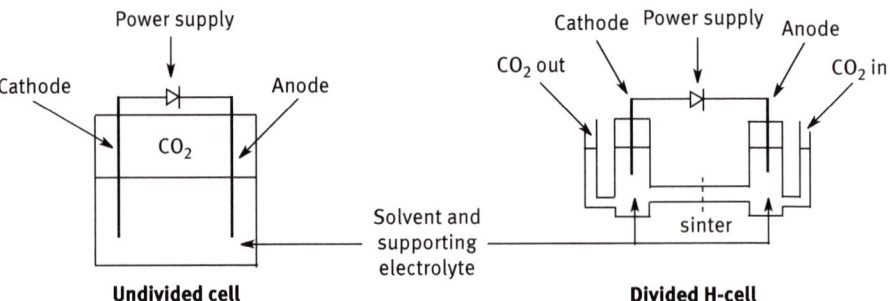

Figure 26.1: Typical cell set-ups used in electrochemical carbon dioxide utilisation.

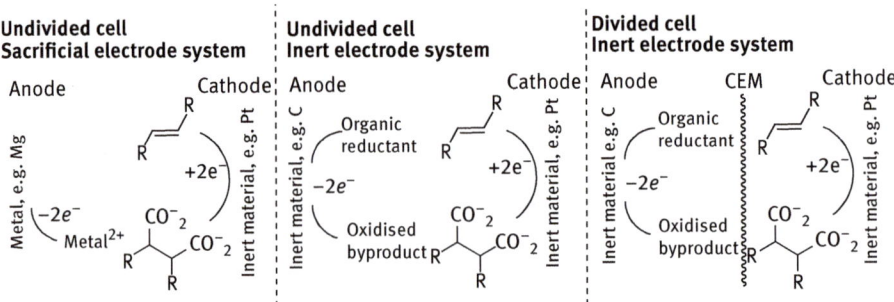

Figure 26.2: Sacrificial and inert electrode systems, highlighted in the decarboxylation of alkenes.

oxidised under the conditions to afford, for example, in the case of Mg^{2+} ions, the electrode is consumed during the reaction and the metal waste would need to be collected and disposed or if possible recycled, probably using a subsequent electro-chemical reaction to regenerate the metal.

With inert electrodes in the undivided cell, an organic reductant can often be employed, a typical example would be an amine in which electrons can be removed under electrolysis. In this case, the organic by-products would need to be separated from the product and again be disposed of, a similar process is observed for divided cells, which are often employed when one of the components of a reaction are destroyed by one of the electrodes.

26.3 Electrochemical carboxylation reactions

As discussed in the introduction, the formation of a thermodynamically and kineti-cally stable C–C bond is one of the most interesting applications for CO$_2$ utilisation, and straightforward procedures for the carboxylation of carbon nucleophiles with CO$_2$ require development. Thus, carboxylation of various types of substrates has been achieved utilising electrochemical processes with varying levels of selectivity; below we examine some of these systems.

26.3.1 Dimerisation of CO$_2$ to oxalate

Oxalic acid has applications in the cleaning industry, in the dyeing processes and in metallurgy and can be prepared through the dimerisation of carbon dioxide. When re-duced directly in aprotic solution, the reduction potential of CO$_2$ is −2.2 V and generally affords mixtures of oxalate, CO, CO$_3^{2-}$ and other C–C coupling products when carried out in the absence of other reactants. Many approaches have been reported in order to optimise the formation of oxalate. For example, Savage reported two routes to oxalate (Figure 26.3): one via direct reduction at constant current in dimethylformamide (DMF) with Bu$_4$NClO$_4$ as electrolyte and an Hg cathode (at low CO$_2$ concentration and high temperature oxalate was favoured over CO + CO$_3^{2-}$) [8]; and the other utilising an indirect approach employing aromatic esters or nitriles as electrocata-lysts [9]. Based on these two types of processes, Savage proposed that oxalate for-mation occurred by the coupling of two radical anions with a second-order rate constant of approximately 5 × 108 M^{-1} s^{-1}.

Oxalate formation using anhydrous conditions in a cell with a sacrificial anode has been reported several times, recently an O$_2$-assisted Al/CO$_2$ electrochemical cell has been developed for CO$_2$ capture/conversion and electric power generation [10]. In addition, nickel catalysis and ionic liquids have been employed in oxalate synthesis [11, 12]. However, a more sustainable approach has been developed in which a stable

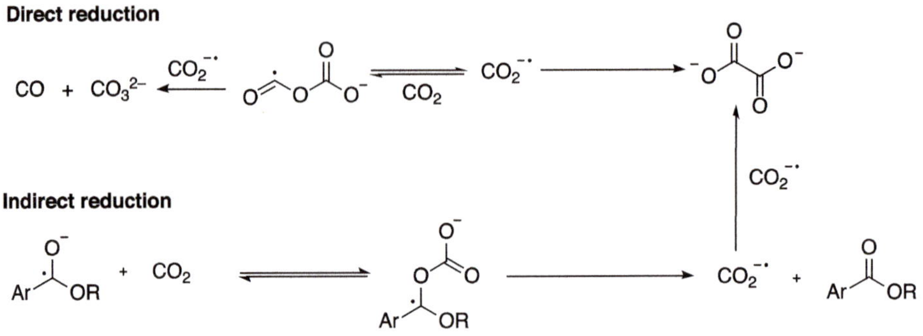

Direct reduction

Indirect reduction

Figure 26.3: Savage's mechanistic rational for oxalate formation through direct and indirect electrochemical reduction.

electrode set-up is employed with current efficiencies over 50% [13]. In addition, oxalate has been formed from CO_2 by solvated electrons produced using an atmospheric pressure plasma electrode, which offers a unique opportunity for CO_2 utilisation [14].

26.3.2 Carboxylation of alkenes (alkyl, aryl)

A wide variety of reports have been published for the carboxylation of alkenes. These processes are dominated by the application of sacrificial anodes and the formation of dicarboxylic acids, below are selected examples highlighting the general processes involved.

26.3.2.1 Dicarboxylation

Vasil'ev and co-workers were one of the first groups to report electrochemical carboxylation of alkenes, through the carboxylation of ethene to produce a non-selective mixture of short- and longer chain dicarboxylic acids (Figure 26.4) [15].

Vasil'ev postulated that oxalate formation proceeded through the CO_2 radical anion and in the presence of an acceptor molecule, such as ethylene, can form the longer chain dicarboxylic acids (Figure 26.5).

This type of process therefore "traps" the radical anion and prevents it from dimerising to form oxalic acid, consequently generating a mixture of succinic acid (C_4), adipic acid (C_6) and suberic acid (C_8) (Figure 26.5).

A wide range of alkenes have been carboxylated. Duñach, a pioneer of electrosynthetic carboxylation chemistry, reported the dicarboxylation of a series of alkenes utilising a sacrificial anode set-up in conjunction with nickel catalysis (Figure 26.6) [16].

Several reports utilising a similar system have appeared. Most recently, Senboku and co-workers reported the electrocarboxylation of aryl-substituted alkenes without

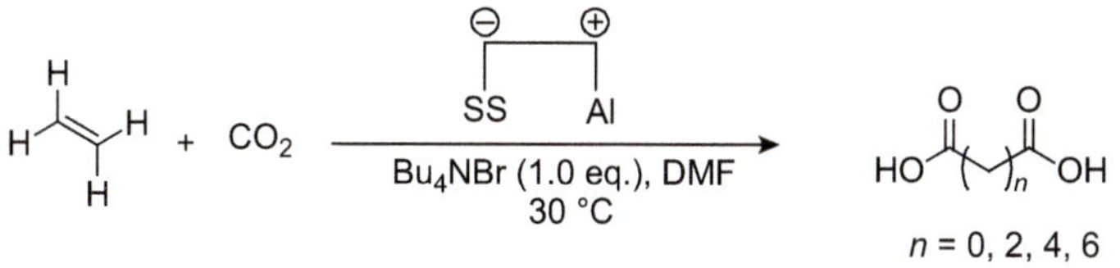

Figure 26.4: Vasil'ev's synthesis of dicarboxylic acids.

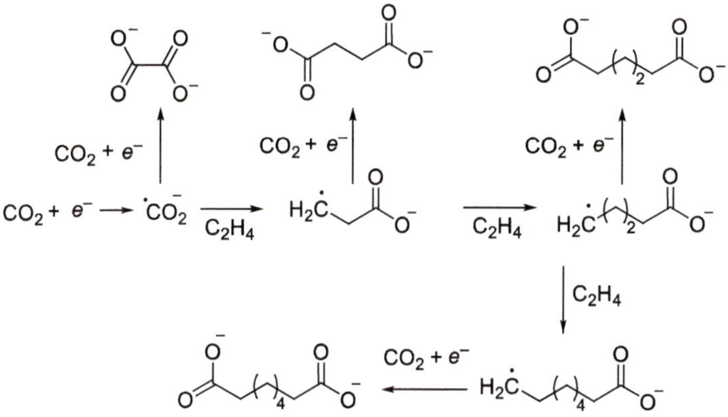

Figure 26.5: Vasil'ev's proposed mechanism for dicarboxylic acid synthesis.

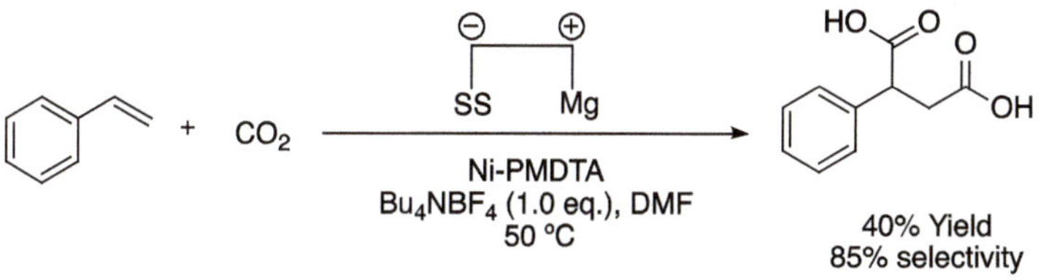

Figure 26.6: Duñach's dicarboxylic acid synthesis.

the need for an additional catalyst; however, the process does rely on the use of a platinum electrode (Figure 26.7) [17]. The group also demonstrated that this set-up could be used for various substituted alkenes, achieving good to excellent yields of the carboxylated product (69–99%; Figure 26.8).

The proposed reaction pathway is similar to the one reported by Vasil'ev. However, cyclic voltammetry studies indicate that CO$_2$ and styrenes have similar reduction potentials: thus, two reaction pathways (path a or path b, Figure 26.8) are possible.

The oxidation of the Mg anode allows for the reduction of CO$_2$ or the alkene depending on which has the lower reduction potential (Figure 26.8). Interestingly both processes can be operative at the same time. The effects of electrodes, electrolytes, temperature and other factors have been examined in the follow-up studies [18].

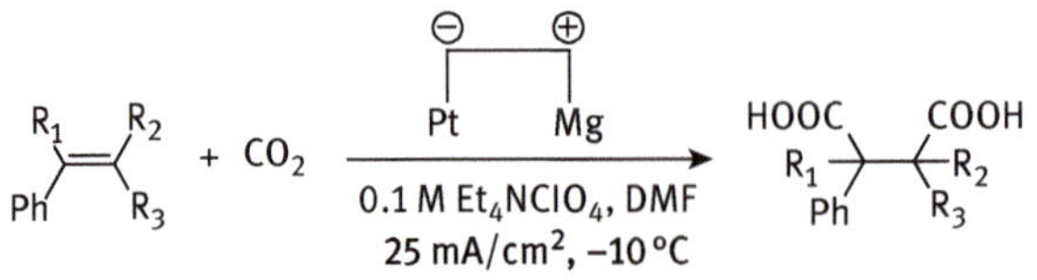

Figure 26.7: Senboku's approach to dicarboxylic acids.

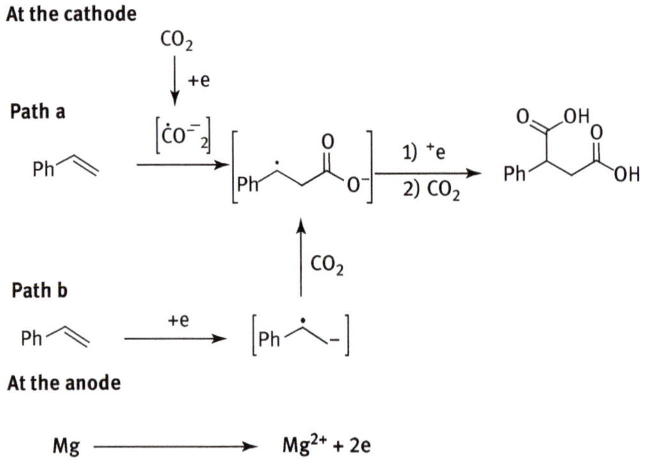

Figure 26.8: Senboku's proposed mechanism for dicarboxylic acid synthesis.

26.3.2.2 Monocarboxylation

Selective mono-carboxylation of alkenes has rarely been reported in electrochemical systems even though this is prevalent in arange of recent publications in the traditional catalysis literature. Under non-electrochemical conditions, the mono-carboxylation of alkenes proceeds to form the Markonikov addition products. This route was pioneered by Hoberg utilising nickel catalysis and modified and enhanced by Rovis, Thomas (Fe catalysis) and more recently by Martin (Figure 26.9) [19–22].

Under the conditions reported earlier, Markonikov's addition dominates. However, photochemical reports have shown that mono-carboxylation can occur, albeit in lower yields and selectivities, to provide anti-Markovnikov addition, thus providing a complimentary route to dicarboxylation alkenes (Figure 26.10) [23–26].

Dunach's reaction system reported for decarboxylation earlier (Figure 26.6) utilising sacrificial electrodes and Ni catalysis did also produce a small amount of anti-Markovnikov product, but this was a mixture with dicarboxylation dominating the process [16].

Tokuda and co-workers have reported the monocarboxylation of bicyclic systems, utilising a platinum plate electrode and a zinc plate sacrificial electrode system [27].

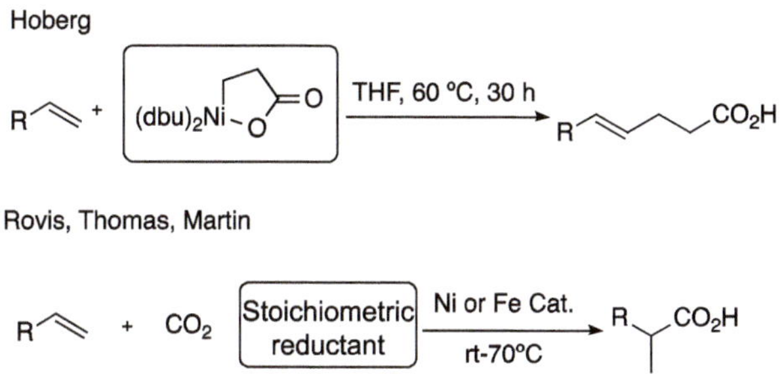

Hoberg

Rovis, Thomas, Martin

Figure 26.9: Selective metal catalysed monocarboxylation of alkenes.

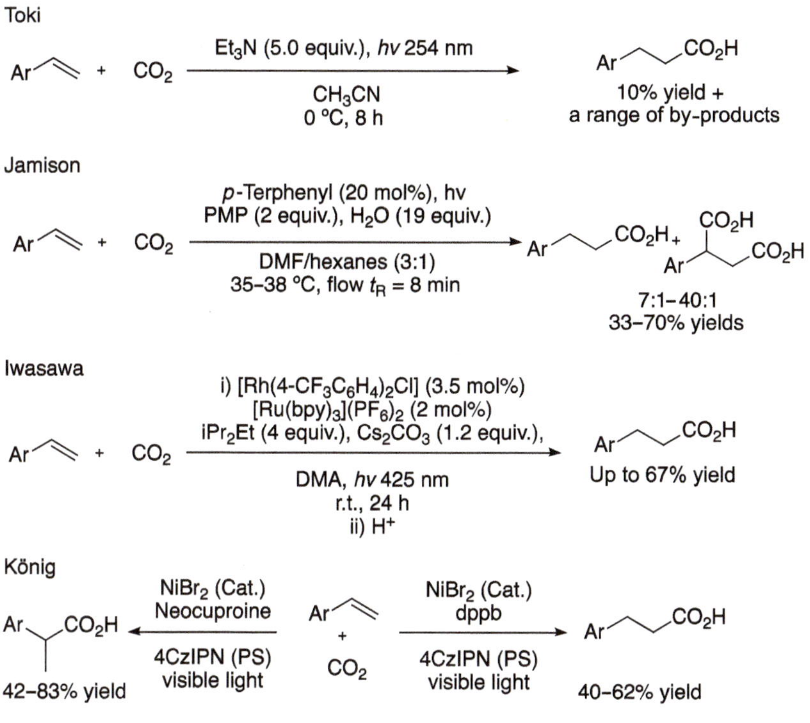

Toki

Jamison

Iwasawa

König

Figure 26.10: Photochemical approaches to monocarboxylic acids.

However, in some cases the yields were low and the product selectivity between mono- and di-carboxylation was poor and unpredictable. For example, in Figure 26.11 when R = Me, the monocarboxylic acid is the major product, but when the methyl ester is replaced with an ethyl ester the dicarboxylic acid product dominates.

Recently, a moderately selective mono-carboxylation of a limited number of substrates was reported that utilised a divided cell and a bespoke 12CaO·7Al$_2$O$_3$ electrode system (Figure 26.12) [28]. Unactivated alkenes such as trans-stilbene as well as

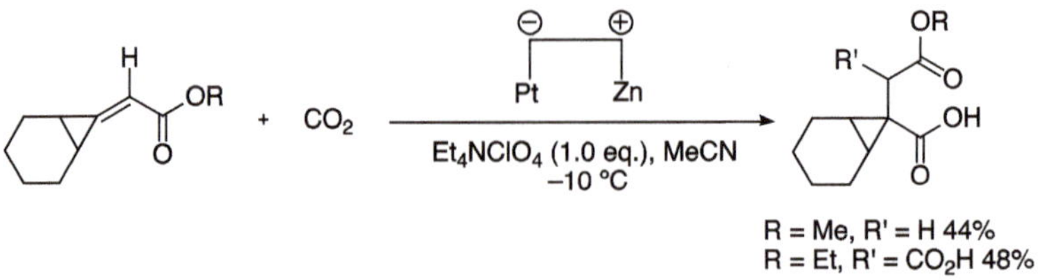

Figure 26.11: Mono- and di-carboxylation of bicyclic alkene substrates.

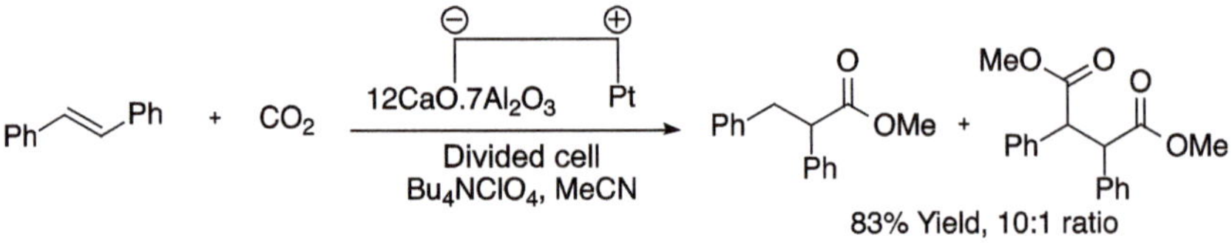

Figure 26.12: Monocarboxylation of *trans*-stilbene reported by Li and co-workers.

activated α,β-unsaturated alkenes such as methyl cinnamate could be selectively monocarboxylated in an anti-Markovnikov fashion. This remains the only report of an electrochemically induced selective anti-Markovnikov monocarboxylation of alkenes. The substrate scope is however limited, and the use of a bespoke electrode may limit the applicability of this system.

26.3.3 Carboxylation of alkynes (alkyl, aryl)

The electrochemical carboxylation of alkynes has also been reported by several re-search groups and like alkenes these processes are dominated by the use of sacrificial anodes. Unlike recent reports in the catalysis literature, there are no current methods for producing monocarboxylated products from addition across the triple bond in al-kyne starting materials with high selectivity. However, some interesting results have been obtained through incorporation of two or more CO_2 molecules when attempting to carboxylate alkynes. In the following section, we review a series of reports directed at electrochemical carboxylation of alkynes.

Duñach and co-workers have reported two approaches to the carboxylation of alkynes; initially, the monocarboxylation of a range of aliphatic and aromatic termi-nal alkynes was achieved (Figure 26.13) [29]. Utilising a carbon–magnesium elec-trode couple and a nickel catalyst, they were able to carboxylate the alkynes with varying levels of selectivity, with the major products undergoing carboxylation at the alpha position. A similar approach was then reported for disubstituted alkynes and again the monocarboxylated product was isolated as the major product [30].

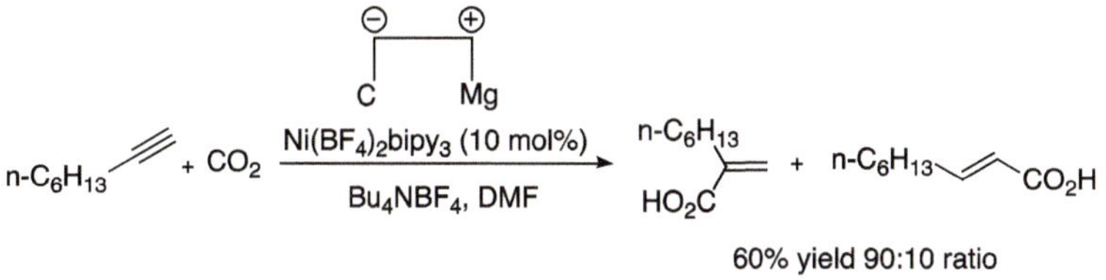

Figure 26.13: Duñach's approach to monocarboxylic acids.

Unsymmetrical disubstituted alkynes proved rather challenging and there was little discrimination between monocarboxylation alpha to alkyl or aryl substituents.

The group then described the carboxylation of terminal alkynes without the reduction of the alkyne bond (Figure 26.14) [31]. Again sacrificial electrode conditions were employed without the use of an additional nickel-based catalyst. Optimum conditions involved the use of a silver cathode and a magnesium sacrificial anode. Up to 95% selectivity was observed for the desired product with up to 90% isolated yield.

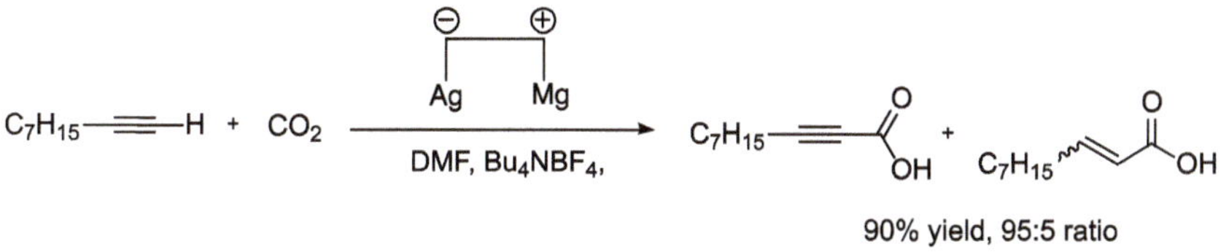

Figure 26.14: Duñach's approach to monocarboxylic acids without reduction of the alkyne bond.

Jiang and co-workers were the first to report the selective dicarboxylation of alkynes using electrochemical methods (Figure 26.15) [32]. Utilising a nickel–aluminium electrode couple, a mixture of the dicarboxylic acid and the corresponding anhydride were produced. The system was able to tolerate both terminal and disubstituted alkynes. However, the ratio of dicarboxylic acid to anhydride varied according to the alkyne employed, with little evidence for a predictable trend.

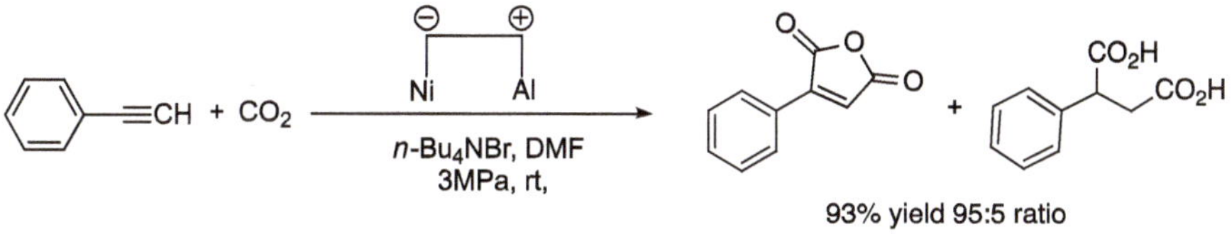

Figure 26.15: Jiang's dicarboxylation of alkynes to afford anhydrides and dicarboxylic acids.

A method for achieving the incorporation of two or three CO$_2$ molecules into an alkyne was reported by Li and co-workers in 2010 (Figure 26.16) [33]. Carboxylation of alkynes in the presence of metal salts was accomplished using a cell set-up similar to

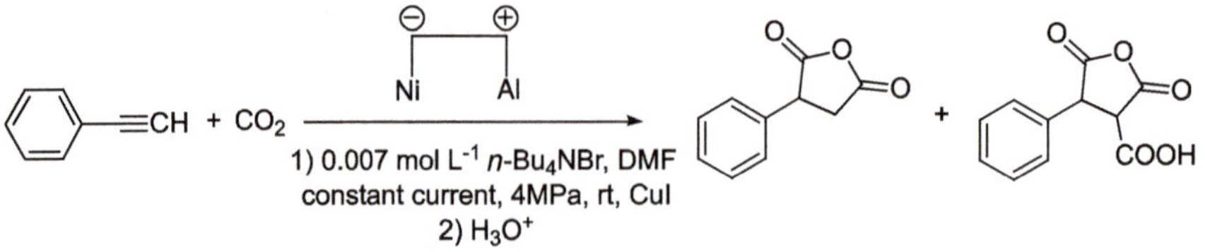

Figure 26.16: Li's dicarboxylation of alkynes to afford fully reduced anhydrides and dicarboxylic acids.

those shown above by Jiang, to afford the corresponding anhydride or anhydride acid. The yields changed according to the cathode material and the effect of substituted groups on the benzene ring. Interestingly, they isolated the reduced anhydride product unlike Jiang but offer no reasoning as to this phenomenon.

A tandem electrochemical radical cyclisation and carboxylation process has been reported through generation of a phenyl centred radical and subsequent intramolecular addition to an alkyne, finally resulting in dicarboxylated products (Figure 26.17) [34].

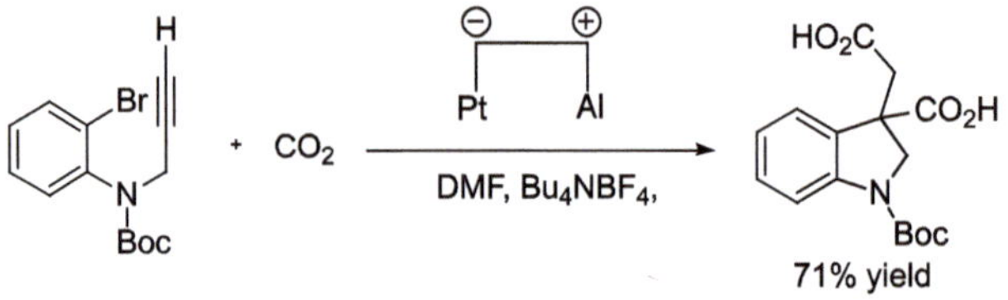

Figure 26.17: Tandem electrochemical radical formation and trapping with carbon dioxide.

26.3.4 Carboxylation of conjugated systems

Conjugated systems, such as 1,3-butadiene, have been extensively investigated for CDU because of the importance of the derived products, such as adipic acid for the polymer industry. A comprehensive review of approaches towards electrochemical carboxylation has been reported by De Vos and co-workers [7]. Very recently, dicarboxylation of dienes has received attention in the catalysis literature; however, like the sacrificial electrode systems highlighted in this chapter the use of a stoichiometric reductant is a mandatory requirement, thus limiting the green credentials of the research.

One of the earliest reports on electrocarboxylation dates back to the 1960s. Loveland patented the electrolytic production of acyclic carboxylic acids from olefins in 1962 [35]. This process was a major step forward for electrochemical carbon dioxide activation. However, several drawbacks to the system such as the use of a mercury electrode prompted further research and development in this area (Figure 26.18).

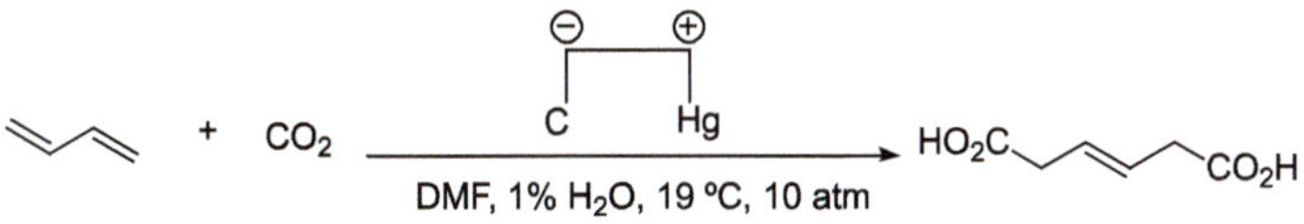

Figure 26.18: Loveland's patented carboxylation of butadienes.

Duñach reported the carboxylation of dienes in 1992 [36] and in 2010 Li and co-workers followed this up with a highly regioselective method to achieve dionic acids from dienes [37]. Their experiments were based around the reaction of 1,3-butadiene with CO_2, which produced 3-hexene-1,6-dionic acid (Figure 26.19).

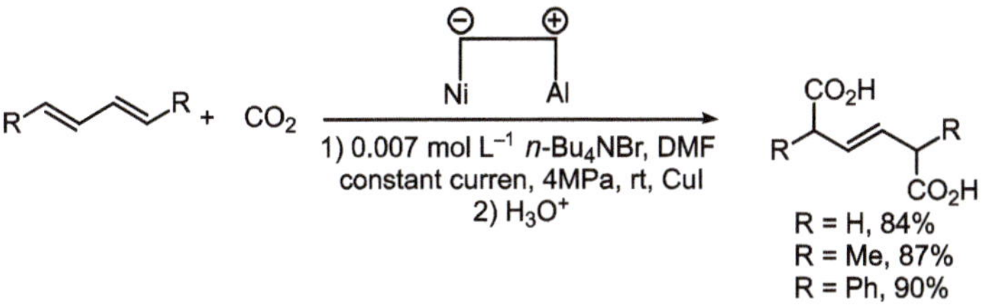

Figure 26.19: Li's modified dicarboxylation of butadienes.

The group used an electrochemical cell with an aluminium anode as the sacrificial electrode, a Ni cathode and Bu_4NBr as the electrolyte. The reaction was carried out at room temperature under a CO_2 pressure of 30 atm. The same experiment was carried out on a variety of substituted 1,3-dienes and the products obtained were exclusively *trans*-substituted. They report that an increase in CO_2 pressure allows for a higher yield. This can be attributed to the solubility of CO_2 in DMF, which increases at higher pressures. In addition, an increase in current also results in higher yields being obtained. This method is a dramatic improvement on the first techniques used by Loveland; however, further improvement could be achieved by removal of the aluminium sacrificial electrode.

Duñach and co-workers have also published the simultaneous activation of carbon dioxide and diynes by electrogenerated LNi(0) complexes (L = bpy, pentamethyldiethylenetriamine: PMDTA; Figure 26.20). They observed the selective incorporation of one molecule of CO_2 into the unsaturated systems [38]. A series of non-conjugated diynes afforded selectively linear or cyclic adducts depending on the ligand employed. Diynes bearing both a terminal and an internal triple bond gave exclusive CO_2 incorporation into the terminal alkynyl group, regioselectively at the 2-position. The electrocarboxylation of 1,3-diynes with the Ni-PMDTA catalytic system yielded, regio- and stereoselectively, (*E*)-2-vinylidene-3-yne carboxylic acids in one step. The electrosyntheses were carried out in a single-compartment cell fitted with a sacrificial magnesium anode, and utilised a catalytic amount of an air-stable Ni(II)

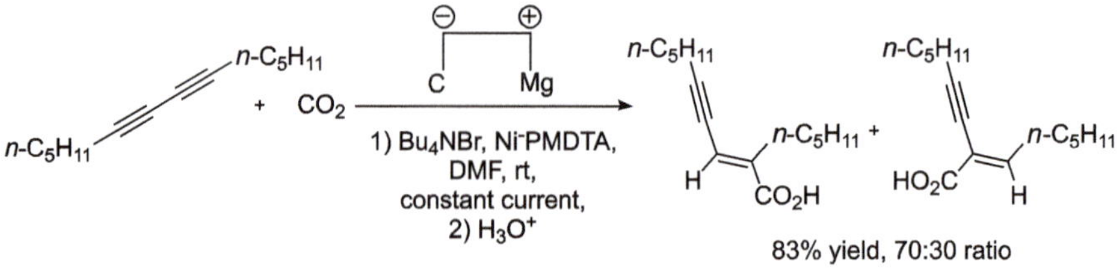

Figure 26.20: Duñach's reaction of CO_2 with diynes.

complex as the catalyst precursor. Cyclic voltammetry studies revealed that both carbon dioxide and the diynes are able to coordinate to LNi(0)-generated species in DMF.

Jiang has also reported the carboxylation of diynes using an undivided electrochemical cell fitted with Ni cathode and Al anode, containing Bu_4NBr electrolyte in DMF with a constant current at 4 MPa pressure of CO_2 (Figure 26.21) [39]. The corresponding alkylidene lactones could be obtained in moderate-to-good yields in the presence of a copper(I) catalyst. In the absence of additional catalyst, the γ-keto carboxylic acids were obtained as the major product.

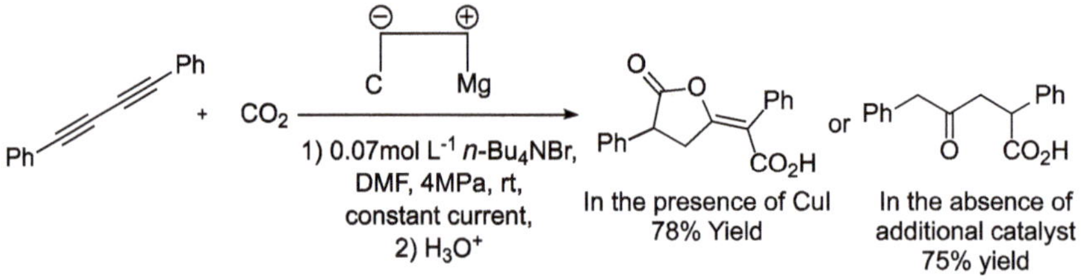

Figure 26.21: Jiang's route to alkylidene lactones and γ-keto carboxylic acids.

26.3.5 Carboxylation of carbonyls and imines

As mentioned in the previous section, electrocarboxylation applies to a variety of substrates. This is also the case for ketones, aldehydes and imines. The scope of the reactions that these substrates have with CO_2 is large with a few even going as far as being applied in semi-industrial processes to produce value-added products such as drug precursors, and precursors for further chemical reactions.

In the 1960s, the first example of electrocarboxylation of ketones was described by Wawzonek and Gundersen [40]. This technique provides a unique way to synthesise α-aryl propionic acids, which are the basis of many of the non-steroidal anti-inflammatory drugs that are now widely used due to their anti-inflammatory and analgesic properties. These drugs include ibuprofen, naproxen and aspirin. In 1994, a research group headed by Chan reported a novel synthesis of hydroxyl naproxen from 2-acetyl–6-methoxynaphthalene (Figure 26.22) [41].

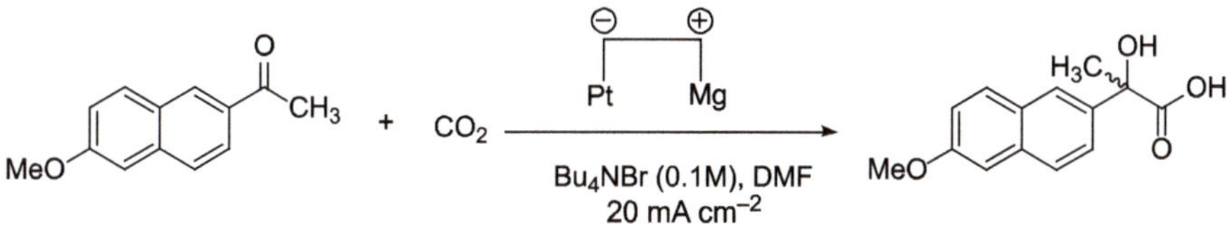

Figure 26.22: Synthesis of hydroxyl naproxen.

The carboxylation process afforded the carboxylated product in high yield (92%) under atmospheric pressure CO$_2$. The groups' investigations also showed that the same reaction conditions and set-up worked for *p*-isobutylacetophenone in even higher yields. This protocol was later applied to a flow cell reactor by Datta et al. [42]. The group started with a 0.2 L scale cell, and various variables were changed to see if they brought about a change in yield. Various changes were implemented, and the most effective one was proven to have an increase in pressure from 138 to 690 kPa, which improved both conversion and selectivity, as did the change in electrolyte to Et$_4$NCl. These changes were applied and a 1 L reactor cell was constructed. At ideal conditions, this reactor cell produced conversions of 92.5% with 96% selectivity. This shows that there is scope for this process to be fully industrialised; however, problems may arise as the groups attempt to increase the cell size to 75 L was met with a significant reduction in yield to 58%.

Methionine hydroxy analogue (MHA) has been used as an industrial feedstock for sometime. The conventional way of synthesising this molecule has been through the use of hydrogen cyanide, which has various health risks as well as extremely high toxicity. Recently, a patent has been filed in the USA and Europe for the electrocarboxylation of 3-methylmercaptopropionaldehyde (MMP) to form MHA [43]. In this process, an electrochemical cell which is separated by a cation exchange membrane with a boron-doped permeable diamond cathode and a platinum-coated non-sacrificial anode is employed. A detailed representation is shown in Figure 26.23.

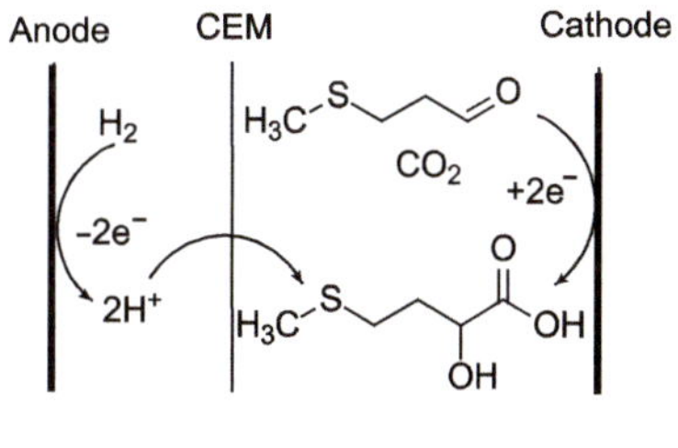

Figure 26.23: Cell set-up employed in the electrocarboxylation of 3-methylmercaptopropionaldehyde.

Due to the fact that the anode is non-sacrificial, the electrons necessary for the reaction to proceed must be provided from a different source; in this case, hydrogen is bubbled into the anodic compartment. The hydrogen is oxidised and this provides the electrons needed for the reduction of MMP to MHA. This process is much safer than the conventional use of hydrogen cyanide; however, at the moment the maximum yields that were obtained by this reaction are only 30% so more development must be carried out before this type of promising technology can replace the current industrial process.

In the 1980s, Silvestri and co-workers conducted extensive research into electrocarboxylation of various substrates. The group was successful with a wide variety of substrates; one of which was the electrocarboxylation of benzalaniline to the corresponding α-amino acid (Figure 26.24) [44]. This reaction was so successful that the group implemented it into a semi-continuous reaction vessel. The cell was shown to be very efficient, achieving a conversion of 92% and a yield of 85% for the desired product.

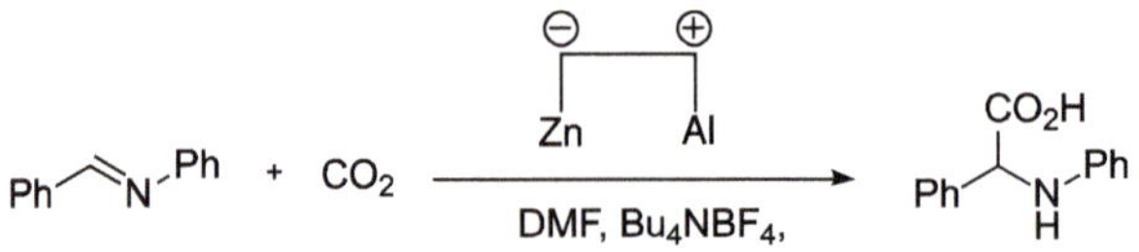

Figure 26.24: Electrocarboxylation of benzalaniline to the corresponding α-amino acid.

26.3.6 Carboxylation of alkyl, benzyl, aryl and alkenyl halides

The carboxylation of benzyl halides has been achieved, again using sacrificial electrodes. Genarro and co-workers reported the carboxylation of benzyl chlorides using a silver cathode and an aluminium anode (Figure 26.25) [45]. It was found that, compared to other cathode materials, silver lowered the reduction potentials of benzyl chlorides while the reduction potential of CO_2 remained relatively unchanged.

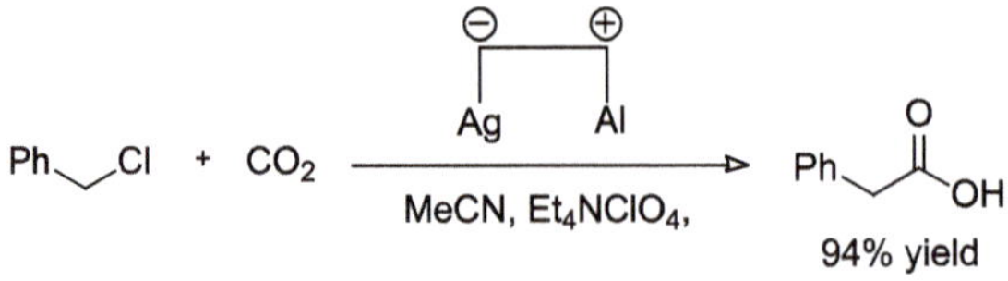

Figure 26.25: Carboxylation of benzyl chloride using a silver cathode and an aluminium anode.

This discovery enabled selective reduction of benzyl halides in the presence of CO_2, and therefore minimised by-products from direct reduction of CO_2. Several other studies on electrochemical carboxylation of benzyl halides using a silver cathode have been published [46–48].

In addition, the cathodic carboxylation of alkyl halides has also been reported; excellent yields were obtained from allyl-chloride-type substrates, with alkyl chlorides providing poor-to-good yields (22-68%) of the corresponding carboxylic acids (Figure 26.26) [47]. The carboxylation of alkyl bromides has also been rendered asymmetric through the use of a chiral auxiliary, delivering good yield (up to 80%) and with high diastereoselectivity (up to >98:2 d.r.) [49].

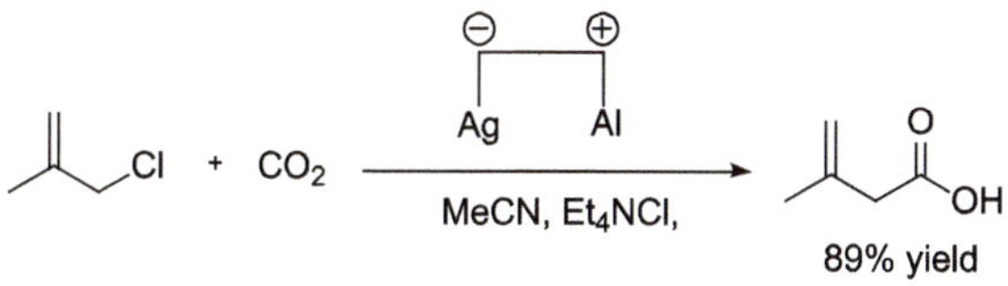

Figure 26.26: Cathodic carboxylation of alkyl halides.

Carboxylation of benzyl fluorides has also been reported to afford the corresponding carboxylic acids in good yield and has been exemplified through the synthesis of fluorinated analogues of non-steroidal anti-inflammatory agents such as α-fluorofenoprofen and α-fluoroloxoprofen (Figure 26.27) [50]. Electrolysis of α,α-difluorotoluene derivatives in the presence of CO$_2$ affords α-fluorinated arylacetic acid derivatives. This approach avoids the use of hazardous fluorinating agents that were used in previous syntheses of these compounds.

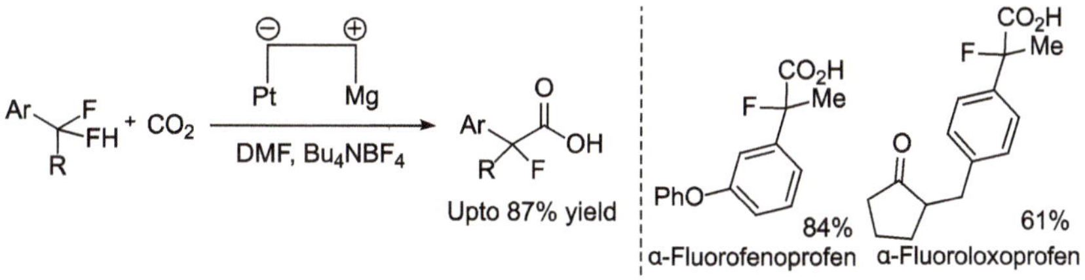

Figure 26.27: Carboxylation of benzyl fluorides.

Alkenyl halides and pseudo-halides have also been carboxylated using a sacrificial electrode approach. The process tends to work well with the addition of a nickel catalyst for alkenylbromides and requires the use of a platinum cathode and magnesium anode [51].

26.4 Summary of carboxylation approaches and future outlook

Clearly, the use of electrosynthesis to enable room temperature carbon dioxide incorporation to a wide variety of substrates has been well demonstrated. The use of these electrons to construct carbon–carbon bonds offers an alternative "reagentless" approach to traditional synthesis/catalysis. On a research laboratory scale, the recent development and availability of commercial electrosynthesis equipment, in both batch and flow modes, should enable more researchers from both an academic and industrial background to attempt electrochemical reactions. The advantage of commercial equipment now being available is the opportunity for the reproducibility of electrochemical reactions across the globe to be increased. The benefits from this standardised approach/equipment are from having a very defined reactor set-up, electrode size, shape and surface area. In time, this may well deliver processes that could be scaled up into industrial processes.

However, significant challenges remain, perhaps the biggest challenge being the removal of the sacrificial electrode. In almost all cases, the use of a magnesium, aluminium or zinc sacrificial electrode enables the electrochemical process to occur under otherwise mild conditions and provides a suitable cation for stabilisation of the resultant carboxylate. These carboxylates are insoluble in most organic solvents and precipitate during the process, thus reducing possible side reactions with the products, but this phenomenon is also a significant drawback, if, for example, one wanted to develop a continuous flow reactor using this technology. Precipitation could potentially block the flow of the solution. The formation of these carboxylates also adds another step in the process in which the products need to be treated with an acidic solution to release the free acid, thus resulting in significant inorganic waste products. The life cycle analysis and technoeconomic analysis of these processes need to be carried out and therefore future studies that aim to provide products from carbon dioxide insertion will need to be carefully studied.

With the recent renaissance in electrosynthesis and the demand for greener processes that include the incorporation of carbon dioxide, it is likely that this area will continue to grow building on the foundation of research carried out most notably by Loveland and Duñach.

References

[1]   Styring P, Jansen D, de Coninck H, Armstrong K, Carbon Capture and Utilisation in the green economy, Report No. 501, The Centre For Low Carbon Futures, ISBN 978-0-9572588-0-8. July 2011.
[2]   Martín C, Fiorani G, Kleij AW. Recent Advances in the Catalytic Preparation of Cyclic Organic Carbonates. ACS Catal 2015;5(2):1353–70.

[3] Ion A, Parvulescu V, Jacobs P, De Vos D. Synthesis of symmetrical or asymmetrical urea compounds from CO$_2$ via base catalysis. Green Chem 2007;9(2):158–61.

[4] Ritter SK. Electrosynthesis gives organic chemists more power. Chemical and Engineering News Mar 13, 2017, pg 23.

[5] Martin R, Kleij AW. Myth or Reality? Fixation of Carbon Dioxide into Complex Organic Matter under Mild Conditions. ChemSusChem 2011;4:1259-63.

[6] Gomes-Jelonek J, Metathesis: the green method of chemical synthesis. Chemistry World, Royal Society of Chemistry June 2018.

[7] For a detailed review on cell set-ups for carboxylation reactions see: Matthessen R, Fransaer J, Binnemans K, De Vos DE. Electrocarboxylation: towards sustainable and efficient synthesis of valuable carboxylic acids. Beilstein Journal of Organic Chemistry 2014;10(1):2484–500.

[8] Gennaro A, Isse AA, Severin M-G, Vianello E, Bhugun I, Savéant J-M. Mechanism of the electrochemical reduction of carbon dioxide at inert electrodes in media of low proton availability. J Chem Soc, Faraday Trans 1996, 92, 3963–8.

[9] Gennaro A, Isse AA, Savéant J-M, Severin M-G, Vianello E. Homogeneous Electron Transfer Catalysis of the Electrochemical Reduction of Carbon Dioxide. Do Aromatic Anion Radicals React in an Outer-Sphere Manner? J Am Chem Soc 1996, 118, 7190–6.

[10] Goodridge F, Presland G. The electrolytic red$_u$ction of carbon dioxide and monoxide f$_o$r the production of carboxylic acids. Journal of Applied Electrochemist6, 2,;14 (6):791–

[11] Rudolph M, Dautz S, Jäger E-G. Macrocyclic [N$_4$$^{2-}$] Coordinated Nickel Complexes as Catalysts for the Formation of Oxalate by Electrochemical Reduction of Carbon Dioxide. J Am Chem Soc 2000, 122, 10821–30.

[12] Sun L, Ramesha GK, Kamat PV, Brennecke JF. Switching the Reaction Course of Electrochemical CO$_2$ Reduction with Ionic Liquids. Langmuir 2014, 30, 6302–8.

[13] Goodridge F, Presland G. The electrolytic reduction of carbon dioxide and monoxide for the production of carboxylic acids. Journal of Applied Electrochemistry 1984, 14, 791–6.

[14] Rumbach P, Xu R, Go DB. Electrochemical Production of Oxalate and Formate from CO$_2$ by Solvated Electrons Produced Using an Atmospheric-Pressure Plasma. J Electrochem Soc 2016, 163, F1157–61.

[15] Shul'zhenko GI, Vasil'ev YB. Electrochemical synthesis of dicarboxylic acids by means of the electroreduction of carbon dioxide in the presence of ethylene and its derivatives. Russ Chem Bull 1991;40(6):1217–20.

[16] Dérien S, Clinet J-C, Duñach E, Périchon J. Electrochemical incorporation of carbon dioxide into alkenes by nickel complexes. Tetrahedron 1992;48(25):5235–48.

[17] Senboku H, Komatsu H, Fujimura Y, Tokuda M. Efficient electrochemical dicarboxylation of phenyl-substituted alkenes: synthesis of 1-phenylalkane-1, 2-dicarboxylic acids. Synlett 2001;2001(03):0418–20.

[18] Ohkoshi M, Michinishi J-Y, Hara S, Senboku H. Tetrahedron 2010;66(39):7732–7.

[19] Hoberg H, Peres Y, Krüger C, Tsay YH. A 1-Oxa-2-nickela-5-cyclopentanone from Ethene and Carbon Dioxide: Preparation, Structure, and Reactivity. Angew Chem Int Ed Engl 1987;26(8): 771–3.

[20] Williams CM, Johnson JB, Rovis T. Nickel-Catalyzed Reductive Carboxylation of Styrenes Using CO$_2$. J Am Chem Soc 2008;130(45):14936–7.

[21] Greenhalgh MD, Thomas SP. Iron-Catalyzed, Highly Regioselective Synthesis of α-Aryl Carboxylic Acids from Styrene Derivatives and CO$_2$. J Am Chem Soc 2012;134(29):11900–3.

[22] Gaydou M, Moragas T, Juliá-Hernández F, Martin R. Site-Selective Catalytic Carboxylation of Unsaturated Hydrocarbons with CO$_2$ and Water. J Am Chem Soc 2017;139(35):12161–4.

[23] Toki S, Hida S, Chemischer ST, Photochemical reaction of styrenes with triethylamine – Photofixation of carbon dioxide by radical anion. Nippon Kagaku Kaishi 1984;(1):152–7.

[24] Seo H, Liu A, Jamison TF. Direct β-Selective Hydrocarboxylation of Styrenes with CO 2Enabled by Continuous Flow Photoredox Catalysis. J Am Chem Soc 2017;139(40):13969–72.

[25] Murata K, Numasawa N, Shimomaki K, Takaya J, Iwasawa N. Construction of a visible light-driven hydrocarboxylation cycle of alkenes by the combined use of Rh(i) and photoredox catalysts. Chem Commun 2017;53(21):3098–101.

26 Meng Q-Y, Wang S, Huff GS, König B. Ligand-Controlled Regioselective Hydrocarboxylation of Styrenes with CO_2 by Combining Visible Light and Nickel Catalysis. J Am Chem Soc 2018;140 (9):3198–201.

[27] Chowdhury MA, Senboku H, Tokuda M. Electrochemical carboxylation of bicyclo[n.1.0] alkylidene derivatives. Tetrahedron 2004;60(2):475–81.

[28] Li J, Inagi S, Fuchigami T, Hosono H, Ito S. Selective monocarboxylation of olefins at 12CaO·7Al2O3 electride cathode. Electrochemistry Communications 2014;44(C):45–8.

[29] Duñach E, Périchon J. Electrochemical carboxylation of terminal alkynes catalyzed by nickel complexes: unusual regioselectivity. Journal of Organometallic Chemistry 1988; 352 (1–2): 239–46.

[30] Duñach E, Dérien S, Périchon J. Nickel-catalyzed reductive electrocarboxylation of disubstituted alkynes. Journal of Organometallic Chemistry 1989;364(3):C33–6.

[31] Köster, F.; Dinjus, E.; Duñach, E. Electrochemical Selective Incorporation of CO_2 Into Terminal Alkynes and Diynes. Eur. J. Org. Chem. 2001, 2001, 2507–2511

[32] Yuan G-Q, Jiang H-F, Lin C. Efficient electrochemical dicarboxylations of arylacetylenes with carbon dioxide using nickel as the cathode. Tetrahedron 2008;64(25):5866–72.

[33] Li C-H, Yuan G-Q, Qi C-R, Jiang H-F. Copper-catalyzed electrochemical synthesis of alkylidene lactones from carbon dioxide and 1,4-diarylbuta-1,3-diynes. Tetrahedron 2013;69(15): 3135–40.

[34] Katayama A, Senboku H, Hara S. Aryl radical cyclization with alkyne followed by tandem carboxylation in methyl 4-tert-butylbenzoate-mediated electrochemical reduction of 2-(2-propynyloxy)bromobenzenes in the presence of carbon dioxide. Tetrahedron 2016;72 (31):4626–36.

[35] Loveland JW, Electrolytic production of acyclic carboxylic acids from hydrocarbons. U.S. Patent 3,032,489, May 1, 1962.

[36] Dérien S, Clinet J-C, Duñach E, Périchon J. Electrochemical incorporation of carbon dioxide into alkenes by nickel complexes. Tetrahedron 1992;48(25):5235–48.

[37] Jiang H, Li C, Yuan G, Qi C, Ji X. Method for synthesizing 3-alkene(cycloalkene)-1, 6-dicarboxylic acid by using electrochemical reaction. Faming Zhuanli Shenqing CN 2009-10192927.

[38] Dérien S, Clinet J-C, Duñach E, Périchon J. Activation of carbon dioxide: nickel-catalyzed electrochemical carboxylation of diynes. J Org Chem 2002;58(9):2578–88.

[39] Li C-H, Yuan G-Q, Qi C-R, Jiang H-F. Copper-catalyzed electrochemical synthesis of alkylidene lactones from carbon dioxide and 1,4-diarylbuta-1,3-diynes. Tetrahedron 2013;69(15): 3135–40.

[40] Wawzonek S, Runner ME. Polarographic Studies in Acetonitrile. J Electrochem Soc 1952;99 (11):457.

[41] Chan ASC, Huang TT, Wagenknecht JH, Miller RE. A Novel Synthesis of 2-Aryllactic Acids via Electrocarboxylation of Methyl Aryl Ketones. J Org Chem 1995;60(3):742–4.

[42] Datta AK, Marron PA, King CJH, Wagenknecht JH. Process development for electrocarboxylationof 2-acetyl-6-methoxynaphthalene. Journal of Applied Electrochemistry 1998;28(6):569–77.

[43] Weckbecker C, Duñach E, Olivero S. Process for the production of 2-hydroxy-4-methylmercaptobutyric acid. EP1309739, 2014.

[44] Silvestri G, Gambino S, Filardo G, Tedeschi F. A filter-press electrolytic cell with semi-continuous renewal of sacrifical electrodes. Journal of Applied Electrochemistry 1989;19(6): 946–8.

[45] Isse AA, Gennaro A. Electrocatalytic carboxylation of benzyl chlorides at silver cathodes in acetonitrile. Chem Commun (Camb) 2002;0(23):2798–9.

[46] Scialdone O, Galia A, Errante G, Isse AA, Gennaro A, Filardo G. Electrocarboxylation of Benzyl Chlorides at Silver Cathode at the Preparative Scale Level. Electrochim. Acta 2008;53:2514–28.

[47] Niu D-F, Xiao L-P, Zhang A-J, Zhang G-R, Tan Q-Y, Lu J-X. Electrocatalytic Carboxylation of Aliphatic Halides at Silver Cathode in Acetonitrile. Tetrahedron 2008;64:10517–20.

[48] Niu D, Zhang J, Zhang K, Xue T. Electrocatalytic Carboxylation of Benzyl Chloride at Silver Cathode in Ionic Liquid BMImBF$_4$. Chin. J. Chem. 2009;27;1041–44.

[49] Feroci M, Orsini M, Palombi L, Sotgiu G, Colapietro M, Inesi, A. Diastereoselective Electrochemical Carboxylation of Chiral α-Bromocarboxylic Acid Derivatives: an Easy Access to Unsymmetrical Alkylmalonic Ester Derivatives. J. Org. Chem. 2004;69:487–94.

[50] Yamauchi Y, Fukuhara T, Hara S, Senboku, H. Electrochemical Carboxylation of α,α-Difluorotoluene Derivatives and Its Application to the Synthesis of α-Fluorinated Nonsteroidal Anti-Inflammatory Drugs. Synlett 2008;2008;438–42.

[51] Senboku H, Kanaya H, Fujimura Y, Tokuda M. Stereochemical Study on Electrochemical Carboxylation of Vinyl Triflates. J. Electroanal. Chem. 2001;507:82–88.

Jean-Marie Fontmorin, Paniz Izadi, Shahid Rasul and Eileen H. Yu

27 Carbon dioxide utilisation by bioelectrochemical systems through microbial electrochemical synthesis

27.1 Introduction

27.1.1 What are bioelectrochemical systems

The gap between energy demand and the availability of fossil fuels has been growing, particularly over the last decades due to the increase in economic growth and social development. Currently, fossil fuels are the main resources for industrial and municipal activities, and their intensive consumption has led to the dreadful consequences of global warming and climate change. Bioelectrochemical systems (BESs) are emerging technologies involving the interaction of microorganisms with solid-state electrodes. The versatility of this technology combining microbiology and electrochemistry has led to various applications such as wastewater treatment, environmental sensing, resource recovery and energy harvesting from waste, as well as converting CO_2 to various products. The primary interest of BES is the ability of microorganisms used as biocatalysts to convert chemical energy present in wastewater into electricity, but also to convert electricity into chemical energy [1]. In the case of electricity generation, oxidation and reduction reactions occur at the anode and cathode, respectively, causing the potential difference and subsequently the driving force between two electrodes for the electric current generation through the external circuit by flowing of the electrons from the anode to the cathode. Microbial fuel cell (MFC) is an important type of BES in which bacteria oxidise organic compounds in the anode compartment and use the anode as the electron acceptor. The common reaction at the cathode is the oxygen reduction reaction (ORR) due to its high redox potential [2], and reactions with other types of electron acceptors, such as metal ions [3], denitrification [4, 5] and disulphate [6] have also been used. Recently, the focus for microbial electrochemical systems has been diversified to various applications on cathodes using the electrons generated from bioanodes decomposing organic waste and generating renewable electric energies.

Jean-Marie Fontmorin, Paniz Izadi, Eileen H. Yu, School of Engineering, Newcastle University
Shahid Rasul, School of Engineering, Newcastle University; Faculty of Engineering and Environment, Northumbria University

https://doi.org/10.1515/9783110665147-027

27.1.2 What is microbial electrochemical synthesis

Besides electricity generation, BESs have the unique capability to produce valuable chemicals and fuels, by changing the electrical energy gained from the cathode by microorganisms to chemical energy. Microbial electrochemical synthesis (MES) is another type of BESs that is able to convert CO_2 to other valuable organic chemicals at the cathode compartment. In other words, bacteria in the cathode chamber are able to use CO_2 as carbon source and the electrons as the energy source to produce a desired organic substance [7]. These types of bacteria are known as autotrophic bacteria, which are normally strictly anaerobic. Therefore, anaerobic conditions are required at the cathodic compartment of the BES. Unlike the ORR in a conventional MFC, the reduction of CO_2 is not thermodynamically favourable. Therefore, external energy is required to drive the reaction that can be supplied by other types of renewable energy such as solar or wind power. Figure 27.1 demonstrates the schematic of basic MES, in which bacteria consume the electrons transferred from the anode and external renewable energy source to convert CO_2 to valuable chemicals.

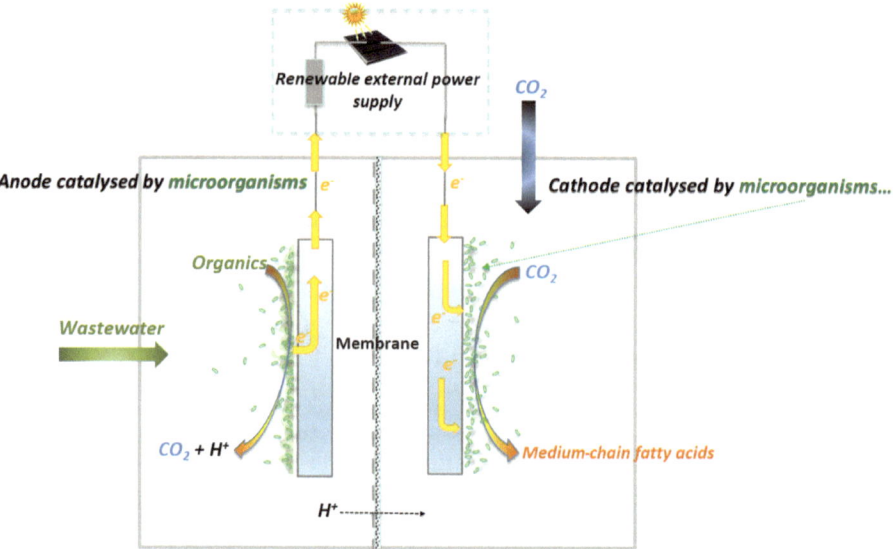

Figure 27.1: Schematic of a typical MES system for conversion of CO_2 to valuable chemicals.

Depending on the operational conditions, types of bacterial catalysts, electron acceptors and possible mediators in the catholyte, products of MES vary. The nature of the biocatalysts involved in the reduction of CO_2 in MES plays a key role in the selectivity of the products due to the different pathways of different bacterial communities. Several studies reported different types of products such as acetate, butyrate, ethanol, butanol, methane and longer chain products such as caproate and caprylate [8] and biopolymer [9, 10].

27.1.3 Advantages and challenges of bioconversion of CO_2

MES, also known as bioelectrosynthesis, is the new perspective of BES and has attracted researchers' attention utilising CO_2 and the electricity driven from the anodic reactions to produce valuable chemicals such as fuels, proteins and bioplastics. In addition, microorganisms as biocatalysts are the main advantage of MES, as compared to chemical catalysts: they are robust, sustainable, compatible with the environment and can reduce CO_2 at lower overpotentials. Moreover, bacteria are able to produce various types of chemicals such as long-chain fatty acids and alcohols, whereas this cannot be achieved by chemical reactions. In addition, the high energy efficiency of MES in the production of chemicals is one of the strengths of this technique [11]. Despite the advantages of the MES, the start-up of bioproduction is slow. It was shown that the development of biocathodes requires longer time than bioanodes, thus also delaying the bioproduction at the cathode. Although CO_2 reduction by biocathodes in BESs has gained much attention recently, little is still known about the biocathode development and molecular mechanisms of electron transfer between the electrode and microorganisms [12, 13]. However, the strong interest in producing valuable chemicals from wastes and particularly CO_2 and the high demand for new sustainable energy sources to reduce the dependence on fossil fuels are the main reasons for the expansion of MES technologies. In this regard, this chapter provides an overview of the principles of MES, while the bacterial pathway for reduction of CO_2 and production of higher value products as well as parameters affecting the production and efficiency of the MES will be discussed. Different reactor designs and the available renewable energy sources for MES will also be presented.

27.2 Principles and reaction pathways for CO_2 utilisation in BES

27.2.1 Electron transfer mechanisms in CO_2-reducing biocathodes

During microbial electrosynthesis (MES), microbial catalysts convert CO_2 into chemical products using electrons derived from the cathode. These electrons can be used by microorganisms to reduce carbon dioxide but also for their own growth and maintenance [14]. Electrons are transferred from the cathode to the microorganisms either through direct electron transfer (DET) or through mediated electron transfer (MET), as illustrated in Figure 27.2. In the case of MET (Figure 27.2, (i) and (ii)), electrons are carried via exogenous shuttles (mediators) or via shuttles secreted/released by microorganisms as metabolic products [15, 16]. Exogenous shuttles can be electrochemically generated, such as H_2, formate, ammonia, sulphide or Fe(II). H_2 and formate are particularly interesting as their redox potentials are negative enough (E^0_{H2} = 0 V (RHE)), and formate E^0 = −0.03 V (RHE), for microorganisms to gain energy to support growth and CO_2 reduction. Growing evidence has been suggesting that the CO_2 reduction by microbes mediated by H_2/H^+ is a major driving force in most biocathodes [17, 18]. Ammonia, sulphide or Fe(II), however, require an electron acceptor with a potential higher than that of CO_2, typically oxygen, to support cell growth. The presence of oxygen would drastically decrease the efficiency of CO_2 conversion, thus requiring physical separation between the sites of electron donor generation and consumption. These shuttles are thus less suitable for CO_2 conversion applications [14]. MET can also be carried out via redox mediators excreted or released by the bacteria (Figure 27.2, (ii)). These mediators can be phenazine, quinone and riboflavins or even other substances released after the cell death, such as vitamin B_{12} or DNA [19].

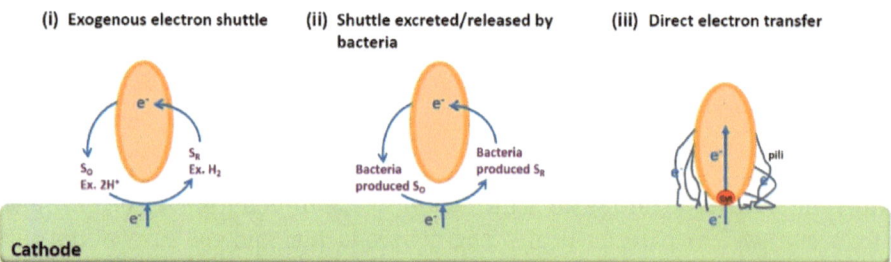

Figure 27 2: Possible extracellular electron transfer mechanisms from the cathode to the microbial catalyst. Indirect electron transfer via (i) an exogenous shuttle or (ii) a shuttle released/excreted by the microbial catalyst. (iii) Direct electron transfer via bacterial outer surface components such as c-type cytochromes or pili. SO is oxidised electron shuttle and SR is reduced electron shuttle [16].

In the case of DET (Figure 27.2, (iii)), no mediator is needed as electrons move from the cathode to the cell via direct contact. Therefore, this mechanism relies on the existence of a biofilm or at least of a single cell layer, that is, a physical contact, at the surface of the electrode [20]. It is suspected that the mechanisms could be similar to those described for bioanodes, such as *Geobacter, Rhodoferax* or *Shewanella* which can evolve electronically conducting molecular pili (nanowires) to reach and utilise more distant solid electrons acceptors. The electron transfer is accomplished, thanks via the connection of these pili to membrane-bound cytochromes [21]. Microorganisms able to take electrons directly from the electrode are known as electrotrophs. It has been suggested that in some MES systems driven by acetogens or methanogens, DET could occur with a cathode poised at less negative potentials than the hydrogen evolution reaction (e.g. –400 mV to –500 mV vs SHE (Standard hydrogen electrode)) [18]. This assumption, however, has not been proven. The standard potential for proton reduction to H_2 ($E_{(2H^+/H_2)}$) is –420 mV versus SHE at 1 bar, 25 °C and pH 7. Thermodynamically, it is preferable for electrons to reduce protons into H_2 than react with other reactants with more negative reduction potentials, which thus require more energy. This can be explained by the equation of change of Gibbs free energy ΔG (kJ mol^{-1}):

$$\Delta G = - zFE \tag{27.1}$$

where z is the number of electrons transferred for 1 mol of reactant; F is the Faraday constant, 96,485 C mol^{-1}; E the redox potential of the reaction, V.

This also suggests the production of H_2 as a mediator for CO_2 reduction with microorganisms [22]. In addition, depending on the pH and hydrogen partial pressure, the reduction potential could occur at much less negative potentials based on the Nernst equation:

$$E = E^0 - (RT/zF) \ln\left(P_{H_2}/[H^+]^2\right) \tag{27.2}$$

where E^0 is the standard reduction potential, V; R is the gas constant, 8.314 J K^{-1} mol^{-1}; T the temperature, K; P_{H_2} the partial pressure of H_2, Pa, which makes DET difficult to distinguish from the electron transfer mediated by hydrogen.

Various types of bacteria can interact with electrodes via DET. For example, in the absence of Fe^{2+} in the medium, iron-oxidising bacteria (IOB) are able to acquire electrons directly from the cathode to generate ATP and fix CO_2. Interestingly, IOB are also known for being able to use anodes as electron sinks and transfer electrons to anodes, which suggests the possible reversibility of the electron transport chain. *Geobacter* and *Shewanella* are also two metal-reducing bacteria able to use anodes as electron sinks and cathodes as electron sources. In both cases, multi-heme c-type cytochrome and filamentous conductive pili are key components of the electron transfer channels. Considering the broad range of redox potential in c-type cytochrome (from –400 to +200 mV vs SHE), the direction of electron flow (microbe → anode or cathode → microbe) would be flexible in an electron transfer channel consisting of cytochrome C [23].

When it comes to conversion of CO_2 into chemical products, both DET and MET present advantages and disadvantages. Indeed, DET with the presence of a biofilm can be limited by the internal (within the biofilm) and the external (between the bulk and the biofilm) diffusion of substrates and products. On the other hand, with the indirect electron transfer mechanism the biocatalyst may be mostly in solution ("planktonic") rather than present as a biofilm. In this case, one limitation could be the lower biomass retention in a continuous system, washing out the biocatalyst in solution.

27.2.2 Microbial communities and reaction pathways in MES

The conversion of CO_2 in MES systems can be driven by two major types of microbial catalysts: mixed cultures and pure cultures. Mixed cultures with various microorganism communities are generally enriched from environmental samples such as wastewater, sludge or sediments. The main advantage is that they eliminate the need to work under sterile conditions as is required with pure cultures. In addition, mixed cultures might benefit from syntrophy, which occurs when different cultures cooperate and depend on each other to perform the metabolic activity observed. An example of syntrophy is the interspecies hydrogen transfer, during which H_2 is transferred from one organism to another. In the context of MES, H_2 could be produced by a microorganism and used as an energy source by another for the reduction of CO_2. Another advantage of systems driven by mixed cultures is that the communities developed are generally more robust [22]. However, mixed cultures are also more difficult to control as different types of bacteria compete in the same environment, which makes the production of desired chemicals more difficult to control. Pure culture is easier to control for chemical production. It is also possible to conduct genetic engineering to produce strains for desired products.

27.2.2.1 Pure cultures and reaction pathways for CO_2 conversion

Methanogens and acetogens are the most studied pure cultures in MES systems for CO_2 conversion. A very direct strategy is to convert carbon dioxide into methane, which is an excellent fuel and easy to separate as it is poorly soluble [14]. However, methane is also a greenhouse gas, which makes the relevance of converting CO_2 into CH_4 questionable. It is assumed that in the presence of electrical current and CO_2, methanogenic bacteria can produce methane based on the following eq. [24]:

$$CO_2 + 8H^+ + 8e^- \rightarrow CH_4 + 2H_2O \tag{27.3}$$

It is however still not clear whether methanogens can accept electrons directly from the electrode or if the mechanism involves molecular hydrogen (eq. (27.4)), in which case methane is formed as follows:

$$2H^+ + 2e^- \rightarrow H_2 \tag{27.4}$$

$$CO_2 + 4H_2 \rightarrow CH_4 + 2H_2O \tag{27.5}$$

Under anaerobic conditions, acetogens can also reduce CO_2 through the Wood–Ljungdahl (W–L) pathway into mainly acetate (Figure 27.3), and more specifically to (i) synthesise acetyl-CoA by the reduction of CO_2, (ii) conserve energy and (iii) fix (assimilate) CO_2 for the synthesis of cell carbon [25]. The W–L pathway is the most energy-efficient non-photosynthetic fixation pathway for the production of acetate

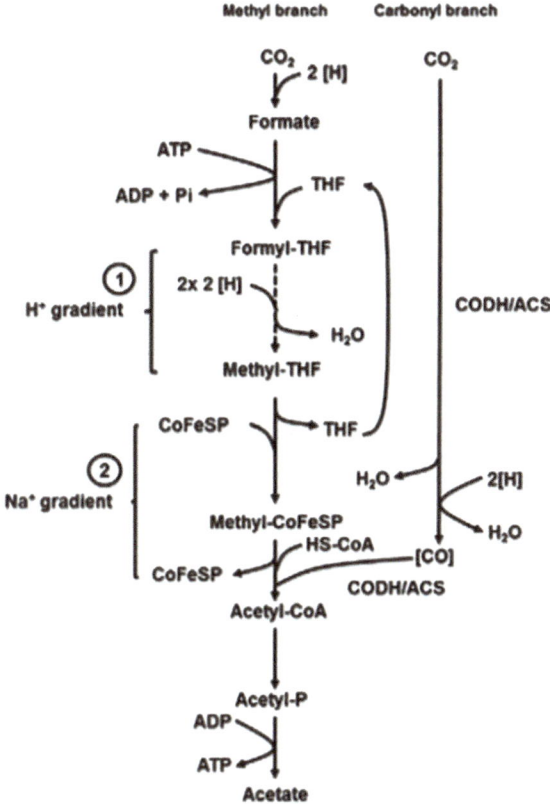

Figure 27.3: Autotrophic (CO_2+H_2) fermentation by acetogens (Wood–Ljungdahl pathway) [25].

and ethanol. In other words, most of the electrons derived from the cathode should end up in simple organic chemical products instead of being used for biomass production [14].

Although acetate as an end-product does not have a high market value, more useful intermediates can be obtained through the W–L pathway for the production of compounds with higher value such as alcohols and longer chain fatty acids [26]. More specifically, acetyl-CoA is an excellent building block for the production of diverse chemicals, and under specific conditions of pH and hydraulic retention time (HRT) and following mechanisms derived from the W–L pathway (see below), butyrate, ethanol, butanol or even bioplastic polyhydroxyalkanoates (PHA) can be produced from acetyl-CoA as a precursor (Figure 27.4). Indeed, in addition to acetate, some bacteria of the genus of *Clostridium* and *Acetobacterium* can produce butyrate, 2,3-butanediol, ethanol and butanol, which are also natural products from acetogens [25]. Ethanol is an important biofuel, used mostly as an additive to gasoline. 2,3-Butanediol is a precursor for the production of various chemical products, and is currently mainly produced from petrochemicals. Developing a sustainable process for the production of

2,3-butanediol is therefore of high interest. Butanol is an important industrial bulk as well as a promising biofuel as it can be used without modifications of car engines [25]. The production of butyrate can occur either via linear extension of the acetyl-CoA to butyral-CoA, or via reverse β-oxidation which is also called microbial chain elongation [27]. As acetic and butyric acids' concentrations increase, pH naturally decreases and a defensive mechanism resulting in a switch in metabolism can occur [28]. In this case, the metabolism switches from acidogenesis [production of volatile fatty acids (VFAs)] to solventogenesis (production of solvents, typically alcohols). As solvento-genic genes are induced as a defensive mechanism, acetic and butyric acids are reassi-milated (Figure 27.3). Acetyl-CoA and butyryl-CoA are regenerated and subsequently converted into butanol and ethanol, respectively [28]. The production of acetone can then occur during an acetone–butanol–ethanol fermentation, and acetone can be fur-ther reduced to isopropanol during an isopropanol–butanol–ethanol fermentation [27]. These reactions can occur in the same reactor; hence, it is important to control operating conditions.

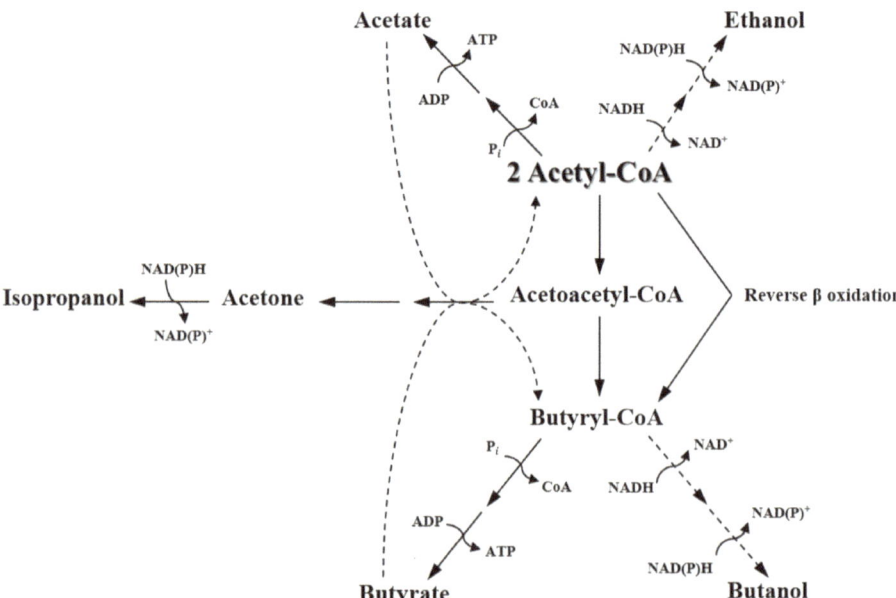

Figure 27.4: Schematic diagram of the suggested metabolic pathways for the production of acetate, ethanol, acetone, isopropanol, butyrate and butanol from acetyl-CoA. In dashed lines are the pathways for solventogenesis after production of acetate and butyrate. Adapted from [28].

2,3-Butanediol is based on pyruvate as a precursor but is also a natural product of bacteria of the genus *Clostridium*. Genetic engineering of acetogens is also a way to specifically target more valuable products such as isopropanol, butanol or ace-tone [14]. This can be achieved by eliminating genes responsible for the production of acetate and ethanol and by introducing or overexpressing genes essential for the production of the targeted commodity in *Clostridium* sp. in order to favour a specific pathway. A wide range of acetogens are able to accept electrons from a cathode and use CO_2 as a terminal electron acceptor, such as *Sporomusa ovata*, *Sporomusa silvacetica*, *Sporomusa sphaeroides*, *Clostridium ljungdahlii*, *Clostridium aceticum* and *Moorella thermoacetica* [29]. The exact mechanisms for the electron transfer (i.e. direct or indirect) are however still hypothetical and highly depend on operating conditions.

Autotrophic Fe(II) oxidising bacteria such as *Acidithiobacillus ferrooxidans* or *Leptospirillum* are also able to reduce CO_2 in BES using Fe(II) as a mediator or by directly accepting electrons from cathode with applied potentials [16]. In these systems, although CO_2 can be used as a carbon source, O_2 is the final electron acceptor which means that more than 90% of the electrons will be used to generate biomass rather than chemical products. Ammonia-oxidising bacteria also fix CO_2 to produce biomass in BES mediated by ammonia [16]. Similarly to IOB, oxygen is also the terminal electron acceptor, which considerably limits the efficiency for the production of multi-carbon compounds.

27.2.2.2 Mixed microbial communities

In MES reactors inoculated with environmental samples, the microbial community quickly becomes dominated by acetogenic bacteria and methanogens [16]. Therefore, in these systems, acetate and methane are the main chemical commodities produced from CO_2 reduction. As acetogenic bacteria and methanogens are competing, methanogens inhibitors are often used to avoid the production of methane (when it is not the desired product) and favour acetate. Autotrophic methanogens use H_2 as electron donor, and it was suggested that they could also directly use the electrode [30, 31]. In addition, acetoclastic methanogens use acetate as electron donor, showing the difficulty to selectively produce a specific compound when mixed communities are present. To date, however, the MES systems with the highest acetate production rate are driven by mixed communities. Acetate production rates of 685 and 21 g m^{-2} day^{-1} were reported in fed-batch and continuous systems, respectively [27, 32].

Lately, the production of higher value, multi-carbon compounds using mixed microbial communities and CO_2 has been targeted. The optimisation of reactor design and operating conditions, for example, continuous versus fed-batch, pH, HRT and applied potentials can lead to the production of C4–C6 products. Indeed, recent studies reported the formation of *n*-butyrate and *n*-caproate at production rates of 3.2 and 0.95 g L^{-1} day^{-1}, respectively [33, 34]. The production of isopropanol was also reported at a rate of 3.3 g m^{-2} day^{-1} [27], showing the potential of mixed community-driven MES systems for the production of valuable chemical commodities from CO_2.

MES also offers the possibility to target more complex compounds than those previously mentioned. Lately, the production of bioplastics such as PHAs has attracted increasing interest because they are biodegradable and mainly produced from renewable sources such as organic acids or wastes [35]. In addition, PHAs can be produced by mixed cultures and with VFAs such as acetate and butyrate as direct metabolic precursors [36]. A recent study has shown the possibility to produce polyhydroxybutyrate (PHB) from acetate and butyrate generated by MES from CO_2. A maximum of 74.4 g PHB per 100 g volatile suspended solids were produced for an equivalent carbon conversion of 0.41 kg C of PHB obtained for 1 kg of carbon as CO_2 inlet to the entire system [35]. However, it should be noted that the production of VFAs and bioplastics was not carried out in the same BES reactor. Indeed, the generation of VFAs is strictly anaerobic, whereas the production of PHAs is aerobic, thus requiring additional steps for VFAs extraction and separated reactors.

27.2.3 Parameters influencing products and efficiency of CO_2 conversion in BES

As described in the previous section, microbial communities and pathways dictate the chemical commodities produced from CO_2 conversion. Nevertheless, operating conditions such as current density or applied potential, mass transfer and pH will impact on both the communities and pathways.

27.2.3.1 Impact of applied potential

Applied potential plays a crucial role in the conversion of CO_2 in BES. It most likely impacts on the nature of the electron transfer with the possible generation of electron carriers such as H_2 at more negative potentials, and most importantly it should be low enough to allow the reduction of CO_2. The role of the cathode is to provide the energy required for a reaction to occur. The energy required for the generation of a product (from reduction of CO_2, for example) is linked to the redox potential of this reaction by the equation for change of Gibbs free energy ΔG (kJ mol^{-1}):

$$\Delta G = -zFE \tag{27.1}$$

where z is the number of electrons transferred for 1 mol of reactant; F is the Faraday constant, 96,485 C mol^{-1}; E the redox potential of the reaction, V, which directly links to the Nernst reaction:

$$E = E^0 - (RT/zF) \ln (a_{Red}/a_{Ox}) \tag{27.6}$$

where E^0 is the standard reduction potential, V; R is the gas constant, 8.314 J K^{-1} mol^{-1}; T the temperature, K; a_{Ox} the activity of the oxidised form (e.g. CO_2); a_{Red} the activity of the reduced form (product of CO_2 reduction, e.g. acetate).

Under standard conditions, CO_2 may be reduced bioelectrochemically to acetate at a potential (E^0) of −280 mV versus SHE (eq. (27.7)), but no electroacetogenesis was reported with a biocathode poised at such potential:

$$CO_2 + 7H^+ + 8e^- \rightarrow CH_3COO^- + 2H_2O; \; E^0 = -280 \, mV \, vs \, SHE \qquad (27.7)$$

It is well understood that potential losses occur in BES. These losses are related to mass transport and kinetics limitations between the medium, the electrode and within the biofilm (when applicable), but also to Ohmic voltage losses associated with the electrolytes, the membrane, the electrodes and connections. Therefore, it has been postulated that in order to overcome potential losses due to the components of BES, microbial acetogenesis requires potentials of −400 mV or much more negative [22]. It was reported that both acetate production and coulombic efficiency were improved when potential was shifted from −600 to −800 mV [37]. At more negative potentials, both the biotic and abiotic production of H_2 increase, which might facilitate the kinetics of electron transfer from the electrode to the bacteria, and also favour the production of acetate through the hydrogen-mediated mechanism (eqs. (27.4) and (27.8)) [26]. More negative potential means more reduction energy:

$$2H^+ + 2e^- \rightarrow H_2 \qquad (27.4)$$

$$CO_2 + 4H_2 \rightarrow CH_3COO^- + H^+ + 2H_2O \qquad (27.8)$$

27.2.3.2 Impact of pH

Most studies focussing on CO_2 conversion in BES are carried out at near neutral pH. The production of acetic and butyric acids naturally decreases the pH. Therefore, replacing the catholyte regularly and constantly supplying CO_2 helps maintain the pH close to neutrality and therefore support acetogenesis [38]. Theoretically, the bioelectrochemical production of acetate is based on eq. (27.7). However, as pH decreases, the hydrogen-mediated mechanism might also be favoured (eqs. (27.4) and (27.8)) [26]. Studies have reported that a pH of 5.2 was optimum for the production of acetate, mainly because of increased substrate and protons availability [26, 39]. However, it was also reported that acetate production with a biocathode poised at −600 mV vs SHE ceased at pH ~5, showing the importance to control the pH [38]. A low pH can also be beneficial for the production of alcohols such as ethanol, butanol and isopropanol. Indeed, acidic pH can induce a switch in metabolism from acidogenesis to solventogenesis as described in the previous section [27].

27.2.3.3 Impact of reactor design: Batch versus continuous

The operation mode (e.g. batch/fed-batch vs continuous) can have a crucial influence on the compounds productivity as it impacts on parameters such as pH, end-product concentration/accumulation and biomass concentration. As mentioned previously, the accumulation of VFAs leads to a gradual decrease of the pH, which can limit their production or lead to product diversification. Operating the BES reactor in continuous mode can help controlling the pH and prevent products' accumulation which can in turn displace the equilibrium of the reaction towards chemical production. Therefore, in continuous mode, optimising the HRT according to the aim of the experiment is necessary, as a possible washout of the biomass from the cathode chamber should also be taken into consideration. A recent study reported that a lower HRT (3.3 days in this specific case) and higher catholyte pH (~7) led to higher acetate productions, whereas a higher HRT (>5 days) and lower pH (<5.5) favoured product diversification, including butyrate and isopropanol [27].

27.2.4 Reactor design and cell configurations

27.2.4.1 MES cell configuration

Depending on the operational conditions and applications, reactors for MES can be designed differently. Aerobic or anaerobic conditions, nature of the products (gas or liquid) and the way of sampling, batch or continuous flow are examples of parameters that can affect the design of the reactors [40]. The common and conventional reactor design is the dual-chamber cell used in many studies. This design, also known as "H-cell", consists of two compartments of anode and cathode usually separated by a membrane such as a cation exchange membrane (CEM) or anion exchange membrane (AEM), bipolar or charge mosaic membranes [41]. Membranes allow the ion movement but separate electrolytes in anode and cathode chamber. Selection of membrane mainly depends on the products targeted from CO_2 conversion and the electrolytes. For instance, as the products of MES are usually anions such as acetate or butyrate, CEM is the most suitable membrane as it limits the diffusion of the products to the anode chamber. In some systems, the anode is performing water oxidation with oxygen evolution reaction (OER). The oxygen produced at the anode would disturb the anaerobic condition required by cathode bacteria for MES. Thus, the utilisation of membranes with low oxygen permeability is preferred [42, 43].

Despite the advantage of the dual-chamber design, membranes are expensive particularly at larger scale. To simplify the BES design for MES, single-chamber reactors can still be considered for CO_2 reduction. In this design, both anode and cathode are placed in one chamber with the same electrolyte. In a study focussing on CO_2 reduction in a half-cell configuration using the pure culture of *S. ovata* [44], an undivided 1 L Pyrex glass bottle was used as the reactor, using two circular graphite disks as anode and cathode. A blunted cannula between two electrodes for the gassing was responsible to keep the oxygen produced at the anode from the diffusion to the cathode.

The solubility of CO_2 in aqueous medium is limited (1.5 g CO_2/per kg water at 25 °C); therefore, the mass transfer of CO_2 could be an issue when purging CO_2 into the electrolyte. To overcome this problem, a gas diffusion electrode (GDE) with an additional gas chamber for CO_2 can be used next to the cathode compartment [45]. Selection of appropriate cathode material for GDE is critical. Carbon paper or cloth with one side coated with polytetrafluoroethylene used for air cathode in MFCs [46] can be applied for this purpose. In a recent study, mass transfer of gaseous CO_2 in the catholyte of GDE reactor was compared to that in the conventional design of purging CO_2 directly in the solution. It was reported in this study that CO_2 mass transfer was enhanced by using the GDE reactor, thus increasing the rate of bioproduction [45]. The schematic of the reactor used in this study is shown in (Figure 27.5)

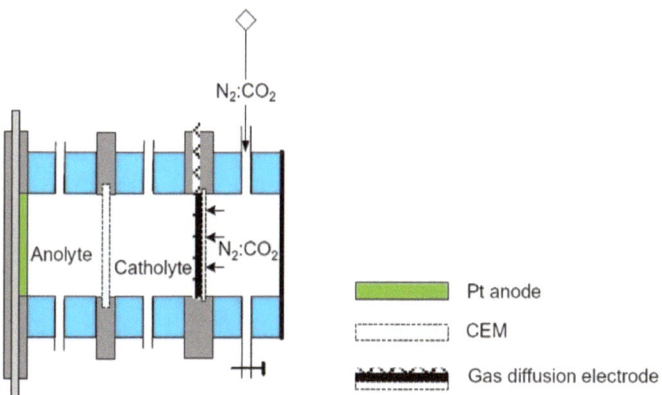

Figure 27.5: Schematic of the reactor used for MES with the gas diffusion biocathode. Reproduced with permission from [45]. Copyright © Springer.

27.2.4.2 Reactor design for product extraction and continuous production

Despite the simple design for two chamber cells, the limitations are the recovery of the products from the catholyte and decrease of pH due to the accumulation of fatty acids such as acetic or butyric acids, which is detrimental for the cathode bacterial community, as well as reaction equilibrium. One method to tackle this problem is to use continuous mode instead of batch mode. By this operational mode, fresh medium will be flushed into the cathode compartment continuously using a pump, providing fresh and essential components for the bacteria and continuously removing products from the chamber [47]. However, choosing a suitable HRT is essential as in the absence of a biofilm and in the presence of a planktonic biocatalyst, a fast flow rate can wash out the bacterial community responsible for the CO_2 reduction and bio-production [48].

Besides continuous mode, another reactor design to extract the products is the three-chamber cell consisting of the anode and cathode compartments and an additional extraction compartment. In this design, the extraction chamber is placed between the anode and cathode compartments, and extraction of products is achieved with the assistance of electro-osmosis force. The cathode chamber is separated from the extraction chamber using an AEM, allowing products of MES (usually negatively charged) to diffuse through the membrane towards the extraction chamber. On the other side, the extraction chamber is separated from the anode chamber using a CEM to let protons diffuse through the membrane from the anode to the extraction and also to prevent the bioproducts from reaching the anode. This design allowed the extraction of overall 17.5 g. L^{-1} of acetate produced through MES in the extraction chamber after 86 days of the experiment [49]. The schematic design of this reactor is shown in (Figure 27.6). This design has the advantage of decreasing the additional

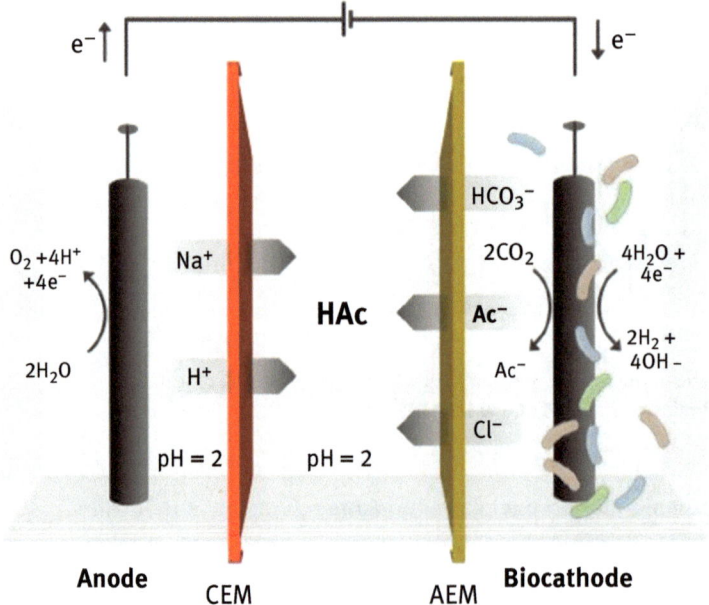

Figure 27.6: Schematic of the three compartments of BES for the simultaneous electrosynthesis and extraction of acetate from the cathode compartment. Reprinted with permission from [49]. Copyright © American Chemical Society.

cost of post-treatment to recover the products from the catholytes, while also stabilising pH in the cathode chamber. It also shifts reaction equilibrium with products removed from the cathode chamber to the extraction chamber.

There are many other reactor designs used in various studies depending on the requirements and applications. The designs described can be modified and fabricated differently or combined for CO_2 conversion in MES.

27.2.5 Integrated and hybrid MES for CO_2 utilisation

Electrical energy is one of the pre-requisites for the MES process. This energy can be provided by two means as demonstrated in Figure 27.7; (1) coupling MES with renewable energy sources, for example, wind and solar systems that allow for a fast switch on/off or adjustments in capacity [50] (Figure 27.7a) and (2) integrate MES with inorganic semiconductor photocatalysts (Figure 27.7b).

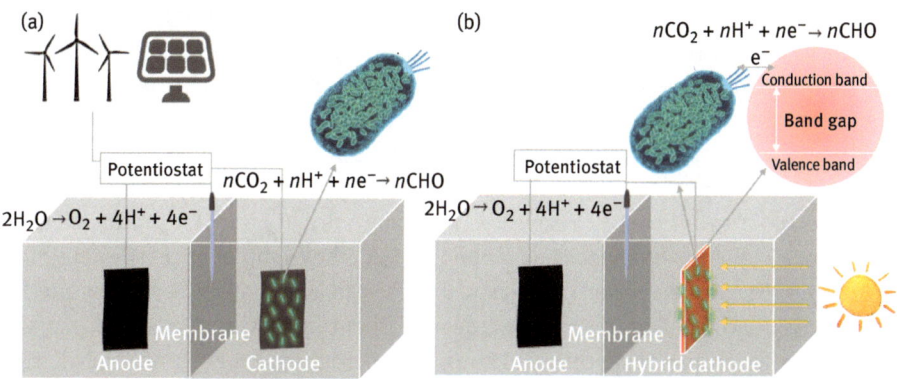

Figure 27.7: (a) Electrical energy supplied through renewable energy systems and (b) aficial photosynthesis utilising hybrid inorganic-MES.

27.2.5.1 MES integrated with renewable electricity

MES integrated with renewable electricity offers a new and efficient photosynthetic technology when coupled to a photovoltaic system, for the production of organic products from CO_2. In this case, MES relies on chemolithoautotrophic bacteria that have the ability to uptake electrons from the cathode of an electrochemical cell to catalyse the reduction of CO_2 into fuels or value-added chemicals at low overpotentials. Nevin et al. integrated MES with a solar-powered system, which comprised a solar panel and a voltage control to directly feed electrons to acetogens through electrodes [51]. They used acetogenic bacteria (*Sporomusa ovata*) to reduce CO_2 to acetate and other multicarbon extracellular products with hydrogen as the electron donor. MES integrated with renewable energy systems has the advantage since it allows better control of the bioprocess and a fast switch on/off or adjustments in capacity could be possible. However, the integrated MES process needs energy storage systems to tackle the intermittency of renewable energy.

27.2.5.2 Artificial photosynthesis utilising hybrid inorganic-MES

The development of photocathodes in integrated MES systems is a sustainable approach to produce solar biofuels from CO_2 where solar light is used to drive the MES process and could be regarded as "artificial photosynthesis". In the artificial photosynthesis process, the photocatalyst captures the photons to excite the electron–hole pairs. The generated electrons are utilised to reduce the CO_2 through the Calvin cycle [52] in two ways: (1) electrons are directly fed to microorganisms for fixing the CO_2 and (2) the generated electrons are utilised to produce H_2 photoelectrochemically through water splitting and generated H_2 and/or O_2 is fed to microorganisms along with CO_2. For the artificial photosynthesis process, several examples have been reported which couple sunlight harvesting semi-conductors with microorganisms to reduce CO_2 to renewable fuels in a hybrid bio-inorganic system. For example, Torella et al. reported that the bacterium *Ralstonia eutropha* could be used to efficiently convert CO_2, along with H_2 and O_2 produced from water splitting, into biomass and alcohols [53]. They employed earth-abundant metals (Co, Ni, Mn, Zn based) catalysts, which split water at low overpotential to generate H_2 and O_2. The authors engineered *R. eutropha* which enabled production of the isopropanol at up to 216 mg/L, the highest bioelectrochemical fuel yield yet reported by >300% [54]. This proposed biosynthetic system exceeded the efficiency of natural photosynthesis demonstrating that catalysts of biotic and abiotic origin can be interfaced to achieve challenging chemical energy-to-fuel transformations [55].

In another example of hybrid inorganic-MES system, an in situ synthesised cadmium sulphide–bacteria hybrid system was developed to produce acetic acid from CO_2 by harvesting light [56]. In this hybrid approach, Sakimoto and co-workers induced the self-photosensitisation of a non-photosynthetic bacterium, *Moorella thermoacetica*, with cadmium sulphide nanoparticles, enabling the photosynthesis of acetic acid from CO_2. This self-augmented biological system selectively produced acetic acid continuously over several days of light–dark cycles at relatively high quantum yields.

Taking together both the routes towards utilisation of renewable energy in MES, it is suggested that integration of photovoltaics, inorganic catalyst, and microbial catalyst for the production of useful chemicals from industrial/atmospheric sourced CO_2 waste has a great potential in the field of renewable energy storage and reduction of CO_2 emission [57].

27.3 Concluding remarks and perspectives

There is an urgent need to develop new sustainable technologies to tackle the accumulation of greenhouse gases, including CO_2 in the atmosphere. As mentioned in this chapter, BESs, and especially MES, are one of the promising technologies for CO_2 utilisation combining renewable electricity and microorganisms as biocatalysts. Although during the last few years, methane and acetate have been the main products from MES, it is now understood that acetyl-CoA can be used as a building block for the production of more valuable compounds such as butyrate, 2,3-butanediol, ethanol, isopropanol, butanol or caproic acid. In addition, the most recent studies have highlighted the importance of understanding the impact of operating parameters such as pH, applied potential or batch mode versus continuous mode to selectively target either acetate or longer chained VFAs, or alcohols. Focus has also been put on optimisation of reactor designs, for example with the utilisation of GDEs to overcome the limited solubility of CO_2 in electrolytes, or with the development of three-chamber cells to increase the production and facilitate the extraction of target compounds. Finally, since CO_2 conversion through MES is not a thermodynamically favourable process, research has also been carried out on integrated and hybrid MES systems using renewable energy sources. Indeed, since typical bioanodes cannot provide the required energy for CO_2 conversion, studies have shown the possibility to power MES with renewable sources such as solar panels, hence making the technology highly sustainable.

Over the last few years, MES has improved considerably to become a highly promising technology for CO_2 conversion into valuable bioproducts. However, a lot of progress is still needed to take the technology to the next level. For example, and similarly to other BES, the very high majority of studies dealing with CO_2 conversion by MES have been carried out at laboratory scale, showing that there is still a long way to go to reach commercialisation. In order to exploit the full potential of MES, prioritising the production of lower volumes of chemicals with higher value is probably the best strategy to adopt at the moment. The versatility of MES is also to be considered, as recent studies have shown the possibility to integrate with other technologies, for example for the production of bioplastics, or with the utilisation of photocathodes to mimic photosynthesis. We are not there yet, as to date MES does not offer a sufficient control of the end-products or requires additional steps for chain elongation, hence the need for optimisation. But considering the current environmental incentives, efforts are well worth being pursued.

References

[1]   K. Rabaey, R.A. Rozendal, Microbial electrosynthesis – revisiting the electrical route for microbial production, Nature Reviews Microbiology, 8 (2010) 706–716.

[2]   V.B. Oliveira, M. Simões, L.F. Melo, A. Pinto, Overview on the developments of microbial fuel cells, Biochemical Engineering Journal, 73 (2013) 53–64.

[3]   D. Ucar, Y. Zhang, I. Angelidaki, An overview of electron acceptors in microbial fuel cells, Frontiers in microbiology, 8 (2017) 643.

[4]   B. Virdis, K. Rabaey, R.A. Rozendal, Z. Yuan, J. Keller, Simultaneous nitrification, denitrification and carbon removal in microbial fuel cells, Water Research, 44 (2010) 2970–2980.

[5]   A. Getachew, F. Woldesenbet, Production of biodegradable plastic by polyhydroxybutyrate (PHB) accumulating bacteria using low cost agricultural waste material, BMC Research Notes, 9 (2016) 509.

[6]   D.-J. Lee, X. Liu, H.-L. Weng, Sulfate and organic carbon removal by microbial fuel cell with sulfate-reducing bacteria and sulfide-oxidising bacteria anodic biofilm, Bioresource technology, 156 (2014) 14–19.

[7]   K.P. Nevin, T.L. Woodard, A.E. Franks, Z.M. Summers, D.R. Lovley, Microbial electrosynthesis: feeding microbes electricity to convert carbon dioxide and water to multicarbon extracellular organic compounds, MBio, 1 (2010) e00103–00110.

[8]   M.C.A.A. Van Eerten-Jansen, A. Ter Heijne, T.I.M. Grootscholten, K.J.J. Steinbusch, T.H.J.A. Sleutels, H.V.M. Hamelers, C.J.N. Buisman, Bioelectrochemical Production of Caproate and Caprylate from Acetate by Mixed Cultures, ACS Sustainable Chemistry & Engineering, 1 (2013) 513–518.

[9]   X. Chen, Y. Cao, F. Li, Y. Tian, H. Song, Enzyme-Assisted Microbial Electrosynthesis of Poly(3-hydroxybutyrate) via CO2 Bioreduction by Engineered Ralstonia eutropha, ACS Catalysis, 8 (2018) 4429–4437.

[10]   S. Bajracharya, M. Sharma, G. Mohanakrishna, X.D. Benneton, D.P. Strik, P.M. Sarma, D. Pant, An overview on emerging bioelectrochemical systems (BESs): technology for sustainable electricity, waste remediation, resource recovery, chemical production and beyond, Renewable Energy, 98 (2016) 153–170.

[11]   P.-L. Tremblay, T. Zhang, Electrifying microbes for the production of chemicals, Frontiers in microbiology, 6 (2015) 201.

[12]   S.A. Patil, C. Hägerhäll, L. Gorton, Electron transfer mechanisms between microorganisms and electrodes in bioelectrochemical systems, Advances in Chemical Bioanalysis, Springer2012, pp. 71–129.

[13]   P.-L. Tremblay, L.T. Angenent, T. Zhang, Extracellular electron uptake: among autotrophs and mediated by surfaces, Trends in biotechnology, 35 (2017) 360–371.

[14]   D.R. Lovley, K.P. Nevin, Electrobiocommodities: powering microbial production of fuels and commodity chemicals from carbon dioxide with electricity, Current Opinion in Biotechnology, 24 (2013) 385–390.

[15]   S.A. Patil, C. Hägerhäll, L. Gorton, Electron transfer mechanisms between microorganisms and electrodes in bioelectrochemical systems, Bioanalytical Reviews, 4 (2012) 159–192.

[16]   P.L. Tremblay, T. Zhang, Electrifying microbes for the production of chemicals, Front Microbiol, 6 (2015).

[17]   F. Kracke, B. Lai, S.Q. Yu, J.O. Kromer, Balancing cellular redox metabolism in microbial electrosynthesis and electro fermentation – A chance for metabolic engineering, Metab Eng, 45 (2018) 109–120.

[18] P.L. Tremblay, L.T. Angenent, T. Zhang, Extracellular Electron Uptake: Among Autotrophs and Mediated by Surfaces, Trends Biotechnol, 35 (2017) 360–371.

[19] M. Rosenbaum, F. Aulenta, M. Villano, L.T. Angenent, Cathodes as electron donors for microbial metabolism: Which extracellular electron transfer mechanisms are involved?, Bioresource Technology, 102 (2011) 324–333.

[20] K. Rabaey, P. Girguis, L.K. Nielsen, Metabolic and practical considerations on microbial electrosynthesis, Current Opinion in Biotechnology, 22 (2011) 371–377.

[21] U. Schröder, Anodic electron transfer mechanisms in microbial fuel cells and their energy efficiency, Phys. Chem. Chem. Phys., 9 (2007) 2619–2629.

[22] H.D. May, P.J. Evans, E.V. LaBelle, The bioelectrosynthesis of acetate, Current Opinion in Biotechnology, 42 (2016) 225–233.

[23] O. Choi, B.I. Sang, Extracellular electron transfer from cathode to microbes: application for biofuel production, Biotechnology for Biofuels, 9 (2016).

[24] A. ElMekawy, H.M. Hegab, G. Mohanakrishna, A.F. Elbaz, M. Bulut, D. Pant, Technological advances in CO2 conversion electro-biorefinery: A step toward commercialization, Bioresource Technology, 215 (2016) 357–370.

[25] B. Schiel-Bengelsdorf, P. Durre, Pathway engineering and synthetic biology using acetogens, Febs Lett, 586 (2012) 2191–2198.

[26] P. Batlle-Vilanova, S. Puig, R. Gonzalez-Olmos, M.D. Balaguer, J. Colprim, Continuous acetate production through microbial electrosynthesis from CO2 with microbial mixed culture, J Chem Technol Biot, 91 (2016) 921–927.

[27] J.B.A. Arends, S.A. Patil, H. Roume, K. Rabaey, Continuous long-term electricity-driven bioproduction of carboxylates and isopropanol from CO2 with a mixed microbial community, J Co2 Util, 20 (2017) 141–149.

[28] S.Y. Lee, Y.S. Jang, J.Y. Lee, J. Lee, Metabolic engineering of Clostridium acetobutylicum ATCC 824 for increased butanol production, J Biotechnol, 150 (2010) S557–S557.

[29] K.P. Nevin, S.A. Hensley, A.E. Franks, Z.M. Summers, J.H. Ou, T.L. Woodard, O.L. Snoeyenbos-West, D.R. Lovley, Electrosynthesis of Organic Compounds from Carbon Dioxide Is Catalyzed by a Diversity of Acetogenic Microorganisms, Appl. Environ. Microbiol., 77 (2011) 2882–2886.

[30] M. Villano, G. Monaco, F. Aulenta, M. Majone, Electrochemically assisted methane production in a biofilm reactor, Journal of Power Sources, 196 (2011) 9467–9472.

[31] C.W. Marshall, D.E. Ross, E.B. Fichot, R.S. Norman, H.D. May, Electrosynthesis of Commodity Chemicals by an Autotrophic Microbial Community, Appl. Environ. Microbiol., 78 (2012) 8412–8420.

[32] L. Jourdin, T. Grieger, J. Monetti, V. Flexer, S. Freguia, Y. Lu, J. Chen, M. Romano, G.G. Wallace, J. Keller, High Acetic Acid Production Rate Obtained by Microbial Electrosynthesis from Carbon Dioxide, Environ. Sci. Technol., 49 (2015) 13566–13574.

[33] S.M.T. Raes, L. Jourdin, C.J.N. Buisman, D.P.B.T.B. Strik, Continuous Long-Term Bioelectrochemical Chain Elongation to Butyrate, Chemelectrochem, 4 (2017) 386–395.

[34] L. Jourdin, S.M.T. Raes, C.J.N. Buisman, D.P.B.T.B. Strik, Critical Biofilm Growth throughout Unmodified Carbon Felts Allows Continuous Bioelectrochemical Chain Elongation from CO2 up to Caproate at High Current Density, Frontiers in Energy Research, 6 (2018).

[35] T. Pepè Sciarria, P. Batlle-Vilanova, B. Colombo, B. Scaglia, M.D. Balaguer, J. Colprim, S. Puig, F. Adani, Bio-electrorecycling of carbon dioxide into bioplastics, Green Chemistry, (2018).

[36] B. Colombo, F. Favini, B. Scaglia, T.P. Sciarria, G. D'Imporzano, M. Pognani, A. Alekseeva, G. Eisele, C. Cosentino, F. Adani, Enhanced polyhydroxyalkanoate (PHA) production from the organic fraction of municipal solid waste by using mixed microbial culture, Biotechnology for Biofuels, 10 (2017).

[37] G. Mohanakrishna, K. Vanbroekhoven, D. Pant, Imperative role of applied potential and inorganic carbon source on acetate production through microbial electrosynthesis, J Co2 Util, 15 (2016) 57–64.

[38] E.V. LaBelle, C.W. Marshall, J.A. Gilbert, H.D. May, Influence of Acidic pH on Hydrogen and Acetate Production by an Electrosynthetic Microbiome, Plos One, 9 (2014).

[39] L. Jourdin, S. Freguia, V. Flexer, J. Keller, Bringing High-Rate, CO2-Based Microbial Electrosynthesis Closer to Practical Implementation through Improved Electrode Design and Operating Conditions, Environ. Sci. Technol., 50 (2016) 1982–1989.

[40] A. Kadier, Y. Simayi, P. Abdeshahian, N.F. Azman, K. Chandrasekhar, M.S. Kalil, A comprehensive review of microbial electrolysis cells (MEC) reactor designs and configurations for sustainable hydrogen gas production, Alexandria Engineering Journal, 55 (2016) 427–443.

[41] R.A. Rozendal, T. Sleutels, H.V.M. Hamelers, C.J.N. Buisman, Effect of the type of ion exchange membrane on performance, ion transport, and pH in biocatalyzed electrolysis of wastewater, Water Science and Technology, 57 (2008) 1757–1762.

[42] S. Bajracharya, R. Yuliasni, K. Vanbroekhoven, C.J.N. Buisman, D.P. Strik, D. Pant, Long-term operation of microbial electrosynthesis cell reducing CO2 to multi-carbon chemicals with a mixed culture avoiding methanogenesis, Bioelectrochemistry, 113 (2017) 26–34.

[43] S.A. Patil, J. Arends, I. Vanwonterghem, J. Van Meerbergen, K. Guo, G.W. Tyson, K. Rabaey, Selective enrichment establishes a stable performing community for microbial electrosynthesis of acetate from CO2, Environmental science & technology, 49 (2015) 8833–8843.

[44] C.G.S. Giddings, K.P. Nevin, T. Woodward, D.R. Lovley, C.S. Butler, Simplifying microbial electrosynthesis reactor design, Frontiers in microbiology, 6 (2015) 468.

[45] S. Bajracharya, K. Vanbroekhoven, C.J.N. Buisman, D. Pant, D.P.B.T.B. Strik, Application of gas diffusion biocathode in microbial electrosynthesis from carbon dioxide, Environmental Science and Pollution Research, 23 (2016) 22292–22308.

[46] E.H. Yu, S. Cheng, K. Scott, B. Logan, Microbial fuel cell performance with non-Pt cathode catalysts, Journal of Power Sources, 171 (2007) 275–281.

[47] L. Jourdin, S.M.T. Raes, C.J.N. Buisman, D.P. Strik, Critical biofilm growth throughout unmodified carbon felts allows continuous bioelectrochemical chain elongation from CO2 up to caproate at high current density, Frontiers in Energy Research, 6 (2018) 7.

[48] J.B.A. Arends, S.A. Patil, H. Roume, K. Rabaey, Continuous long-term electricity-driven bioproduction of carboxylates and isopropanol from CO2 with a mixed microbial community, Journal of CO2 Utilization, 20 (2017) 141–149.

[49] S. Gildemyn, K. Verbeeck, R. Slabbinck, S.J. Andersen, A. Prévoteau, K. Rabaey, Integrated Production, Extraction, and Concentration of Acetic Acid from CO2 through Microbial Electrosynthesis, Environmental Science & Technology Letters, 2 (2015) 325–328.

[50] K. Rabaey, R.A. Rozendal, Microbial electrosynthesis – revisiting the electrical route for microbial production, Nature Reviews Microbiology, 8 (2010) 706.

[51] K.P. Nevin, T.L. Woodard, A.E. Franks, Z.M. Summers, D.R. Lovley, Microbial Electrosynthesis: Feeding Microbes Electricity To Convert Carbon Dioxide and Water to Multicarbon Extracellular Organic Compounds, mBio, 1 (2010).

[52] J.H. Kim, D.H. Nam, C.B. Park, Nanobiocatalytic assemblies for artificial photosynthesis, Current Opinion in Biotechnology, 28 (2014) 1–9.

[53] J.P. Torella, C.J. Gagliardi, J.S. Chen, D.K. Bediako, B. Colón, J.C. Way, P.A. Silver, D.G. Nocera, Efficient solar-to-fuels production from a hybrid microbial–water-splitting catalyst system, Proceedings of the National Academy of Sciences, 112 (2015) 2337–2342.

[54] C. Liu, B.C. Colón, M. Ziesack, P.A. Silver, D.G. Nocera, Water splitting–biosynthetic system with CO_2 reduction efficiencies exceeding photosynthesis, Science, 352 (2016) 1210–1213.

[55] R.E. Blankenship, D.M. Tiede, J. Barber, G.W. Brudvig, G. Fleming, M. Ghirardi, M.R. Gunner, W. Junge, D.M. Kramer, A. Melis, T.A. Moore, C.C. Moser, D.G. Nocera, A.J. Nozik, D.R. Ort, W.W. Parson, R.C. Prince, R.T. Sayre, Comparing Photosynthetic and Photovoltaic Efficiencies and Recognizing the Potential for Improvement, Science, 332 (2011) 805–809.

[56] K.K. Sakimoto, A.B. Wong, P. Yang, Self-photosensitization of nonphotosynthetic bacteria for solar-to-chemical production, Science, 351 (2016) 74–77.

[57] D.F. Savage, J. Way, P.A. Silver, Defossiling fuel: How synthetic biology can transform biofuel production, ACS Publications, 2008.

Part VI: **Photo- and plasma induced reactions of CO_2**

Annemie Bogaerts, Xin Tu, Gerard van Rooij
and Richard van de Sanden

28 Plasma-based CO_2 conversion

28.1 Introduction

Plasma technology is gaining increasing interest for CO_2 conversion. Plasma is an ionised gas, consisting of a variety of different species, including electrons, various types of radicals, ions, excited species, photons, besides neutral gas molecules. This reactive cocktail makes it useful for a myriad of applications [1]. Furthermore, as plasma is generated by electrical power, and can easily be switched on/off, this combination makes it suitable for using intermittent renewable electricity. Hence, it may provide a solution for the current challenges on efficient storage and transport of renewable electricity, that is, peak shaving and grid stabilisation.

So let us consider in more detail what plasma is and which promises it carries for chemical transformations in general and of CO_2 in particular. Plasma as ionised gas is generally sustained by the application of electric fields, as depicted in the cartoon in Figure 28.1. Energy transfer from the external electric field starts with acceleration of the free electrons. Subsequent collisions with (blue) feedstock molecules pass their kinetic energy on. However, the large mass difference between electrons and molecules makes momentum transfer extremely inefficient. Instead, energy transfer occurs predominantly via excitation of internal degrees of freedom, such as molecular vibration. On the microscopic scale, it means that the free electron modifies the configuration of the bound electrons of the atom or molecule. Internal energy is subsequently transferred to translational and rotational degrees of freedom, of which the rates are highly dependent on molecular properties, cross sections, pressure and temperature. In effect, a hierarchy in excitation of different degrees of freedom of the system is typically found. The free plasma electrons are hottest, typically 1–3 eV. Rotational and translational degrees are coldest whilst molecular vibration temperatures are necessarily intermediate. It goes without saying that at all time energy might be consumed (or released) in chemical reactions, which is the overall purpose and hence to be optimised.

Annemie Bogaerts, Department of Chemistry, University of Antwerp, Antwerp, Belgium
Xin Tu, Department of Electrical Engineering and Electronics, University of Liverpool, Liverpool, UK
Gerard van Rooij, Richard van de Sanden, Dutch Institute for Fundamental Energy Research, Eindhoven, The Netherlands

https://doi.org/10.1515/9783110665147-028

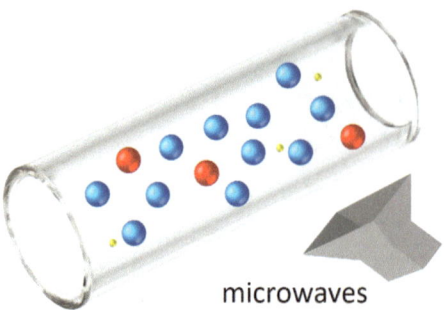

microwaves

Figure 28.1: Plasma is ionised gas and generally sustained by acceleration of the light electrons in an electric field (here indicated as microwaves). Ionisation is indicated by positively charged atoms/molecules (in red) and free electrons (in yellow) and is a small fraction compared to the neutral particle density.

In the present context of plasma-based CO_2 transformations, especially those cases in which the strongest non-equilibrium between the different modes is found are highly interesting. These are generally referred to as non-thermal plasma (NTP). It is under the far from thermodynamic equilibrium conditions that it is possible to intensify traditional chemical processes and to achieve the highest values of energy efficiency [2]. In the most ideal situation, one would have room temperature rotation and translation (or perhaps even lower), whilst high vibration temperature still drives strongly endothermic reactions. Simply speaking, this saves energy that is otherwise to be invested in these modes and likely to be lost as heat to the environment. It has the additional advantage of inherent quenching of the reaction products. Both aspects makes the approach particularly advantageous for thermodynamically unfavourable or energy-intensive chemical reactions, such as CO_2 splitting or dry reforming of methane (DRM), to proceed in an energy-efficient way. The strong non-equilibrium situation is opposed to thermal plasma (TP), in which all degrees of freedom are in thermal equilibrium.

The nature of the excitation process depends on the energy of the electrons. In the tail of the electron energy distribution function, the energy is high enough to excite the heavy gas particles into higher electronic states or even induce ionisation, as shown in Figure 28.2. Obviously, ionisation is required for sustaining the plasma discharge. For efficient CO_2 reforming it should not become a dominant pathway as it is an energetically inefficient way of initiating chemical reactions. After all, ionisation of CO_2 requires ca. 14 eV/molecule, whereas its "net" dissociation energy is ca. 3 eV (considering the "net" reaction $CO_2 \rightarrow CO + \frac{1}{2}O_2$). This simple consideration implies a maximum energy efficiency of at most 20% for each dissociation event via ionisation. In practice, due to the fact that only the high energy tail would drive dissociation and all other energy input would be "lost", it would limit efficiencies to even lower values of ca. 5%.

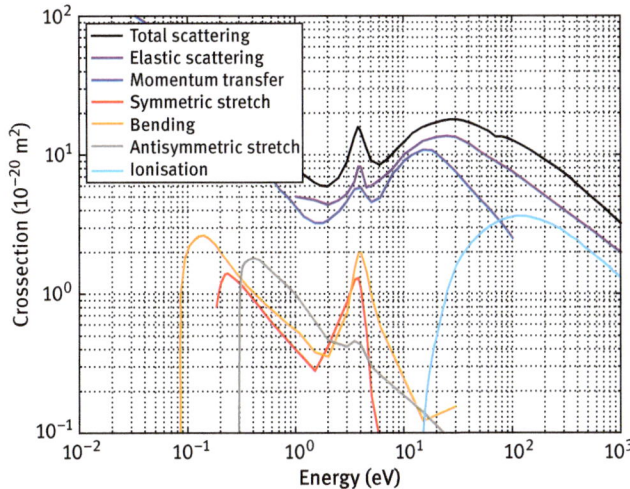

Figure 28.2: Cross sections for electron collisions with CO2, obtained from [3]. Average plasma electron energies are usually of few eV, exactly where the cross section of vibrational excitation peaks. This confirms the hand waving picture of preferential vibrational excitation in low temperature plasma. Ionisation to replenish plasma losses requires energies over 10 eV and is still possible by electrons in the high-energy tail of the electron-energy distribution function. The different approaches to plasma generation vary all in shape and mean of the distribution, which determines the balance between power deposited in vibration versus power consumed by ionisation and other high-energy excitations.

The majority of the electrons are however at lower energy, typically a few eV. These are responsible for collisions that predominantly excite vibrational modes in the molecule. The resulting vibrationally excited molecules will further interact with each other and exchange vibrational energy or convert vibrational energy into translational energy. Of special interest here is the asymmetric stretch vibration of CO_2, which carries two important properties. Firstly, the vibrational quanta are too large to be easily converted into translational energy in a low energy collision. Secondly, the vibration is anharmonic, which means that the vibrational level spacing of highly excited molecules is smaller than that of molecules at a lower level. This results in a slight preference of highly excited molecules gaining additional quanta compared to losing it to (the majority) molecules at the first levels.

In effect, it is the asymmetric stretch vibrational mode that can be brought to the highest degree of non-equilibrium and in which vibrational energy can be driven up along the energy scale to reach the dissociation limit. In this ladder-climbing scheme, illustrated in Figure 28.3, the electrons, that were energetically "expensive" to create, are used many times to deliver energy to overpopulate the lower asymmetric stretch levels and essentially to the bond that is to be broken up to the point where dissociation of the molecule is achieved. It is this qualitative

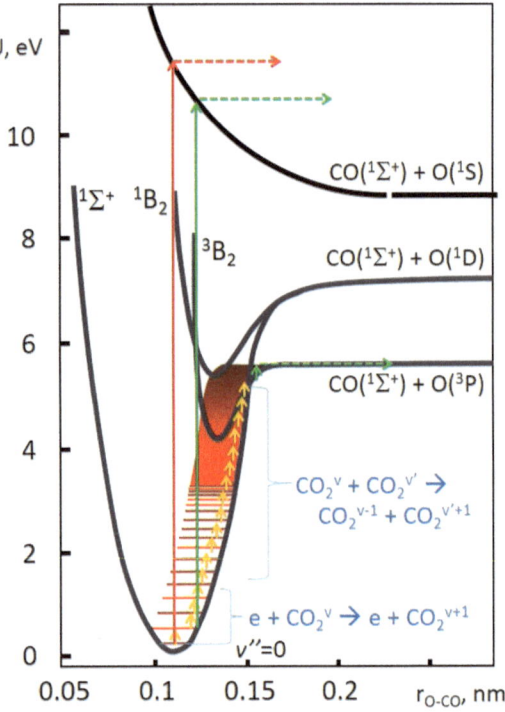

Figure 28.3: Schematic illustration of some CO_2 electronic and vibrational levels. Illustrated is stepwise vibrational excitation, that is, the so-called ladder-climbing process. It is initiated by plasma electron excitation and carried by vibrational exchange up to the point of dissociation. Opposed to energetically advantageous ladder climbing is dissociative excitation, which involves a large activation barrier. Reproduced from [10] with permission.

mechanism that has been put forward to explain the ultimate energy efficiencies that have been demonstrated in the former Soviet Union in the 1970s [4–9] for the net reaction $CO_2 \rightarrow CO + O_2$, that is, with over 80% energy efficiency.

The fraction of charge is usually small, often 10^{-5} or even less compared to the neutral species. Ionisation is therefore not significant in the power balance and the plasma acts as a *power transfer medium*, converting electric energy into internal energy of molecules.

Having explained why plasma is promising for CO_2 conversion, we will briefly present the most common types of plasma reactors with their characteristic features in the next section. We refer here to the non-equilibrium nature of the discharges to illustrate why some plasma types exhibit better energy efficiency than others. Subsequently, we will discuss the state of the art on plasma-based CO_2 conversion, including the combined conversion of CO_2 with CH_4, H_2O or H_2. Finally, we will discuss the major limitations and steps to be taken for further improvement.

28.2 Plasma reactor types for CO_2 conversion

Plasma is created, in its simplest form, by applying an electric potential difference between two electrodes, positioned in a gas. The gas pressure can range from a few Torr up to (above) 1 atm. The potential difference can be direct current (DC), alternating current (AC), ranging from 50 Hz over kHz to MHz (radio-frequency; RF), or pulsed. Furthermore, the electrical energy can also be supplied in other ways, for example, by a coil (inductively coupled plasma [ICP]) or as microwaves (MW).

Three types of plasma reactors are most often studied for CO_2 conversion: dielectric barrier discharges (DBDs), MW plasmas and gliding arc (GA) discharges. Below, we will briefly present their working principles and typical operating conditions, to explain why they are particularly interesting and what their current limitations are. Furthermore, besides these three major types of plasma reactors, other plasma types are being explored as well for CO_2 conversion, and they will also be very briefly discussed. Finally, we will introduce the concept of plasma catalysis, for the selective production of value-added chemicals.

28.2.1 Dielectric barrier discharge (DBD)

A dielectric barrier discharge is created by applying an AC potential difference between two electrodes, of which at least one is covered by a dielectric barrier. The latter limits the amount of charge transported between both electrodes, and thus it prevents that the discharge would undergo a transition into a TP, which is a less efficient regime for CO_2 conversion. The electrodes can be two parallel plates, but a more common design for CO_2 conversion is based on two concentric cylindrical electrodes (cf. Figure 28.4(a)), in which the inner electrode is surrounded by a dielectric tube with a mesh or foil electrode wrapped around it. The gap between inner electrode and dielectric tube is in the order of a few millimetres. One of the electrodes is connected to a power supply, while the other electrode is grounded. The gas flows in from one side, and is gradually converted along its way through the gap between inner electrode and dielectric tube, and flows out from the other side.

A DBD typically operates at atmospheric pressure, which is beneficial for industrial applications. Furthermore, it has a very simple design, making is suitable for upscaling, and thus industrial implementation, as demonstrated already for ozone synthesis, by placing a large number of DBD reactors in parallel [11].

On the other hand, a DBD has only a limited energy efficiency for CO_2 conversion, typically around 10%, with some exceptions up to 20% [12]; see also next section. The reason is that the reduced electric field (i.e. ratio of electric field over gas number density) is typically above 100–200 Td (1 Td = 10^{-21} V m^2), creating high-energy electrons, which mainly give rise to electronic excitation, ionisation and

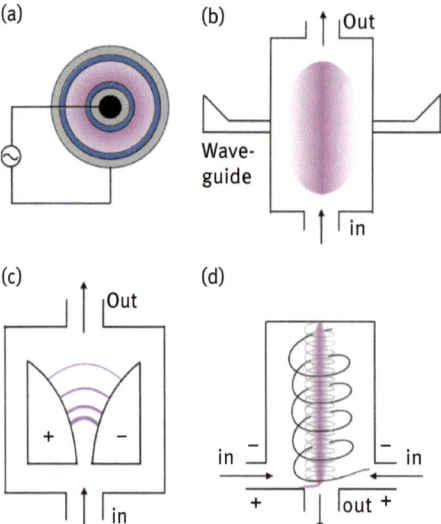

Figure 28.4: Schematic illustration of the three plasma reactors most often used for CO_2 conversion, that is, dielectric barrier discharge (a), MW plasma (b) and GA discharge, in classical configuration (c) and cylindrical geometry, called gliding arc plasmatron (GAP) (d). Reproduced from [16] with permission.

dissociation of CO_2 molecules in the ground state, and this is not the most energy-efficient CO_2 dissociation pathway (see below).

By introducing a packing of dielectric material in the discharge gap, the energy efficiency can in principle be improved, because of polarisation of the dielectric packing beads resulting from the applied potential difference. Indeed, this will enhance the electric field near the contact points of the packing beads, and thus the electron energy [13], causing more electron impact excitation, ionisation and dissociation, and thus more CO_2 conversion for the same applied power. In addition, such a packed bed DBD is very suitable for plasma catalysis, as will be discussed in Section 28.2.5. However, it should be noted that the CO_2 conversion efficiency is not always enhanced in a packed-bed DBD [14, 15], because of the competing effect of reduced residence time in the smaller discharge volume, when comparing at the same gas flow rate, as well as the loss of electrons and reactive plasma species at the surface of the packing material.

28.2.2 Microwave plasma

In a MW plasma, electromagnetic radiation with frequency between 300 MHz and 10 GHz is applied to a gas, without using electrodes. Depending on the configuration, there exist different types of MW plasmas, that is, cavity-induced plasmas, free expanding atmospheric plasma torches, electron cyclotron resonance plasmas and surface wave discharges. The latter are most frequently used for CO$_2$ conversion. In this configuration, the gas flows through a quartz tube, which is transparent to MW radiation, intersecting with a rectangular waveguide, to initiate the discharge (see Figure 28.4(b)). The MWs propagate along the interface between the quartz tube and the plasma column, and the wave energy is absorbed by the plasma.

MW plasmas can operate in a wide pressure regime, ranging from very low pressure (e.g. 10 mTorr) up to atmospheric pressure. The low-pressure regime yields very efficient CO$_2$ conversion. Energy efficiencies up to 90% were reported for very specific conditions, that is, supersonic gas flow and pressures around 100–200 Torr [17]. This is attributed to the role of the vibrational kinetics (discussed above in section 28.2) [2, 10, 12, 18]. Indeed, a MW plasma is characterised by typical reduced electric fields below 100 Td. This yields electron energies around 1 eV, which are most beneficial for vibrational excitation of CO$_2$ [2, 10, 12]. Hence, the electrons populate the lower vibrational levels of CO$_2$, which collide with each other in so-called vibrational–vibrational (VV) relaxation, gradually populating the higher levels. This so-called ladder-climbing process requires 5.5 eV for CO$_2$ dissociation, which is exactly the C=O bond dissociation energy, while electronic excitation to a dissociative level, which is the main process in a DBD (see above), would require 7–10 eV. As the latter is much more than the C=O bond dissociation energy, this extra energy is just waste of energy. This explains why the energy efficiency in a DBD is much more limited (see above).

Note, however, that the vibrational levels can also get lost by vibrational-translational (VT) relaxation. This becomes especially important at high gas temperature, as revealed by computer simulations [19], and it results in a vibrational distribution function (VDF) in (near) thermal equilibrium with the gas temperature. When the MW plasma operates at atmospheric pressure, it exhibits a quite high gas temperature (in the order of several 1,000 K), resulting in a VDF that is indeed close to thermal [19, 20]. Deviation from a thermal distribution can be realised by increasing the power density, reducing the pressure and the gas temperature [19]. At atmospheric pressure, it is not straightforward to realise a low gas temperature. A solution could be to apply a pulsed power, so that the gas can cool down in between the applied pulses, or to apply a supersonic gas flow, as demonstrated by Azivov et al. [17], so that the gas has not enough time to be heated. On the other hand, the gas must have a sufficiently long residence time for the conversion to take place as well.

28.2.3 Gliding arc discharge

A GA discharge is a transient type of arc discharge. A classical (two-dimensional) GA discharge is created between two flat diverging electrodes (see Figure 28.4(c)). The arc is initiated at the shortest interelectrode distance, and it "glides" towards larger interelectrode distance under influence of the gas, which flows along the electrodes, until it extinguishes and a new arc is created at the shortest interelectrode distance.

The classical GA discharge yields only limited CO_2 conversion, because only a limited fraction of the gas passes through the arc. Therefore, other types of (three-dimensional) GA discharges have been designed, such as a GA plasmatron and a rotating GA, operating between cylindrical electrodes. Figure 28.4(d) schematically illustrates the operating principle of a GAP. The cylindrical reactor body acts as cathode (powered electrode), while the reactor outlet is the anode (grounded). The gas enters tangentially between both cylindrical electrodes. When the outlet diameter is (significantly) smaller than the diameter of the reactor body, the gas flows in an outer vortex towards the upper part of the reactor body, followed by a reverse inner vortex towards the outlet, with a smaller diameter because it has lost some speed, and therefore it can leave the reactor through the outlet. The arc is again initiated at the shortest interelectrode distance, and expands till the upper part of the reactor, rotating around the axis of the reactor until it stabilises in the centre after about 1 ms, because of the vortex gas flow. Ideally, the inner gas vortex passes completely through this stabilised arc, allowing most of the gas to be converted. However, the fraction of gas passing through the arc is still too limited, thereby limiting the CO_2 conversion [21, 22].

The GA discharge operates at atmospheric pressure, which makes it suitable for industrial implementation. Moreover, it shows a good energy efficiency, that is, around 30% for CO_2 splitting [21] and 60% for DRM [22]. The reason is the same as in the MW plasma, that is, because of the favourable reduced electric field, creating electrons of about 1 eV, which mainly give rise to vibrational excitation of CO_2, and thus, the vibrational pathway of CO_2 dissociation is again promoted. Nevertheless, the gas temperature is also fairly high (typically a few 1,000 K), which limits the energy efficiency, because of VT relaxation, yielding a VDF too close to a thermal distribution, just like in a MW plasma (see above). More efforts are thus needed to better exploit the non-equilibrium behaviour of a GA plasma, by reducing the gas temperature.

28.2.4 Other plasma types used for CO_2 conversion

Besides these three types of plasma reactors explained above, other plasma types are also increasingly being used for CO_2 conversion, such as nanosecond (ns)-pulsed discharges [23], spark discharges [24], corona discharges [25] and atmospheric pressure glow discharges (APGDs) [26].

Ns-pulsed discharges are basically generated by repetitive ns-pulsed excitation, leading to a high non-equilibrium with very high plasma densities for relatively low power consumption because of the short pulse duration. The short pulses offer good control of the electron energy, depending on the pulse length, so that more energy can be directed towards the desired dissociation channels.

Spark discharges consist of an initiation of streamers between two electrodes, developing into highly energetic spark channels, which extinguish and reignite periodically, just as lightning, even without pulsed power supply.

Corona discharges are created near sharp edges or thin wires used as electrode. Either a negative or a positive voltage can be applied to the electrode, yielding a negative or positive corona discharge. Corona discharges are non-uniform discharges, with a strong electric field, ionisation and luminosity close to the sharp electrode, while the charged particles are dragged to the other electrode by a weak electric field. Their performance towards CO$_2$ conversion is similar as for DBDs.

The name "APGD" stands for a collection of several types of plasmas, including miniaturised DC glow discharges, microhollow cathode DC discharges, RF discharges as well as DBDs. They typically operate at not too elevated temperature, and can exist either in stable homogeneous glow or filamentary glow mode. They can exhibit a typical electron temperature around 2 eV, thus still suitable for vibrational excitation of CO$_2$, while the gas temperature is limited to about 900–1,000 K, hence lower than for GA and MW plasmas. This guarantees more pronounced thermal non-equilibrium, and makes them promising for CO$_2$ conversion.

28.2.5 Principle of plasma catalysis

As explained above, plasma on its own is very reactive, because of the cocktail of chemical species (electrons, various types of molecules, atoms, radicals, ions and excited species), but for the same reason, it is not selective in the production of targeted compounds. This problem can be solved by so-called plasma catalysis, which combines the high reactivity of plasma with the selectivity of a catalyst [27–29]. Plasma catalysis is most straightforward in a DBD plasma, more specifically in a packed bed DBD, because the packing beads can be covered by a catalytic material or they can have catalytic properties from their own. This is called one-stage plasma catalysis, but the catalyst can also be placed after the plasma reactor, in so-called two-stage plasma catalysis. In the first case, short-lived plasma species, such as excited species, radicals, photons and electrons, can interact with the catalyst, providing more possibilities for synergy than in the latter case, where only long-lived species can interact with the catalyst. On the other hand, the two-stage configuration can also be applied to other plasma types, such as MW and GA discharge, where one-stage plasma catalysis is not so straightforward, among others because of the high gas temperature in the plasma (cf. above). Nevertheless, the latter may also provide other opportunities; it can open the way for thermal activation of catalysts, either inside the discharge zone (if the temperature could be somewhat reduced, and when using thermally stable catalysts) but also downstream, when the gas leaving the MW or GA reactor is still hot, in two-stage plasma catalysis.

Although plasma catalysis is a quite promising combination, not only to improve the selectivity of product formation, but also to enhance the overall plasma performance in terms of conversion and energy efficiency, the underlying mechanisms, especially in one-stage plasma catalysis, are very complicated and far from understood.

On the one hand, the plasma can affect the catalyst performance in several ways:

a) changes in the physicochemical properties of the catalyst, that is, a higher adsorption probability [30], a higher surface area [31], because of reduced metal particle size and enhanced dispersion of metal particles at the catalyst surface [32], a change in the oxidation state [33], reduced coke formation [34] and a change in the work function because of the presence of a voltage and current (or charge accumulation) at the catalyst surface [35];

b) the formation of hot spots, modifying the local plasma chemistry [36];

c) lower activation barriers, because of the existence of short-lived active species, such as radicals and vibrationally excited species [33].

On the other hand, the catalyst will also affect the plasma performance, by:

a) enhancement of the local electric field in the plasma, because the catalyst is mostly present in a structured packing (e.g. pellets, beads and honeycomb; so-called packed-bed reactor), or simply because of porosity of the catalyst surface [36–38];

b) change of the discharge type from streamers inside the plasma to streamers along the catalyst surface, resulting in more intense plasma around the contact points [39–42];

c) formation of microdischarges in the catalyst pores, resulting in more discharge per volume, increasing the mean energy density of the plasma [36, 43];

d) adsorption of plasma species on the catalyst surface, affecting the residence time and hence the concentration of species in the plasma [44], while new reactive species might be formed at the catalyst surface.

Figure 28.5 presents a schematic overview of some of these plasma–catalyst interaction processes, in one-stage plasma catalysis [45]. Roughly speaking, we can distinguish two types of effects, that is, physical and chemical effects. While the physical effects, such as enhanced electric field, are mainly responsible for gaining a better energy efficiency, the chemical effects can lead to improved selectivity towards value-added products. In case of CO_2 splitting, mainly CO and O_2 are formed, so the primary added value of the catalyst is to increase the energy efficiency, although the conversion can also be improved by chemical effects, such as enhanced dissociative chemisorption because of catalyst acid/basic sites. When adding a co-reactant (e.g. CH_4, H_2O and H_2), the catalyst allows to modify the selectivity towards value-added products.

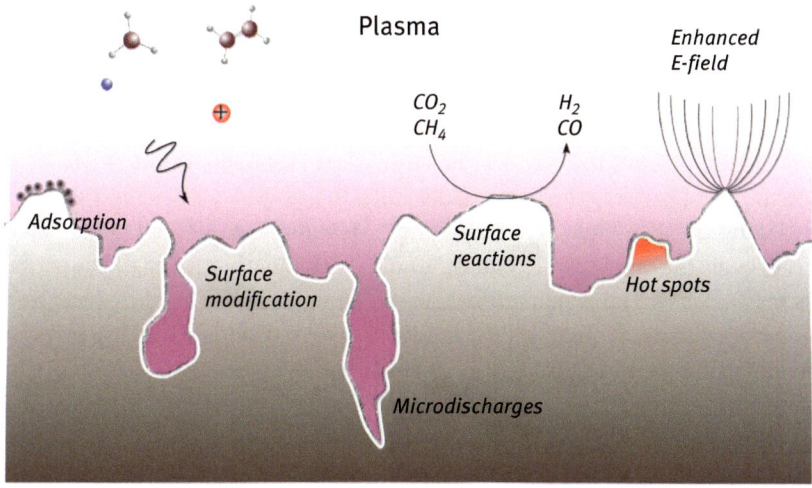

Figure 28.5: Schematic illustration of some plasma–catalyst interaction mechanisms. Modified from [45].

The plasma–catalyst interactions can lead to synergy in plasma catalysis, when the combined effect is larger than the sum of the two separate effects under the same operating conditions, but this is not always realised up to now. Indeed, a lot of research, by combined experiments and computer modelling, will be needed to understand all these mechanisms and to fully exploit the possible synergy. Furthermore, more dedicated research is needed to effectively design catalysts tailored for the plasma environment, which make profit of the typical plasma conditions, instead of using commercial catalysts typically used in thermal catalysis. Indeed, nowadays this is still too often the case, limiting the real potential of plasma catalysis in selectively producing the desired products. Examples of successful plasma catalytic CO$_2$ conversion will be given later in this chapter.

28.3 CO$_2$ conversion processes: Reactions, reactors and performance

28.3.1 CO$_2$ splitting

CO$_2$ splitting in the plasma phase was pioneered around the 70s for a two-step hydrogen production process in the former Soviet Union. It remained largely unknown to the rest of the world until it was summarised in the book by A. Fridman in 2008 [2].

The potential of vibrational excitation to intensify chemical reactions is a recurring theme in this work and CO_2 reduction forms its showcase. Since that moment, the promise to address the current global challenges regarding CO_2 emissions has been well recognised by the international plasma chemistry community and a number of groups started investigating the maximally achievable energy efficiency for the reduction of CO_2 in plasma.

A graph summarising the great promise of plasma chemistry on the basis of work from the 80s [2] and some first results since the revival of the field [18, 46] is shown in Figure 28.6. It is clear that the unprecedented high energy efficiency of ca. 85% has been the food for inspiration and is extremely promising for opening pathways to CO_2 reuse. The explanation of the ultimate energy efficiencies has been the preferential excitation of vibrational modes that drives ladder climbing of vibrational quanta all the way to dissociation, as explained in section 28.2.

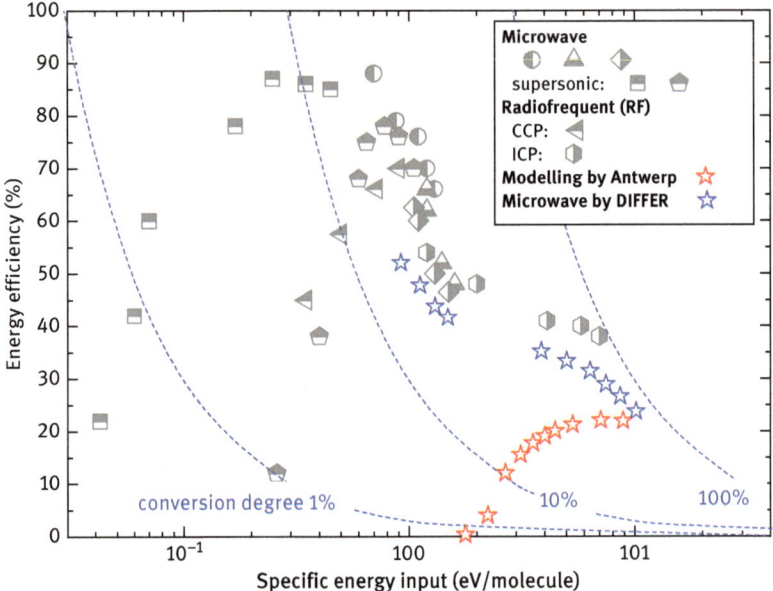

Figure 28.6: Energy efficiency of plasma assisted CO_2 reduction as a function of the specific energy input. The black and white markers are from the summary by Fridman [2]. The red markers show initial modelling results of vibrational ladder climbing by Kozak [18] and the orange markers first experimental microwave results by Bongers [46]. Dashed lines indicate contours of conversion degree.

Other remarkable features of the early data are the superior performance of MW discharges compared to other approaches (here only RF shown) and an apparent trade-off between energy efficiency and conversion. Both observations align with the mechanism of vibrational ladder climbing. MW discharges are recognised to have an average electron energy that is optimal for preferential vibrational excitation, although it requires sub-atmospheric pressures. Degrading efficiency at higher input power levels can be expected due to gas heating that quenches the vibrational non-equilibrium and reduces performance to thermal values [2].

Some first results since the revival have also been included in the figure and are topical for many more findings since. Firstly, zero-dimensional modelling of vibrational kinetics in MW plasma that included 25 vibrational levels (of the asymmetric stretch mode) up to the dissociation limit could not reproduce the record efficiencies but predicted a maximum efficiency of ca. 25% [18]. Secondly, experimental characterisation of MW plasma achieved higher efficiencies of ca. 50% [46], however, temperature measurements in similar configurations revealed gas temperatures of typically 3,500 K [47]. The latter means that thermal decomposition must have been of importance too whilst vibrational ladder climbing cannot be expected to dominate at such high temperatures [19]. Noteworthy is that also for the early experiments that yielded 85% efficiency a high temperature core was observed and that vibrational dynamics were assumed to be important in the colder surroundings [2]. In other words, ultimate thermal conversion performance of 50% energy efficiency has been achieved in recent experiments. However, the older record of 85% has not been reproduced yet and it seems that schemes in which vibrational excitation is dominant are probably non-uniform and involve transport of power and species.

28.3.1.1 Mechanisms of CO₂ dissociation

Let us summarise the main pathways that lead to dissociation in a plasma. Here, we start with the electron driven processes. Important for the present is that these differentiate largely from each other in their threshold energy E_{th}:

i. Electron impact ionisation followed by dissociative recombination,

$$e + CO_2 \rightarrow CO_2^+ + 2e \rightarrow CO + O + e \qquad E_{th} = E_{ion} = 13.8 \text{ eV} \qquad (28.1)$$

ii. Electron impact dissociative excitation,

$$e + CO_2 \rightarrow CO + O(^1S) + e \qquad E_{th} = 11.5 \text{ eV} \qquad (28.2)$$

iii. Vibrationally enhanced electron impact dissociative excitation, including vibrational ladder climbing,

$$e + CO_2(v \geq 1) \rightarrow CO + O(^3P) + e \qquad 0 < E_{th} < 11.5 \text{ eV} \qquad (28.3)$$

It is clear that the first two of these are *a priori* not beneficial for achieving high-energy efficiency. The high threshold energies have to be compared with the reaction enthalpy of the net elemental dissociation reaction,

$$CO_2 \rightarrow CO + O(^3P) \qquad \Delta H = 5.5\,eV \qquad (28.4)$$

Thus, (i) dissociative recombination of ions as well as (ii) dissociative excitation are highly inefficient dissociation mechanisms and convert 6–8.3 eV into heat and/or internal energy per event. In fact, more than one excited state will probably contribute to dissociative excitation, with (slightly) lower threshold energy, and ^1D or ^3P oxygen atoms being created. These excited states have bent structures of which still little is known, which means that their Franck–Condon overlap with the electronic ground state is unknown and their electron impact energy threshold cannot be predicted.

The third mechanism is evidently favourable, providing potentially the smallest threshold energy. Here, we include within (iii) vibrationally enhanced electron impact dissociative excitation also the aforementioned vibrational ladder-climbing mechanism. The latter was explained in detail in Figure 28.3, invoking the potential energy diagram of CO_2 along one O–CO coordinate. A subtle addition here is the decrease of the activation barrier of dissociative excitation by more favourable Franck–Condon overlap of a vibrationally excited level. Dissociative excitation from the ground state produced in this example roughly 2.5 eV of kinetic energy in the fragments and an ^1S oxygen radical. A molecule excited in the first vibrational level benefits not only from a lower threshold energy because of its initial vibrational energy (0.3 eV for the asymmetric stretch vibration), but also from Franck–Condon overlap with reduced energy of the upper state. In the schematic representation of Figure 28.3, it means a reduction in threshold energy of ca. 1 eV and in released kinetic energy of ca. 0.7 eV.

The energy threshold vanishes as molecules get into the highest vibrational (asymmetric stretch) levels, close to the dissociation limit. The non-adiabatic transition $^1\Sigma^+ \rightarrow\ ^3B_2$ in the point of crossing of the $^1B_2$ and $^3B_2$ terms opens the most effective dissociation pathway $CO_2(^1\Sigma^+) \rightarrow CO\ (^1\Sigma^+) + O(^3P)$. As was briefly touched upon before, this can become a significant channel by virtue of the vibrational ladder anharmonicity under strongly non-equilibrium conditions in which the kinetic energy of the molecules remains low. Because of the anharmonicity, vibration–vibration (VV)-exchange is no longer resonant and Treanor has shown (although neglecting effects of dissociation and VT relaxation [19], a boundary condition that determines the exact shape of the vibrational distribution [Diomede, 2017]) that this results into strong deviation from a Boltzmann distribution, of overpopulation of high vibrational levels. Such strong overpopulation is indeed observed in the aforementioned state-to-state modelling of the asymmetric stretch manifold, which is shown in Figure 28.7. The depopulation of the highest levels is because of dissociative excitation.

Finally, atomic oxygen created in the plasma should be able to create a second CO molecule and molecular oxygen in order to optimise the overall efficiency and to explain the observed ultimate energy efficiencies close to 90% [2]. Again, this requires a vibrationally excited CO_2^* molecule as the reaction is endothermic:

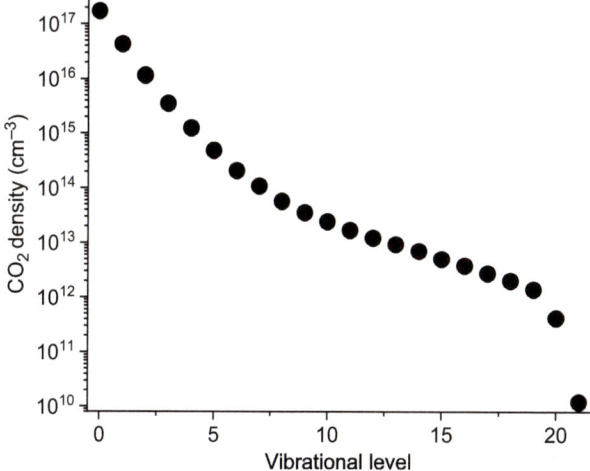

Figure 28.7: Vibrational distribution functions of the asymmetric mode vibrational levels of CO$_2$ in a MW discharge after 8.0 ms of power input at a rate of 20, 25 and 30 W/cm^3, reproduced from [18] with permission. The plateau behaviour around $10 < v < 17$ reflects the Treanor-like overpopulation. Dissociative excitation of the levels v>19 causes strong depopulation of these highest levels and produces CO most efficiently.

$$O + CO_2^* \rightarrow CO + O_2 \qquad \Delta H = 0.3\,eV,\ E_{th} = 0.5 - 3\,eV \qquad (28.5)$$

One should notice the large range that is given for the activation energy of the reaction. Its consequence is that it may well be limiting the overall efficiency, as has been discussed in [48, 49, 50]. At sufficiently high power density, the neutral gas temperature in the plasma reactor can become high enough for thermal decomposition of CO$_2$ to set in. This requires temperatures exceeding ca. 1,700 K, as is seen from the calculated equilibrium composition of a carbon dioxide mixture at 100 mbar in Figure 28.8. The thermal conversion optimum shown in the same graph is just over 50% at 3,200 K, which requires ideal quenching of the reaction products. In fact, also here the plasma phase can help by providing a vibrational non-equilibrium and thus quenching atomic oxygen by producing additional CO in reaction (28.5). This is referred to as super-ideal quenching and would bring the efficiency limit of thermal conversion in plasma up to at least 60% [2].

28.3.1.2 Dissociation performance in different plasma approaches

Recently, Snoeckx and Bogaerts reviewed the state of the art of plasma chemistry concerning CO$_2$ conversion [12]. Figure 28.9 summarises the performance of the different plasma approaches in terms of combinations of efficiency and conversion.

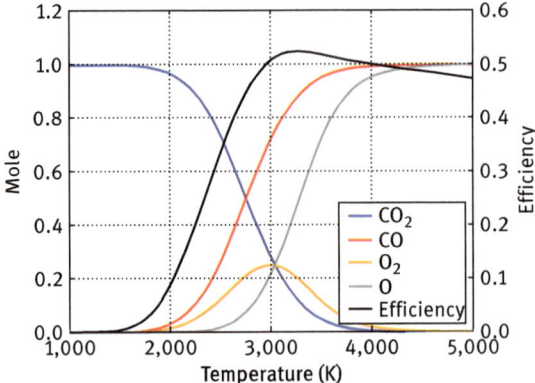

Figure 28.8: The equilibrium composition of CO_2 and its dissociation products as a function of temperature at a pressure of 100 mbar. Instantaneous quenching preserving all CO formed is assumed to calculate the efficiency. Reproduced from [47] with permission.

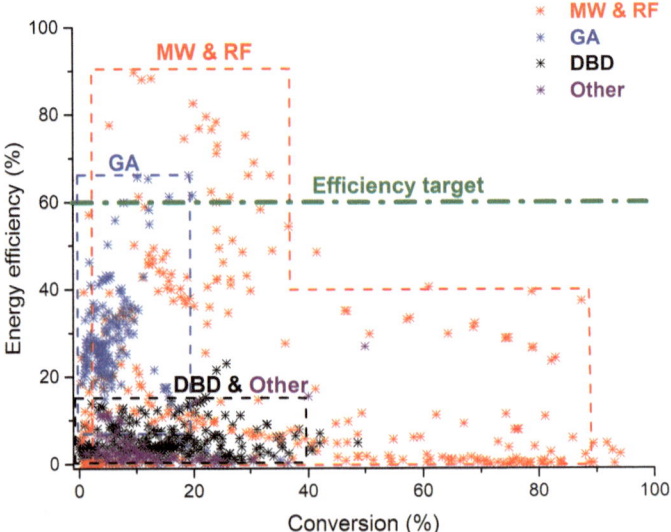

Figure 28.9: Comparison of all the data collected from the literature for CO_2 splitting in the different plasma types as collected by Snoeckx and Bogaerts [12]. It shows combinations of energy efficiency and conversion grouping the data per discharge type.

Although DBDs have been widely researched in view of their strong non-equilibrium character, their successful commercial application for O_3 production and relatively ease of operation at atmospheric pressure, their performance stays significantly behind that of the other types. Although conversion can reach up to 40%, efficiency appears to be limited to ca. 18% (except for the DBD record of 23% by [51]). The effect of changing the applied frequency, power, gas flow rate, discharge length, discharge gap, reactor temperature, dielectric material, electrode material, mixing with gases, that is, Ar, He, N_2 and by introducing (catalytic) packing materials has been studied extensively, as well as numerically modelling has been applied to gain understanding in the underlying reaction pathways. In general, it appears that conversion can well be controlled via specific energy input (SEI) (or, equivalently, the residence time of the gas in the reactor); however, this goes on the expense of energy efficiency. The limited efficiency seems to be because of unfavourable plasma electron energy distribution (or E/n), which causes *vibrational* non-equilibrium effects to be insignificant and reactions (1) and (2) the main dissociation pathways. This is despite the fact that gas temperatures remain low and in this respect the DBD discharge being non-equilibrium par excellence. The low gas temperatures, however, also cause thermal conversion insignificant.

MW discharges clearly span the largest parameter range and reach the highest energy efficiencies. This discharge type is known to be best suitable to channel most of the discharge power to vibrational modes and is thus best equipped to benefit from vibrational excitation. However, the best results (efficiency >60%) date from the early work [4–9] and have not been reproduced in recent years. All recent work is within the range of thermal equilibrium conversion and gas temperature measurements have indeed shown that 50% efficiency is well achievable up to high conversion values [47, 52–57]. As thermal regions have also been observed in the early work, it seems that a combination of thermal conversion with non-equilibrium chemistry in the periphery is an interesting route to further optimisation. It means that transport of power and particles in a complex 3D geometry is to be optimised.

A drawback that is often put forward to MW discharges is their preference to operate at reduced pressure. On the one hand, the lower operational pressure is compensated by high flow rates so that in effect the reactor power density is of the highest possible. On the other hand, also atmospheric pressure operation is well possible, but likely limits performance to thermal operational space.

GA discharges seem to succeed in exploiting vibrational excitation enhanced dissociation channels while operating at atmospheric pressure. Energy efficiencies up to 50% are common [58] and also record values of 65% have been reported [59]. Model calculations (e.g. [20, 60–63]) revealed that also GA discharges induce elevated temperatures. Just like for MW discharges, preventing gas heating to operate at lower gas temperatures might be the key for benefitting fully from the potential of the vibrational excitation pathways to dissociation. At the same time, it is very much likely that also in GA thermal conversion plays a significant role [20, 60–63].

28.3.2 Plasma conversion of CO_2 with CH_4

28.3.2.1 Plasma conversion

$$CO_2(g) + CH_4(g) \rightarrow 2CO\,(g) + 2H_2(g) \quad \Delta H° = 247\,\text{kJ}\,\text{mol}^{-1} \qquad (28.6)$$

The conversion of CO_2 with CH_4, known as DRM has received significant interest as this reaction uses two abundant greenhouse gases CO_2 and CH_4 in the form of different sources (e.g. landfill gas, biogas and shale gas) to produce value-added fuels and chemicals, with syngas (H_2+CO) being the most common target product (28.6). Syngas is a vital chemical feedstock that can be used to produce a variety of platform chemicals and synthetic fuels, including via the Fischer-Tropsch process. However, both CO_2 and CH_4 are highly stable, therefore high temperature (> 700 °C) is always required for thermal catalytic activation of CO_2 and CH_4 with reasonable conversions and syngas production because of the thermodynamic barrier of this process, resulting in high-energy consumption. In addition, the rapid deactivation of reforming catalysts at high temperatures because of sintering and coke deposition remains a major challenge for the use of this process at a commercial scale. NTPs provide a promising alternative to the thermal catalytic process for the conversion of CO_2 with CH_4 into higher value chemicals and fuels at low temperatures and ambient pressure. Significant efforts have been devoted to the synthesis of syngas using different plasma systems with or without catalyst [64–68]. In addition to syngas production, noticeable amounts of higher hydrocarbons are often produced in the plasma DRM process, especially in the presence of a catalyst (e.g. zeolite) [69]. CH_x radicals initially formed in the dissociation of CH_4 play a key role in the production of higher hydrocarbons [10, 19, 70]. Thus, the content of CH_4 in the CO_2/CH_4 mixture is of primary importance for the synthesis of higher hydrocarbons in the plasma DRM reaction. Eliasson et al. investigated the synthesis of higher hydrocarbons from CO_2 and CH_4 using a DBD plasma. The selectivity of C_{5+} and oxygenates was up to 41.2% at a discharge power of 500 W and a CO_2/CH_4 molar ratio of 1:2[69]. A mixture containing mainly C_2H_2 and synthesis gas with a H_2/CO ratio of 2:1 was produced in the plasma reaction using a point-to-point pulsed discharge at a CO_2/CH_4 ratio of 1:2[71].

In addition, a few groups have reported the formation of trace oxygenates (e.g. alcohols and acids) along with the production of syngas and hydrocarbons in plasma-based DRM [72, 73]. Zhang et al. [72] reported the production of acetic acid, propanoic acid, ethanol and methanol in the plasma DRM using a DBD reactor. Acetic acid was the major liquid product with the highest selectivity of 5.2% achieved at CH_4 and CO_2 conversion of 64.3% and 43.1%, respectively [72]. Acetic, formic, butanoic and propanoic acids were also formed, along with methanol and ethanol, in the plasma oxidation of CH_4 with CO_2 using a DBD [74]. Zhou et al. [75] developed a starch-enhanced plasma process for the conversion of CO_2 and CH_4 into a range of oxygenates, including primarily formaldehyde, methanol, ethanol, formic acid and acetic acid. The total selectivity of oxygenates was about 10–40% with the conversion of CH$_4$ and CO$_2$ of about 20%. They found that a lower methane concentration was favourable for the production of oxygenates, and a higher feed flow rate led to higher selectivity of oxygenates in the presence of starch [75]. The direct conversion of CH$_4$ with CO$_2$ using DBD plasmas was carried out by Li et al. The product includes syngas, gaseous hydrocarbons (C$_2$ to C$_4$), liquid hydrocarbons (C$_5$ to C$_{11+}$) and oxygenates [76]. Bogaerts et al. developed a 1D fluid model to understand the plasma chemistry of the DRM process in a DBD reactor. Their modelling results showed that oxygenates, including methanol, ethanol, acetaldehyde and ketene, can be formed in the plasma DRM reaction [10, 77]. Very recently, Wang et al. have developed a water-electrode DBD plasma reactor for the direct, one-step reforming of CO$_2$ with CH$_4$ into oxygenates (e.g. acetic acid, methanol, ethanol and acetone) at atmospheric pressure (1 bar) and room temperature (30 °C). The total selectivity to oxygenates was approximately 50–60% without a catalyst, with acetic acid being the major liquid product at 40.2% selectivity [78]. Two possible reaction pathways could contribute to the formation of acetic acid in this process (Figure 28.10). CO can react with a CH$_3$ radical to form an acetyl radical (CH$_3$CO) with a low-energy barrier of 28.77 kJ mol^{-1}, followed by recombination with OH to produce acetic acid with no energy barrier. Direct coupling of CH$_3$ and carboxyl radicals (COOH) could also form acetic acid based on density functional theory (DFT) modelling [74].

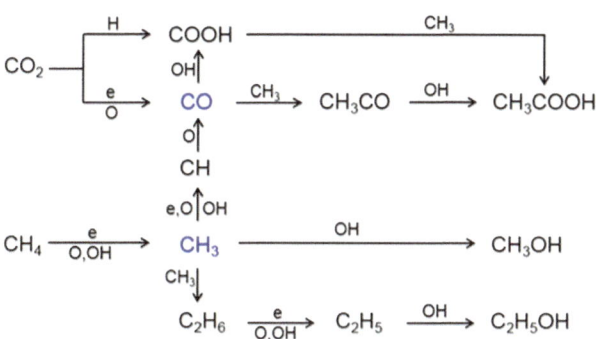

Figure 28.10: Possible reaction pathways for the formation of oxygenates in plasma DRM using a DBD. Reproduced from [78] with permission.

Carbon nanomaterials are often produced as a by-product in the plasma dry reforming reaction. Tu and Whitehead reported the production of multi-wall carbon nanotubes (MWCNTs) and spherical carbon nanoparticles (NPs) with a diameter of 40–50 nm in the DRM reaction using an AC GA discharge with knife-shaped Al electrodes (Figure 28.11(a)) [66]. Chung and Chang reported the synthesis of MWCNTs via plasma DRM using a spark discharge. They found that the stainless-steel electrodes of the spark discharge acted as a substrate for the deposition of MWCNTs

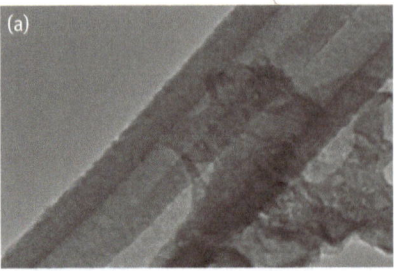

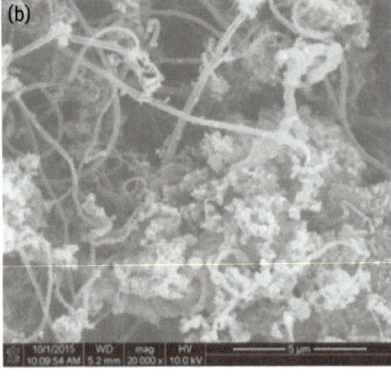

Figure 28.11: (a) The formation of carbon nanotubes using a gliding arc plasma; (b) the formation of MWCNTs using a spark discharge. Reproduced from [66, 79] with permission.

(Figure 28.11(b)) [79]. Carbon nanomaterials have a variety of applications and are higher value products in the plasma DRM process, which can further reduce the energy cost of the overall plasma DRM process and make this process more attractive.

The conversion of CO_2 with CH_4 has been explored using different plasma systems. Most of previous works have mainly focused on the production of syngas via plasma DRM [80–85, 86]. The reaction performance of the plasma dry reforming process has been affected by a range of operating parameters, such as plasma input power, total gas flow rate, SEI, CH_4/CO_2 molar ratio and dielectric material. The plasma power is one of the most important parameters determining the effectiveness of the plasma DRM process. Increasing discharge power enhances the conversion of CO_2 and CH_4 regardless of the type of plasma system used [65–67, 87]. A higher discharge power generates more energetic electrons and reactive species (e.g. O and OH radicals), which can activate the reactants and promote the conversion [69]. In addition, increasing discharge power would increase the temperature of the plasma reaction, which also contributes to the enhanced conversion of CO_2 and CH_4. In a DBD plasma reactor, increasing discharge power by changing the applied voltage at a fixed frequency increases the number of microdischarge and creates more reaction channels for chemical reactions, resulting in higher conversion of CO_2 and CH_4. This effect can be demonstrated by the increased magnitude and number of current pulses of the DBD plasma at a higher plasma power [88].

However, the discharge power can also affect the distribution of gas products produced in the plasma DRM process. Previous results showed that increasing discharge power decreases the selectivity of lower hydrocarbons (e.g. C$_2$) but increases the selectivity of higher hydrocarbons (e.g. C$_4$ and C$_5$) [89]. By contrast, the change of discharge power has a limited effect on the selectivity of syngas and the H$_2$/CO molar ratio, although the yield of syngas is enhanced at a higher discharge power [90].

Increasing the total feed flow rate decreases the conversion of CO$_2$ and CH$_4$ because of the decrease of the residence time of the reactants in the discharge region, which reduces the possibility of the reactant molecules colliding with energetic electrons and reactive species [66]. A lower gas flow rate is beneficial for producing more syngas and reducing the selectivity of higher hydrocarbons [89]. The increase of the residence time resulting at a lower feed gas flow rate increases the chance for C$_2$–C$_4$ hydrocarbons to be further dissociated via electron impact reactions and converted to produce more CO and H$_2$ [91]. By contrast, a high total feed flow rate is preferred for the production of C$_2$–C$_4$ hydrocarbons. In addition, the change of the feed flow rate does not significantly change the H$_2$/CO molar ratio [64, 82]. Although the conversion of the reactants decreases when increasing the feed flow rate, the energy efficiency of the plasma process increases as the total amount of reactants converted increases and more electric energy could be converted to chemical energy stored in the products [66].

SEI is a major determining factor for the conversion and energy efficiency in plasma chemical processes, as it combines the effect of power and gas flow rate. The variation of the SEI can be achieved by changing the discharge power and/or gas flow rate. However, previous findings showed that manipulating the SEI by changing the gas flow rate has a more pronounced effect on the conversion of the reactants compared to the change of discharge power [88]. Increasing the SEI at a constant gas ratio and frequency results in a higher conversion of CO$_2$ and CH$_4$ but with a decreased energy efficiency of the plasma process. The trade-off between the conversion and energy efficiency was often reported in previous studies [67, 87]. Therefore, both discharge power and gas flow rate should be considered when pursuing a suitable SEI to achieve higher conversion and energy efficiency simultaneously.

The reactant conversion and the H$_2$/CO molar ratio, along with the product yields and selectivities, are significantly affected by the molar ratio of CO$_2$/CH$_4$ in the feed [81, 92]. Increasing the CO$_2$/CH$_4$ molar ratio significantly enhances the conversion of CH$_4$ but only weakly decreases the conversion of CO$_2$. At a higher CO$_2$ content in the feed, oxygen atoms generated from the dissociation of CO$_2$ can also react with CH$_4$, enhancing the CH$_4$ conversion. The CO$_2$/CH$_4$ molar ratio also significantly affects the yield of CO and H$_2$. Mei reported that the yield of H$_2$ and CO was more than doubled when increasing the CO$_2$/CH$_4$ molar ratio from 1:4 to 4:1 in the plasma DRM using a DBD (Figure 28.12) [93]. Zhang et al. found that increasing the CO$_2$/CH$_4$ molar ratio from 2:3 to 3:1 significantly increased the H$_2$ yield from 11.4 to 20.4% and the CO yield from 7.3 to 31.3% in a DBD reactor [64]. The CO$_2$/CH$_4$ molar

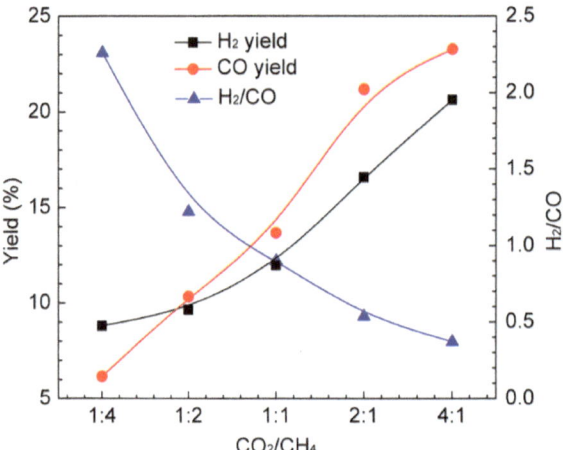

Figure 28.12: Effect of CO_2/CH_4 molar ratio on the yield of syngas and H_2/CO ratio in the plasma DRM using a DBD reactor (discharge power 50 W, total flow rate 50 ml/min). Reproduced from [93] with permission.

ratio plays a key role in determining the H_2/CO molar ratio in the produced syngas. Thus, syngas with a desired H_2/CO molar ratio for the further synthesis of chemicals or fuels can be controlled by tuning the CO_2/CH_4 molar ratio in the feed.

Higher CO_2 content in the CO_2/CH_4 mixture leads to higher CO selectivity. In addition to direct CO_2 dissociation to CO, more C_2–C_4 hydrocarbons generated by CH_4 dissociation could be oxidised by O atoms from CO_2 dissociation, resulting in the enhanced CO selectivity and decreased selectivity to C_2–C_4. For instance, the CO selectivity was increased from ca. 20% to over 80% when changing the CO_2/CH_4 molar ratio from 1:4 to 4:1 [93]. On the other hand, the lower content of CO_2 in the feed gas leads to a higher selectivity of C_2–C_4 hydrocarbons. Zhang et al. suggested that lower CO_2/CH_4 ratio decreased the availability of O radicals in the reaction, which enhanced the possibility of recombination of CH_x (x=1–3) species to form C_2–C_4 hydrocarbons compared with that of direct CH_4 oxidation to form CO [64]. This explanation is consistent with the decreasing trend in CO selectivity as a result of decreasing the CO_2 content in the feed gas. Wang et al. investigated the effect of CO_2/CH_4 molar ratio on the synthesis of oxygenates via DRM using a water-cooled DBD system. The selectivity of acetic acid and methanol increased initially and then decreased when changing the CO_2/CH_4 molar ratio from 3:1 to 1:2, with the highest selectivity achieved at a CO_2/CH_4 molar ratio of 1:1. By contrast, the selectivity of ethanol decreased continuously when decreasing the CO_2/CH_4 molar ratio [78]. Zhang et al. also reported that there exists an optimum CH_4/CO_2 molar ratio for the maximum selectivity of target oxygenates [72].

Other process parameters also affect the performance of the plasma DRM process. Khoja et al. evaluated the effect of discharge gap (1–4 mm) on the plasma DRM at a constant SIE in a DBD reactor. The highest conversion of CH_4 and CO_2 and H_2 selectivity was achieved at a discharge gap of 3 mm using quartz as a dielectric material [94]. In most of the previous works, a discharge gap between 1 and 5 mm was used. The most appropriate discharge gap may be 2–3 mm for adequate residence time and effective collision between electron molecules, as studied in many cases. Enlarging the discharge gap can increase the residence time of the reactants in the discharge zone, which can have a positive effect on the conversion. However, increasing the gap distance at a constant input power decreases the power density because of the increased discharge volume, which in turn negatively affects the conversion. The balance between these opposite effects determines whether the change of the gap distance has a positive or negative effect on the conversion [88]. A partial discharge is more likely to form at a larger discharge gap, resulting in reduced conversion of the reactants. Li et al. found that a wider discharge gap (1.8 mm) is more favourable for the formation of methanol and ethanol in the plasma reforming of CO_2 with CH_4 at a lower CO_2/CH_4 feed ratio, while a smaller discharge gap (1.1 mm) produced more acetic acid [76]. In the first reactor, the roughness of the inner electrodes was demonstrated to play an important role in the conversion and efficiency levels of methane [95]. Zhu et al. investigated the effect of pressure on plasma-based DRM using a kHz spark discharge plasma. Their results showed that increasing the pressure from 1 to 2 bars enhanced the conversion of CO_2 and CH_4 by 7–14.8% and reduced the energy costs by 7.7–15.2% for the conversion of the reactants [80].

Considerable effort has been devoted to further improving the performance of the dry reforming process to maximise the conversion of CO_2 and CH_4 while reducing the energy consumption of the plasma process through the development of new plasma reactor designs. Wu et al. designed a novel rotating GA co-driven by a magnetic field and tangential flow for the conversion of CO_2 with CH_4. A total conversion of 39% with an energy cost of 1 eV per molecule was achieved in this process [96]. Very recently, Cleiren et al. applied a novel GA plasmatron for the reforming of CH_4 with CO_2 with syngas being the major product. The CO_2 and CH_4 conversions reached their highest values of approximately 18 and 10%, respectively, at 25% CH_4 in the gas mixture, which corresponded to an energy efficiency of 66%. This value was above the required energy efficiency target (i.e. 60%) reported in literature to be competitive with thermal catalytic DRM processes [22]. Modification of a plasma reactor design has been carried out to manipulate the product distribution with enhanced selectivity of target products. Wang et al. proposed a multi-stage ionisation design to enhance the conversion of reactants and syngas production. It was found that the multi-stage ionisation process favoured a higher conversion of CO_2, but lowered the conversion of CH_4. Meanwhile, the selectivity to CO and H_2 was increased, while the selectivity to the by-products (C_2–C_6) was decreased [92]. Ozkan et al. developed a new geometry of a DBD reactor with multiple electrodes for the processing of high gas flow rates in DRM. In their work, the main products were syngas, C_2H_4 and C_2H_6 when Ar or He was used as the carrier gas [97]. Wang et al. developed a specially designed coaxial DBD reactor using water as both the ground electrode and cooling for the direct synthesis of a range of oxygenates, with acetic acid being the dominant liquid chemical via plasma DRM reaction [78].

28.3.2.2 Plasma catalysis

The combination of NTP with heterogeneous catalysis has been demonstrated as a promising solution to further enhance the conversion and energy efficiency of the plasma process, as well as the selectivity towards target products (e.g. syngas). Plasma-catalytic DRM has been carried out using different plasma systems, including DBD, corona discharge, glow discharge and GA plasma. Those catalysts that have successfully demonstrated their activities in thermal catalytic dry reforming are generally used as a starting point in the plasma-catalytic DRM reaction.

Catalysts can be coupled with NTP in different ways, which in turn affects the interaction between the plasma and catalyst and the formation of a plasma-catalytic synergy. Tu et al. found that a fully packed Ni/Al_2O_3 catalyst into the entire discharge region of a DBD reactor decreased the conversion of CH_4 and CO_2 in comparison to the plasma reaction without a catalyst [40]. Packing catalyst pellets into the discharge area was found to shift the discharge mode from typical microdischarges in the gas to a combination of surface discharge and weak microdischarges, which could partly contribute to the negative effect of the plasma-catalyst coupling. A similar negative effect from the integration of plasma and catalyst was also reported in previous studies [90, 98, 99]. These results suggest that the generation of plasma-catalytic synergy at low temperatures (without extra heating) in the DRM reaction is conditional and depends on the balance between the change in discharge properties induced by the catalyst and the catalyst activity generated by the plasma [40]. The former strongly depends on how the catalysts are packed into the discharge volume and the packing geometry significantly affects the interactions between the plasma and the catalyst. Tu and Whitehead compared the influence of three different catalyst packing methods on the plasma–catalyst interactions and the resulting plasma-catalytic synergy in the DRM reaction (Figure 28.13) [65]. They found that partially packing a $Ni/y\text{-}Al_2O_3$ catalyst in flake form into the bottom of the discharge gap significantly enhanced the reaction performance with a doubled CH_4 conversion and hydrogen yield in comparison to a fully packed reactor. This is because the discharge in the partially packed-bed reactor retains the strong filamentary discharge, whereas the reduction in discharge volume in the fully packed bed DBD reactor strongly suppresses the formation of microdischarges and changes the discharge mode to surface discharge and spatially limited microdischarges [65].

Ray et al. also investigated the effect of catalyst packing volume (0%, 25%, 50% and 100%) on the plasma dry reforming reaction in a DBD reactor [100]. They found that 25% catalyst packing showed the highest conversion of CH_4 and CO_2,

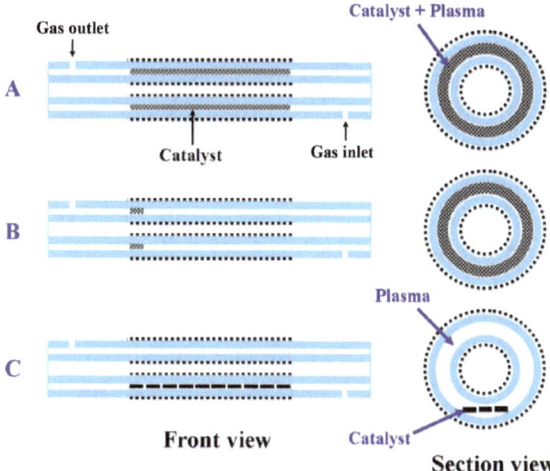

Figure 28.13: Different catalyst packing methods in a DBD plasma reactor. Reproduced from [65] with permission.

while fully (100%) packing the catalyst into the discharge zone decreased the conversion of CH$_4$ and CO$_2$ compared to the plasma DRM reaction without packing. Wang et al. studied the synergistic effect of catalyst and NTP on DRM in fluidised bed and packed-bed DBD reactors with a Ni/y-Al$_2$O$_3$ catalyst. They concluded that both interaction modes between the plasma and catalytic particles could promote the reaction at relatively low temperatures (e.g. 673 K) [101].

The use of zeolites has been shown to be effective for enhancing selectivities towards higher hydrocarbons, particularly C$_2$–C$_4$ species. Zeolites are known for their adsorbent properties, which are beneficial to plasma reactions, because they allow species to be adsorbed onto the zeolite surface or inside the pore structure, which can increase the residence time of the reactant species in the plasma discharge. This can lead to an increased probability of successful collisions with active plasma species. Eliasson et al. reported the direct formation of higher hydrocarbons with reduced carbon formation in the plasma dry reforming of CH$_4$ when zeolite NaX was used. They found that the presence of zeolite NaX in the discharge zone reduced the overall conversion but increased the concentration of C$_2$ to C$_4$ hydrocarbons in the products [69]. Zhang et al. compared the effects of zeolite X, zeolite HY and zeolite NaY on the plasma DRM reaction at ambient conditions. Zeolite NaY was found to be the most promising catalyst for the production of syngas and liquid hydrocarbons (C$_{5+}$) [89], while Zeolite HY showed the best performance in the generation of syngas and C$_4$ hydrocarbons (C$_4$H$_8$, n-C$_4$H$_{10}$ and i-C$_4$H$_{10}$) with high selectivity [102]. Jiang et al. reported that the coupling of a DBD with zeolite A inhibited the formation of carbon black and polymers and resulted in a higher selectivity towards valuable hydrocarbons (C$_2$–C$_4$) compared to the use of zeolite X in the DBD

[103]. Li et al. reported the formation of phenol in the plasma DRM combined with a HZSM-5 catalyst using a corona discharge [98].

Supported metal catalysts have been extensively used in thermal catalytic dry reforming, with transition metals being prevalent because of their low cost and wide availability. Catalysts with high activity in the thermal catalytic process have been used as the starting point for the plasma-catalytic DRM process. Ni/y-Al$_2$O$_3$ [64, 65, 68, 81, 40, 101, 104], Ag/Al$_2$O$_3$ [99], Pd/Al$_2$O$_3$ [99], Cu-Ni/Al$_2$O$_3$ [102], Cu/Al$_2$O$_3$ [82, 105], Co/y-Al$_2$O$_3$ [82], Mn/y-Al$_2$O$_3$ [82], Fe/Al$_2$O$_3$ [106], La$_2$O$_3$/y-Al$_2$O$_3$ [107] have been tested in the plasma DRM, with Ni/Al$_2$O$_3$ catalysts being the most commonly used. Song et al. reported that the presence of Ni/y-Al$_2$O$_3$ in a DBD reactor enhanced the CO selectivity by 22% but had a weak effect on the CO$_2$ conversion. In addition, they found that Ni loading (2–10 wt. %) had no effect on the conversion of CO$_2$ and CH$_4$, the selectivity of gas products and H$_2$/CO molar ratio [90]. Mahammadunnisa et al. also investigated the effect of Ni loading (10, 20 and 30 wt.%) on the plasma DRM over partially packed Ni/Al$_2$O$_3$ catalysts in a DBD reactor. The coupling of the DBD with 20 wt.% Ni/Al$_2$O$_3$ showed the highest reactant conversion and syngas selectivity with doubled hydrogen yield and H$_2$/CO molar ratio [108]. Zhu et al. reported that increasing the Ni loading of Ni/Al$_2$O$_3$ from 6 wt.% to 10 wt.% enhanced the conversion of CH$_4$ with maximum CH$_4$ conversion of 58.5% in the plasma DRM using a rotating GA plasma, which could be attributed to the increased catalytic effect because of the decreased Ni particle size and enhanced Ni dispersion on the catalyst surface at a higher Ni loading (10 wt.%) [84]. Tu and Whitehead evaluated the influence of calcination temperature (300–800 °C) of Ni/Al$_2$O$_3$ on the plasma-catalytic DRM at low temperatures (ca. 250 °C) in a DBD reactor. The results showed a synergistic effect from the combination of the DBD with partially packed Ni/Al$_2$O$_3$ catalyst calcined at 300 °C, which almost doubled the conversion of CH$_4$ (56%) and hydrogen yield (17.5%) compared to the plasma reaction without a catalyst [65]. Long et al. carried out the plasma DRM reaction using a cold plasma jet coupled with a 12 wt.% Ni/Al$_2$O$_3$ catalyst placed in the downstream of the plasma jet. Compared to the reaction using plasma only, the combination of the plasma jet and Ni/Al$_2$O$_3$ enhanced the conversion of CO$_2$ and CH$_4$ by 6–14% and the yield of hydrogen and CO by 11–18% at a discharge power of 700 W [109]. Zeng et al. carried out the plasma DRM reaction over different supported metal catalysts, that is, M/y-Al$_2$O$_3$ (M = Ni, Co, Cu and Mn), in a DBD reactor. They found that the combination of the plasma with Ni/y-Al$_2$O$_3$ or Mn/y-Al$_2$O$_3$ significantly enhanced the conversion of CH$_4$ in comparison to the reaction without catalyst. The presence of Ni/y-Al$_2$O$_3$ in the plasma showed the highest activity for syngas production [82]. However, the use of these catalysts did not improve the CO$_2$ conversion [82]. Sentek et al. found that the presence of a Pd/Al$_2$O$_3$ catalyst in a DBD reactor slightly decreased the conversion of CO$_2$ and CH$_4$ compared to the reaction without packing, but significantly changed the distribution of C$_2$–C$_4$ hydrocarbons with the formation of more C$_2$ and less C$_3$–C$_4$ [99].

The catalyst support is of primary importance, as the support, along with its interactions with the active metal, can affect the reaction performance. Mei et al. investigated the use of a Ni catalyst supported on y-Al$_2$O$_3$, TiO$_2$, MgO and SiO$_2$ in plasma-catalytic DRM [110]. The results of this experiment concluded that the y-Al$_2$O$_3$ support was most beneficial on the reaction performance, giving the highest CO$_2$ (26.2%) and CH$_4$ (44.1%) conversions, as well as the maximum energy efficiency and highest yields of CO and H$_2$ (Figure 28.14). This was attributed to the increased reducibility of the Ni/y-Al$_2$O$_3$ catalyst and the number of stronger basic sites present on its surface (which facilitate CO$_2$ chemisorption and activation), along with its higher specific surface area and greater dispersion of smaller NiO particles [110]. Carbon deposition also occurred to a lower extent on this catalyst, as the increase in CO$_2$ chemisorption and activation may have resulted in adsorbed CO$_2$ undergoing gasification by surface adsorbed oxygen [110]. Weaker interactions between the catalyst and support are favourable as this increases the reducibility of the catalyst, increasing its activity [65].

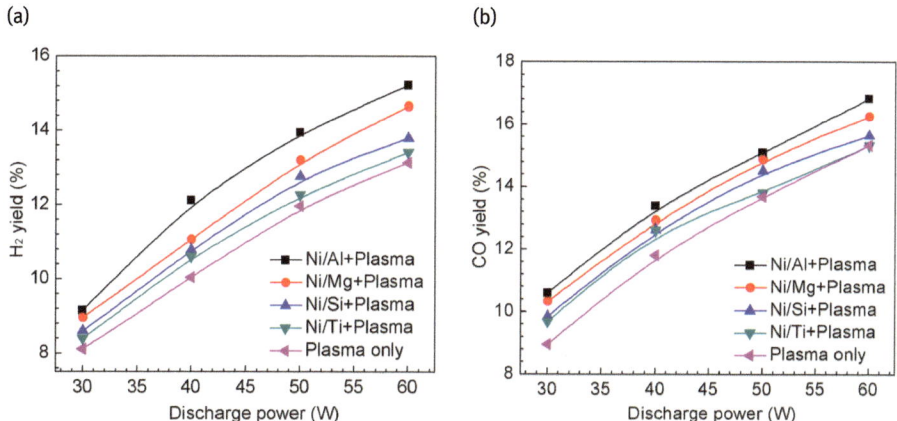

Figure 28.14: Effect of catalyst support on the yield of H$_2$ (a) and CO (b) as a function of discharge power in the plasma-catalytic DRM reaction (total flow rate 50 mL/min, CO$_2$/CH$_4$ molar ratio 1:1). Reproduced from [110] with permission.

The addition of dopants and use of bimetallic catalysts has also been studied in the plasma DRM reaction [111]. Zhang et al. investigated the effect of Cu/Ni ratio in Cu-Ni /y-Al$_2$O$_3$ catalysts and found that the 12 wt.%Cu–12 wt. % Ni/y-Al$_2$O$_3$ catalyst gave the optimum results for both CH$_4$ and CO$_2$ conversion and showed a synergistic effect of plasma-catalysis at 450 °C (Figure 28.15) [64]. This catalyst also achieved the maximum CO selectivity of 75%. However, this selectivity was also achieved when using the 5 wt.% Ni–12 wt.% Cu/y-Al$_2$O$_3$ catalyst, whereas the maximum selectivity to H$_2$ was achieved with 16 wt.% Ni–12 wt.% Cu/y-Al$_2$O$_3$ and 20 wt.% Ni–12 wt.% Cu/y-Al$_2$O$_3$ catalysts [64]. Ray et al. found that the addition of Mn to a Ni/Al$_2$O$_3$ catalyst

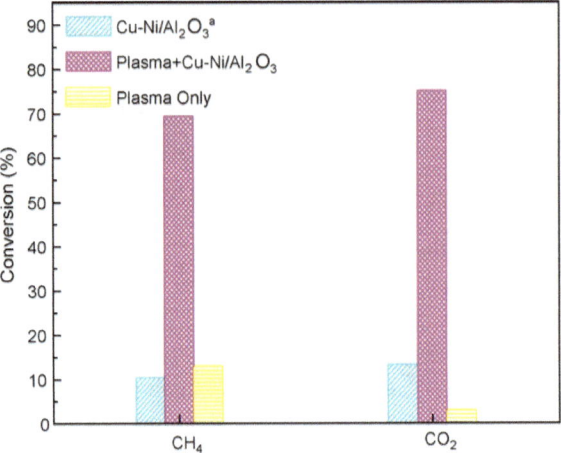

Figure 28.15: DRM using plasma only, Cu-Ni/Al₂O₃ catalyst only and plasma-catalysis at 450 °C (total flow rate 60 mL/min, argon flow rate 30 mL/min, CO₂/CH₄ molar ratio 1:1, discharge power 60 W, GHSV 1800 h⁻¹, ᵃ 60 mL/min, 50% Ar in the feed, CO₂/CH₄ molar ratio 1:1, GHSV 1800 h⁻¹, 0.1 g catalyst). Reproduced from [64] with permission.

enhanced the conversion of CO₂ and CH₄, and the yield of H₂ and CO in the plasma DRM [100]. In addition, the coupling of the DBD with the Ni-Mn/Al₂O₃ bimetallic catalyst reduced the carbon formation on the catalyst surface compared to Ni/Al₂O₃ [100]. Zhang et al. investigated the effect of La₂O₃/Al₂O₃ catalysts on the production of C₂ hydrocarbons in the plasma DRM using a pulsed corona discharge. They found that the La₂O₃/Al₂O₃ catalysts with different La loadings (5–12 wt. %) gave a C₂ hydrocarbon selectivity of more than 60% with C₂H₂ being the major C₂ product, and maintained the methane conversion of ca. 24%. Note the La₂O₃/Al₂O₃ catalysts with different La loadings showed no change in the distribution of C₂ products. Adding 0.01 wt.% Pd to 5 wt.% La₂O₃/Al₂O₃ still gave a high C2 selectivity of 70% but significantly changed the distribution of C₂ hydrocarbons with C₂H₄ being the major C₂ product (65 vol.%) [112]. Kado et al. reported that packing a Ni₀.₀₃Mg₀.₉₇O catalyst into a flow-type tubular pulsed discharge reactor significantly changed the selectivity of gas products compared to the plasma reaction without a catalyst: the selectivity of C₂ drastically decreased from 33.6% to 1%, and the CO selectivity increased from 65.4% to 99%, at a CO₂/CH₄ molar ratio of 1:1 [113]. More recently, K-, Mg- and Ce-promoted Ni/Al₂O₃ catalysts have also been evaluated in plasma-catalytic DRM at 160 °C [85]. The addition of promoters (K, Mg and Ce) into the Ni/Al₂O₃ catalyst enhanced the conversion of CH₄, the yield of H₂ and the energy efficiency of the plasma process. The highest conversion of CO₂ (22.8%) and CH₄ (31.6%) was achieved by placing the K-promoted Ni/Al₂O₃ catalyst in the plasma reforming process. In addition, compared to the unpromoted Ni/Al₂O₃ catalyst, although the use of the promoted catalysts increased the carbon deposition on the surface of the spent catalysts by 22–26%, the total amount of deposited carbon was still less than that reported in high temperature catalytic dry reforming processes. More than 80% of the increased carbonaceous species was in the form of reactive carbon species, which can be easily oxidised by CO$_2$ and O atoms and maintain the stability of the catalysts during the reforming reaction [85]. In this study, the behaviour of K, Mg and Ce promoters in the low temperature plasma-catalytic DRM reforming was opposite to that in high temperature thermal catalytic DRM process in terms of the conversion of CH$_4$ and carbon deposition, which could be ascribed to the temperature-dependent character of the promotors [85]. These results also suggest that those catalysts that have shown poor catalytic activity (e.g. conversion) in thermal catalytic reactions might work well in low temperature plasma-catalytic processes, and vice versa.

Core–shell structured catalysts have attracted significant interest in DRM as the metallic active sites could be uniformly distributed within the shells. The strong interaction between the cores and shells is ascertained to be highly capable of preventing metallic NPs from carbon deposition and sintering even at high temperatures. Zheng et al. reported that the combination of a DBD plasma with LaNiO$_3$–SiO$_2$ core–shell nanoparticle catalysts showed a better catalytic performance in plasma-based DRM with higher reactant conversion, product selectivity and catalytic stability, compared to the traditional Ni-based catalysts (Ni/SiO$_2$, LaNiO$_3$/SiO$_2$ and LaNiO$_3$) [114, 115]. The conversion of CH$_4$ and CO$_2$ reached 88.3 and 77.8%, and the selectivity of CO and H$_2$ was 92.4 and 83.7%, respectively. Their results suggest that the SiO$_2$ shell is capable of preventing Ni from sintering and mitigating carbon deposition in the plasma-catalytic reaction (Figure 28.16) [114]. Compared to the supported Ni-based catalysts (Ni/y-Al$_2$O$_3$, NiFe/y-Al$_2$O$_3$, NiFe/SiO$_2$ and NiFe$_2$O$_4$), the use of spinel nickel ferrite NPs (NiFe$_2$O$_4$ NPs) embedded in silica (NiFe$_2$O$_4$–SiO$_2$) also showed excellent catalytic performance and high resistance to carbon formation in the plasma dry reforming under ambient conditions without the involvement of extra heat. The results indicated that the special structure of the as-synthesised NiFe$_2$O$_4$–SiO$_2$ catalyst was capable of restraining the aggregation of NiFe alloy and suppressing the carbon formation in the plasma reforming process [83].

In addition, the catalytic effect of electrode materials on the plasma DRM reaction has also been investigated. Li et al. evaluated the influence of different electrode materials (Ti, Al, Fe and Cu) on the production of syngas and higher hydrocarbons in the plasma DRM using a DBD reactor. They found that the Ti electrode showed the highest conversion of CH$_4$ and CO$_2$, while the other electrode materials showed a similar performance [116]. Scapinello et al. reported that nickel and copper electrodes are more efficient than stainless steel in producing carboxylic acids, in particular formic acid in the plasma DRM using a DBD reactor [117]. However, no major catalytic effects of the metal surface on the conversion of reactants (CO$_2$ and CH$_4$) and the production of H$_2$ and CO were observed [117].

The energy efficiency is higher in GA discharges in comparison to other types of discharges, and catalysts can increase this even more [81]. Placing a NiO/Al$_2$O$_3$

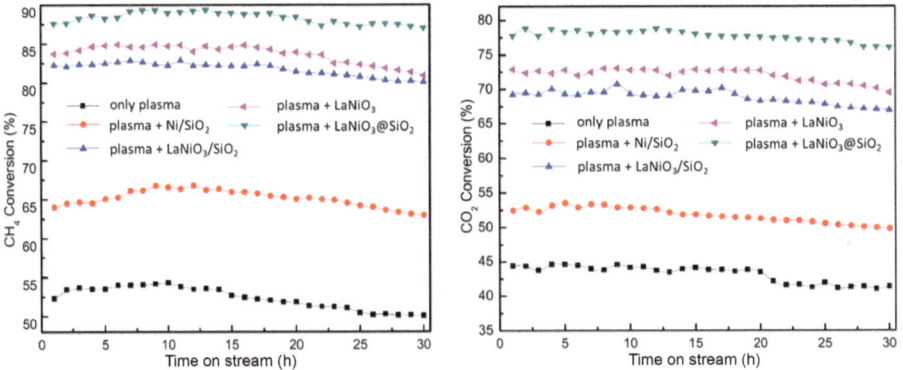

Figure 28.16: The conversion of CH_4 and CO_2 with time on stream over different Ni catalysts. Adopted from [114] with permission.

catalyst in the afterglow of the GA reactor was found to increase the energy efficiency by over 20% in comparison to that achieved using plasma only [81]. The H_2 yield, along with the CO_2 and CH_4 conversions, was also increased. The concentration of active metal was found to influence the reaction performance, as a 33 wt.% NiO/Al_2O_3 catalyst resulted in a decrease in reaction performance in comparison to an 18 wt. % NiO/Al_2O_3 catalyst, whilst a smaller catalyst diameter was found to be beneficial [81]. Goujard et al. investigated the influence of the type of plasma power supply on the plasma-catalytic synergy for DRM. Their experiments were performed in a DBD reactor packed with a cordierite honeycomb monolith and excited by two different power supplies: a pulsed excitation and a sinusoidal excitation. In the absence of a Ni catalyst, higher CO_2 and CH_4 conversion was achieved using the pulsed power supply. However, when using a 2 wt.% Ni catalyst in the plasma, the reactive species generated by the AC power supply promoted the activation of CO_2 and CH_4 on the Ni catalyst surface, leading to a significant increase of CH_4 and CO_2 conversion [118].

28.3.3 Plasma CO_2 hydrogenation

CO_2 hydrogenation for the synthesis of higher value fuels and chemicals has provided an attractive route for CO_2 conversion and utilisation, as this process has a lower thermodynamic limitation compared to direct CO_2 decomposition and DRM. One of the key challenges facing this process is the cost and source of hydrogen. In order for this process to be both economically viable and sustainable, hydrogen must be generated using a low cost, environmentally friendly and sustainable process, such as from water electrolysis using wind or solar power or from bioenergy. The overall process should be CO_2 neutral, which means that CO_2 hydrogenation must convert a greater amount of CO_2 than renewable hydrogen production pathways generate. Although CO_2 reduction with H_2 using heterogeneous catalysis has been extensively investigated in the past few years, there are still significant challenges in developing active, selective and stable catalysts suitable for large-scale commercialisation. In addition, it is key to lower the operating temperature of the CO_2 hydrogenation to minimise the energy consumption of the process.

28.3.3.1 Plasma conversion

The direct hydrogenation of CO_2 mainly produces three types of C1 chemicals: CO via reverse water gas shift reaction (RWGS, (28.7), CH_4 via CO_2 methanation (28.8), and CH_3OH via CO_2 hydrogenation (28.9). Up until now, very limited research has been concentrated on CO_2 hydrogenation using NTPs [119–123]. The majority of this research reports CO as the dominant chemical, with CH_4 formed as a minor product and no or trace CH_3OH detected [124–126].

$$CO_2(g) + H_2(g) \leftrightarrow CO\,(g) + H_2O\,(g) \qquad \Delta H° = +40.9\,\text{kJ mol}^{-1} \qquad (28.7)$$

$$CO_2(g) + 4H_2(g) \rightarrow CH_4(g) + 2H_2O(g) \qquad \Delta H° = -165.3\,\text{kJ mol}^{-1} \qquad (28.8)$$

$$CO_2(g) + 3H_2(g) \rightarrow CH_3OH(g) + H_2O(g) \qquad \Delta H° = -49.9\,\text{kJ mol}^{-1} \qquad (28.9)$$

The reverse water-gas shift reaction converts CO_2 and H_2 to produce CO and H_2O. CO is an important chemical feedstock for Fischer–Tropsch synthesis (FTS) to produce higher hydrocarbons such as liquefied petroleum gas (LPG), naphtha, gasoline and diesel; or for the synthesis of valorised chemicals such as acetic acid, phosgene and formic acid. Recently, Porosoff et al. has proposed the combination of the RWGS reaction with FTS for the synthesis of hydrocarbons [127].

Zeng and Tu investigated the influence of H_2/CO_2 ratio on the RWGS reaction using a DBD reactor [128]. They found that the conversion of CO_2 increased almost linearly with the increase of the H_2/CO_2 ratio from 1:1 to 4:1. Increasing the H_2/CO_2 ratio significantly enhanced the yield of CO, while the CO selectivity was only slightly increased [128]. The dependence of CO selectivity on the gas flow rate at a fixed H_2/CO_2 ratio of 4:1 was investigated using a low-pressure radio-frequency discharge [129]. The selectivity of CO increased gradually when increasing the total flow rate. This phenomenon is most likely due to the decreased residence time associated with the increase of the flow rate, resulting in suppression of the recombination of CO and O. The effect of argon on the plasma RWGS was evaluated in a DBD plasma reactor at 150 °C. In the absence of a catalyst, the CO_2 conversion increased from 18.3 to 38% at a discharge power of 30 W and a fixed H_2/CO_2 ratio of 4:1 when increasing the Ar content from 0 to 60% in the gas mixture [121]. The presence of metastable argon species in the DBD creates new reaction pathways for the dissociation of CO_2, resulting in enhanced CO_2 conversion.

In the CO_2 methanation reaction, CO_2 reacts with hydrogen to produce methane and water. This reaction was first discovered by Sabatier and Senderens in 1902. The CO produced during methanation has been recognised as an important intermediate in the CO_2 methanation pathways (28.10).

$$CO\,(g) + 3H_2(g) \rightarrow CH_4(g) + H_2O\,(g) \quad \Delta H^\circ = -249.8\,kJ\,mol^{-1} \quad (28.10)$$

However, limited efforts have been devoted to the use of NTPs for CO_2 methanation, especially in the plasma reaction without catalyst. CO is the major product with CH_4 being the minor one in CO_2 methanation. Zeng and Tu showed that the selectivity of CH_4 (2–5%) was significantly lower than that of CO (>90%) in the plasma processing of CO_2 with H_2 at low temperature (150 °C) [121]. In the plasma CO_2 methanation process, a higher H_2/CO_2 ratio is desirable as this increases the conversion of CO_2 and the selectivity of CH_4, which has been demonstrated both experimentally [128, 130] and through the use of a 1D fluid model [131]. Optimising the total flow rate can also maximise the CH_4 selectivity and CO_2 conversion. A very low total flow rate can lead to reverse reactions occurring, reforming CO from CH_4 according to (28.11), because of the longer residence time, increasing the interactions between the CO_2 hydrogenation products and the reactive species in the plasma [129].

$$CH_4(g) + H_2O\,(g) \rightarrow CO(g) + 3H_2(g) \quad \Delta H^\circ = 206\,kJ\,mol^{-1} \quad (28.11)$$

Zeng and Tu reported that adding argon up to 60% in the CO_2/H_2 mixture significantly enhanced the CH_4 selectivity by 85%, which suggests that the presence of metastable argon species in the discharge creates new reaction routes for the production of methane [121]. For DBD plasmas, the use of alumina as a dielectric material instead of quartz is beneficial on CO_2 methanation, which might be attributed to the relatively higher dielectric constant of alumina [130]. Addition of a magnetic field to a plasma system enhanced the CO_2 conversion and CH_4 selectivity by over 10% at a discharge power of 30 W, whilst the energy efficiency of the process was tripled [124]. This process however employed low pressure (200 Pa), reducing simplicity of design and requiring extra energy input. Increasing input power generally results in a higher selectivity to CH_4 because of the increased power density [124, 130]. However, it has been found that at high power input (>160 W), energy is transferred to the electrodes through heating rather than being used for plasma chemical reactions, resulting in the saturation of CH_4 selectivity [124]. A smaller discharge gap is beneficial on the CO_2 conversion and CH_4 selectivity because of the rise in input power density and enhanced electric field [124]. In fact, a smaller discharge gap can achieve the same CH_4 selectivity at a lower input power than when using a larger discharge gap [124]. A reduction in discharge gap can also increase the production efficiency of the plasma process.

CO$_2$ hydrogenation to liquid fuels (e.g. methanol and ethanol) is one of the most attractive routes for CO$_2$ conversion and utilisation. CH$_3$OH is a valuable fuel substitute and additive, and is also a key feedstock for the synthesis of other higher value chemicals. In addition, methanol is considered a promising hydrogen carrier, suitable for storage and transportation [132]. In the late 1990s, Eliasson and cow-orkers investigated CO$_2$ hydrogenation to CH$_3$OH using a DBD plasma reactor at pressures up to 10 bar [133]. However, the major products were CO and H$_2$O with a CO selectivity of over 90%. Only trace amounts of CH$_3$OH were produced in the plasma CO$_2$ hydrogenation without a catalyst, with a maximum CH$_3$OH yield of 0.06% (selectivity 0.4–0.5%) obtained at 8 bars, a relatively high plasma power of 400 W, a total flow rate of 250 mL/min and a H$_2$/CO$_2$ ratio of 3:1 [133]. Increasing the wall temperature from 100 to 220 °C had a limited effect on the selectivity and yield of methanol [133]. The formation of trace amounts of CH$_3$OH in the plasma CO$_2$ hydrogenation was also reported using a radio frequency (RF) impulse discharge at low pressures (1–10 Torr) [129]. Very recently, Wang et al. reported that the methanol production via plasma-assisted CO$_2$ hydrogenation was strongly dependent on the structure of the DBD plasma reactor (Figure 28.17). The proposed single dielectric DBD reactor with a special design using water as a ground electrode and cooling significantly enhanced the production of methanol with the highest methanol selectivity of 54% achieved in the plasma hydrogenation of CO$_2$ without a catalyst at atmospheric pressure (1 bar) and room temperature (30 °C). The concentration and yield of CH$_3$OH, as well as the conversion of CO$_2$ were affected by the H$_2$/CO$_2$ molar ratio. Increasing the H$_2$/CO$_2$ molar ratio from 1:1 to 3:1 increased the yield of CH$_3$OH from 6.0 to 7.2%, while the selectivity of CO decreased from 40.0 to 30.0%[123].

28.3.3.2 Plasma catalysis

Catalysts are the key to manipulate the selectivity of target products in the plasma hydrogenation of CO$_2$. Zeng and Tu investigated the influence of Al$_2$O$_3$ supported metal catalysts (Mn, Cu and Cu–Mn) on the plasma RWGS reaction in a DBD reactor at atmospheric pressure (1 bar). Compared to the reaction using plasma alone, the presence of these catalysts in the DBD enhanced the conversion of CO$_2$ by up to 36%, with the Mn/y-Al$_2$O$_3$ catalyst showing the best activity for CO$_2$ conversion at a H$_2$/CO$_2$ molar ratio of 1:1. The coupling of the DBD with Mn/y-Al$_2$O$_3$ also significantly enhanced the yield of CO by 114%, followed by Cu-Mn/y-Al$_2$O$_3$ (91%) and Cu/y-Al$_2$O$_3$ (71%). As a result, the energy efficiency for CO production was significantly enhanced by up to 116%[128]. However, only the Cu/y-Al$_2$O$_3$ catalyst was found to enhance the selectivity of CH$_4$ compared to the reaction using plasma alone [128]. The weaker activity of Cu/y-Al$_2$O$_3$ in CO$_2$ conversion in comparison to the Mn/y-Al$_2$O$_3$ catalyst might be attributed to the increased prevalence of the water

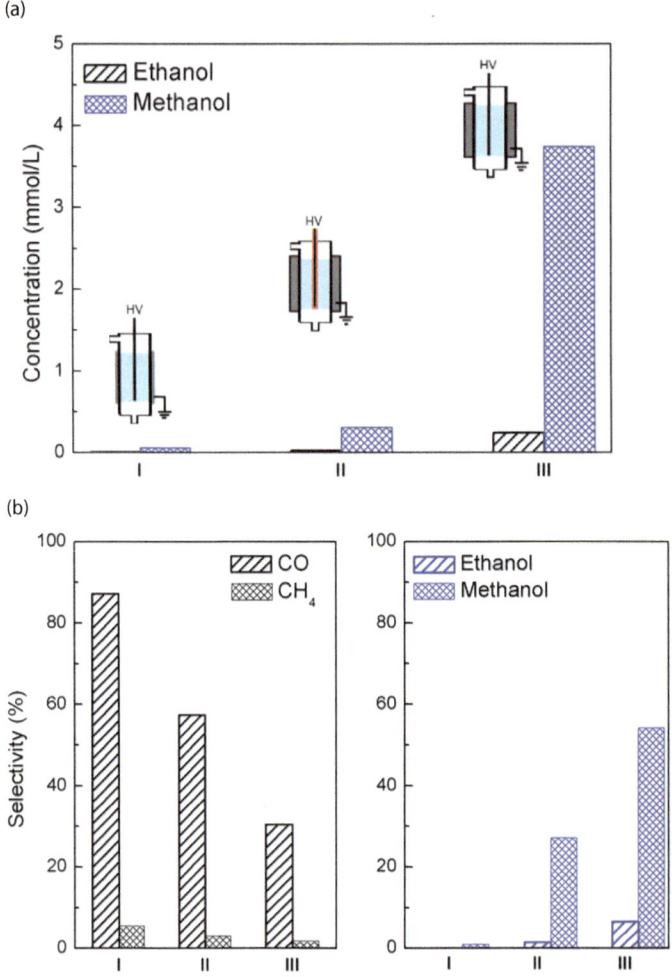

Figure 28.17: Effect of DBD reactor structure (Reactor I, II and III) on plasma CO_2 hydrogenation to oxygenates without a catalyst at a fixed discharge power of 10 W and a H_2/CO_2 molar ratio 3:1 (a) concentration of oxygenates; (b) selectivity of gas products and oxygenates. Reproduced from [123] with permission.

gas shift (WGS) reaction in the presence of Cu/y-Al_2O_3 as Cu catalysts are often used for catalysing the WGS reaction [128]. It is therefore important to select a catalyst that will supress the WGS reaction and simultaneously increase the CO_2 conversion and the selectivity to CH_4.

The combination of plasma and suitable catalysts enables the CO_2 methanation reaction to occur at much lower temperatures than those required in the thermal catalytic process [125]. Nizio et al. evaluated the activity of ceria and zirconia-promoted Ni-containing hydrotalcite-derived catalysts in the plasma-catalytic hydrogenation of CO$_2$ to methane using a low temperature DBD reactor. Below 250 °C, negligible CO$_2$ conversion occurs for the catalytic process using a Ce–Zr supported Ni catalyst. However, when combined with a DBD plasma the CO$_2$ conversion reached 80%, with 90% selectivity to CH$_4$ even at 110 °C [125]. This is because of the creation of excited species in the plasma, which generate new pathways for CO$_2$ dissociation; hence, the reaction is not limited by the rate of CO$_2$ dissociation at the catalyst surface as it is in the thermal catalytic process [126]. The use of nickel containing hydrotalcite catalysts also showed promising results in the plasma-catalytic CO$_2$ methanation reaction, with a CO$_2$ conversion of 80% and a CH$_4$ selectivity of nearly 100% achieved [134].

Eliasson et al. investigated the plasma-catalytic hydrogenation of CO$_2$ to methanol over a Cu/ZnO/Al$_2$O$_3$ catalyst in a DBD reactor at a relatively high pressure (up to 10 bar). Compared to the plasma reaction without catalyst (see previous section), the presence of the Cu/ZnO/Al$_2$O$_3$ catalyst in the DBD increased the methanol yield (from 0.1 to 1.0%), methanol selectivity (from 0.4 to 10.0%), and CO$_2$ conversion (from 12.4% to 14.0%) at 8 bars, 100 °C and a H$_2$/CO ratio of 3:1. However, the methanol yield and selectivity were still significantly lower than those reported in thermal catalytic CO$_2$ hydrogenation processes [133]. Wang et al. has successfully demonstrated the synthesis of methanol with a high selectivity via plasma-catalytic CO$_2$ hydrogenation using a water-cooled DBD reactor at room temperature and atmospheric pressure (Figure 28.18). Packing Cu/y-Al$_2$O$_3$ or Pt/y-Al$_2$O$_3$ into the DBD significantly enhanced the CO$_2$ conversion and methanol yield compared to the plasma hydrogenation of CO$_2$ without a catalyst (Figure 28.19). The maximum methanol yield of 11.3% and methanol selectivity of 53.7% were achieved using the Cu/y-Al$_2$O$_3$ catalyst with a CO$_2$ conversion of 21.2% in the

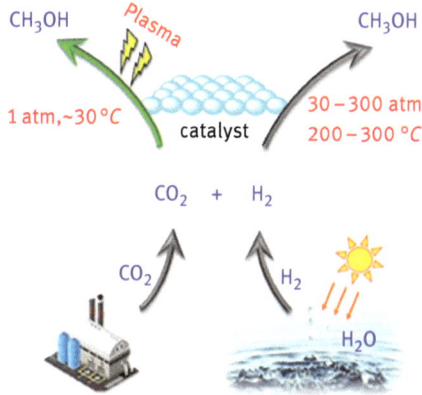

Figure 28.18: Scheme of CO$_2$ hydrogenation to methanol. Reproduced from [123] with permission.

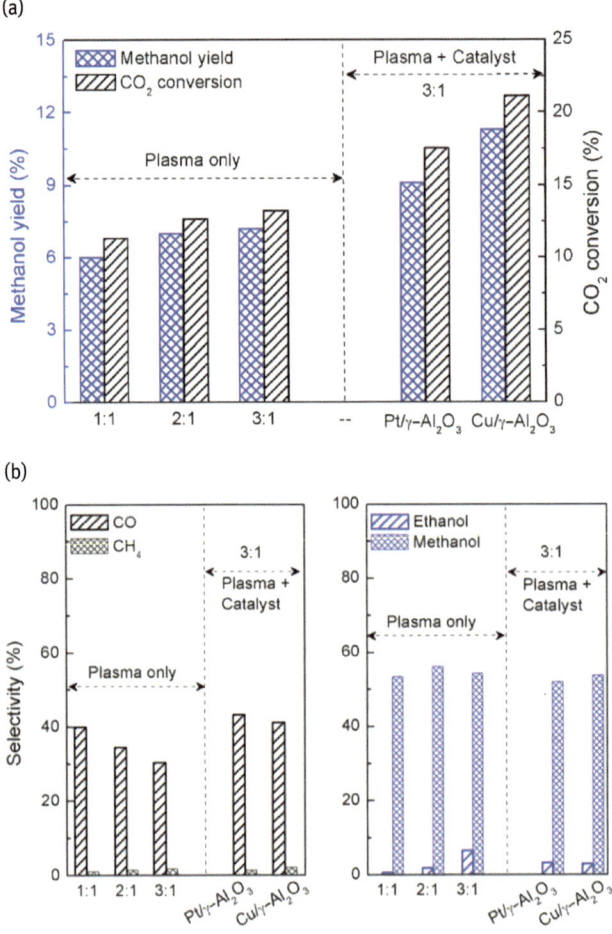

Figure 28.19: Effect of H_2/CO_2 ratio and catalysts on the plasma-catalytic CO_2 hydrogenation to oxygenates at a discharge power of 10 W. (a) methanol yield and CO_2 conversion; (b) selectivity of gas products and oxygenates. Reproduced from [123] with permission.

plasma-catalytic process, while no reaction occurred at ambient conditions without using plasma [123].

In addition, the production of dimethyl ether (DME) from plasma CO_2 hydrogenation was reported using an atmospheric pressure surface discharge, with a CO_2 conversion of 15% and a H_2/CO_2 molar ratio of 1:1 [122]. Compared to thermal catalytic CO_2 hydrogenation to value-added fuels and chemicals, which has been carried out using a wide range of catalysts for a range of target products, very limited catalysts that are active for the thermal catalytic process have been examined in plasma hydrogenation of CO_2 at low temperatures.

28.3.4 CO_2 with water

28.3.4.1 Plasma conversion

Compared to the large amount of work performed for plasma-based DRM, only limited research has been performed for the simultaneous conversion of CO_2 and H_2O, that is, so-called artificial photosynthesis.

$$CO_2(g) + H_2O(g) \leftrightarrow CO(g) + H_2(g) + O_2(g) \quad \Delta H° = 525\, kJ\, mol^{-1} \tag{28.12}$$

$$CO_2(g) + 2H_2O(g) \leftrightarrow CH_3OH(g) + 3/2\,O_2(g) \quad \Delta H° = 676\, kJ\, mol^{-1} \tag{28.13}$$

Futamura et al. [135] investigated a CO_2/H_2O mixture diluted to 0.5–2.5% in N_2 in a DBD reactor, and reported a CO_2 conversion of only 0.5%, with product yields of 0.7% for H_2, 0.5% for CO and no O_2, and also no mention of oxygenated products. Mahammadunnisa et al. [136] obtained a CO_2 conversion of 12–25% in a DBD reactor, depending on the SEI, with a selectivity of 18–14% for H_2 and 97–99% for CO, yielding a syngas ratio of 0.55–0.18.

   Snoeckx et al. [137] performed a combined experimental and computational study for CO_2/H_2O conversion in a DBD. Adding a few % of H_2O to the CO_2 plasma was found to cause a steep drop in the CO_2 conversion, and both the CO_2 and H_2O conversion were quite low. CO, H_2, O_2 and H_2O_2 (up to 2%) were the major products and no oxygenates were detected. The experimental data could be explained by a chemical kinetics model (see Figure 28.20). The main reactive species created were OH, CO, O and H, and the model reveals that the OH radicals quickly recombine with CO into CO_2, which explains the limited CO_2 conversion upon H_2O addition. In addition, the O and H atoms recombine in a few subsequent reactions to form H_2O again, explaining why also the H_2O conversion was limited. Finally, the fast reaction between O/OH and H atoms explains why no oxygenated products were formed.

   Ihara et al. [138] investigated a 1:1 CO_2/H_2O mixture in a MW plasma, and detected low yields of oxalic acid and H_2O_2. They also assumed that H_2 and O_2 are generated, but these products were not measured. In their follow-up study [139] they varied the CO_2/H_2O gas mixing ratio from 1:4 to 1:1, and detected methanol instead of H_2O_2 and oxalic acid, albeit again in very low concentrations < 0.01%. However, a rise in the methanol yield by a factor 3.5 was observed upon increasing the pressure from 240 to 400 Pa. They suggested two pathways for methanol formation, that is, (i) direct formation from CO_2 and H_2O in the plasma, and (ii) the reformation of deposited polymeric material on the walls during the plasma reaction with H_2O. Chen et al. [140] detected syngas (in a ratio close to 1, for a 1:1 CO_2/H_2O ratio) and O_2 in a surface-wave MW plasma, but no hydrocarbons or oxygenates. In a follow-up study [141], they reported that adding 10% H_2O to a CO_2 MW plasma yielded a higher CO_2 conversion (i.e. 31% vs. 23%) and a lower energy cost (i.e. 22.4 eV/molec vs. 30.2 eV/molec),

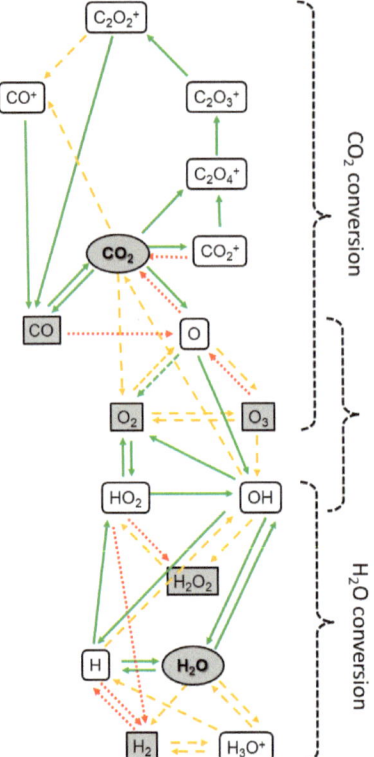

Figure 28.20: Reaction scheme to illustrate the main pathways for CO_2 and H_2O conversion and their interactions. The arrow lines represent the formation rates of the species, with full green lines being formation rates over 10^{17} cm^{-3}·s^{-1}, orange dashed lines between 10^{17} and 10^{16} cm^{-3}·s^{-1} and red dotted lines between 10^{16} and 10^{15} cm^{-3}·s^{-1}. Adopted from [137] with permission.

along with a lower gas temperature, which was attributed to the higher heat capacity of water and the induced endothermic reactions. This lower gas temperature can explain the higher conversion and lower energy cost, because of less VT relaxation (causing vibrational depopulation; see section 28.2.2 above) and a lower reaction rate of the recombination of CO back into CO_2.

Indarto et al. [142] reported that H_2O addition (in the range of 5 to 31%) to a CO_2 plasma in a classical GA yielded a drop in CO_2 conversion (around 7.1–3.0%, compared to 13.4% in pure CO_2) and a higher energy cost (around 89–189 eV/molec, compared to 53 eV/molec in pure CO_2), which they attributed to instabilities in the plasma upon adding H_2O. Nunnally et al. [58] also found a higher energy cost when adding 1% H_2O to a GAP, that is, 14.8 eV/molec vesus 9.5 eV/molec in pure CO_2, but they did not observe arc instabilities. Instead, they attributed the higher energy cost to VT relaxation, causing depopulation of the CO_2 vibrational levels, which is much faster for collisions with H_2O than with CO_2 [2, 58].

Hayashi et al. [122] compared the conversion and product formation for a 1:1 mixture of CO_2/H_2O and CO_2/H_2 in a surface discharge, and reported much lower values in the CO_2/H_2O mixture than in the CO_2/H_2 mixture (i.e. 5% vs. 15%), but the same products were detected, that is, CO, CH$_4$, DME and formic acid. Finally, Guo et al. [143] studied the combined CO_2/H_2O conversion in a negative DC corona discharge, for H_2O contents between 10 and 43% and pressures between 1 and 4 bar. Again, a drop in CO_2 conversion upon increasing H_2O content was observed. The main products formed were H_2 and CO, as well as ethanol and methanol, in roughly a 3:1 ratio, with a total molar yield up to 4.7%, increasing with pressure.

In general, it is clear from the above literature overview that adding even small amounts of H_2O (1–2%) yields a significant drop in CO_2 conversion, followed by a further drop upon addition of even more H_2O. For a DBD, this can be explained by the chemical pathways presented in [137] (see Figure 28.20 above). For a GA, the fast quenching of the CO_2 vibrational levels by VT relaxation with H_2O molecules is the most probable explanation [2, 58]. This would be expected in a MW plasma as well, but Chen et al. [141] reported a higher CO_2 conversion upon H_2O addition. However, this MW setup operates at low pressures (30–60 Torr), where VT relaxation and thus quenching is less important, and cooling upon H_2O addition might be the dominant effect, which reduces VT relaxation, and thus enhances the CO_2 conversion and corresponding energy efficiency.

The main products formed by the combined CO_2/H_2O conversion are H_2 and CO, like for DRM (see Section 28.3.2), as well as O_2, but some papers also report the production of hydrogen peroxide (H_2O_2) [137, 138], oxalic acid ($C_2H_2O_4$) [138], formic acid (CH_2O_2) [122], methane (CH_4) [122, 136], dimethyl ether (C_2H_6O, DME) [122], methanol (CH_3OH) [136, 139, 143], ethanol (C_2H_5OH) al. [143], acetylene (C_2H_2) [136], propadiene (C_3H_4) [136] and even carbon nanofibres (CNFs) [136]. However, most of these data are only qualitative and mainly incomplete, so we cannot deduce a general trend on product yields or selectivities. Nevertheless, the formation of oxygenates, and other value-added compounds, in a one-step process seems limited, in the absence of a catalyst. The reason is that too many steps are involved in creating these oxygenates, and all of them involve H atoms, which will rather recombine quickly with OH into H_2O or with O_2 into HO_2, which also reacts further with O into OH – and hence H_2O. In other words, the interactions of H atoms with oxygen species (i.e. OH, O_3, O_2 or O atoms) are too fast and their tendency to form H_2O is too strong.

On the other hand, the H_2/CO ratio produced in CO_2/H_2O plasmas can be very high (even up to 8.5, for sufficient amounts of H_2O addition), and they can be easily controlled, as revealed by computer modelling [137], which might be useful for the production of value-added chemicals in a two-step process.

28.3.4.2 Plasma catalysis

It is clear from above that the direct production of value-added compounds in CO_2/H_2O plasmas requires suitable catalysts. Futamura et al. [135] investigated the potential of a ferroelectric packed bed reactor with $BaTiO_3$ pellets for the diluted CO_2/H_2O mixture mentioned above. They obtained a CO_2 conversion of 12.3%, with product yields of 12.4% for H_2, 11.8% for CO and 2.8% for O_2, hence much higher than in the non-packed DBD reactor (cf. above), but nothing was mentioned on the formation of oxygenates. Likewise, adding a Ni/y-Al_2O_3 catalyst, in both unreduced and partially reduced form, Mahammadunnisa et al. [136] obtained a higher conversion and syngas ratio than without catalyst (see above), that is, 18–28% and 0.95–0.45 for the unreduced catalyst (NiO/y-Al_2O_3), and 24–36% and 0.66–0.35 for the partially reduced catalyst (Ni/y-Al_2O_3). The NiO catalyst was found to yield a reduction of the produced CO to CH_4, CH_3OH, C_2H_2, propadiene, while the Ni catalyst also gave rise to carbon nanofibres.

Chen et al. [141] studied the effect of NiO/TiO_2 catalysts (treated with an Ar plasma) in a MW plasma. Compared to their results without catalyst (see above), the CO_2 conversion was further enhanced to 48%, with an energy cost of 14.5 eV/molec, but no oxygenated products were detected. The authors suggested that CO_2 is adsorbed at oxygen vacancies on the catalyst surface, reducing the threshold for dissociative electron attachment into CO, adsorbed O atoms at the vacancies and electrons. The adsorbed O atoms may subsequently recombine with gas phase O atoms or OH radicals into O_2 (and H atoms). Hence, the catalyst seems to have a beneficial effect in tuning O/OH from the plasma into O_2, by means of adsorbed O atoms at the vacancies, before they recombine again into CO_2 and H_2O, which seems the limiting step in CO_2/H_2O conversion, at least in a DBD, as revealed by computer modelling [137] (see above).

Although the above papers use catalysts, and report beneficial effects, the production of oxygenates or other value-added compounds is very limited, if reported at all. Hence, clearly more research is needed for tailored catalyst design. These catalysts should allow the plasma-generated CO and H_2 to selectively react into oxygenates, such as methanol, and subsequently they should separate the methanol from the mixture. As mentioned in previous section, Eliasson et al. [133] applied a CuO/ZnO/Al_2O_3 catalyst in a CO_2/H_2 DBD, which gave a ten times higher methanol yield and selectivity. Other possible candidates could be Ni–zeolite catalysts, for which methanation is reported [126], a Rh_{10}/Se catalyst, yielding an ethanol selectivity up to 83% [144], and a Ni–Ga catalyst for the conversion into methanol [145].

28.4 Summary and steps to be taken for further improvement

It is clear from the above sections that plasma-based CO_2 conversion is promising, but more research is needed before it can be implemented in industry. Although plasma creates a very reactive environment, and is thus chemically not selective in producing value-added compounds, plasma is selective in another way, in the sense that it can selectively populate the vibrational levels of CO_2, without activating the other degrees of freedom, that is, without the need to heat the gas. This selectivity induces thermal non-equilibrium, and explains the good energy efficiency compared to thermal conversion, at least for some plasma types, like MW and GA plasmas (and maybe APGDs).

To compete with classical and other emerging technologies, Snoeckx and Bogaerts [12] stated that an energy efficiency of 60% (or an energy cost below 4.27 eV/molec for DRM) would be required (see also Figures 28.6 and 28.9 in sections 28.3.1). MW and GA plasmas already reach energy efficiencies close to, or above, this defined efficiency target. This shows their great potential, attributed indeed to the important role of vibrational excitation for energy-efficient CO$_2$ dissociation. This is especially true for MW plasmas at reduced pressure for CO$_2$ conversion, which clearly exhibit a thermal non-equilibrium between the vibrational and gas temperature. However, for DRM, very limited results have been reported in MW plasmas, while many promising results are published for GA discharges, and also ns-pulsed discharges, spark discharges and APGDs show potential.

DBDs are most frequently studied, both for CO$_2$ conversion and DRM, but we believe that even with further improvements the energy efficiency will remain too low for industrial implementation. Indeed, they have typical energy efficiencies of 5–10%, with some exceptions up to 20%, so significant improvements, by a factor 3–5, would still be needed to reach the defined efficiency target of 60%, in order to make them competitive with other emerging technologies.

It should be realised, however, that the efficiency target of 60% was defined for the production of syngas by DRM, which is indeed the major product in plasma-based DRM. Nevertheless, also other higher value products are formed in the plasma, although not selectively, but when suitable catalysts can be found, plasma catalysis can produce these value-added compounds, such as higher hydrocarbons or oxygenates, in a one-step process at low temperatures and ambient pressure. This would significantly reduce the energy efficiency target to be competitive with other technologies, if the latter would need a two-step process, because indeed, the subsequent Fischer–Tropsch or methanol synthesis from syngas is quite energy intensive.

In this sense, even DBD could become suitable, especially because of their simple design, allowing easy upscaling and straightforward implementation of catalysts. However, much more research is needed to design catalysts, tailored to the plasma environment, to directly produce such value-added chemicals with high selectivity. The latter does not only apply to DRM, but certainly also to CO$_2$ hydrogenation (CO$_2$/H$_2$) and artificial photosynthesis (CO$_2$/H$_2$O mixtures), where the formation of value-added compounds without catalyst seems even less straightforward.

To improve the capabilities of plasma-based CO$_2$ conversion, first of all the energy efficiency should be further improved. We believe this should be realised by a low enough reduced electric field (order of 5–100 Td), in combination with high enough plasma power for sufficient vibrational excitation, which is the most energy-efficient CO$_2$ dissociation pathway, and with a low gas temperature, to minimise vibrational losses upon collision with other gas molecules (so-called VT relaxation), that is, strong thermal non-equilibrium. MW and GA plasmas already make use of this most energy-efficient dissociation pathway, but at atmospheric pressure, the VT relaxation is quite significant, and hence, the VDF is too close to thermal, so for further improvement, the non-equilibrium should be further exploited.

In addition, the conversion should be further improved, along with the product yield/selectivity. Indeed, the major disadvantage of plasma-based CO_2 conversion is in our opinion the need for a post-reaction separation step, as the gas conversion is typically (far) below 100%, and many different reaction products can be formed. In case of CO_2 splitting, the separation of CO and O_2 was calculated to yield the largest energy cost [146]. When a H-source is added (either CH_4, H_2 or H_2O), mainly syngas is formed, but some minor side products are observed as well. The syngas mixture does not really pose a problem, when it is subsequently used for Fischer–Tropsch or methanol synthesis. Moreover, plasma technology can deliver a wide variety of syngas ratios, depending on the initial feed gas mixing ratio. However, when higher hydrocarbons or oxygenates could be directly produced with sufficient selectivity and yields, the post-reaction separation step would not be so critical, as these (liquid) compounds can more easily be separated. This brings us back, however, to the crucial need for tailored catalysts, specifically designed to the plasma environment.

In spite of the further improvement of energy efficiency or selectivity being required, we believe that plasma-based CO_2 conversion is very promising, especially because of its overall flexibility, in terms of (i) feed gas (i.e. pure CO_2 splitting, but also mixtures with any H-source are possible), (ii) energy source (solar, wind, hydro, wave and tidal power, as well as nuclear power) and (iii) fast on/off switching and modular upscaling capabilities. This flexibility makes plasma very useful as a so-called "turnkey" process, which might be able to use renewable electricity in a flexible way, following its intermittency, and convert it into fuels or chemicals. An example of how a (MW) plasma-based process for pure CO production could be implemented at industrial scale has been evaluated in terms of its economics [146]. It indicates that CO production cost price translates into 0.22 $/kWh stored in CO at an electricity price of 0.05 $/kWh based on present day technology and performance.

References

[1] Bogaerts, A., et al., *Gas discharge plasmas and their applications*. Spectrochimica Acta. Part B: Atomic Spectroscopy, 2002. 57(4): 609–658.
[2] Fridman, A., *Plasma chemistry*. 2008: Cambridge university press.
[3] Itikawa, Y., *Cross sections for electron collisions with carbon dioxide*. Journal of Physical and Chemical Reference Data, 2002. 31(3): 749–767.
[4] Semiokhin, I., Y.P. Andreev, and G. Panchenkov, *Dissociation of Carbon Dioxide Circulating in a Silent Electric Discharge*. Russian Journal of Physical Chemistry, 1964. 38(8): 1126.

[5]   Azizov, R., et al. *The nonequilibrium plasmachemical process of decomposition of CO2 in a supersonic SHF discharge*. in *Soviet Physics Doklady*. 1983.

[6]   Maltsev, A., E. Eremin, and V. Ivanter, *Dissociation of carbon dioxide in glow discharge*. 1967, Interperiodica PO Box 1831, Birmingham, Al 35201-1831. 633--.

[7]   Andreev, Y., et al., *Dissociation of carbon-dioxide in a pulse discharge*. Russian Journal of Physical Chemistry, USSR, 1971. 45(11): 1587-&.

[8]   Rusanov, V., A. Fridman, and G. Sholin, *The physics of a chemically active plasma with nonequilibrium vibrational excitation of molecules*. Physics-Uspekhi, 1981. 24(6): 447–474.

[9]   Asisov, R.F., AA; Givotov, VK; Krasheninnikov, EG; Petrushev, BI; Potapkin, BV; Rusanov, VD; Krotov, MF and Kurchatov, IV, *Carbon dioxide dissociation in non-equilibrium plasma*, in *5th Int. Symp. on Plasma Chemistry* 1981: Edinburgh, Scotland, 774.

[10] De Bie, C., J. van Dijk, and A. Bogaerts, *The Dominant Pathways for the Conversion of Methane into Oxygenates and Syngas in an Atmospheric Pressure Dielectric Barrier Discharge*. Journal of Physical Chemistry C, 2015. 119(39): 22331–22350.

[11] Kogelschatz, U., *Dielectric-barrier discharges: their history, discharge physics, and industrial applications*. Plasma Chemistry and Plasma Processing, 2003. 23(1): 1–46.

[12] Snoeckx, R. and A. Bogaerts, *Plasma technology–a novel solution for CO2 conversion?* Chemical Society Reviews, 2017. 46(19): 5805–5863.

[13] Van Laer, K. and A. Bogaerts, *Fluid modelling of a packed bed dielectric barrier discharge plasma reactor*. Plasma Sources Science and Technology, 2015. 25(1): 015002.

[14] Michielsen, I., et al., *CO2 dissociation in a packed bed DBD reactor: First steps towards a better understanding of plasma catalysis*. Chemical Engineering Journal, 2017. 326: 477–488.

[15] Uytdenhouwen, Y., et al., *A packed-bed DBD micro plasma reactor for CO2 dissociation: Does size matter?* Chemical Engineering Journal, 2018. 348: 557–568.

[16] Bogaerts, A. and E.C. Neyts, *Plasma Technology: An Emerging Technology for Energy Storage*. ACS Energy Letters, 2018. 3(4): 1013–1027.

[17] Azizov, R., et al. *Nonequilibrium plasmachemical process of CO2 decomposition in a supersonic microwave discharge*. in *Akademiia Nauk SSSR Doklady*. 1983.

[18] Kozák, T. and A. Bogaerts, *Splitting of CO2 by vibrational excitation in non-equilibrium plasmas: a reaction kinetics model*. Plasma Sources Science and Technology, 2014. 23(4): 045004.

[19] Berthelot, A. and A. Bogaerts, *Modeling of CO2 splitting in a microwave plasma: how to improve the conversion and energy efficiency*. The Journal of Physical Chemistry C, 2017. 121(15): 8236–8251.

[20] Sun, S., et al., *CO2 conversion in a gliding arc plasma: Performance improvement based on chemical reaction modeling*. Journal of CO2 Utilization, 2017. 17: 220–234.

[21] Ramakers, M., et al., *Gliding Arc Plasmatron: providing an alternative method for carbon dioxide conversion*. ChemSusChem, 2017.

[22] Cleiren, E., et al., *Dry Reforming of Methane in a Gliding Arc Plasmatron: Towards a Better Understanding of the Plasma Chemistry*. ChemSusChem, 2017. 10(20): 4025–4036.

[23] Scapinello, M., et al., *Conversion of CH4/CO2 by a nanosecond repetitively pulsed discharge*. Journal of Physics D: Applied Physics, 2016. 49(7): 075602.

[24] Zhu, B., et al., *Kinetics study on carbon dioxide reforming of methane in kilohertz spark-discharge plasma*. Chemical Engineering Journal, 2015. 264: 445–452.

[25] Indarto, A., et al., *Effect of additive gases on methane conversion using gliding arc discharge*. Energy, 2006. 31(14): 2986–2995.

[26] Li, D., et al., *CO2 reforming of CH4 by atmospheric pressure glow discharge plasma: a high conversion ability*. Int. J. Hydrog. Energ., 2009. 34(1): 308–313.

[27] Neyts, E.C., et al., *Plasma catalysis: synergistic effects at the nanoscale.* Chemical Reviews, 2015. 115(24): 13408–13446.

[28] Whitehead, J.C., *Plasma–catalysis: the known knowns, the known unknowns and the unknown unknowns.* Journal of Physics D: Applied Physics, 2016. 49(24): 243001.

[29] Tu, X., J.C. Whitehead, and T. Nozaki, *Plasma-catalysis: Fundamentals and applications.* 2018: Springer.

[30] Blin-Simiand, N., et al., *Removal of 2-Heptanone by Dielectric Barrier Discharges–The Effect of a Catalyst Support.* Plasma Processes and Polymers, 2005. 2(3): 256–262.

[31] Hong, J., et al., *Cobalt species and cobalt-support interaction in glow discharge plasma-assisted Fischer–Tropsch catalysts.* Journal of Catalysis, 2010. 273(1): 9–17.

[32] Zou, J.-J., Y.-p. Zhang, and C.-J. Liu, *Reduction of Supported Noble-Metal Ions Using Glow Discharge Plasma.* Langmuir, 2006. 22(26): 11388–11394.

[33] Demidyuk, V. and J.C. Whitehead, *Influence of temperature on gas-phase toluene decomposition in plasma-catalytic system.* Plasma Chemistry and Plasma Processing, 2007. 27(1): 85–94.

[34] Shang, S., et al., *Research on Ni/γ-Al2O3 catalyst for CO2 reforming of CH4 prepared by atmospheric pressure glow discharge plasma jet.* Catal. Today, 2009. 148(3–4): 268–274.

[35] Liu, C.-j., R. Mallinson, and L. Lobban, *Nonoxidative methane conversion to acetylene over zeolite in a low temperature plasma.* Journal of Catalysis, 1998. 179(1): 326–334.

[36] Holzer, F., F. Kopinke, and U. Roland, *Influence of ferroelectric materials and catalysts on the performance of non-thermal plasma (NTP) for the removal of air pollutants.* Plasma Chemistry and Plasma Processing, 2005. 25(6): 595–611.

[37] Guaitella, O., et al., *C2H2 oxidation by plasma/TiO2 combination: influence of the porosity, and photocatalytic mechanisms under plasma exposure.* Applied Catalysis B: Environmental, 2008. 80(3–4): 296–305.

[38] Liu, C.-j., et al., *Floating double probe characteristics of non-thermal plasmas in the presence of zeolite.* Journal of Electrostatics, 2002. 54(2): 149–158.

[39] Kim, H.-H., J.-H. Kim, and A. Ogata, *Microscopic observation of discharge plasma on the surface of zeolites supported metal nanoparticles.* Journal of Physics D: Applied Physics, 2009. 42(13): 135210.

[40] Tu, X., et al., *Dry reforming of methane over a Ni/Al2O3 catalyst in a coaxial dielectric barrier discharge reactor.* Journal of Physics D: Applied Physics, 2011. 44(27): 274007.

[41] Nozaki, T., et al., *Dissociation of vibrationally excited methane on Ni catalyst: Part 2. Process diagnostics by emission spectroscopy.* Catal. Today, 2004. 89(1–2): 67–74.

[42] Mizuno, A., *Generation of non-thermal plasma combined with catalysts and their application in environmental technology.* Catal. Today, 2013. 211: 2–8.

[43] Hensel, K., *Microdischarges in ceramic foams and honeycombs.* The European Physical Journal D, 2009. 54(2): 141.

[44] Rousseau, A., et al., *Combination of a pulsed microwave plasma with a catalyst for acetylene oxidation.* Applied Physics Letters, 2004. 85(12): 2199–2201.

[45] Neyts, E. and A. Bogaerts, *Understanding plasma catalysis through modelling and simulation –a review.* Journal of Physics D: Applied Physics, 2014. 47(22): 224010.

[46] Bongers, W., et al., *Plasma-driven dissociation of CO2 for fuel synthesis.* Plasma processes and polymers, 2017. 14(6): 1600126.

[47] den Harder, N., et al., *Homogeneous CO2 conversion by microwave plasma: Wave propagation and diagnostics.* Plasma Processes and Polymers, 2017. 14(6): 1600120.

[48] Berthelot, A. and A. Bogaerts, *Pinpointing energy losses in CO2 plasmas–Effect on CO2 conversion.* Journal of CO2 Utilization, 2018. 24: 479–499.

[49] Kozák, T. and A. Bogaerts, *Evaluation of the energy efficiency of CO2 conversion in microwave discharges using a reaction kinetics model.* Plasma Sources Science and Technology, 2014. 24(1): 015024.

[50] Bogaerts, Annemie, et al. "Plasma-based conversion of CO2: current status and future challenges." Faraday discussions 183 (2015): 217–232.

[51] Ozkan, A., A. Bogaerts, and F. Reniers, *Routes to increase the conversion and the energy efficiency in the splitting of CO2 by a dielectric barrier discharge.* Journal of Physics D: Applied Physics, 2017. 50(8): 084004.

[52] Tsuji, M., et al., *Decomposition of CO2 into CO and O in a microwave-excited discharge flow of CO2/He or CO2/Ar mixtures.* Chemistry Letters, 2001. 30(1): 22–23.

[53] Vesel, A., et al., *Dissociation of CO2 molecules in microwave plasma.* Chemical Physics, 2011. 382(1–3): 127–131.

[54] Spencer, L. and A. Gallimore, *CO2 dissociation in an atmospheric pressure plasma/catalyst system: a study of efficiency.* Plasma Sources Science and Technology, 2012. 22(1): 015019.

[55] Goede, A.P.H., et al., *Production of solar fuels by CO2 plasmolysis.* EPJ Web of Conferences, 2014. 79: 01005.

[56] Van Rooij, G., et al., *Taming microwave plasma to beat thermodynamics in CO2 dissociation.* Faraday discussions, 2015. 183: 233–248.

[57] Chen, G., et al., *Plasma assisted catalytic decomposition of CO2.* Applied Catalysis B: Environmental, 2016. 190: 115–124.

[58] Nunnally, T., et al., *Dissociation of CO2 in a low current gliding arc plasmatron.* Journal of Physics D: Applied Physics, 2011. 44(27): 274009.

[59] Kim, S.C., M.S. Lim, and Y.N. Chun, *Reduction characteristics of carbon dioxide using a plasmatron.* Plasma Chemistry and Plasma Processing, 2014. 34(1): 125–143.

[60] Wang, W., et al., *CO2 conversion in a gliding arc plasma: 1D cylindrical discharge model.* Plasma Sources Science and Technology, 2016. 25(6): 065012.

[61] Wang, W., et al., *Gliding arc plasma for CO2 conversion: better insights by a combined experimental and modelling approach.* Chemical Engineering Journal, 2017. 330: 11–25.

[62] Heijkers, S. and A. Bogaerts, *CO2 Conversion in a Gliding Arc Plasmatron: Elucidating the Chemistry through Kinetic Modeling.* The Journal of Physical Chemistry C, 2017. 121(41): 22644–22655.

[63] Trenchev, G., et al., *CO2 Conversion in a Gliding Arc Plasmatron: Multidimensional Modeling for Improved Efficiency.* The Journal of Physical Chemistry C, 2017. 121(44): 24470–24479.

[64] Zhang, A.-J., et al., *Conversion of greenhouse gases into syngas via combined effects of discharge activation and catalysis.* Chemical Engineering Journal, 2010. 156(3): 601–606.

[65] Tu, X. and J.C. Whitehead, *Plasma-catalytic dry reforming of methane in an atmospheric dielectric barrier discharge: Understanding the synergistic effect at low temperature.* Applied Catalysis B-Environmental, 2012. 125: 439–448.

[66] Tu, X. and J.C. Whitehead, *Plasma dry reforming of methane in an atmospheric pressure AC gliding arc discharge: Co-generation of syngas and carbon nanomaterials.* Int. J. Hydrog. Energ., 2014. 39(18): 9658–9669.

[67] Snoeckx, R., et al., *Plasma-based dry reforming: improving the conversion and energy efficiency in a dielectric barrier discharge.* RSC Adv., 2015. 5(38): 29799–29808.

[68] Mei, D.H., S.Y. Liu, and X. Tu, *CO 2 reforming with methane for syngas production using a dielectric barrier discharge plasma coupled with Ni/γ-Al 2 O 3 catalysts: Process optimization through response surface methodology.* Journal of CO2 Utilization, 2017. 21: 314–326.

[69] Eliasson, B., C.J. Liu, and U. Kogelschatz, *Direct conversion of methane and carbon dioxide to higher hydrocarbons using catalytic dielectric-barrier discharges with zeolites.* Industrial & Engineering Chemistry Research, 2000. 39(5): 1221–1227.

[70] Snoeckx, R., et al., *Plasma-based dry reforming: a computational study ranging from the nanoseconds to seconds time scale.* The Journal of Physical Chemistry C, 2013. 117(10): 4957–4970.

[71] Yao, S.L., et al., *Plasma reforming and coupling of methane with carbon dioxide.* Energy & Fuels, 2001. 15(5): 1295–1299.

[72] Zhang, Y.P., et al., *Plasma methane conversion in the presence of carbon dioxide using dielectric-barrier discharges.* Fuel Processing Technology, 2003. 83(1–3): 101–109.

[73] Kozlov, K.V., P. Michel, and H.E. Wagner, *Synthesis of organic compounds from mixtures of methane with carbon dioxide in dielectric-barrier discharges at atmospheric pressure.* Plasmas and Polymers, 2001. 5(3–4): 129–150.

[74] Martini, L.M., et al., *Oxidation of CH4 by CO2 in a dielectric barrier discharge.* Chemical Physics Letters, 2014. 593: 55–60.

[75] Zou, J.-J., et al., *Starch-Enhanced Synthesis of Oxygenates from Methane and Carbon Dioxide Using Dielectric-Barrier Discharges.* Plasma Chemistry and Plasma Processing, 2003. 23(1): 69–82.

[76] Li, Y., et al., *Synthesis of oxygenates and higher hydrocarbons directly from methane and carbon dioxide using dielectric-barrier discharges: Product distribution.* Energy & Fuels, 2002. 16(4): 864–870.

[77] A. Bogaerts, C. De Bie, R. Snoeckx and T. Kozák, Plasma Process. Polym. 14, e1600070 (2017).

[78] Wang, L., et al., *One-Step Reforming of CO2 and CH4 into High-Value Liquid Chemicals and Fuels at Room Temperature by Plasma-Driven Catalysis.* Angew Chem Int Ed Engl, 2017. 56 (44): 13679–13683.

[79] Chung, W.-C. and M.-B. Chang, *Simultaneous Generation of Syngas and Multiwalled Carbon Nanotube via CH4/CO2 Reforming with Spark Discharge.* ACS Sustainable Chemistry & Engineering, 2017. 5(1): 206–212.

[80] Zhu, B., et al., *Pressurization effect on dry reforming of biogas in kilohertz spark-discharge plasma.* Int. J. Hydrog. Energ., 2012. 37(6): p. 4945–4954.

[81] Abd Allah, Z. and J.C. Whitehead, *Plasma-catalytic dry reforming of methane in an atmospheric pressure AC gliding arc discharge.* Catal. Today, 2015. 256: 76–79.

[82] Zeng, Y.X., et al., *Plasma-catalytic dry reforming of methane over gamma-Al2O3 supported metal catalysts.* Catal. Today, 2015. 256: 80–87.

[83] Zheng, X.G., et al., *Plasma-assisted catalytic dry reforming of methane: Highly catalytic performance of nickel ferrite nanoparticles embedded in silica.* Journal of Power Sources, 2015. 274: 286–294.

[84] Zhu, F., et al., *Plasma-catalytic reforming of CO 2 -rich biogas over Ni/γ-Al 2 O 3 catalysts in a rotating gliding arc reactor.* Fuel, 2017. 199: 430–437.

[85] Zeng, Y.X., et al., *Low temperature reforming of biogas over K-, Mg- and Ce-promoted Ni/Al 2 O 3 catalysts for the production of hydrogen rich syngas: Understanding the plasma-catalytic synergy.* Applied Catalysis B: Environmental, 2018. 224: 469–478.

[86] Chung, Wei-Chieh, and Moo-Been Chang. "Dry reforming of methane by combined spark discharge with a ferroelectric." Energy conversion and management 124 (2016): 305–314.

[87] Scapinello, M., et al. "Conversion of CH_4/CO_2 by a nanosecond repetitively pulsed discharge." Journal of Physics D: Applied Physics 49.7 (2016): 075602.

[88] Mei, D. and X. Tu, *Conversion of CO2in a cylindrical dielectric barrier discharge reactor: Effects of plasma processing parameters and reactor design.* Journal of CO2 Utilization, 2017. 19: 68–78.

[89] Zhang, K., U. Kogelschatz, and B. Eliasson, *Conversion of greenhouse gases to synthesis gas and higher hydrocarbons.* Energy & Fuels, 2001. 15(2): 395–402.

[90] Song, H.K., et al., *Synthesis gas production via dielectric barrier discharge over Ni/gamma-Al 2O3 catalyst*. Catal. Today, 2004. 89(1–2): 27–33.

[91] Aziznia, A., et al., *Comparison of dry reforming of methane in low temperature hybrid plasma-catalytic corona with thermal catalytic reactor over Ni/gamma-Al2O3*. Journal of Natural Gas Chemistry, 2012. 21(4): 466–475.

[92] Wang, Q., et al., *Investigation of Dry Reforming of Methane in a Dielectric Barrier Discharge Reactor*. Plasma Chemistry and Plasma Processing, 2009. 29(3): 217–228.

[93] Mei, D., *Plasma-catalytic conversion of greenhouse gas into value-added fuels and chemicals*, in *Department of Electrical Engineering and Electronics*. 2016, University of Liverpool: Liverpool.

[94] Khoja, A.H., M. Tahir, and N.A.S. Amin, *Dry reforming of methane using different dielectric materials and DBD plasma reactor configurations*. Energy Conversion and Management, 2017. 144: 262–274.

[95] Rico, V.J., et al., *Evaluation of Different Dielectric Barrier Discharge Plasma Configurations As an Alternative Technology for Green C-1 Chemistry in the Carbon Dioxide Reforming of Methane and the Direct Decomposition of Methanol*. Journal of Physical Chemistry A, 2010. 114(11): 4009–4016.

[96] Wu, A., et al., *Study of the dry methane reforming process using a rotating gliding arc reactor*. Int. J. Hydrog. Energ., 2014. 39(31): 17656–17670.

[97] Ozkan, A., et al., *CO2-CH4 conversion and syngas formation at atmospheric pressure using a multi-electrode dielectric barrier discharge*. Journal of Co2 Utilization, 2015. 9: 74–81.

[98] Li, M.W., et al., *Effects of catalysts in carbon dioxide reforming of methane via corona plasma reactions*. Energy & Fuels, 2006. 20(3): 1033–1038.

[99] Sentek, J., et al., *Plasma-catalytic methane conversion with carbon dioxide in dielectric barrier discharges*. Appl. Catal. B-Environmental, 2010. 94(1–2): 19–26.

[100] Ray, D., P.M.K. Reddy, and C. Subrahmanyam, *Ni/Mn/γ-Al 2 O 3 assisted plasma dry reforming of methane*. Catal. Today, 2018. 309: 212–218.

[101] Wang, Q., Y. Cheng, and Y. Jin, *Dry reforming of methane in an atmospheric pressure plasma fluidized bed with Ni/gamma-Al2O3 catalyst*. Catal. Today, 2009. 148(3–4): 275–282.

[102] Zhang, K., B. Eliasson, and U. Kogelschatz, *Direct conversion of greenhouse gases to synthesis gas and C-4 hydrocarbons over zeolite HY promoted by a dielectric-barrier discharge*. Industrial & Engineering Chemistry Research, 2002. 41(6): 1462–1468.

[103] Jiang, T., et al., *Plasma methane conversion using dielectric-barrier discharges with zeolite A*. Catal. Today, 2002. 72(3–4): 229–235.

[104] Bo, Z., et al., *Plasma assisted dry methane reforming using gliding arc gas discharge: Effect of feed gases proportion*. Int. J. Hydrog. Energ., 2008. 33(20): 5545–5553.

[105] Kroker, T., et al., *Catalytic Conversion of Simulated Biogas Mixtures to Synthesis Gas in a Fluidized Bed Reactor Supported by a DBD*. Plasma Chemistry and Plasma Processing, 2012. 32(3): 565–582.

[106] Krawczyk, K., et al., *Methane conversion with carbon dioxide in plasma-catalytic system*. Fuel, 2014. 117: 608–617.

[107] Pham, M.H., et al., *Activation of methane and carbon dioxide in a dielectric-barrier discharge-plasma reactor to produce hydrocarbons-Influence of La2O3/gamma-Al2O3 catalyst*. Catal. Today, 2011. 171(1): 67–71.

[108] Mahammadunnisa, S., et al., *Catalytic Nonthermal Plasma Reactor for Dry Reforming of Methane*. Energy & Fuels, 2013. 27(8): 4441–4447.

[109] Long, H., et al., *CO2 reforming of CH4 by combination of cold plasma jet and Ni/gamma-Al 2O3 catalyst*. Int. J. Hydrog. Energ., 2008. 33(20): 5510–5515.

[110] Mei, Danhua, et al. "Plasma-catalytic reforming of biogas over supported Ni catalysts in a dielectric barrier discharge reactor: Effect of catalyst supports." Plasma Processes and Polymers 14.6 (2017): 1600076

[111] Ray, Debjyoti, P. Manoj Kumar Reddy, and Ch Subrahmanyam. "Ni-Mn/γ-Al$_2$O$_3$ assisted plasma dry reforming of methane." Catalysis Today 309 (2018): 212–218.

[112] Zhang, X.L., et al., *The simultaneous activation of methane and carbon dioxide to C-2 hydrocarbons under pulse corona plasma over La2O3/gamma-Al2O3 catalyst*. Catal. Today, 2002. 72(3–4): 223–227.

[113] Kado, S., et al., *Low temperature reforming of methane to synthesis gas with direct current pulse discharge method*. Chem. Comm., 2001(5): 415–416.

[114] Zheng, X., et al., *LaNiO3@SiO2 core-shell nano-particles for the dry reforming of CH4 in the dielectric barrier discharge plasma*. Int. J. Hydrog. Energ., 2014. 39(22): 11360–11367.

[115] Zheng, X.G., et al., *Silica-coated LaNiO3 nanoparticles for non-thermal plasma assisted dry reforming of methane: Experimental and kinetic studies*. Chem. Eng. J., 2015. 265: 147–156.

[116] Li, Y., et al., *Co-generation of syngas and higher hydrocarbons from CO2 and CH4 using dielectric-barrier discharge: Effect of electrode materials*. Energy & Fuels, 2001. **15**(2): 299–302.

[117] Scapinello, M., L.M. Martini, and P. Tosi, *CO2 Hydrogenation by CH4 in a Dielectric Barrier Discharge: Catalytic Effects of Nickel and Copper*. Plasma Processes and Polymers, 2014. 11 (7): 624–628.

[118] Goujard, V., J.M. Tatibouet, and C. Batiot-Dupeyrat, *Influence of the Plasma Power Supply Nature on the Plasma-Catalyst Synergism for the Carbon Dioxide Reforming of Methane*. Ieee Transactions on Plasma Science, 2009. 37(12): 2342–2346.

[119] Zou, J.-J. and C.-J. Liu, *Utilization of carbon dioxide through nonthermal plasma approaches*. Carbon dioxide as chemical feedstock, 2010: 267–290.

[120] Amouroux, J., S. Cavadias, and A. Doubla. *Carbon dioxide reduction by non-equilibrium electrocatalysis plasma reactor*. in *IOP conference series: materials science and engineering*. 2011. IOP Publishing.

[121] Zeng, Y. and X. Tu, *Plasma-catalytic hydrogenation of CO2 for the cogeneration of CO and CH4 in a dielectric barrier discharge reactor: effect of argon addition*. J. Phys. D: Appl. Phys., 2017. 50(18): 184004.

[122] Hayashi, N., T. Yamakawa, and S. Baba, *Effect of additive gases on synthesis of organic compounds from carbon dioxide using non-thermal plasma produced by atmospheric surface discharges*. Vacuum, 2006. 80(11–12): 1299–1304.

[123] Wang, L., et al., *Atmospheric Pressure and Room Temperature Synthesis of Methanol through Plasma-Catalytic Hydrogenation of CO2*. ACS Catal., 2018. 8(1): 90–100.

[124] Arita, K. and S. Iizuka, *Production of CH4 in a low-pressure CO2/H2 discharge with magnetic field*. Journal of Materials Science and Chemical Engineering, 2015. 3(12): 69.

[125] Nizio, M., et al., *Hybrid plasma-catalytic methanation of CO2 at low temperature over ceria zirconia supported Ni catalysts*. Int. J. Hydrog Energ, 2016. 41(27): 11584–11592.

[126] Jwa, E., et al., *Plasma-assisted catalytic methanation of CO and CO2 over Ni–zeolite catalysts*. Fuel Process. Technol., 2013. 108: 89–93.

[127] Porosoff, M.D., B. Yan, and J.G. Chen, *Catalytic reduction of CO 2 by H 2 for synthesis of CO, methanol and hydrocarbons: challenges and opportunities*. Energy & Environmental Science, 2016. 9(1): 62–73.

[128] Zeng, Y. and X. Tu, *Plasma-catalytic CO2 hydrogenation at low temperatures*. IEEE Transactions on Plasma Science, 2015. 44(4): 405–411.

[129] Kano, M., G. Satoh, and S. Iizuka, *Reforming of Carbon Dioxide to Methane and Methanol by Electric Impulse Low-Pressure Discharge with Hydrogen*. Plasma Chem. Plasma P., 2012. 32(2): 177–185.

[130] Mora, E.Y., A. Sarmiento, and E. Vera, *Alumina and quartz as dielectrics in a dielectric barrier discharges DBD system for CO 2 hydrogenation*. J. Phys.: Confer Ser., 2016. 687(1): 012020.

[131] De Bie, C., J. van Dijk, and A. Bogaerts, *CO $_2$ Hydrogenation in a Dielectric Barrier Discharge Plasma Revealed*. J. Phys. Chem. C, 2016. 120(44): 25210–25224.

[132] Wang, W., et al., *Recent advances in catalytic hydrogenation of carbon dioxide*. Chem. Soc. Rev., 2011. 40(7): 3703–3727.

[133] Eliasson, B., et al., *Hydrogenation of Carbon Dioxide to Methanol with a Discharge-Activated Catalyst*. Ind. Eng. Chem. Res., 1998. 37(8): 3350–3357.

[134] Nizio, M., et al., *Low temperature hybrid plasma-catalytic methanation over Ni-Ce-Zr hydrotalcite-derived catalysts*. Catal. Commun., 2016. 83: 14–17.

[135] Futamura, S. and H. Kabashima, *Synthesis gas production from CO2 and H2O with nonthermal plasma, in Studies in Surface Science and Catalysis*. 2004, Elsevier. 119–124.

[136] Mahammadunnisa, S., et al., *CO2 reduction to syngas and carbon nanofibres by plasma-assisted in situ decomposition of water*. Int. J. Greenh Gas Control, 2013. 16: 361–363.

[137] Snoeckx, R., et al., *The Quest for Value-Added Products from Carbon Dioxide and Water in a Dielectric Barrier Discharge: A Chemical Kinetics Study*. ChemSusChem, 2017. 10(2): 409–424.

[138] Ihara, T., M. Kiboku, and Y. Iriyama, *Plasma reduction of CO2 with H2O for the formation of organic compounds*. Bulletin of the Chemical Society of Japan, 1994. 67(1): 312–314.

[139] Ihara, T., et al., *Formation of methanol by microwave-plasma reduction of CO2 with H2O*. Bulletin of the Chemical Society of Japan, 1996. 69(1): 241–244.

[140] Chen, G., et al., *Simultaneous dissociation of CO2 and H2O to syngas in a surface-wave microwave discharge*. Int. J. Hydrog Energ, 2015. 40(9): 3789–3796.

[141] Chen, G., et al., *An overview of CO2 conversion in a microwave discharge: the role of plasma-catalysis*. J. Phys. D: Appl. Phys, 2017. 50(8): 084001.

[142] Indarto, A., et al., *Gliding arc plasma processing of CO2 conversion*. J. Hazard. Mater, 2007. 146(1–2): 309–315.

[143] Guo, L., et al., *A novel method of production of ethanol by carbon dioxide with steam*. Fuel, 2015. 158: 843–847.

[144] Centi, G. and S. Perathoner, *Opportunities and prospects in the chemical recycling of carbon dioxide to fuels*. Catal. Today, 2009. 148(3–4): 191–205.

[145] Studt, F., et al., *Discovery of a Ni-Ga catalyst for carbon dioxide reduction to methanol*. Nat. Chem., 2014. 6(4): 320.

[146] Van Rooij, G., et al., *Plasma for electrification of chemical industry: a case study on CO2 reduction*. Plasma Phys Control Fusion, 2017. 60(1): 014019.

Andreas Kafizas

29 Photocatalytic approaches for converting CO_2 into fuels and feedstocks

29.1 Introduction

The excessive use of fossil fuels, and release of CO_2 into the atmosphere, is the primary cause of global warming. This has caused global average temperatures to increase, polar ice to melt, deserts to expand and sea levels to rise, and is damaging the Earth's climate and ecosystems [1, 2]. To mitigate the potentially catastrophic results of climate change, the Intergovernmental Panel on Climate Change (IPCC) has set a target of restricting warming to 2 °C above pre-industrial era temperatures [3]. In order to meet this target, immediate and extensive reductions in CO_2 emissions must occur, which can only be achieved if new technologies that exploit carbon neutral, and ideally renewable, sources of energy are adopted. This problem is on a very large scale (*ca.* 36 GT total CO_2 emissions, 2015), so there is a need to develop upscalable and cost-effective technologies, which ideally, are compatible with existing infrastructure.

Sunlight is our largest source of renewable energy, the amount reaching the Earth in 1 h being more than the total energy we consume annually from all sources combined (fossil, nuclear, hydroelectric, etc) [4]. Consequently, photovoltaics (i.e. solar cells), which convert sunlight into electricity, is the fastest growing renewable technology [5]. However, one of the biggest barriers to the growth of photovoltaics is the intermittency of sunlight. In addition to this, there is a mismatch between when photovoltaics generate electricity the most (typically midday) and when we consume electricity the most (typically mornings and evenings) [6]. This is a major barrier to their implementation, as the electrical energy they produce must be used at the point of generation (in the absence of energy storage infrastructure) [7]. To circumvent this problem, photovoltaics need to be coupled with energy storage technologies, such as Li-ion rechargeable batteries for short-term storage or pumped hydroelectric, compressed air, redox-flow batteries, and so on for inter-seasonal storage [8]; however, it should be noted that many of these technologies require substantial infrastructure spending and do not address heating energy demand in cold climates. Another issue is transportation, which is currently primarily powered by fossil fuels and accounts for *ca.* 30% of global energy use. Although there has been substantial progress in the adoption of electrically powered vehicles, current battery technologies are not sufficiently energy dense to meet the power requirements of freight (e.g. ships and planes

Andreas Kafizas, Molecular Science Research Hub, Imperial College, White City Campus, London, W12 0BZ, United Kingdom

https://doi.org/10.1515/9783110665147-029

that currently account for *ca.* 5.3% of global CO_2 emissions) [9]. Therefore, technological solutions that address the need for both: (i) energy storage and (ii) renewable fuels for transportation need to be developed.

Currently all of those strategies exploit sunlight, and can loosely be separated into two approaches: (i) biofuels and (ii) solar fuels. By mimicking photosynthesis in plants, solar energy is used to drive the synthesis of chemical bonds (i.e. fuels) that can be indefinitely be stored and used as and when required. Biofuels exploit photosynthesis in plants to generate fuels. In Brazil, sugarcane is used as a feedstock for producing biofuel ethanol, and is produced on an incredibly large scale in Brazil (*ca.* 6.2 billion gallons, 2014). However, relatively low solar-to-biomass energy conversion efficiencies (*ca.* < 0.2%) [10] have constrained biofuel production to economies with vast amounts of arable land (e.g. Brazil and USA produced *ca.* 83% of the world's biofuel in 2014) [11]. Alternatively, semiconductor materials can be used to drive the synthesis of fuels using sunlight (often deemed solar fuels or artificial photosynthesis) [12]. Various reactions have been studied, including water splitting, where water is converted into hydrogen fuel [12], and CO_2 reduction, where CO_2 is converted into carbon-based fuels (e.g. CH_4, CH_3OH) and/or useful chemical feedstocks (e.g. CO, syngas) [13]. Methanol can be converted into dimethyl ether for use as a fuel in transportation [14]. Feedstocks can be upgraded using the Fischer–Tropsch process to produce larger hydrocarbons for applications in freight.

Solar-driven water splitting has received greater attention, and since its first demonstration by Fujishima and Honda in 1972 [19], has resulted in *ca.* 14,000 publications. Various device architectures and semiconducting materials have been explored [20–22]. The highest solar-to-hydrogen efficiency to date of *ca.* 30% was achieved by coupling a photovoltaic (an InGaP/GaAs/GaInNAsSb triple junction) to electrolysers (Pt and Ir) [15]. High efficiencies have also been observed in photoelectrochemical systems, where Turner et al. recently achieved solar-to-hydrogen efficiency of *ca.* 16% using a GaInP/GaInAs buried junction coated with Pt and Ru surface catalysts [16]. Although high solar-to-hydrogen efficiencies have been achieved, such devices typically composed of silicon and III–V semiconductor light absorber layers, are not considered economically viable and have not been commercialised. Interestingly, the only semiconductor to be applied commercially for photocatalytic applications[1] is the transition metal oxide, TiO_2. Given their durability, low-cost and ease of synthesis, a wide range of transition metal oxides have been studied for applications in water splitting [23], and have achieved solar-to-hydrogen efficiencies of *ca.* 1% in an all oxide-based device (La,Rh: $SrTiO_3$/ Mo: $BiVO_4$ coated with Ru and RuO_x surface

---

[1] Windows, tiles, concretes, paints and indoor air purifiers that can oxidise a wide range of pollutants (*e.g.* NO_x and VOCs to remediate air). The global market for photocatalytic compositions was worth $1.6 billion in 2015 and is forecast to reach *ca.* $2.9 billion in 2020 [117].

catalysts) [17, 24]. The energy conversion efficiencies for various biofuels and semiconductor materials for producing fuels are compared in Table 29.1.

Table 29.1: A summary of the typical energy conversion efficiencies for biofuels production and state-of-the-art semiconductor materials for solar fuels.

Approach	Device	Fuel	Energy conversion efficiency (%)	Reference
Biofuels	n/a	Oil palm	0.16–0.18	Gust et al. [10]
		Sugarcane	0.08	
		Castor oil	0.05–0.09	
Inorganic semiconductors	PV + electrolyser	H_2 from H_2O	*ca.* 30	Jia et al. [15]
	Photoelectrode	H_2 from H_2O	*ca.* 16	Turner et al. [16]
	Nanopowders	H_2 from H_2O	*ca.* 1	Domen et al. [17]
	Nanopowders	CH_3OH from CO_2	*ca.* 2.5	Tseng et al. [18]

Source: PV = photovoltaic

Solar-driven CO_2 reduction has received comparatively less attention than water splitting, but is a field growing in importance [25, 26]. In order to realise a sustainable carbon neutral economy, a global-scale transition needs to occur in which CO_2 is treated as a recyclable feedstock for making fuels (rather than simply treated as a waste product from fossil fuel combustion) [27]. Solar-driven CO_2 reduction can be split into two main approaches: (i) photocatalytic [28–37] and (ii) photothermal [38–40] (also known as solar thermochemical).

The photocatalytic approach applies semiconductors that can operate at near ambient temperatures. A wide range of semiconductors have been studied, including binary and ternary transition metal oxides, Si and III–V semiconductors, transition metal chalcogenides and nitrides, organic semiconductors (e.g. C_3N_4) and porous materials (e.g. MOFs, COFs and zeolites) [28–37]. For the photocatalytic approach, ultrabandgap light is absorbed by the semiconductor, resulting in the photoexcitation of electrons from the valence to the conduction band (leaving behind a concomitant hole in the valence band). Photogenerated electrons reduce CO_2 in reactions at the material surface. Typically, the reaction is carried out in the presence of H_2O vapour, which is oxidised by photogenerated holes into O_2, and thereby provides a source of protons for CO_2 hydrogenations reactions.[2] Photocatalytic reactions can be carried

---

2 All photocatalytic CO_2 reactions covered in this chapter were carried out in the presence of CO_2 gas and H_2O vapour unless otherwise stated.

out in the liquid or gas phase. In liquid-phase reactions, CO_2 is dissolved in a protic solution[3]; in gas-phase reactions, CO_2 and the proton source are both in the gas phase. A wide range of reaction products have been observed (e.g. CO, CH_4, CH_3OH), which depends strongly on the photocatalyst and conditions employed. The valence and conduction band potentials for various semiconductor materials that have been explored for photocatalytic CO_2 reduction are shown in Figure 29.1, and the redox potentials for driving CO_2 reduction are shown in Table 29.2.

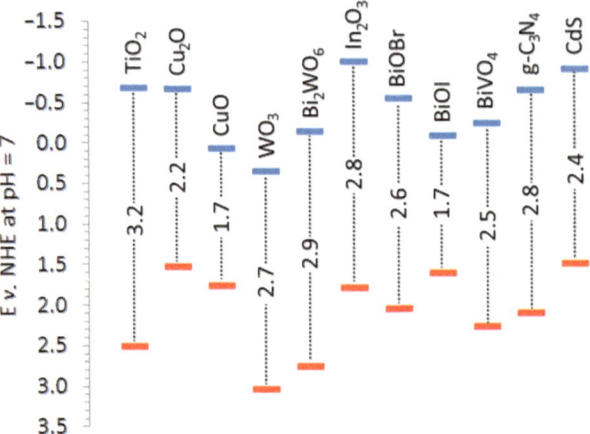

Figure 29.1: Optical gaps (eV), and valence and conduction band potentials, for the various semiconductor materials that have been explored for photocatalytic CO_2 reduction reviewed herein [41–45] Energies are versus the normal hydrogen electrode (NHE) at pH = 7.

Unlike the photocatalytic approach, the photothermal approach does not require the use of semiconducting light absorber materials. In a typical photothermal reaction, concentrated sunlight is used to heat the catalyst in the presence of CO_2 and H_2O. A wide range of transition metal oxides and carbides have been studied at various temperatures [38–40]. Depending on the catalyst used, the process temperature ranges widely from *ca.* 300 to 1,300 °C. Given the high temperature requirements, all photothermal reactions are conducted in the gas phase (i.e. the substrate CO_2 and proton source are in the gas phase). Again, a wide range of reaction products have been observed (e.g. CO, CH_4, CH_3OH). Using the photothermal approach, a benchmark solar-to-fuel energy conversion efficiency of 5.25% was achieved using a two-step CeO_{2-x} redox cycle; however, this process required solar concentration (*ca.* 3,000 suns) to reach the high temperatures required to drive this reaction (*ca.* 1,300 °C) [46].

---

3 Typically an aqueous solution, although ionic liquids are increasingly in popularity [118]. Moreover, liquid phase reactions facilitate the use of photoelectrochemical methods [28].

Table 29.2: Standard redox potentials related to CO_2 reduction and water splitting. Energies are versus the normal hydrogen electrode (NHE) at pH = 7 [29].

Process	Reaction	E^0 vs NHE at pH = 7
CO_2 reduction	$CO_2 + 2e^- + 2H^+ \rightarrow HCOOH$	−0.61V
	$CO_2 + 2e^- + 2H^+ \rightarrow CO + H_2O$	−0.53V
	$CO_2 + 4e^- + 4H^+ \rightarrow HCHO + H_2O$	−0.48V
	$CO_2 + 6e^- + 6H^+ \rightarrow CH_3OH + H_2O$	−0.38V
	$CO_2 + 8e^- + 8H^+ \rightarrow CH_4 + 2H_2O$	−0.24V
water splitting	$2H^+ + 2e^- \rightarrow H_2$	−0.41 V
	$2H_2O \rightarrow O_2 + 4e^- + 4H^+$	+0.82 V

In this chapter, we cover a range of semiconductor materials that drive photo-catalytic[4] CO_2 reduction. It should be noted that this coverage is not exhaustive, but will highlight key materials and studies, and provide a discussion on avenues of future research in this promising field.

29.2 Materials

29.2.1 TiO$_2$

TiO$_2$ was amongst the first semiconductors studied for photocatalytic CO_2 reduction [47]. A wide range of TiO$_2$-based materials have been examined [13, 31] which can produce a wide range of majority products, including CO [48–53], CH$_4$ [54–67], and CH$_3$OH [18, 68, 69]. TiO$_2$ exhibits three polymorphs: anatase, rutile and brookite (where the anatase phase typically shows higher photocatalytic activity than rutile or brookite alone) [70]. Li et al. studied the activity of the various polymorphs of TiO$_2$ for photocatalytic CO_2 reduction [50]. By introducing surface oxygen vacancies, they found that they could enhance the activity of anatase and brookite *ca.* tenfold. Interestingly, brookite showed marginally higher activity than anatase (*ca.* 3 and *ca.* 0.5 μmol/g$_{cat}$/h of CO and CH$_4$ produced respectively; ~90 mW/cm^2 solar simu-lated light). Šolcová et al. studied the effect of TiO$_2$ particle size on photocatalytic CO_2 reduction activity [57]. Anatase TiO$_2$ particles of various average sizes (*ca.* 5 to 30 nm) were prepared by precipitation and sol–gel methods. The best performing materials possessed an average particle size of 14 nm, and produced *ca.* 0.05, *ca.* 0.04 and *ca.* 0.045 μmol/g$_{cat}$/h of CH$_3$OH, CH$_4$ and CO, respectively (Hg lamp[5]).

---

4 This review does not cover photoelectrocatalytic and molecular photocatalytic approaches, nor does it cover photothermal approaches in depth.
5 Incident power not specified.

Jaroniec et al. studied the effect of altering the ratio of {001} and {101} facets of ana-tase TiO_2 nanocrystals on the photocatalytic CO_2 reduction performance [54] The {101} and {001} facets exhibited different band structures and band edge positions, and thus co-exposed {101} and {001} facets can form a surface heterojunction within a single TiO_2 particle, which is beneficial for the transfer of photogenerated elec-trons and holes to {101} and {001} facets, respectively. Particles that possessed *ca.* 40% {101} and *ca.* 60% {001} facets showed the highest activity, producing 1.35 µmol/g_{cat}/h of CH_4, which was almost four times higher than a commercial standard, P25 Evonik (Xe lamp[5]). Oymak et al. studied the mechanism of photocata-lytic CO_2 reduction on TiO_2, grown using a sol–gel method [61]. The accumulation of carbonaceous intermediates on the surface indicated the rate limiting step is the H_2O oxidation reaction, similar to the natural photosynthetic systems. However, when H_2 was introduced into the system, the carbonaceous intermediates were con-verted into CH_4 at a far higher rate.

29.2.2 CuO_x and other composites with TiO_2

Using a commercial TiO_2 anatase: rutile composite (P25 Evonik), Biswas et al. synthes-ised Cu_2O/TiO_2 composites on high surface area mesoporous SiO_2 substrates (> 300 m^2/g) using a one-pot sol–gel method [48]. Peak production rates of *ca.* 60 and *ca.* 10 µmol/g_{cat}/h of CO and CH_4 were observed respectively, which corresponded to an overall peak quantum yield of *ca.* 1.4% (considering only ultra-bandgap light ab-sorption by TiO_2 from the Xe lamp light source, which amounted to 2.4 mW/cm^2). Using P25, Li et al. also synthesised Cu/TiO_2 composites using a simple precipitation and calcination method [53]. The as-prepared composites predominantly contained CuO; however, thermal treatment in He and H_2 atmospheres resulted in the formation of predominantly Cu_2O and mixed Cu_2O/Cu, respectively. The H_2 reduced sample showed highest photocatalytic activity, with respective peak production rates of *ca.* 7 and *ca.* 1.5 µmol/g_{cat}/h of CO and CH_4 (*ca.* 90 mW/cm^2 solar simulated light). Li et al. also studied the effect of co-modifying TiO_2 with Cu and I [51, 52]. Iodine entered the TiO_2 lattice as I^{5+} and I^- states, and Cu_2O formed at the surface. They observed visible light activity in their material (*ca.* 3 µmol/g_{cat}/h of CO, λ > 400 nm with a power *ca.* four times greater than sunlight), which was attributed to the introduction of inter-band states. Wu et al. synthesised a range of Cu_2O/anatase TiO_2 materials using a one-pot sol–gel method [62]. Optimal CH_4 production was observed in the sample with 0.03 wt.% Cu_2O, with a tenfold enhancements over the analogous unloaded ana-tase (3.25 mW/cm^2 UVA light). Tseng et al. studied the aqueous photocatalytic reduc-tion of CO_2 using Cu_2O/TiO_2 synthesised by a sol–gel method [18]. With an optimal Cu loading of 2.0 wt.%, they observed the formation of CH_3OH in high yield (*ca.* 20 µmol/g_{cat}/h, Hg lamp, 254 nm peak wavelength, *ca.* 140 µW/cm^2), which was three times higher than a P25 Evonik standard, and corresponded to a quantum efficiency of *ca.*

10% and energy efficiency of *ca.* 2.5%. The turnover frequency (TOF) per catalytic Cu site was estimated to be 40 s^{-1}. In another study, Tseng et. al. also investigated the effect of light source on CO_2 reduction activity for their Cu_2O/TiO_2 materials [69]. Using UVC light, a far higher CH_4 yield was observed (*ca.* 20 µmol/g$_{cat}$/h, 254 nm light[5]) compared with UVA light (*ca.* 0.5 µmol/g$_{cat}$/h, 365 nm light[5]).

Zhou et al. synthesised CeO_2/TiO_2 composites using a nano-casting route [49]. The composites, composed of cubic fluorite CeO_2 and anatase TiO_2, exhibited highly ordered 2D hexagonal mesostructures and possessed high specific surface area (> 130 m^2/g). The highest CO_2 reduction activity was observed in their 1: 1, Ce: Ti composite, which produced *ca.* 18 and *ca.* 3 µmol/g$_{cat}$/h of CO and CH_4 respectively (Xe lamp irradiation[5]). Hersam et al. investigated graphene/P25 Evonik TiO_2 composites [65]. In the photocatalytic reduction of CO_2, their best performing composite produced CH_4 at a rate of *ca.* 8 µmol/m^2/h under UVA light[5] and *ca.* 4 µmol/m^2/h[1] under visible light[5], which in both cases was at least four times higher than a P25 standard. Zou et al. developed a novel reduction-hydrolysis method to produce graphene/TiO_2 composites [60]. In the photocatalytic reduction of CO_2, their best performing composites produced CH_4 and C_2H_6 at respective rates of *ca.* 10 and *ca.* 17 µmol/g$_{cat}$/h (Xe lamp[5]). A P25 standard produced CH_4 at a rate of *ca.* 0.7 µmol/g$_{cat}$/h and negligible C_2H_6. Ma et al. synthesised multi-walled carbon nanotube (MWCNT) composites with anatase or rutile TiO_2 [68]. In the photocatalytic reduction of CO_2 (UVA light, 365 nm[5]), their best performing anatase composite produced CH_3CH_2OH as the main product (*ca.* 30 µmol/g$_{cat}$/h) and their best performing rutile composite produced HCOOH as the main product (*ca.* 25 µmol/g$_{cat}$/h).

29.2.3 Noble and platinum-group metals on TiO_2

Garcia et al. loaded P25 Evonik TiO_2 with Au–Cu alloy nanoparticles using a stepwise precipitation method [55]. Photocatalytic reduction of CO_2 produced CH_4 with high selectivity (*ca.* 2,000 µmol/g$_{cat}$/h of CH_4, 97% selectivity, 1,000 mW.cm^{-2} solar simulated light *i.e.* 10 suns). Using a UV filter (> 400 nm), their photocatalysts were able to photocatalytically reduce CO_2 using visible light alone (*ca.* 600 µmol/m^2/h of CH_4, *ca.* 90% selectivity, Xe lamp[5]), which was attributed to the plasmon resonance absorption of the Au–Cu alloy nanoparticles. Šolcová et al. synthesised Ag/TiO_2 powders using a sol–gel process [56]. The best performing sample for photocatalytic CO_2 reduction contained 7 wt.% Ag (*ca.* 0.4 and *ca.* 0.1 µmol/g$_{cat}$/h of CH_4 and CH_3OH respectively, Hg lamp[5]).

Li et al. synthesised AgBr/P25 Evonik TiO_2 composites using a precipitation method [66]. They examined their photocatalytic CO_2 reduction activity using visible light (> 420 nm, Xe lamp[5]). They found that a *ca.* 23 wt.% AgBr loading produced the most active photocatalyst (*ca.* 26, *ca.* 16, *ca.* 6.4 and *ca.* 2.7 µmol/g$_{cat}$/h of

CH_4, CH_3OH, CO and CH_3CH_2OH, respectively) and showed good stability over five photocatalytic cycles.

Pan et al. studied N:TiO_2 loaded with Pt, Au or Ag [63]. They found that Pt loading produced the most active materials for photocatalytic CO_2 reduction, with the optimum loading being *ca.* 0.2 wt.% Pt and the optimum N-doping level being *ca.* 0.8% in replacing oxygen sites. They evaluated their performance under UV (Xe lamp, *ca.* 35 mW.cm^{-2}) and visible light (Xe lamp, > 420 nm, *ca.* 38 mW.cm^{-2}). Under UV irradiation, they observed a CH_4 production rate of *ca.* 4 µmol/g_{cat}/h, and under visible light they observed a CH_4 production rate of *ca.* 0.5 µmol/g_{cat}/h. Yu et al. studied Pt-loaded TiO_2 nanoparticles and nanotubes [59]. The Pt particles were 1 to 2 nm in diameter. Optimal photocatalytic CO_2 reduction was observed at a Pt loading of 0.12 wt. % (*ca.* 0.06 µmol g_{Ti}^{-1} hr^{-1} of CH_4, Hg lamp[5]). The CH_4 yield showed no dependence on the H_2O/CO_2 partial pressure in Pt/TiO_2 nanoparticles; however, in Pt/TiO_2 nanotubes, significantly higher CH_4 yields were observed at higher H_2O/CO_2 partial pressures. In addition, a stronger dependence on reaction temperature was also observed in Pt/TiO_2 nanotubes, showing higher CH_4 yields at higher temperatures. Biswas et al. grew nanostructured Pt-loaded TiO_2 films grown using a gas-phase deposition method [64]. The Pt particles were 0.5 to 2 nm in diameter. At an optimal Pt loading, they observed high yields of CH_4 (*ca.* 1,360 µmol/g_{cat}/h) and CO (*ca.* 190 µmol/g_{cat}/h) from the photocatalytic reduction of CO_2, which corresponded to a quantum yield of *ca.* 2.4% (considering only ultra-bandgap light absorption by TiO_2 from the Xe lamp light source, which amounted to *ca.* 20 mW/cm^2). Time-resolved laser spectroscopy measurements showed that electron-hole recombination was suppressed in their Pt-loaded TiO_2 compared with TiO_2 alone. Grimes et al. studied Pt and Cu-loaded N: TiO_2 nanotubes [58]. The N:TiO_2 nanotubes were grown by anodising titanium foil, and the Pt and Cu were deposited by sputtering. All photocatalytic CO_2 reduction reactions were conducted outdoors using natural sunlight (power varied between *ca.* 100 and 75 mW/cm^2). The predominant product was CH_4, with their best performing sample producing hydrocarbons at a rate of *ca.* 32 nmol cm^{-2} hr^{-1}.

Ishitani et al. examined the effect of Pd loading on the photocatalytic CO_2 reduction activity of P25 Evonik [67]. On P25 Evonik alone, the majority product was CO (with small amounts of CH_4 formed). In contrast, on Pt-loaded TiO_2, CH_4 was the majority product (with small amounts of CO formed). Isotopic labelling studies showed that CO_2 and CO_3^{2-} are the main carbon sources for CH_4 formation, which proceeds via Pd sites. However, prolonged irradiation resulted in de-activation of the catalyst because of oxidation of Pd to PdO.

29.2.4 WO_3 and Bi_2WO_6

Ozin et al. examined the photocatalytic CO_2 reduction activity of WO_3 nanowires decorated with palladium nanocrystals [71]. Experiments were conducted in a H_2:CO_2

atmosphere at 250 °C and *ca.* 2 bar of pressure. CO was the majority product (99% selectivity) and was produced at a rate of *ca.* 3,000 µmol/g_{cat}/h (Xe lamp, 1.36 W. cm^{-2}). This corresponded to a quantum efficiency of *ca.* 0.006%. Ozin et al. carried out a rate law analysis for this reaction, by conducting a series of experiments at various partial pressures. In dark experiments (i.e. thermally activated), the activation energy for CO production was 54 kJ/mol. With illumination, this decreased to 31 kJ/mol. Concomitantly, the rate law changed from r = $k[CO_2]^{0.7}[H_2]^{0.1}$ in the dark to r = $k[CO_2]^{0.3}[H_2]^{0.6}$ under illumination. Xie et al. studied oxygen-deficient and oxygen-rich WO_3 grown using a hydrothermal process [72]. Interestingly, they found that oxygen-deficient WO_3 could drive photocatalytic CO_2 reduction using IR light (a Si_3N_4 lamp, > 800 nm[5]). Under these conditions, oxygen-rich WO_3 was not active. The majority product was CO, and was produced at a rate of *ca.* 2.7 µmol/g_{cat}/h. With 800 nm light, an apparent quantum yield of *ca.* 0.03% was observed. Zou et al. investigated ultrathin, single-crystal WO_3 nanosheets, grown using an arc discharge method [73]. During photocatalytic CO_2 reduction measurements, CH_4 was produced at a rate of *ca.* 1.1 µmol/g_{cat}/h (Xe lamp, > 420 nm[5]). A commercial WO_3 powder showed negligible activity. This was attributed to size-quantization effects within their ultrathin nanostructure (*ca.* 4–5 nm) that altered the band properties of the material. Liu et al. studied graphene/WO_3 composites, grown using a hydrothermal method [74]. During photocatalytic CO_2 reduction, this composite produced CH_4 at a rate of *ca.* 1.1 µmol/g_{cat}/h (Xe lamp, > 400 nm[5]). This activity was at least ten times higher than WO_3 and P25 TiO_2 standards. They attributed this enhanced activity to the role of graphene, which elevates the conduction band of WO_3, and thereby increases the thermodynamic potential to drive CO_2 reduction.

Wu et al. studied Bi_2WO_6 nanoplates, grown using a hydrothermal method [75]. Photocatalytic CO_2 reduction produced CO as the majority product at a rate of *ca.* 0.4 µmol/g_{cat}/h (Xe lamp, 420 nm < λ < 620 nm, 100 mW/cm^2). Over time, the CO yield decreased because of the accumulation of reaction intermediates on the surface of the nanoplates; however, the activity could be restored by washing the catalyst with water. Zou et al. investigated ultrathin Bi_2WO_6 square nanoplates, grown using a hydrothermal method [76]. The nanoplates were ~9.5 nm thick and possessed well-defined {001} facets. Photocatalytic CO_2 reduction measurements showed that the nanoplates produced the majority product CH_4 at a rate of *ca.* 1.7 µmol/g_{cat}/h (Xe lamp, > 420 nm[5]). Xie et al. studied atomically thin Bi_2WO_6 materials (a single unit cell thick) [77]. During photocatalytic CO_2 reduction, the majority product CH_3OH was produced at a rate of *ca.* 75 µmol/g_{cat}/h (Xe lamp[5]). This rate was more than 100 times higher than bulk Bi_2WO_6, and was attributed to an enhanced charge carrier lifetime, which increased from *ca.* 15 ns in bulk Bi_2WO_6 to *ca.* 80 ns in atomically thin Bi_2WO_6. Dai et al. examined Bi_2WO_6 hollow microspheres, grown using an anion exchange method [78]. Photocatalytic CO_2 reduction produced CH_3OH as the majority product at a rate of *ca.* 16 µmol/g_{cat}/h (visible light[5]).

Yang et al. grew composites of hierarchical Bi_2WO_6 hollow microspheres with various conducting polymers [79]. They found that composites with polythiophene exhibited the highest performance. Photocatalytic CO_2 reduction produced CH_3OH and CH_3CH_2OH at rates of *ca.* 14 and *ca.* 5 µmol/g_{cat}/h, respectively (Xe lamp > 420 nm[5]). The total hydrocarbon yield was almost three times higher than Bi_2WO_6 alone. The apparent quantum efficiency in the polythiophene composite was *ca.* 0.009% (475 nm light). Luo et al. grew hierarchical flower-like Bi_2WO_6 microspheres coated with MoS_2 [80]. At an optimal loading of 0.4 wt.% MoS_2, photocatalytic CO_2 reduction produced CH_3OH and CH_3CH_2OH each at an equivalent rate of *ca.* 9 µmol/g_{cat}/h (Xe lamp, > 420 nm[5]). The total hydrocarbon yield was nearly double that of pure Bi_2WO_6.

29.2.5 In_2O_3 and $InNbO_4$

Ozin et al. investigated hydroxylated In_2O_3 nanocrystals grown by a thermal treatment method [81]. Using CO_2:H_2 gas mixtures, they studied the photocatalytic reduction of CO_2 and observed the formation of the majority product CO at a rate of *ca.* 150 µmol/g_{cat}/h, which was four times higher than the dark reaction (Xe lamp, 100 mW/cm^2, 190 °C). The activity of this highly defective material was attributed to the formation of frustrated Lewis pairs, identified by computation [82] and experiment [83]. Ozin et al. also studied hydroxylated In_2O_3 nanorods [84]. CO was again the majority product in photocatalytic CO_2 reduction tests (*ca.* 1.2 µmol/g_{cat}/h, metal halide lamp[5]). Activity increased with nanorod length, which was attributed to the observed increase in charge carrier lifetime. A subsequent study showed that increased defect concentrations in In_2O_3 resulted in enhanced charge carrier lifetimes [85]. Ozin et al. studied the effect of precursor on the observed photocatalytic activity [86]. Photocatalytic CO_2 reduction tests were carried out in a batch reactor at equal partial pressures of H_2:CO_2 (*ca.* 2 bar total pressure, 150 °C). They found that In_2O_3, grown from indium hydroxide by thermal decomposition, was most active (majority product CO produced at a rate of *ca.* 1.4 µmol/g_{cat}/h, metal halide lamp, 80 mW/cm^2). The choice of precursor had a substantial impact on the population of surface defect sites, which in turn, was highly correlated with CO_2 reduction activity. Goodenough et al. synthesised carbon-coated In_2O_3 nanobelts using a hydrothermal method, and subsequently deposited a Pt co-catalyst by photodeposition [87]. Photocatalytic CO_2 reduction was carried out in an aqueous environment, yielding CO and CH_4 at rates of *ca.* 630 and *ca.* 140 µmol/g_{cat}/h, respectively (Xe lamp[5]).

Chen et al. investigated $InNbO_4$ prepared by a solid-state reaction, and the effect of NiO and Co_3O_4 co-catalysts deposited using an incipient wetness method [88]. The NiO co-catalyst possessed a core–shell Ni–NiO structure. A 0.5 wt.% loading showed the highest activity in aqueous photocatalytic CO_2 reduction tests, producing *ca.* 1.6 µmol/g_{cat}/h of CH_3OH (halogen lamp, 0.14 mW/cm^2).

29.2.6 BiOBr, BiOI and BiVO$_4$

Zhang et. al. synthesised BiOBr microspheres using a solvothermal method in an ionic liquid [89]. The optical bandgap was *ca.* 2.7 eV. Photocatalytic CO_2 reduction measurements yielded CO as the majority product (> 95% selectivity) at a rate of *ca.* 30 μmol/g$_{cat}$/h (Xe lamp[5]). Wong et al. studied ultrathin (< 4 nm), bismuth-rich BiOBr, grown by a two-step solvothermal, hydrolysis method [90]. These materials possessed a substantially lower optical bandgap (*ca.* 2.4 eV) than the bulk material (*ca.* 2.9 eV). Photocatalytic CO_2 reduction produced CO and CH_4 at rates of *ca.* 2.6 and *ca.* 2.0 μmol/g$_{cat}$/h, respectively (Xe lamp[5]), where the CH_4 production rate was substantially higher than bulk or ultrathin BiOBr (< 0.2 μmol/g$_{cat}$/h).

Liu et al. studied the effect of facet control on the activity of BiOI nanosheets [91]. By varying the temperature of the synthesis, they were able to grow nanosheets with exposed {001} and {100} facets. The optical bandgaps were similar (*ca.* 1.9 eV); however, the activity of the nanosheets with exposed {001} facets was higher, producing CO and CH_4 at respective rates of *ca.* 5.0 and *ca.* 1.5 μmol/g$_{cat}$/h (Xe lamp[5]). This was attributed to enhanced charge carrier separation and a more negative conduction band position. Liu et al. also studied ultrathin BiOI (< 3 nm thick), grown by a hydrothermal method [92]. Ultrathin BiOI was more widely spaced in the {001} plane (*ca.* 4% expansion), and possessed a widened optical bandgap (*ca.* 2.2 eV) compared with the bulk material (*ca.* 1.9 eV). Ultrathin BiOI showed higher photocatalytic CO_2 reduction activity than bulk BiOI, producing CO at a rate of *ca.* 4.2 μmol/g$_{cat}$/h using visible light (Xe lamp, > 440 nm[5]) and *ca.* 0.8μmol/g$_{cat}$/h using red light (Xe lamp, > 700 nm[5]). The apparent quantum yields at 420 and 700 nm were 0.14% and 0.02%, respectively.

Whangbo et al. studied the effect of crystal phase on the activity of BiVO$_4$ [93]. They found that monoclinic BiVO$_4$ is substantially more active than tetragonal BiVO$_4$, which showed negligible activity. The photocatalytic reduction of CO_2 was examined in an aqueous environment. Monoclinic BiVO$_4$ produced CH_3CH_2OH as the majority product at a rate of *ca.* 110 μmol/g$_{cat}$/h with a UV filter (Xe lamp, > 420 nm[5]) and *ca.* 550 μmol/g$_{cat}$/h without a UV filter (Xe lamp[5]). Zan et al. studied BiVO$_4$ microspheres prepared by a hydrothermal method [94]. The microspheres adopted the monoclinic BiVO$_4$ phase, and showed high selectivity for the formation of CH_3OH (*ca.* 25 μmol/g$_{cat}$/h, Xe lamp[5]) during the aqueous photocatalytic reduction of CO_2 (apparent quantum efficiency of *ca.* 0.24%). Xie et al. synthesised single unit cell orthorhombic-BiVO$_4$ layers (*ca.* 1.3 nm thick), and studied the effect of controlling the vanadium vacancy concentration [95]. They found that vanadium-poor orthorhombic-BiVO$_4$ possessed superior electronic conductivity and charge carrier lifetime than its vanadium-rich counterpart. The materials showed high selectivity for producing CH_3OH in the aqueous photocatalytic reduction of CO_2, producing up to 400 μmol/g$_{cat}$/h (Xe lamp, 100 mW/cm^2). Moreover, at 350 and 550 nm, the vanadium-poor material drove photocatalytic CO_2 reduction with an apparent quantum efficiency of *ca.* 6.0 and *ca.* 1.0%, respectively. Kudo et al. developed a Z-scheme photocatalyst for CO_2 reduction (and water splitting) [96]. One side of the Z-scheme was formed of CoO_x-coated $BiVO_4$, which drove water oxidation, and the other side was formed of $CuGaS_2$, which reduced CO_2. The two halves of the Z-scheme were interlinked by reduced graphene oxide. Photocatalytic CO_2 reduction was carried out in an aqueous environment, producing CO at a rate of *ca.* 2.9 μmol/g_{cat}/h (Xe lamp, > 420 nm, 100 mW/cm^2).

29.2.7 Metal sulphides

Reisner et al. investigated self-assembled nickel terpyridine complexes supported on CdS quantum dots [97]. Photocatalytic CO_2 reduction was carried out in an aqueous environment using the sacrificial electron donor, triethanolamine. Their photocatalysts reduced CO_2 to CO with high selectivity (> 90%) with an average external quantum efficiency of 0.28 ± 0.04% (400 nm light). Ye et. al. studied the effect of anchoring Co(II) species on the surface of CdS [98]. They found that a 0.7 wt.% loading resulted in optimal aqueous photocatalytic CO_2 reduction activity, producing CO at a rate of *ca.* 400 μmol/g_{cat}/h (Xe lamp, > 420 nm[5]). Ho et al. synthesised WO_3/CdS composites using a precipitation method [99]. Optimal aqueous photocatalytic CO_2 reduction was observed at 5 mol% CdS loading, with CH_4 produced as the majority product at a rate of *ca.* 1.0 μmol/g_{cat}/h (Xe lamp, > 420 nm, 150 mW/cm^2). This activity in the composite was more than ten times higher than either parent material.

Zhu et al. coated TiO_2 nanotubes with either CdS or Bi_2S_3 using a direct precipitation method [100]. Photocatalytic CO_2 reduction testing was carried out in an aqueous environment. They found that Bi_2S_3-modified nanotubes possessed higher activity than CdS-modified nanotubes, and produced CH_3OH at a rate of *ca.* 45 μmol/g_{cat}/h (Xe lamp, *ca.* 1 sun), which was more than double the activity of the nanotubes alone.

Lou et al. studied hierarchical In_2S_3/$CdIn_2S_4$ heterostructured nanotubes [101]. The two components, In_2S_3 and $CdIn_2S_4$, possess similar optical bandgaps of 2.15 and 2.22 eV, respectively. Photocatalytic CO_2 reduction was measured in an aqueous environment using the sacrificial electron donor, triethanolamine. Their hierarchical heterostructures produced CO in high yield at a rate of *ca.* 825 μmol/g_{cat}/h (> 400 nm[5]), which was about 12 times higher than In_2S_3 alone. Moreover, their hierarchical heterostructures showed excellent stability to repeat photocatalytic testing.

29.2.8 C₃N₄, MOFs and zeolites

Various graphitic C_3N_4-based photocatalysts have been investigated for photocatalytic CO_2 reduction [102, 103]. Wu et al. studied the effect of precursor on the activity of g-C_3N_4 [104]. They found that g-C_3N_4 grown from urea (*ca.* 1.0 μmol/g_{cat}/h) showed higher visible light CO_2 reduction activity than g-C_3N_4 grown from thiourea (*ca.* 0.67 µmol/g$_{cat}$/h) for total CO$_2$ conversion to CO, CH$_4$ and CH$_3$CHO (Xe lamp, > 420 nm[5]). However, g-C$_3$N$_4$ grown from urea (*ca.* 1.6 µmol/g$_{cat}$/h) was less active under UV light than g-C$_3$N$_4$ grown from thiourea (*ca.* 2.3 µmol/g$_{cat}$/h) for total CO$_2$ conversion to CO, CH$_4$ and CH$_3$CHO (Xe lamp, > 200 nm[5]). A P25 Evonik standard showed significantly lower activity (*ca.* 0.3 µmol/g$_{cat}$/h). In addition, g-C$_3$N$_4$ grown from thiourea showed good stability to repeat photocatalytic testing.

Wong et al. investigated the effect of phosphoric acid post-treatment on mesoporous g-C$_3$N$_4$ [105]. In the photocatalytic reduction of CO$_2$, CO and CH$_4$ were produced at respective rates of *ca.* 40 and *ca.* 80 µmol/g$_{cat}$/h (Xe lamp[5]). Moreover, apparent quantum efficiencies of *ca.* 1.3 and *ca.* 0.8% were observed at 365 and 420 nm, respectively. Cui et al. prepared Mo-doped g-C$_3$N$_4$ with worm-like mesostructures using a pyrolysis method [106]. They found that their Mo-doped materials exhibited superior photocatalytic CO$_2$ reduction activity than their undoped materials, producing CO and CH$_4$ at respective rates of *ca.* 110 and *ca.* 15 µmol/g$_{cat}$/h (Hg lamp, 110,000 Lux).

In order to overcome the limitation of g-C$_3$N$_4$ not possessing a sufficiently positive and oxidising valence band to drive water oxidation, Ohno et al. studied g-C$_3$N$_4$/WO$_3$ composite photocatalysts [107]. Their composites were prepared by three methods (solid state, ball milling and impregnation). Their most active composites for driving aqueous photocatalytic CO$_2$ reduction were produced by ball milling, forming the majority product CH$_3$OH at a rate of *ca.* 14 µmol/g$_{cat}$/h (435 nm LED, *ca.* 3.0 mW/cm^2). Photodeposition of gold nanoparticles on g-C$_3$N$_4$ in the composite resulted in further enhancements in activity, producing CH$_3$OH at a rate of *ca.* 35 µmol/g$_{cat}$/h. Shi et al. synthesised g-C$_3$N$_4$/ Bi$_2$WO$_6$ composites through a facile hydrothermal method [108]. In the photocatalytic reduction of CO$_2$, their best performing composite produced CO, with 100% selectivity, at a rate of *ca.* 5.2 µmol/g$_{cat}$/h (Xe lamp[5]). This was *ca.* 20 and 6 times higher than the parent materials g-C$_3$N$_4$ and Bi$_2$WO$_6$, respectively. Petit et al. studied TiO$_2$/C$_3$N$_4$ nanosheet composites [109]. TiO$_2$ crystal formation was controlled to favour growth of {001} facets, where the composite with more {001} facets showed improved interfacial charge transfer and activity. Photocatalytic reduction CO$_2$ reduction was measured in a batch reactor at equal partial pressures of CO$_2$ and H$_2$. Highly selective reduction of CO$_2$ to CO was observed, with the best performing composite producing CO at a rate of *ca.* 2.5 µmol/g$_{cat}$/h (Xe lamp, > 325 nm[5]). This activity was more than tenfold higher than the parent materials. Time-resolved laser spectroscopy studies showed that hole transfer from TiO$_2$ to C$_3$N$_4$ occurred on the sub-µs, resulting in the formation of long-lived electrons that could drive CO$_2$ reduction.

Various MOFs have been studied for photocatalytic CO$_2$ reduction [110]; however, many do not show sufficient stability in water or under irradiation for practical use. Petit et. al. studied highly stable MOF (NH$_2$-UiO-66)/TiO$_2$ composites [111]. The composites were bi-functional, with high CO$_2$ adsorption capacity present in the MOF and high photocatalytic activity present in TiO$_2$. Photocatalytic CO$_2$ reduction activity was measured in a batch reactor with equal partial pressures of CO$_2$ and H$_2$ (1.15 bar). The majority product was CO and could be produced at a rate of > 4 µmol/g$_{cat}$/h in their most active materials (Xe lamp[5]). The higher activity in the composite was attributed to longer lived charge carriers, observed through time-resolved laser spectroscopy studies. Wang et al. investigated CdS/MOF composites (MOF was a zeolitic imidazolate framework) [112]. Photocatalytic CO_2 reduction was measured in an aqueous environment alongside the sacrificial electron donor, triethanolamine. CO was produced as the majority product in high yield (*ca.* 2,500 $\mu mol/g_{Cds}/h$, Xe lamp, > 420 nm[5]). An apparent quantum yield of *ca.* 1.9% was observed using 420 nm light.

Tasumi et al. investigated the mesoporous zeolites, Ti-MCM-41 and Ti-MCM-48, which contain titanium oxide species within their framework [113]. The zeolites were prepared by a hydrothermal process. In the photocatalytic reduction of CO_2, CH_4 and CH_3OH was produced at respective rates of *ca.* 7.5 and *ca.* 3 $\mu mol/g_{cat}/h$ (Hg lamp > 280 nm[5]). Pt-loading increased the CH_4 yield (*ca.* 12.5 $\mu mol/g_{cat}/h$), but decreased the CH_3OH yield (< 1 $\mu mol/g_{cat}/h$). Mul et al. studied the zeolite Ti-SBA-15 and found that photocatalytic CO_2 reduction was higher when using H_2O as the electron donor, as opposed to H_2 [114].

29.3 Summary and future outlook

Anthropogenic climate change, primarily caused by the excessive use of fossil fuels, has made the need to develop renewable, carbon-neutral sources of energy a priority. Since the first demonstration of photocatalytic CO_2 reduction to produce carbon-based fuels in 1979 [47], there has been a great number of studies on a wide range of materials. Although some materials have shown promise, we are yet to see an economically feasible device enter the market.

In this chapter, we covered key materials and studies of photocatalytic CO_2 reduction. The traditional calculation of activity in this field is in moles/mass$_{cat}$/time (typically $\mu mol/g_{cat}/h$). However, given the numerous ways in which photocatalytic CO_2 reduction was measured, it is not possible to make any fair comparisons of activity. This is a significant barrier for this field to progress, where ideally, some standardised protocol for measuring activity should be introduced. This would enable one to make rational comparisons, and would better guide future materials focus and development. Moreover, very few studies determine apparent quantum efficiency, which is now standard in the field of photocatalytic water splitting. In addition, $^{13}CO_2$ labelling studies should be carried out more frequently to ensure reaction products are formed from the reduction of CO_2, rather than the decomposition of the photocatalyst (or impurities within the photocatalyst). There is also a growing trend to use sacrificial electron donors, such as triethanolamine or hydrogen, that enhance photocatalytic activity by more rapidly scavenging photogenerated hole carriers (and thereby circumvent the need to oxidise water, a kinetically slow process [23]). However, the use of such sacrificial electron donors in an applied device would be impractical (unless these scavengers can be produced renewably and economically). In order to increase commercial viability, value-added oxidation re-actions should also be explored alongside CO_2 reduction; examples include glycerol [115] and chloride [116] oxidation to form the value-added chemicals glyceraldehyde and chlorine, respectively.

In the photocatalytic reduction of CO_2, a wide range of reaction products have been observed, where there is no obvious correlation between the material used and the products formed (e.g. TiO_2 can form CO, CH_4, CH_3OH). For such heteroge-neous reactions, more work needs to be carried out to determine the factors that control product selectivity. Moreover, in the parallel field of photocatalytic water splitting, the charge carrier dynamics of the most popular materials, and the rate-limiting steps for driving this reaction are well studied [23]. However, in the field of photocatalytic CO_2 reduction, such studies are lacking. In order to guide future ma-terials development in this field, similar in-depth studies should be carried out.

One vision for renewably producing fuels is to photocatalytically reduce CO_2: (i) locally at source, or (ii) alongside CO_2 capture and transport technologies. However, if such fuels are then used in transportation, the CO_2 emissions produced would be highly delocalised, and difficult to capture and recycle. One solution is to develop ma-terials that can drive photocatalytic CO_2 reduction at ambient concentrations (*ca.* 400 ppm), similar to plants, and thereby circumvent the need for CO_2 capture and storage.

To achieve a carbon-neutral economy, and avoid the potentially catastrophic impacts of climate change, all fossil fuel-dependant technologies need to be re-placed. In the case of freight (i.e. large transport vehicles such as trucks, ships and planes), current battery technologies are not sufficiently dense to power them. Therefore, renewable fuels, for the purpose of freight, need to be generated on a large scale to meet our demands. For photocatalytic CO_2 reduction technologies to have a future role, devices need to be produced at a scale commensurate to demand using ideally low-cost and earth abundant components. More work should be car-ried out on prototyping and scaling state-of-the-art materials for this purpose, and include techno-economic analyses.

References

[1] Broecker WS. Climatic change: are we on the brink of a pronounced global warming? Science (80-) [Internet] 1975 [cited 2015 Aug 21];189 (4201):460–3. Available from: http://www.scien cemag.org/content/189/4201/460.abstract

[2] Gruber N. Warming up, turning sour, losing breath: ocean biogeochemistry under global change. Philos Trans A Math Phys Eng Sci [Internet] 2011 [cited 2015 Jul 15];369(1943): 1980–96. Available from: http://rsta.royalsocietypublishing.org/content/369/1943/1980

[3] New M, Liverman D, Schroeder H, Schroder H, Anderson K. Four degrees and beyond: the potential for a global temperature increase of four degrees and its implications. Philos Trans A Math Phys Eng Sci [Internet] 2011 [cited 2015 Aug 21]; 369 (1934) : 6–19. Available from: http://rsta.royalsocietypublishing.org/content/369/1934/6

[4]   Morton O. Solar energy: a new day dawning? Silicon Valley sunrise. Nature [Internet] 2006
        [cited 2015 Jul 9];443(7107):19–22. Available from: http://dx.doi.org/10.1038/443019a

[5]   Time to shine: Solar power is fastest-growing source of new energy | Environment | The
        Guardian [Internet]. [cited 2018 Oct 26];Available from: https://www.theguardian.com/
        environment/2017/oct/04/solar-power-renewables-international-energy-agency

[6]   Sovacool BK. The intermittency of wind, solar, and renewable electricity generators:
        Technical barrier or rhetorical excuse? Util Policy [Internet] 2009 [cited 2015 Apr 10];
        17(3–4):288–96. Available from: http://www.sciencedirect.com/science/article/pii/
        S0957178708000611

[7]   Beaudin M, Zareipour H, Schellenberglabe A, Rosehart W. Energy storage for mitigating the
        variability of renewable electricity sources: An updated review. Energy Sustain Dev [Internet]
        2010 [cited 2014 Jul 10];14(4):302–14. Available from: http://www.sciencedirect.com/sci
        ence/article/pii/S0973082610000566

[8]   Dunn B, Kamath H, Tarascon J-M. Electrical energy storage for the grid: a battery of choices.
        Science (80-) [Internet] 2011 [cited 2019 Jan 26];334(6058):928–35. Available from: http://
        www.ncbi.nlm.nih.gov/pubmed/22096188

[9]   Lee DS, Lim L, Owen B. Shipping and aviation emissions in the context of a 2 °C emission
        pathway [Internet]. [cited 2018 Oct 26]. Available from: https://www.transportenvironment.
        org/sites/te/files/publications/Shipping and aviation emissions and 2 degrees v1-6.pdf

[10]   Gust D, Kramer D, Moore A, Moore TA, Vermaas W. Engineered and Artificial Photosynthesis:
        Human Ingenuity Enters the Game. MRS Bull [Internet] 2008 [cited 2019 Jan 26];33(04):
        383–7. Available from: http://www.journals.cambridge.org/abstract_S0883769400004905

[11]   Hall CAS, Dale BE, Pimentel D. Seeking to Understand the Reasons for Different Energy
        Return on Investment (EROI) Estimates for Biofuels. Sustainability [Internet] 2011 [cited 2015
        Aug 21]; 3 (12):2413–32. Available from: http://www.mdpi.com/2071-1050/3/12/2413/htm

[12]   Tachibana Y, Vayssieres L, Durrant JR. Artificial photosynthesis for solar water-splitting. Nat.
        Photonics. [Internet] 2012 [cited 2014 Jul 14];6(8):511–8. Available from: http://www.nature.
        com/doifinder/10.1038/nphoton.2012.175

[13]   Habisreutinger SN, Schmidt-Mende L, Stolarczyk JK. Photocatalytic reduction of CO_2 on TiO2
        and other semiconductors. Angew. Chemie. – Int. Ed. 2013;52(29):7372–408.

[14]   Álvarez A, Bansode A, Urakawa A, et al. Challenges in the Greener Production of Formates/
        Formic Acid, Methanol, and DME by Heterogeneously Catalyzed CO_2 Hydrogenation
        Processes. Chem. Rev. 2017;117(14):9804–38.

[15]   Jia J, Seitz LC, Benck JD, et al. Solar water splitting by photovoltaic-electrolysis with a solar-
        to-hydrogen efficiency over 30%; Nat. Publ. Gr. [Internet] 2016 7(May):1–6. Available from:
        http://dx.doi.org/10.1038/ncomms13237

[16]   Young JL, Steiner MA, Döscher H, France RM, Turner JA, Deutsch TG. Direct solar-to-hydrogen
        conversion via inverted metamorphic multi-junction semiconductor architectures. Nat. energ.
        2017;17028(March):1–8.

[17]   Wang Q, Hisatomi T, Suzuki Y, et al. Particulate Photocatalyst Sheets Based on Carbon
        Conductor Layer for Efficient Z-Scheme Pure-Water Splitting at Ambient Pressure. J. Am.
        Chem. Soc. [Internet] 2017 [cited 2019 Jan 26];139(4):1675–83. Available from: http://pubs.
        acs.org/doi/10.1021/jacs.6b12164

[18]   Tseng I, Chang W, Wu JCS. Photoreduction of CO_2 using sol – gel derived titania and titania-
        supported copper catalysts. Appl. Catal. B. Environ 2002;37:37–48.

[19]   Fujishima A, Honda K. Electrochemical photolysis of water at a semiconductor electrode.
        Nature [Internet] 1972 [cited 2011 Jul 22];238(5358):37–8. Available from: http://dx.doi.org/
        10.1038/238037a0

[20] Osterloh FE. Inorganic nanostructures for photoelectrochemical and photocatalytic water splitting. Chem. Soc. Rev. 2013;42:2294–320.

[21] Osterloh FE. Inorganic Materials as Catalysts for Photochemical Splitting of Water. Chem. Mater. [Internet] 2008;20(1):35–54. Available from: http://pubs.acs.org/doi/abs/10.1021/cm7024203

[22] Walter MG, Warren EL, Mckone JR, et al. Solar Water Splitting Cells. Chem. Rev. 2010;110:6446.

[23] Andreas Kafizas, Robert Godin JRD. Charge Carrier Dynamics in Metal Oxide Photoelectrodes for Water Oxidation. Semicond. Semimetals. 2017;97:3–46.

[24] Wang Q, Hisatomi T, Jia Q, et al. Scalable water splitting on particulate photocatalyst sheets with a solar-to-hydrogen energy conversion efficiency exceeding 1%. Nat Mater 2016;15 (June):611.

[25] Ozin GA. You can't have an energy revolution without transforming advances in materials, chemistry and catalysis into policy change and action. Energy. Environ. Sci. [Internet] 2015;8 (6):1682–4. Available from: http://xlink.rsc.org/?DOI=C5EE00907C

[26] Ozin GA. Throwing New Light on the Reduction of CO_2. Adv. Mater. 2015;27(11):1957–63.

[27] Jia J, Qian C, Dong Y, et al. Heterogeneous catalytic hydrogenation of CO_2 by metal oxides: defect engineering – perfecting imperfection. Chem. Soc. Rev. [Internet] 2017;46(15): 4631–44. Available from: http://xlink.rsc.org/?DOI=C7CS00026J

[28] Kalamaras E, Maroto-valer MM, Shao M, Xuan J, Wang H. Solar carbon fuel via photoelectrochemistry. Catal. Today. [Internet] 2018;317(October 2017):56–75. Available from: https://doi.org/10.1016/j.cattod.2018.02.045

[29] Li K, Peng B, Peng T. Recent Advances in Heterogeneous Photocatalytic CO_2 Conversion to Solar Fuels. ACS Catal. 2016;6(11):7485–527.

[30] White JL, Baruch MF, Pander JE, et al. Light-Driven Heterogeneous Reduction of Carbon Dioxide: Photocatalysts and Photoelectrodes. Chem. Rev. 2015;115(23):12888–935.

[31] Dhakshinamoorthy A, Navalon S, Corma A, Garcia H. Photocatalytic CO_2 reduction by TiO2 and related titanium containing solids. Energ. Environ. Sci. [Internet] 2012;5 (11) : 9217. Available from: http://xlink.rsc.org/?DOI=c2ee21948d

[32] Navalón S, Dhakshinamoorthy A, Álvaro M, Garcia H. Photocatalytic CO_2 reduction using non-itanium metal oxides and sulfides. ChemSusChem 2013;6(4):562–77.

[33] Hendon CH, Tiana D, Fontecave M, et al. Engineering the Optical Response of the Titanium-MIL-125 Metal–Organic Framework through Ligand Functionalization. J. Am. Chem. Soc. [Internet] 2013 [cited 2019 Jan 26];135(30):10942–5. Available from: http://pubs.acs.org/doi/10.1021/ja405350u

[34] Xu H-Q, Hu J, Wang D, et al. Visible-Light Photoreduction of CO_2 in a Metal–Organic Framework: Boosting Electron–Hole Separation via Electron Trap States. J. Am. Chem. Soc. [Internet] 2015 [cited 2019 Jan 26];137(42):13440–3. Available from: http://pubs.acs.org/doi/10.1021/jacs.5b08773

[35] Chambers MB, Wang X, Ellezam L, et al. Maximizing the Photocatalytic Activity of Metal–Organic Frameworks with Aminated-Functionalized Linkers: Substoichiometric Effects in MIL-125-NH $_2$. J. Am. Chem. Soc. [Internet] 2017 [cited 2019 Jan 26];139(24):8222–8. Available from: http://pubs.acs.org/doi/10.1021/jacs.7b02186

[36] Yang S, Hu W, Zhang X, et al. 2D Covalent Organic Frameworks as Intrinsic Photocatalysts for Visible Light-Driven CO_2 Reduction. J. Am. Chem. Soc. [Internet] 2018 [cited 2019 Jan 26];140 (44):14614–8. Available from: http://pubs.acs.org/doi/10.1021/jacs.8b09705

[37] Vyas VS, Haase F, Stegbauer L, et al. A tunable azine covalent organic framework platform for visible light-induced hydrogen generation. Nat. Commun. [Internet] 2015 [cited 2019 Jan 26];6(1):8508. Available from: http://www.nature.com/articles/ncomms9508

[38]   Kho ET, Tan TH, Lovell E, Wong RJ, Scott J, Amal R. A review on photo-thermal catalytic conversion of carbon dioxide. Green. Energy. Environ. [Internet] 2017;2(3):204–17. Available from: http://linkinghub.elsevier.com/retrieve/pii/S246802571730064X

[39]   Scheffe JR, Steinfeld A. Oxygen exchange materials for solar thermochemical splitting of H2O and CO_2: A review. Mater. Today [Internet] 2014;17(7):341–8. Available from: http://dx.doi.org/10.1016/j.mattod.2014.04.025

[40]   Smestad GP, Steinfeld A. Review: Photochemical and thermochemical production of solar fuels from H2O and CO_2 using metal oxide catalysts. Ind. Eng. Chem. Res. 2012;51(37):11828–40.

[41]   Chen S, Wang L-W. Thermodynamic Oxidation and Reduction Potentials of Photocatalytic Semiconductors in Aqueous Solution. Chem. Mater. [Internet] 2012;24(18):3659–66. Available from: http://pubs.acs.org/doi/abs/10.1021/cm302533s

[42]   Xu Y, Schoonen AA. The absolute energy positions of conduction and valence bands of selected semiconducting minerals. Am. Mineral. 2000;85:543.

[43]   Bhachu DS, Moniz SJA, Sathasivam S, et al. Bismuth oxyhalides: synthesis, structure and photoelectrochemical activity. Chem. Sci. [Internet] 2016 [cited 2019 Jan 26];7(8):4832–41. Available from: http://xlink.rsc.org/?DOI=C6SC00389C

[44]   Saison T, Gras P, Chemin N, et al. New Insights into Bi_2WO_6 Properties as a Visible-Light Photocatalyst. J. Phys. Chem. C. [Internet] 2013 [cited 2019 Jan 26];117(44):22656–66. Available from: http://pubs.acs.org/doi/10.1021/jp4048192

[45]   Bai Y, Ye L, Wang L, et al. g-C3N4/Bi4O5I2 heterojunction with I3-/I-redox mediator for enhanced photocatalytic CO_2 conversion. Appl. Catal. B. Environ. [Internet] 2016;194:98–104. Available from: http://dx.doi.org/10.1016/j.apcatb.2016.04.052

[46]   Marxer D, Furler P, Takacs M, Steinfeld A. Solar thermochemical splitting of CO_2 into separate streams of CO and O_2 with high selectivity, stability, conversion, and efficiency. Energ. Environ. Sci. [Internet] 2017 [cited 2019 Jan 26];10(5):1142–9. Available from: http://xlink.rsc.org/?DOI=C6EE03776C

[47]   T Inoue, A Fujishima SK and KH. Photoelectrocatalytic reduction of carbon dioxide in aqueous suspensions of semiconductor powders.. Nature 1979;27:637–638.

[48]   Li Y, Wang WN, Zhan Z, Woo MH, Wu CY, Biswas P. Photocatalytic reduction of CO_2 with H2O on mesoporous silica supported Cu/TiO2catalysts. Appl. Catal. B. Environ. [Internet] 2010;100(1–2):386–92. Available from: http://dx.doi.org/10.1016/j.apcatb.2010.08.015

[49]   Wang Y, Li B, Zhang C, et al. Ordered mesoporous CeO2-TiO2 composites: Highly efficient photocatalysts for the reduction of CO_2 with H2O under simulated solar irradiation. Appl. Catal. B. Environ. 2013;130–131:277–84.

[50]   Liu L, Zhao H, Andino JM, Li Y. Photocatalytic CO_2 reduction with H2O on TiO2 nanocrystals: Comparison of anatase, rutile, and brookite polymorphs and exploration of surface chemistry. ACS Catal. 2012;2(8):1817–28.

[51]   Zhang Q, Gao T, Andino JM, Li Y. Copper and iodine co-modified TiO2 nanoparticles for improved activity of CO_2 photoreduction with water vapor. Appl. Catal. B. Environ. [Internet] 2012;123–124:257–64. Available from: http://dx.doi.org/10.1016/j.apcatb.2012.04.035

[52]   Zhang Q, Li Y, Ackerman EA, Gajdardziska-Josifovska M, Li H. Visible light responsive iodine-doped TiO2 for photocatalytic reduction of CO_2 to fuels. Appl. Catal. A. Gen. [Internet] 2011;400(1–2):195–202. Available from: http://dx.doi.org/10.1016/j.apcata.2011.04.032

[53]   Liu L, Gao F, Zhao H, Li Y. Tailoring Cu valence and oxygen vacancy in Cu/TiO2 catalysts for enhanced CO2photoreduction efficiency. Appl. Catal. B. Environ. [Internet] 2013; 134–135:349–58. Available from: http://dx.doi.org/10.1016/j.apcatb.2013.01.040

[54]   Yu J, Low J, Xiao W, Zhou P, Jaroniec M. Enhanced photocatalytic CO_2-Reduction activity of anatase TiO2 by Coexposed {001} and {101} facets. J. Am. Chem. Soc. 2014;136(25):8839–42.

[55] Neațu Ș, Maciá-Agulló JA, Concepción P, Garcia H. Gold–Copper Nanoalloys Supported on TiO2 as Photocatalysts for CO$_2$ Reduction by Water. J. Am. Chem. Soc. 2014;136:15969–76.

[56] Kočí K, Matějů K, Obalová L, et al. Effect of silver doping on the TiO2 for photocatalytic reduction of CO$_2$. Appl. Catal. B. Environ. 2010;96(3–4):239–44.

[57] Kočí K, Obalová L, Matějová L, et al. Effect of TiO2 particle size on the photocatalytic reduction of CO$_2$. Appl. Catal. B. Environ. 2009;89(3–4):494–502.

[58] Varghese OK, Paulose M, Latempa TJ, et al. High-Rate Solar Photocatalytic Conversion of CO$_2$ and Water Vapor to Hydrocarbon Fuels. Nano. Lett. 2009;9(2):731–7.

[59] Zhang QH, Han WD, Hong YJ, Yu JG. Photocatalytic reduction of CO$_2$ with H2O on Pt-loaded TiO2 catalyst. Catal. Today 2009;148(3–4):335–40.

[60] Tu W, Zhou Y, Liu Q, et al. An in situ simultaneous reduction-hydrolysis technique for fabrication of TiO2-graphene 2D sandwich-like hybrid nanosheets: Graphene-promoted selectivity of photocatalytic-driven hydrogenation and coupling of CO 2 into methane and ethane. Adv Funct Mater 2013;23(14):1743–9.

[61] Uner D, Oymak MM. On the mechanism of photocatalytic CO$_2$ reduction with water in the gas phase. Catal. Today [Internet] 2012;181(1):82–8. Available from: http://dx.doi.org/10.1016/j.cattod.2011.06.019

[62] Liu D, Fernández Y, Ola O, et al. On the impact of Cu dispersion on CO$_2$ photoreduction over Cu/TiO2. Catal. Commun 2012;25:78–82.

[63] Li X, Zhuang Z, Li W, Pan H. Photocatalytic reduction of CO$_2$ over noble metal-loaded and nitrogen-doped mesoporous TiO2. Appl. Catal. A. Gen. [Internet] 2012;429–430:31–8. Available from: http://dx.doi.org/10.1016/j.apcata.2012.04.001

[64] Wang WN, An WJ, Ramalingam B, et al. Size and structure matter: Enhanced CO$_2$ photoreduction efficiency by size-resolved ultrafine Pt nanoparticles on TiO2 single crystals. J. Am. Chem. Soc. 2012;134(27):11276–81.

[65] Liang YT, Vijayan BK, Gray KA, Hersam MC. Minimizing graphene defects enhances titania nanocomposite-based photocatalytic reduction of CO$_2$ for improved solar fuel production. Nano. Lett. 2011;11(7):2865–70.

[66] Asi MA, He C, Su M, et al. Photocatalytic reduction of CO$_2$ to hydrocarbons using AgBr/TiO2 nanocomposites under visible light. Catal. Today [Internet] 2011;175(1):256–63. Available from: http://dx.doi.org/10.1016/j.cattod.2011.02.055

[67] Yui T, Kan A, Saitoh C, Koike K, Ibusuki T, Ishitani O. Photochemical reduction of CO$_2$ using TiO2: Effects of organic adsorbates on TiO2 and deposition of Pd onto TiO2. ACS Appl. Mater. Interfaces. 2011;3(7):2594–600.

[68] Xia XH, Jia ZJ, Yu Y, Liang Y, Wang Z, Ma LL. Preparation of multi-walled carbon nanotube supported TiO2 and its photocatalytic activity in the reduction of CO$_2$ with H2O. Carbon N Y 2007;45(4):717–21.

[69] Tseng IH, Wu JCS, Chou HY. Effects of sol-gel procedures on the photocatalysis of Cu/TiO2 in CO$_2$ photoreduction. J. Catal. 2004;221(2):432–40.

[70] Moss B, Lim KK, Beltram A, et al. Comparing photoelectrochemical water oxidation, recombination kinetics and charge trapping in the three polymorphs of TiO2. Sci. Rep. [Internet] 2017;7(1):2938. Available from: http://www.nature.com/articles/s41598-017-03065-5

[71] Li YF, Soheilnia N, Greiner M, et al. Pd@HyWO3−x Nanowires Efficiently Catalyze the CO$_2$ Heterogeneous Reduction Reaction with a Pronounced Light Effect. ACS Appl. Mater. Interfaces. 2018;

[72] Liang L, Li X, Liang L, et al. Infrared Light-Driven CO 2 Overall Splitting at Room Temperature Infrared Light-Driven CO 2 Overall Splitting at Room Temperature. Joule [Internet] 2018;2(5):1004–16. Available from: https://doi.org/10.1016/j.joule.2018.02.019

[73] Chen X, Zhou Y, Liu Q, Li Z, Liu J, Zou Z. Ultrathin, single-crystal WO3 nanosheets by two-dimensional oriented attachment toward enhanced photocatalystic reduction of CO_2 into hydrocarbon fuels under visible light. ACS Appl. Mater. Interfaces. 2012;4(7):3372–7.

[74] Wang P-Q, Bai Y, Luo P-Y, Liu J-Y. Graphene–WO3 nanobelt composite: Elevated conduction band toward photocatalytic reduction of CO_2 into hydrocarbon fuels. Catal. Commun. [Internet] 2013;38:82–5. Available from: http://linkinghub.elsevier.com/retrieve/pii/S156673671300157X

[75] Sun Z, Yang Z, Liu H, Wang H, Wu Z. Visible-light CO_2 photocatalytic reduction performance of ball-flower-like Bi2WO6 synthesized without organic precursor: Effect of post-calcination and water vapor. Appl. Surf. Sci. [Internet] 2014;315(1):360–7. Available from: http://dx.doi.org/10.1016/j.apsusc.2014.07.153

[76] Zhou Y, Tian Z, Zhao Z, et al. High-Yield Synthesis of Ultrathin and Uniform Bi2WO6 Square Nanoplates Benefitting from Photocatalytic Reduction of CO_2 into Renewable Hydrocarbon Fuel under Visible Light. ACS Appl. Mater. Interfaces. [Internet] 2011;3(9):3594–601. Available from: http://pubs.acs.org/doi/abs/10.1021/am2008147

[77] Liang L, Lei F, Gao S, et al. Single Unit Cell Bismuth Tungstate Layers Realizing Robust Solar CO_2 Reduction to Methanol. Angew Chemie – Int Ed 2015;54(47):13971–4.

[78] Cheng H, Huang B, Liu Y, et al. An anion exchange approach to Bi2WO6 hollow microspheres with efficient visible light photocatalytic reduction of CO_2 to methanol. Chem. Commun. [Internet] 2012;48(78):9729. Available from: http://xlink.rsc.org/?DOI=c2cc35289c

[79] Dai W, Xu H, Yu J, et al. Photocatalytic reduction of CO_2 into methanol and ethanol over conducting polymers modified Bi2WO6 microspheres under visible light. Appl. Surf. Sci. 2015;356:173–80.

[80] Dai W, Yu J, Deng Y, Hu X, Wang T, Luo X. Facile synthesis of MoS2/Bi2WO6 nanocomposites for enhanced CO_2 photoreduction activity under visible light irradiation. Appl. Surf. Sci. 2017;403:230–9.

[81] Ghuman KK, Wood TE, Hoch LB, Mims CA, Ozin GA, Singh CV. Illuminating CO_2 reduction on frustrated Lewis pair surfaces: investigating the role of surface hydroxides and oxygen vacancies on nanocrystalline $In_2O_{3-x}(OH)_y$. Phys. Chem. Chem. Phys. [Internet] 2015;17(22):14623–35. Available from: http://xlink.rsc.org/?DOI=C5CP02613J

[82] Ghuman KK, Hoch LB, Szymanski P, et al. Photoexcited Surface Frustrated Lewis Pairs for Heterogeneous Photocatalytic CO_2 Reduction. J. Am. Chem. Soc. [Internet] 2016;138(4):1206–14. Available from: http://pubs.acs.org/doi/10.1021/jacs.5b10179

[83] Ghoussoub M, Yadav S, Ghuman KK, Ozin GA, Singh CV. Metadynamics-Biased ab Initio Molecular Dynamics Study of Heterogeneous CO_2 Reduction via Surface Frustrated Lewis Pairs. ACS Catal. [Internet] 2016;6(10):7109–17. Available from: http://pubs.acs.org/doi/10.1021/acscatal.6b01545

[84] He L, Wood TE, Wu B, et al. Spatial Separation of Charge Carriers in In2O3-x(OH)y Nanocrystal Superstructures for Enhanced Gas-Phase Photocatalytic Activity. ACS Nano. 2016;10(5):5578–86.

[85] Hoch LB, Szymanski P, Ghuman KK, et al. Carrier dynamics and the role of surface defects: Designing a photocatalyst for gas-phase CO_2 reduction. Proc. Natl. Acad. Sci. U S A [Internet] 2016 [cited 2017 Sep 1];113(50):E8011–20. Available from: http://www.ncbi.nlm.nih.gov/pubmed/27911785

[86] Hoch LB, He L, Qiao Q, et al. Effect of Precursor Selection on the Photocatalytic Performance of Indium Oxide Nanomaterials for Gas-Phase CO_2Reduction. Chem. Mater. 2016;28(12):4160–8.

[87] Pan Y-X, You Y, Xin S, et al. Photocatalytic CO_2 Reduction by Carbon-Coated Indium-Oxide Nanobelts. J. Am. Chem. Soc. [Internet] 2017;139(11):4123–9. Available from: http://pubs.acs.org/doi/10.1021/jacs.7b00266

[88] Lee DS, Chen HJ, Chen YW. Photocatalytic reduction of carbon dioxide with water using InNbO4 catalyst with NiO and Co3O4 cocatalysts. J Phys Chem Solids [Internet] 2012;73(5): 661–9. Available from: http://dx.doi.org/10.1016/j.jpcs.2012.01.005

[89] Wang P, Yang P, Bai Y, et al. Synthesis of 3D BiOBr microspheres for enhanced photocatalytic CO_2reduction. J. Taiwan. Inst. Chem. Eng. [Internet] 2016;68:295–300. Available from: http:// dx.doi.org/10.1016/j.jtice.2016.09.013

[90] Ye L, Jin X, Liu C, et al. Thickness-ultrathin and bismuth-rich strategies for BiOBr to enhance photoreduction of CO_2into solar fuels. Appl. Catal. B. Environ. [Internet] 2016;187:281–90. Available from: http://dx.doi.org/10.1016/j.apcatb.2016.01.044

[91] Ye L, Jin X, Ji X, et al. Facet-dependent photocatalytic reduction of CO_2 on BiOI nanosheets. Chem. Eng. J. [Internet] 2016;291:39–46. Available from: http://dx.doi.org/10.1016/j. cej.2016.01.032

[92] Ye L, Wang H, Jin X, et al. Synthesis of olive-green few-layered BiOI for efficient photoreduction of CO_2 into solar fuels under visible/near-infrared light. Sol. Energy. Mater. Sol. Cells. [Internet] 2016;144:732–9. Available from: http://dx.doi.org/10.1016/j. solmat.2015.10.022

[93] Liu Y, Huang B, Dai Y, et al. Selective ethanol formation from photocatalytic reduction of carbon dioxide in water with BiVO4 photocatalyst. Catal. Commun. [Internet] 2009;11(3): 210–3. Available from: http://dx.doi.org/10.1016/j.catcom.2009.10.010

[94] Mao J, Peng T, Zhang X, Li K, Zan L. Selective methanol production from photocatalytic reduction of CO_2 on BiVO4 under visible light irradiation. Catal. Commun. [Internet] 2012;28: 38–41. Available from: http://dx.doi.org/10.1016/j.catcom.2012.08.008

[95] Gao S, Gu B, Jiao X, et al. Highly Efficient and Exceptionally Durable CO_2 Photoreduction to Methanol over Freestanding Defective Single-Unit-Cell Bismuth Vanadate Layers. J. Am. Chem. Soc. 2017;139(9):3438–45.

[96] Iwase A, Yoshino S, Takayama T, Ng YH, Amal R, Kudo A. Water Splitting and CO_2 Reduction under Visible Light Irradiation Using Z-Scheme Systems Consisting of Metal Sulfides, CoOx-Loaded BiVO4, and a Reduced Graphene Oxide Electron Mediator. J. Am. Chem. Soc. 2016;138(32):10260–4.

[97] Kuehnel MF, Orchard KL, Dalle KE, Reisner E. Selective Photocatalytic CO_2Reduction in Water through Anchoring of a Molecular Ni Catalyst on CdS Nanocrystals. J. Am. Chem. Soc. 2017;139(21):7217–23.

[98] Zhao G, Zhou W, Sun Y, et al. Efficient photocatalytic CO_2 reduction over Co(II) species modified CdS in aqueous solution. Appl Catal B Environ 2018;226(November2017):252–7.

[99] Jin J, Yu J, Guo D, Cui C, Ho W. A Hierarchical Z-Scheme CdS-WO3 Photocatalyst with Enhanced CO_2 Reduction Activity. Small [Internet] 2015;(39):n/a-n/a. Available from: http:// doi.wiley.com/10.1002/smll.201500926

[100] Li X, Liu H, Luo D, et al. Adsorption of CO_2 on heterostructure CdS(Bi2S3)/TiO2 nanotube photocatalysts and their photocatalytic activities in the reduction of CO_2 to methanol under visible light irradiation. Chem. Eng. J. [Internet] 2012;180:151–8. Available from: http://dx. doi.org/10.1016/j.cej.2011.11.029

[101] Wang S, Guan BY, Lu Y, Lou XW "David." Formation of Hierarchical In $_2$ S $_3$ –CdIn $_2$ S $_4$ Heterostructured Nanotubes for Efficient and Stable Visible Light CO_2 Reduction. J. Am. Chem. Soc. [Internet] 2017;(step I):jacs.7b10733. Available from: http://pubs.acs.org/doi/10. 1021/jacs.7b10733

[102] Shen M, Zhang L, Shi J. Converting CO_2 into fuels by graphitic carbon nitride-based photocatalysts. Nanotechnology 2018;29(41).

[103] Ong WJ, Tan LL, Ng YH, Yong ST, Chai SP. Graphitic Carbon Nitride (g-C3N4)-Based Photocatalysts for Artificial Photosynthesis and Environmental Remediation: Are We a Step Closer to Achieving Sustainability? Chem. Rev. 2016;116(12):7159–329.

[104] Wang H, Sun Z, Li Q, Tang Q, Wu Z. Surprisingly advanced CO_2 photocatalytic conversion over thiourea derived g-C3N4 with water vapor while introducing 200–420 nm UV light. J. CO_2 Util. [Internet] 2016;14:143–51. Available from: http://dx.doi.org/10.1016/j.jcou.2016.04.006

[105] Ye L, Wu D, Chu KH, et al. Phosphorylation of g-C3N4 for enhanced photocatalytic CO_2 reduction. Chem. Eng. J. [Internet] 2016;304:376–83. Available from: http://dx.doi.org/10.1016/j.cej.2016.06.059

[106] Wang Y, Xu Y, Wang Y, et al. Synthesis of Mo-doped graphitic carbon nitride catalysts and their photocatalytic activity in the reduction of CO_2 with H2O. Catal. Commun. 2016;74:75–9.

[107] Ohno T, Murakami N, Koyanagi T, Yang Y. Photocatalytic reduction of CO_2 over a hybrid photocatalyst composed of WO3 and graphitic carbon nitride (g-C3N4) under visible light. J. CO_2 Util. [Internet] 2014;6:17–25. Available from: http://linkinghub.elsevier.com/retrieve/pii/S2212982014000134

[108] Li M, Zhang L, Fan X, Zhou Y, Wu M, Shi J. Highly selective CO_2 photoreduction to CO over g-C3N4/Bi 2WO6 composites under visible light. J. Mater. Chem. A. [Internet] 2015;3(9): 5189–96. Available from: http://xlink.rsc.org/?DOI=C4TA06295G

[109] Crake A, Christoforidis KC, Godin R, et al. Titanium dioxide/carbon nitride nanosheet nanocomposites for gas phase CO_2 photoreduction under UV-visible irradiation. Appl. Catal. B. Environ. 2019;242:369–78.

[110] Chen Y, Wang D, Deng X, Li Z. Metal-organic frameworks (MOFs) for photocatalytic CO_2reduction. Catal. Sci. Technol. 2017;7(21):4893–904.

[111] Crake A, Christoforidis KC, Kafizas A, Zafeiratos S, Petit C. CO_2 capture and photocatalytic reduction using bifunctional TiO2/MOF nanocomposites under UV–vis irradiation. Appl. Catal. B. Environ. [Internet] 2017 [cited 2017 Apr 3];210:131–40. Available from: http://www.sciencedirect.com/science/article/pii/S0926337317302527

[112] Wang S, Wang X. Photocatalytic CO_2reduction by CdS promoted with a zeolitic imidazolate framework. Appl. Catal. B. Environ. [Internet] 2015;162:494–500. Available from: http://dx.doi.org/10.1016/j.apcatb.2014.07.026

[113] Anpo M, Yamashita H, Ikeue K, et al. Photocatalytic reduction of CO_2 with H2O on Ti-MCM-41 and Ti-MCM-48 mesoporous zeolite catalysts. Catal. Today [Internet] 1998;44(1–4):327–32. Available from: http://linkinghub.elsevier.com/retrieve/pii/S0920586198002065

[114] Yang CC, Vernimmen J, Meynen V, Cool P, Mul G. Mechanistic study of hydrocarbon formation in photocatalytic CO_2 reduction over Ti-SBA-15. J. Catal. [Internet] 2011;284(1):1–8. Available from: http://dx.doi.org/10.1016/j.jcat.2011.08.005

[115] Maurino V, Bedini A, Minella M, Rubertelli F, Pelizzetti E, Minero C. Glycerol Transformation Through Photocatalysis: A Possible Route to Value Added Chemicals. J. Adv. Oxid. Technol. 2008;11:184–192.

[116] McCafferty L, Rourke CO, Mills A, Kafizas A, Parkin I. P, Darr JA. Light-driven generation of chlorine and hydrogen from brine using highly selective Ru / Ti oxide redox catalysts. Sustain. Energ. Fuels 2017;1:254–7.

[117] Gagliardi M. Photocatalysts: Technologies and Global Markets: AVM069B | BCC Research [Internet]. 2015 [cited 2017 Dec 12];Available from: https://www.bccresearch.com/market-research/advanced-materials/photocatalysts-technologies-markets-report-avm069b.html

[118] Jinliang Lin, Zhengxin Ding YH& XW. Ionic Liquid Co-catalyzed Artificial Photosynthesis of CO . Sc.i Rep. 2013;3:1056.

Simon C. Parker, Andrew J. Sadler, James D. Shipp
and Julia A. Weinstein

30 Photochemical reduction of CO_2 with metal-based systems

30.1 Introduction

Carbon fixation is the process by which organisms convert CO_2, into organic compounds. Photosynthesis is the archetypal process by which carbon fixation occurs. CO_2 fixation via photosynthesis typically results in complex organic molecules such as sugars that would be challenging to achieve in a simple catalytic system. When designing systems for CO_2 fixation, simpler chemical processes are considered that require lesser energy input (Scheme 30.1).

$$CO_2 + 2H^+ + 2e^- \rightarrow CO + H_2O \quad E^0 = -0.53\,V \qquad \text{(Eq.1)}$$

$$CO_2 + 2H^+ + 2e^- \rightarrow HCO_2H \quad E^0 = -0.61\,V \qquad \text{(Eq.2)}$$

$$CO_2 + 4H^+ + 4e^- \rightarrow HCHO + H_2O \quad E^0 = -0.48\,V \qquad \text{(Eq.3)}$$

$$CO_2 + 6H^+ + 6e^- \rightarrow CH_3OH + H_2O \quad E^0 = -0.38\,V \qquad \text{(Eq.4)}$$

$$CO_2 + 8H^+ + 8e^- \rightarrow CH_4 + 2H_2O \quad E^0 = -0.24\,V \qquad \text{(Eq.5)}$$

$$CO_2 + e^- \rightarrow CO_2^- \quad E^0 = -1.90\,V \qquad \text{(Eq.6)}$$

Scheme 30.1: The reduction potentials for various reactions of CO_2, recorded at pH 7 in aqueous solution versus the normal hydrogen electrode (NHE), 25 °C, 1 atm gas pressure and 1 M for the other solutes [1].

Some of the reduction products of CO_2, for example methanol, or ether, are promising as liquid fuels [1, 2], whilst CO is an industrially important chemical as a component of syngas used in the Fischer–Tropsch process for the production of hydrocarbons [3]. All of the processes of CO_2 reduction are multielectron, energy-demanding reactions (the one-electron process leads to CO_2-anion, which has different geometry to that of CO_2, hence requires large reorganisation energy) and require either multielectron catalysts, or a panel of catalysts [4].

Photochemical reduction of CO_2 aims to utilise the energy of light to drive this energy-demanding process. A photocatalytic system for CO_2 reduction should

Simon C. Parker, Andrew J. Sadler, James D. Shipp, Julia A. Weinstein, Department of Chemistry, The University of Sheffield, Sheffield, U. K.

https://doi.org/10.1515/9783110665147-030

include a light-absorbing unit, a catalyst and an electron donor to replenish the catalyst after electron transfer to CO_2 takes place.

Several approaches to photochemical CO_2 reduction include, for example, a light-absorbing photocatalyst in a homogeneous system; a photosensitiser that activates a catalyst in a bimolecular process; a photosensitiser that activates a catalyst in an intramolecular process; or a heterogeneous system whereby the light is absorbed by a semiconductor, which transfers electrons to the immobilised catalyst. The catalysts can be, for example, molecular catalysts, enzymes or semiconductor catalysts. The photosensitisers could be molecular complexes, semiconductors, nanostructures (quantum dots, nanoparticles), polymers or metal organic frameworks (MOFs). Various combinations of these approaches and components are being developed, including the most promising at present, a photoelectrochemical approach. The challenge remains to combine reduction of CO_2 with an oxidation process, to complete the cycle – otherwise, each reduction of CO_2 will be accompanied by an irreversible oxidation of a sacrificial donor. In the following sections, different classes of catalysts and photosensitisers are outlined briefly.

30.2 Rhenium-containing catalysts

The first published example of a complex in the form *fac*-[Re(NN)(CO)$_3$X] is [Re(bpy)(CO)$_3$I] (bpy = 2,2′-bipyridine), Figure 30.1, which appeared in a 1959 paper from Abel and Wilkinson [5]. In 1974, Wrighton and Morse published [6] the first report on metal carbonyls that exhibited luminescence in solution at room temperature, [Re(phen)(CO)$_3$Cl] (phen = 1,10-phenanthroline), which was followed by extensive research on photochemistry of Re(I) complexes. In 1983 and 1984, Jean-Marie Lehn and co-workers reported that [Re(NN)(CO)$_3$X] complexes can photocatalytically [7] or electrocatalitically [8] reduce CO_2 to CO – the discovery that started the field of catalytic applications of Re(I) diimine complexes [9–14].

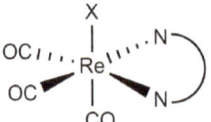

Figure 30.1: The general structure for the complexes of the type *fac*-[Re(NN)(CO)$_3$X], NN = polypyridyl ligand, X = Halide.

The advantages of Re(I) diimine complexes as photocatalysts include their photostability, absorption of light in visible region and the ease of synthesis [15]. The rhenium(I) tricarbonyl complexes possess a low-energy excited state of mainly MLCT character [16], which is usually long-lived (hundreds of nanoseconds) and strongly phosphorescent because of strong spin-orbit coupling resulting in enhanced singlet-triplet mixing [9]. The light-absorbing properties of [Re(α-diimine)(CO)$_3$X] complexes can be tuned by modifications of the α-diimine ligand, such as the extension of the planar aromatic system and incorporation of electron donating or withdrawing substituents to the aromatic rings [6, 10, 17–19].

Ishitani et al. [20] reported the effect of substituting the chloride ligand by phosphorus-containing ligands PR$_3$ (R = alkyl, alkoxy or allyl group) on the CO$_2$ reduction reaction, also suggesting a mechanism of this process. The mechanism presented agrees with that of Fujita et al. [21] in that after the initial photoexcitation, the ligand P(OEt)$_3$ dissociates, and the solvated complex [Re(bpy$^{•-}$)(CO)$_3$(solv)], is being formed; however, the formation of a dinuclear bridge was also considered. The mechanism proposed by Kubiak et al. [22] and expanded upon by Inoue et al. [23], involves the Re(bpy)(CO)$_3$COOH species that interacts with H$^+$, followed by release of a water molecule, and finally, followed by CO release.

The metal-to-ligand charge transfer (MLCT) transition in the archetypal [Re(bpy)(CO)$_3$]Cl complex and its immediate derivatives occurs below 450 nm. In order to extend absorption of the catalyst further into visible region, more extended ligands such as bis(arylimino)acenaphthene (aryl-BIAN), have been explored (Figure 30.2). Whilst no photocatalytic activity has been observed (which the authors attributed to energy of the excited state being insufficient to perform CO$_2$ reduction), the compounds containing sterically bulky substituents R=1,2,3,5-tetramethylbenzene and 2-methyl-1,3-bis(1-methylethyl)benzene, proved to be good electrocatalysts. The presence of sterically demanding groups in close proximity to the active rhenium tricarbonyl site appears to have a strong influence of the catalytic properties of these systems [24]. This influence is likely be linked to the loss of the halide as a key step in the catalytic cycle forming the active species, the five-coordinate anion species with a vacant coordination site [25].

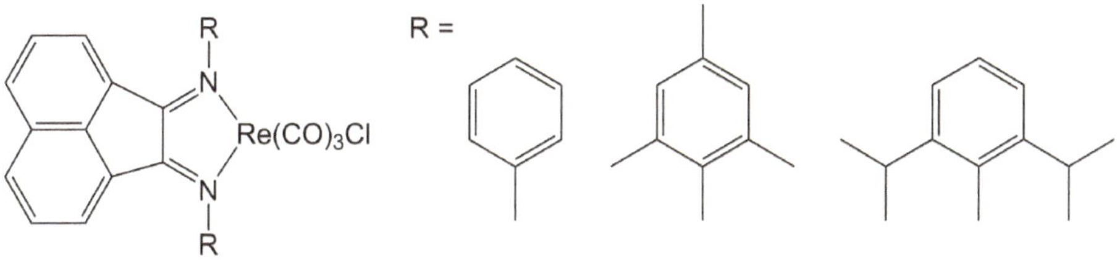

Figure 30.2: Chemical structures of rhenium(I) aryl-BIAN complexes investigated by Portenkirchner and co-workers for CO$_2$ reduction [24].

In a photocatalytic process, the [Re(bpy)(CO)$_3$]$^-$ active catalyst can be formed through reductive quenching of the excited state by a sacrificial donor, most commonly a tertiary amine, such as triethylamine (TEA) or triethanolamine (TEOA), see Figure 30.3. In the presence of a coordinating solvent, such as acetonitrile, the vacant coordination site is transiently occupied by solvent molecules until either CO$_2$ or the free halide coordinates to the rhenium centre [2].

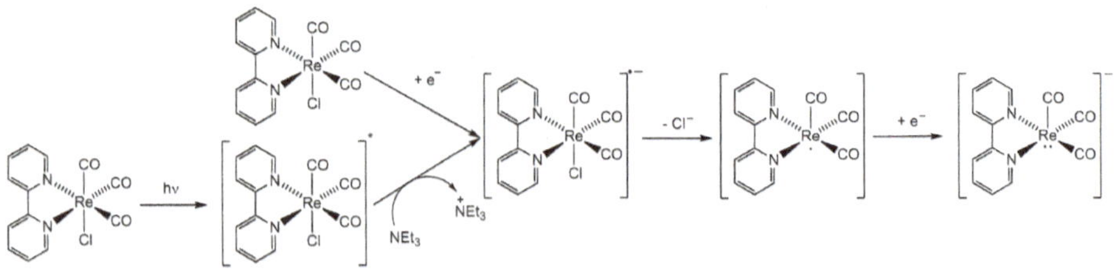

Figure 30.3: One of the proposed mechanisms of [Re(bpy)(CO)$_3$]$^-$ formation [25].

The mechanism for the formation of CO by the rhenium catalysts is still under debate, partly because of difficulty of performing mechanistic studies involving multiple species in the reaction mixture [25–27]. The state of the field with regards to photosensitisation of Re(I) catalysts with the use of various photosensitisers, in both mono- and bimolecular processes, has been comprehensively described by Perutz et al. [28].

Whilst Re(I) catalysts are effective at CO$_2$ reduction, they are not practical for large-scale application because of the scarcity of rhenium in the Earth's crust. Rhenium has an abundance of 2.6 ppb, with a total industrial production per year of 47.2 tonnes, making it one of the least available elements on Earth [29]. Therefore, analogues of the Lehn-type catalysts that employ earth-abundant elements are required for CO$_2$ reduction to be industrially relevant. Suitable elements for this purpose are the first row transition metals, such as iron, cobalt and manganese.

30.3 Manganese-based electrocatalysts

Manganese-based catalysts would present an excellent alternative to their Re(I) counterparts, as manganese abundance is over 365,000 times greater than that of rhenium [30].

Electrocatalytic activity of Mn(I) complexes for CO$_2$ reduction was first demonstrated in 2011, on the example of [Mn(bpy)(CO)$_3$Br]. The selectivity for CO production over formate production and the Faradaic efficiency observed were very similar to the rhenium analogue. A key difference between the two types of catalysts is that the Mn(I)-based analogue requires the presence of a weak Brønsted acid, such as water, to ensure significant catalytic activity [31]. The studies by cyclic voltammetry and spectroelectrochemistry determined that a key intermediate in the CO$_2$ reduction is the Mn–Mn dimer formed from the radical species initially produced by electrolysis.

Density functional theory (DFT) demonstrated that the manganese catalysts could undergo an additional reduction pathway, different to that of the rhenium species. It is proposed that the manganese analogue would first be reduced to [Mn(bpy)(CO)$_3$]^0, this species can then either dimerise to form a Mn^0–Mn^0 dimer or be reduced again to form the five coordinate anion required for catalysis (Figure 30.4) [32]. Whilst

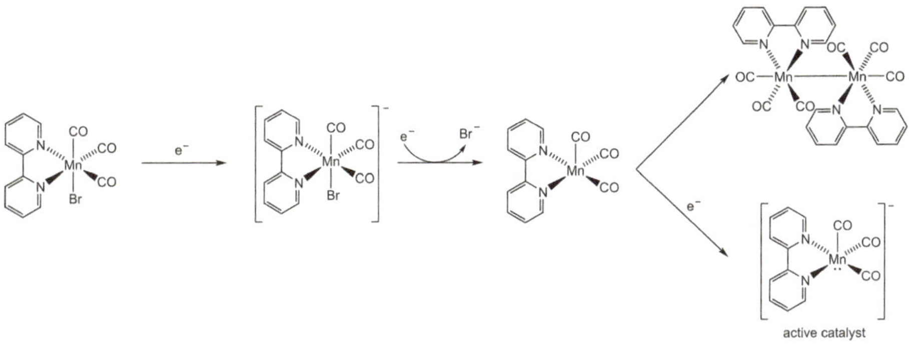

Figure 30.4: The thermodynamically preferred reduction pathways for [Mn(bpy)(CO)$_3$Br] to form the active catalyst [32].

[Re(bpy)(CO)$_3$Cl] can undergo the same pathway as the Mn catalyst, the neutral five-coordinate species is very unstable; hence, this path is thermodynamically disfavoured.

Both the Re and Mn catalysts can form a M^0–M^0 dimer. The Re^0–Re^0 species shows no reactivity to CO$_2$ when the diimine ligand is 2,2′-bipyridine; however, the manganese analogue has been shown to reduce CO$_2$ when photosensitised by [Ru(dmbpy)$_3$]$^{2+}$, where dmbpy = 4,4′-dimethyl-2,2′-bipyridine [33]. Time-resolved IR studies have shown that the rate of dimer formation is significantly higher for the Mn species compared to Re [34].

Introduction of bulky substituents on the diimine ligand in [Mn(mesbpy)(CO)$_3$Br] (mesbpy = 5,5′-dimesityl-2,2′-bipyridine) prevents the dimerisation of the complex and enables mechanistic studies (Figure 30.5) [35].

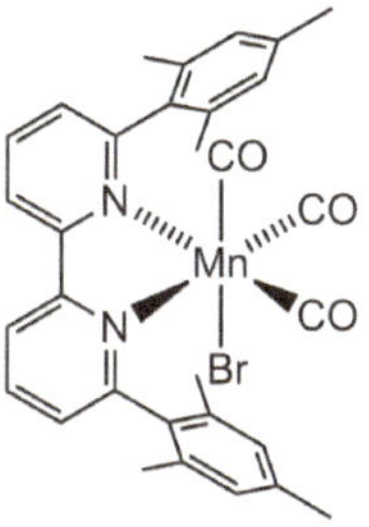

Figure 30.5: The structure of [Mn(mesbpy)(CO)$_3$Br].

The active catalyst [Mn(mesbpy)(CO)$_3$]$^-$ was formed at 300 mV lower overpotential than was required for the [Mn(bpy)(CO)$_3$Br] analogue, showing that preventing the dimerisation of the manganese [Mn(L$_2$)(CO)$_3$]^0 catalyst precursor significantly reduces the potential required to produce the active catalyst. In 2016, the effect of adding a Lewis acid to the catalytic process was investigated: addition of Mg^{2+} to [Mn(mesbpy)(CO)$_3$Br] led to CO$_2$ reduction to form CO and MgCO$_3$ through a reductive disproportionation mechanism (Figure 30.6) [36]. This new pathway required a significantly lower overpotential than any other catalytic system at the time.

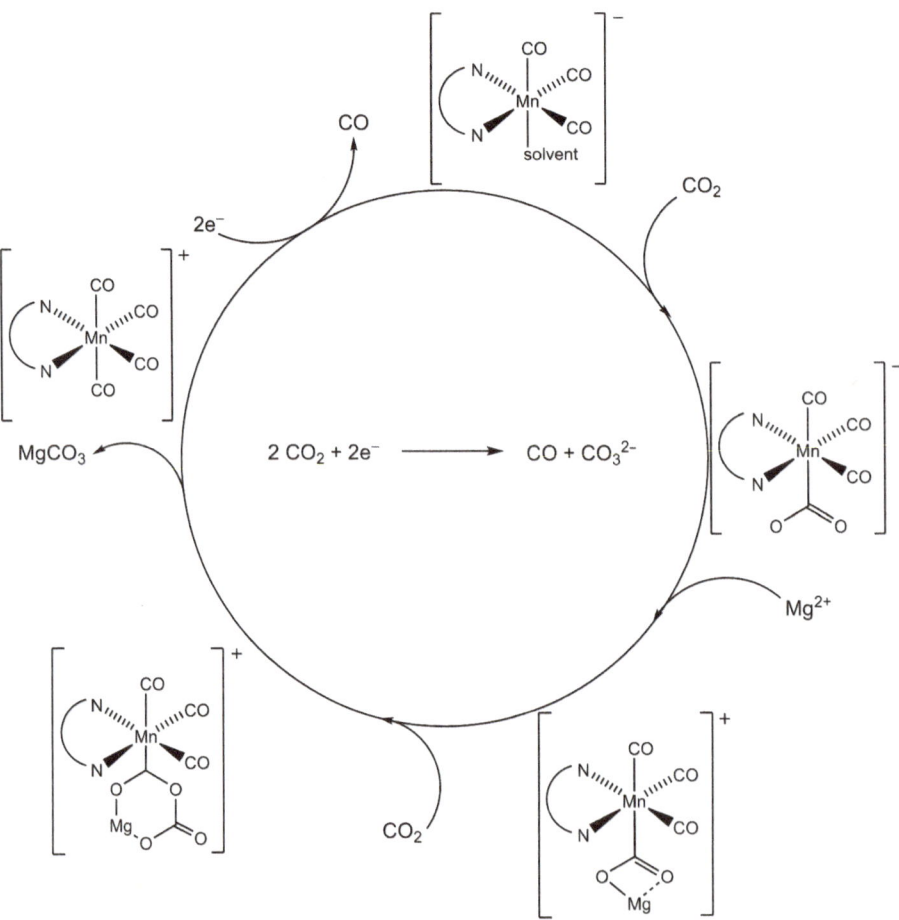

Figure 30.6: The reductive disproportionation mechanism for CO_2 reduction catalysed by [Mn(mesbpy)(CO)$_3$Br] and Mg(OTf)$_2$.

This Lewis acid catalysed process is an example of a bimetallic catalytic system where the manganese centre initiates CO_2 reduction, and the magnesium facilitates C–O bond cleavage. Hence, the Mg(OTf)$_2$ acts as a co-substrate in this system, rather than as a co-catalyst. A drawback of this system is that the Mg^{2+} cation binds strongly to the carbonate ion; hence, the magnesium cation is not regenerated. The use of softer Lewis acids must be explored in future to allow for release of the carbonate ion and regeneration of the Lewis acid. The use of other Lewis acidic species, such as $AlCl_3$ could also prove to be a promising area of research, as acids with different coordination numbers may alter the CO_2 reduction mechanism.

The Weinstein and Hartl groups [37] investigated Mn(I) catalysts with asymmetric diimine ligands, imino-pyridyl (IP) (Figure 30.7). The Mn(IP) complexes proved to be efficient electrocatalysts of CO$_2$ reduction into CO, whilst the partial separation of the steric and electronic effects allowed for detailed mechanistic studies.

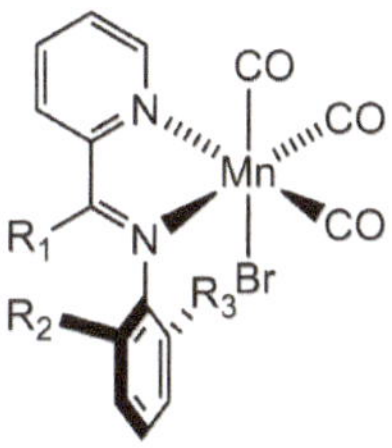

Figure 30.7: The structure of [Mn(IP)(CO)$_3$Br] (R$_1$, R$_2$, R$_3$ = H). The R groups can be substituted for more sterically hindering or electron donating groups to alter the catalyst behaviour [37].

UV/vis/IR-spectroelectrochemical studies under inert atmosphere found that introduction of sterically demanding substituents resulted in the direct formation of the five-coordinate anion [Mn(CO)$_3$(IP)]$^-$. Less sterically demanding IP ligands resulted in a stepwise reduction mechanism where the anion is produced from the Mn–Mn dimer. An IP ligand with "moderate" steric bulk and electron density was found to have an intermediate behaviour, where both the anion and the Mn–Mn dimer were detected at the same time. Experiments under a CO$_2$ atmosphere revealed that for most of the complexes the CO$_2$ coordination outcompetes the dimerisation pathway, as the Mn–Mn dimer was not detected *in situ* [37].

A comprehensive review of electrocatalytic and photocatalytic CO$_2$ reduction with Mn-based complexes was recently completed by Grills et al. [38]. This includes a discussion of computational and experimental methods used to help unravel the mechanisms of reduction.

30.4 Ruthenium-containing catalysts

One of the first known examples of a ruthenium complex used for photochemical reduction of CO$_2$ was [Ru(2,2′-bipyridine)$_3$]$^{2+}$, ([Ru(bpy)$_3$]$^{2+}$), reported in 1982 [39]. Many other Ru(II) catalysts for the production of CO or formate have been investigated since. For example, Ishitani et al. [20]. examined the system of [Ru(bpy)$_3$]$^{2+}$ photosensitiser and [Ru(bpy)$_2$(CO)$_2$]$^{2+}$ in the presence of CO$_2$ and a light source. They postulate that a one electron reduction (OER) species is being produced in the form of [Ru(II)(bpy$^{•-}$)(bpy)$_2$]$^+$. Reduction of CO$_2$ then proceeds via electron capture by the catalyst from the OER species, in which a CO$_2$ molecule coordinates to an open site on the ruthenium centre produced by dissociation of one of the CO ligands after the reduction of [Ru(bpy)$_2$(CO)$_2$]$^{2+}$. By controlling the pH of the solution, the outcome of the CO$_2$ reduction reaction was controlled.

30.5 Photocatalytic systems with Earth-abundant components

The development of Mn catalysts for photo-driven reactions is very appealing – the only problem is that manganese catalysts such as [Mn(bpy)(CO)$_3$Br] are photosensitive.

Thus, photosensitisers are required – which usually contain noble metals. For example, when [Mn(bpy)(CO)$_3$(CN)] was photosensitised by [Ru(dmbpy)$_3$]$^{2+}$ and irradiated with monochromatic 470 nm light for 15 h, this led to the production of both CO and formate, with the ratio between the two products changing between acetonitrile and DMF solutions, indicating differing stabilities of the singly reduced catalyst in acetonitrile and DMF [40]. A general schematic for the process can be seen in Figure 30.8.

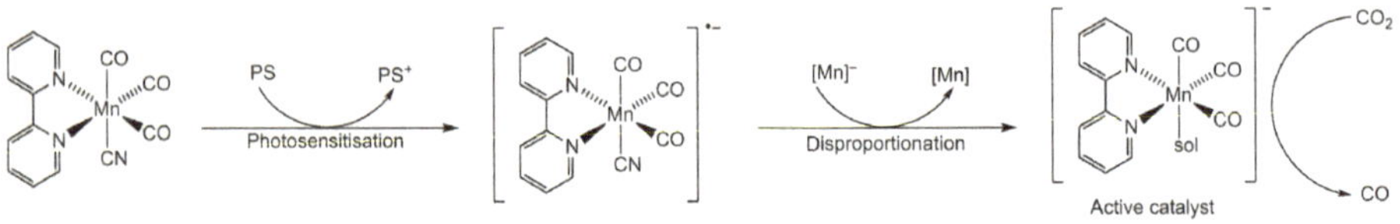

Figure 30.8: The formation of the active catalyst, [Mn(bpy)(CO)$_3$]$^-$, from [Mn(bpy)(CO)$_3$(CN)], in a light-driven process [41].

An attractive alternative to [Ru(bpy)$_3$]$^{2+}$ are copper-based photosensitisers[42]. Specifically, sterically hindered bis-diimine copper(I) complexes (Figure 30.9) have been found to exhibit room temperature luminescence with a long lifetime, allowing for their potential application as photosensitisers for dye-sensitised solar cells (DSSCs), and in photocatalysis [43].

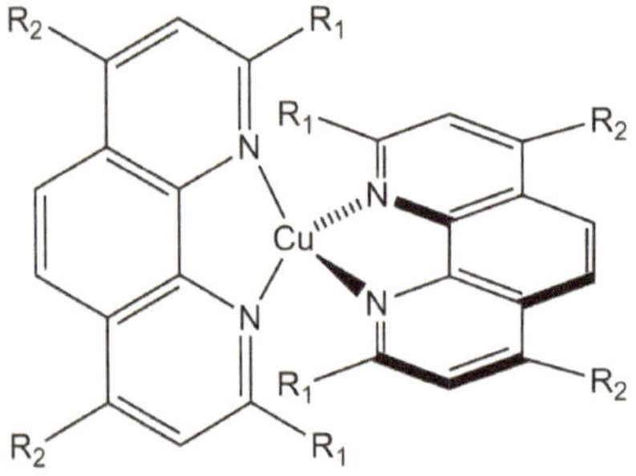

Figure 30.9: The structure of [Cu(dmphen)$_2$]$^+$ (dmphen = 2,9-dimethyl-1,10-phenanthroline) [43].

There are only two examples of a photocatalytic CO$_2$ reducing system employing only earth-abundant elements. The first one is employing tetraphenylporphyrin zinc(II) (ZnTPP) to photosensitise [Mn(phen)(CO)$_3$Br] in the presence of a TEA sacrificial donor (Figure 30.10) [44], the system produced CO with a modest turnover number (TON$_{CO}$ = 119). The second system, which was published at the end of 2018, uses a Cu-based photosensitiser and has TON of >1,000 [45].

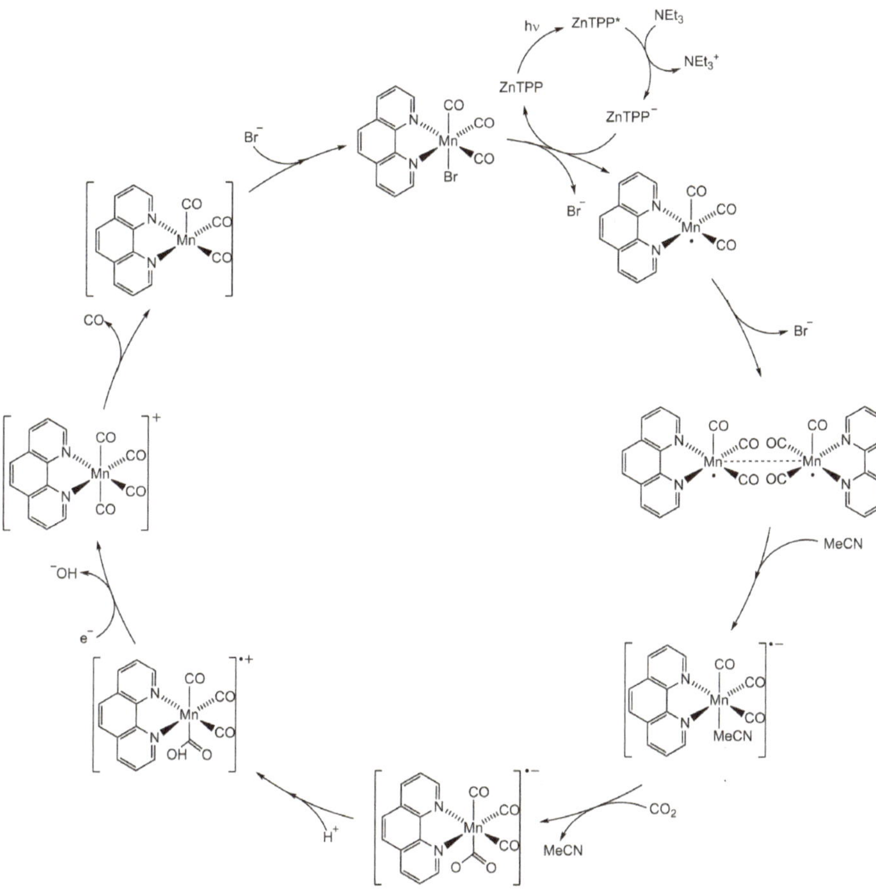

Figure 30.10: The proposed catalytic cycle for CO$_2$ to CO reduction catalysed by [Mn(phen)(CO)$_3$Br], which was photosensitised by ZnTPP [44].

It is also possible to photosensitise a catalyst with an organic chromophore [46]: the advantage of organic chromophores is their photostability, the drawbacks are that they mainly absorb light in the UV region, and/or have short excited state lifetimes, which rules out intermolecular photosensitisation.

Porphyrins, metalloporphyrins and derivatives thereof are one of the most widespread photosensitisers owing to their strong absorption of light in the visible region of the spectrum. The singlet excited states of porphyrins are usually shortlived, populating long-lived triplet excited states with high yields; the triplet states are then engaged in the sensitisation process [2].

30.6 Iron-, Cobalt-, Nickel-, Zinc-containing catalysts and photosensitisers

Initial studies on cobalt and iron corroles [47] (Figure 30.11) demonstrated their ability to reduce CO_2 to CO or produce H_2 in the presence of a sacrificial donor (Et_3N) and a photosensitiser (p-terphenyl). However, the systems suffered from high overpotentials and catalyst decomposition.

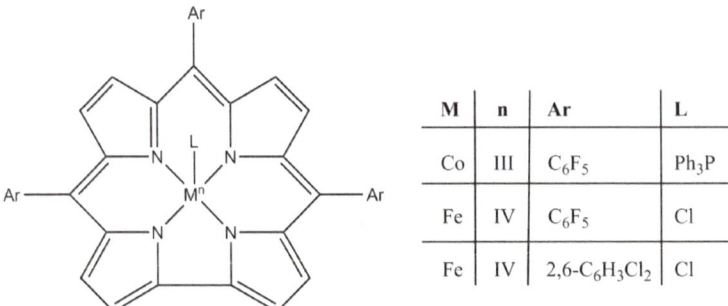

M	n	Ar	L
Co	III	C_6F_5	Ph_3P
Fe	IV	C_6F_5	Cl
Fe	IV	$2,6\text{-}C_6H_3Cl_2$	Cl

Figure 30.11: The metal corrole complexes for the photochemical reduction of CO_2 to CO [47].

To design iron complexes capable of acting as photosensitisers is challenging, as most complexes of iron do not have long-lived excited states. Recently, an iron(II) centre substituted with bis(1,2,3-triazol-5-ylidene) (btz) ligands with a MLCT lifetime of over 100 ps (Figure 30.12) [48, 49] has been reported. Whilst this is too short a lifetime for diffusion controlled intermolecular photosensitisation, it does show that Fe^{II} excited states can be stabilised by ligands with strong σ-donating and π-accepting properties [50].

Iron porphyrin complexes, such as tetraphenylporphyrin iron(II), are some of the most effective molecular catalysts for the reduction of CO_2 [51]. Unlike other metalloporphyrins such as tetraphenylporphyrin zinc(II), the iron centre possesses vacant coordination sites, which allow for the binding of CO_2 to initiate the reduction process. Electrochemical activation of an iron porphyrin requires the reduction of the Fe centre to Fe^0. The resulting species is nucleophilic enough to attack CO_2 and form an $Fe^I\text{-}CO_2^{\bullet-}$ intermediate. Similar to manganese catalysts, iron porphyrins require weak Brønsted acids or Lewis acids to function efficiently (water, trifluoroethanol or phenol). A study of three iron porphyrin electrocatalysts for CO_2 reduction (Figure 30.13) revealed that these compounds have a lower overpotential and greater turnover frequency than $[Mn(dmbpy)(CO)_3Br]$ and $[Ni(cyclam)]^{2+}$ [52].

As with many other electrocatalysts, iron porphyrins also have the potential to act as photocatalysts. This can be achieved through direct excitation of the porphyrin Soret or Q-bands, where the active Fe^0TPP catalyst is generated through

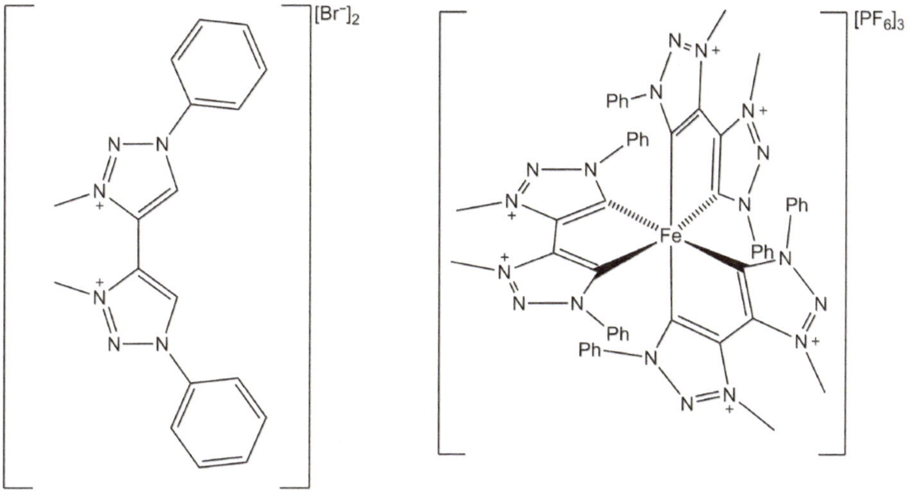

Figure 30.12: The structures of the btz ligand and [Fe(btz)$_3$][PF$_6$]$_3$.

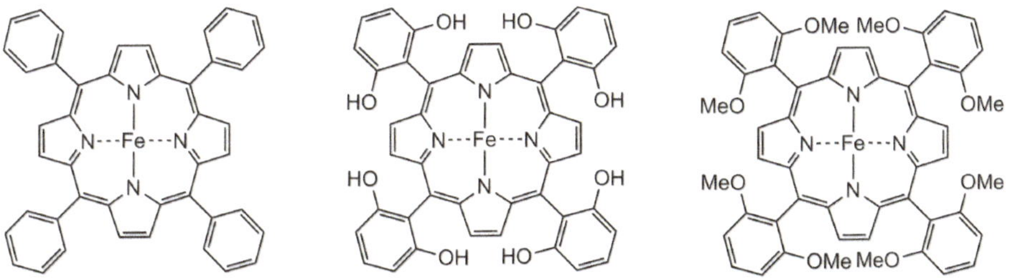

Figure 30.13: The structures of *meso*-tetraphenylporphyrin iron, *meso*-tetrakis (2′,6′-dihydroxyphenyl)porphyrin, and *meso*-tetrakis(2′,6′-dimethoxyphenyl)porphyrin [52].

photolysis of the Fe–Cl bond of the water soluble catalyst (WSCAT) porphyrin (see Figure 30.14), and three subsequent reductions from a sacrificial donor [53]. The system was reported to have TON of 30 if TEA (E$_{ox}$ = 0.69 V vs. SCE) is used as a sacrificial donor; the use of more strongly reducing 1*H*-benzoimidazole (BIH) (E$_{ox}$ = 0.33 V vs. SCE) improved the performance to TON >70 for CO production [54]. Importantly, the introduction of the trimethylannilinium groups in WSCAT seemed to have avoided the photodegredation observed with other iron porphyrins that had prevented their application as photocatalysts when irradiated with near-UV light. It also increases the selectivity for CO formation over hydrogen production, overcoming the competitive iron hydride formation pathway typical for porphyrin-based photocatalysts [55]. A second finding was that WSCAT did not require

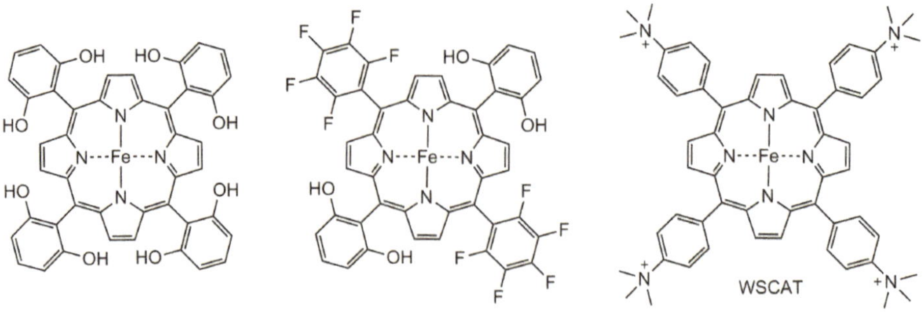

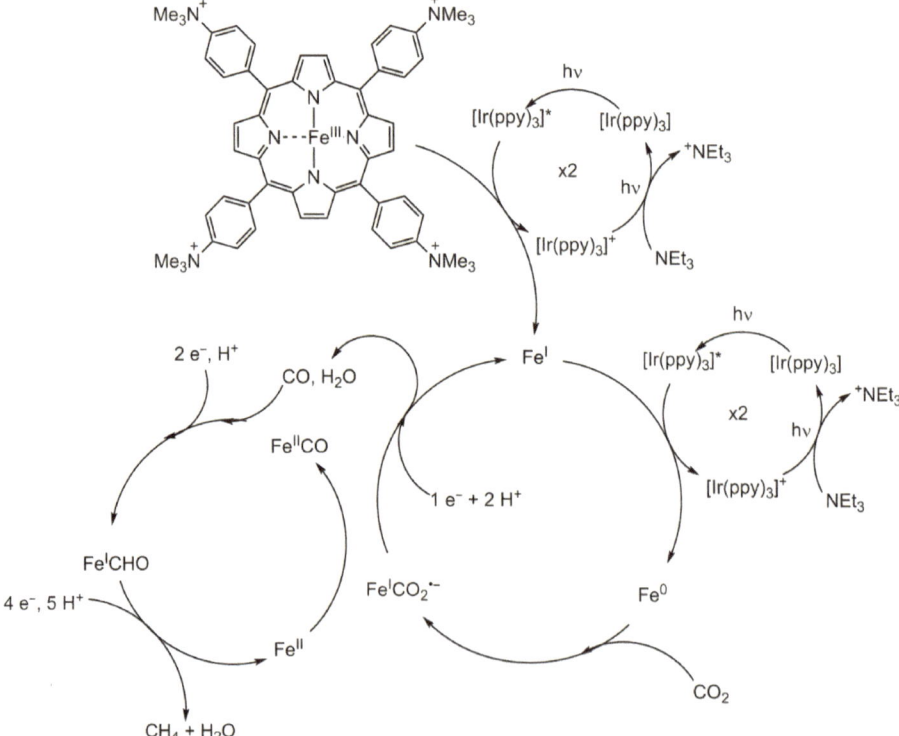

Figure 30.14: Three examples of iron porphyrins with CO_2 ligand-stabilising groups [53].

photosensitisation to work under visible light irradiation, unlike previous porphyrin photocatalysts which were photosensitised by 9-cyanoanthracene [56]. These systems required photosensitisation as near-UV excitation was required to activate the porphyrin complex; hence, these catalysts cannot operate under visible light irradiation.

Figure 30.15: The proposed mechanism for CO_2 to CH_4 reduction by WSCAT, TEA and [Ir(PPy)₃] [57].

Introduction of an Ir(PPy)$_3$ photosensitiser [57] radically altered the products of CO$_2$ reduction by WSCAT, from CO to methane. This was the first case of methane production from an iron-molecular catalyst by CO$_2$ reduction; a preliminary mechanism is outlined in Figure 30.15.

Recently, *meso*-tetramesitylporphyrin zinc (II) was found to be electrocatalytically active towards CO$_2$ reduction. Interestingly, the porphyrin ligand acted as the redox centre for CO$_2$ reduction to CO instead of the metal centre [58], making the mechanism of the reduction process by this species distinct from that of the other porphyrin molecular catalysts.

Some of the Fe- and Co-macrocyclic complexes are also effective catalysts (Figure 30.16) [59], operating both as electro- and photocatalysts. The Fe catalyst is selective with regard to production of formic acid from CO$_2$ at a low overpotential, whilst the Co-analogue catalyses production of CO with high selectivity.

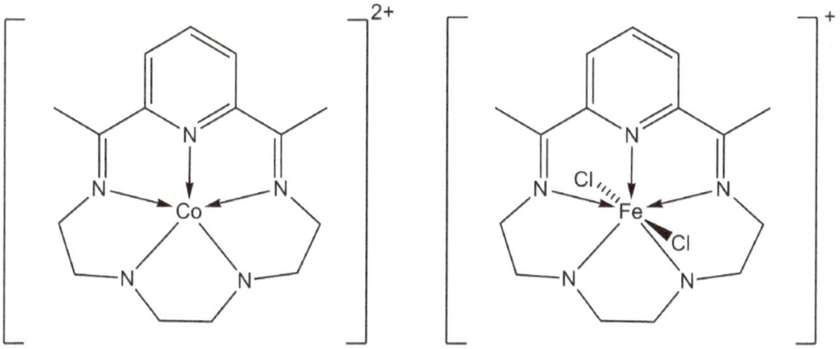

Figure 30.16: Co^{II} and Fe^{III} CO$_2$ reduction catalysts with pentadentate macrocyclic ligands [59].

Macrocyclic nickel complexes have been demonstrated as efficient CO$_2$ to CO reduction catalysts in both aqueous and aprotic solvents [60, 61]. For example, [Ni(cyclam)]$^{2+}$ can reduce CO$_2$ to CO with high efficiency and selectivity in pH 4–5 water at a mercury electrode (Figure 30.17).

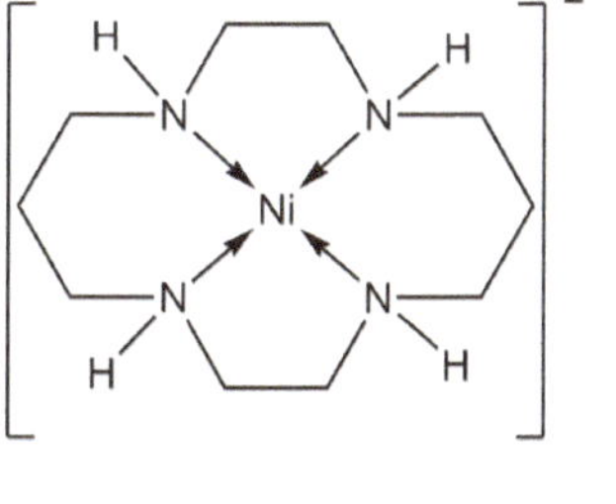

Figure 30.17: The structure of [Ni(cyclam)]$^{2+}$ [60].

A $[Ni(cyclam)]_2^{4+}$ dimer has also been used for photocatalytic CO_2 reduction, using $[Ru(bpy)_3]^{2+}$ as a photosensitiser [62]. A problem with these photocatalytic systems is the large yield of hydrogen, rendering them poor CO_2 reduction catalysts compared to other species.

30.7 Heterogeneous catalytic systems

The applicability of the previously described homogeneous CO_2 reduction catalysts is limited by their relatively poor stability under irradiation, and difficulty in separating the catalyst from the products. Heterogeneous catalysts present an attractive alternative as they are easily separated from liquid products, such as formic acid, and are simple to use. However, there are problems with heterogeneous catalysts such as poorly defined active sites and lack of selectivity that molecular species do not have. Therefore, a good compromise is a hybrid catalytic system where the molecular reduction catalyst is covalently bound to a solid support. The support can be inert or play a role in the reduction, for example, TiO_2 semiconductors can be used as an electron relay and solid support in a photoelectrochemical cell.

Many hybrid catalysts use combinations of metal complexes as the active catalyst bound to inorganic semiconductors in which absorption of light in visible region promotes an electron to the conduction band, which can then be donated to a molecular catalyst.

There are many suitable semiconductors, such as metal oxides (TiO_2, ZrO_2, Ta_2O_5, to name a few), metal sulphides (cadmium sulphide [CdS]) and carbon-based materials such as graphene and its derivatives [63]. The bandgap energies and relative positions of the valence and conduction bands for several semiconductors are shown in Figure 30.18.

The "trade-off" is that narrow bandgap semiconductors have good visible light absorption but suffer from photocorrosion over time, whereas wide bandgap semiconductors are more stable but absorb a limited range of the solar spectrum [64]. The band gap will also determine the rate of electron transfer to the molecular catalyst; if the band gap energy closely matches the LUMO energy of the catalyst then electron transfer will be facile. Therefore, a balance must be struck between optimising stability, visible light harvesting and electron transfer rates when selecting a material for this purpose.

Anchoring the catalyst onto a semiconductor (TiO_2 [65, 66], N-Ta_2O_5 [67, 68]) can be achieved by modifying the diimine ligands with a group such as –COOH or –PO_3H. One example is $[Ru(dcbpy)_2(CO)_2]^{2+}$ (dcbpy = 4,4′-dicarboxy-2,2′-bipyridine) coupled to a N–Ta_2O_5 (N = nitrogen doped) semiconductor [67, 69] which acts as a photosensitiser, injecting electrons into the catalyst to initiate reduction of CO_2 to formic acid. A schematic diagram of the system is shown in Figure 30.19, with the accompanying energy level diagram shown in Figure 30.20.

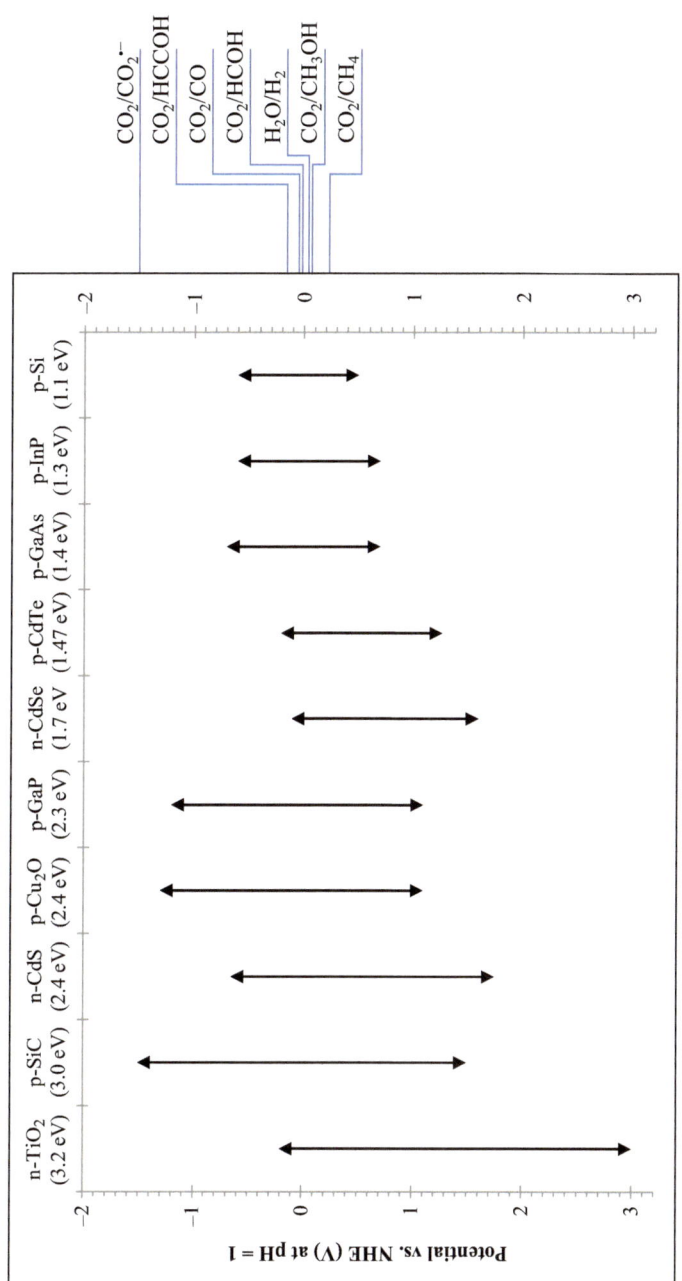

Figure 30.18: The position of the conduction and valence bands of several semiconductors at pH = 1 versus a normal hydrogen electrode (NHE). Thermodynamic potentials for CO_2 reduction to different products at pH = 1 versus a NHE are shown beside the band edge positions of semiconductors [22]. (see also Figure 29.1).

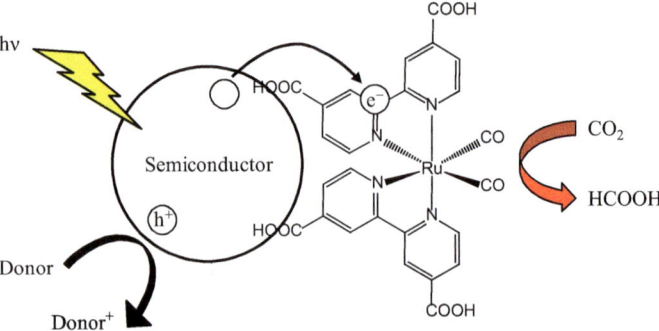

Figure 30.19: A schematic of hybrid photocatalysis under visible light irradiation utilising a semiconductor and a metal complex catalyst [67].

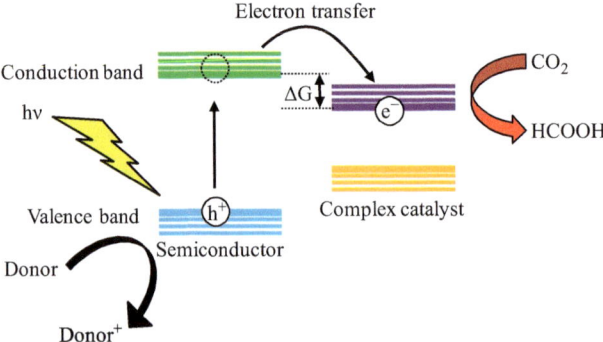

Figure 30.20: The energy level diagram of hybrid photocatalysis under visible light irradiation utilising a semiconductor and a metal complex catalyst [67].

An additional requirement for these hybrid systems to be effective is that the catalyst must be strongly bound to the semiconductor, preferably through a covalent linkage. An example of such a system is manganese catalyst [Mn(dpbpy)(CO)$_3$Br] (dpbpy = 4,4′-diphosphonate-2,2′-bipyridine) immobilised on a mesoporous TiO$_2$ electrode, with a CdS photoanode chromophore to prevent photodegradation of the light-sensitive catalyst [70].

The semiconductive material does not have to be a bulk solid, it is possible to bind molecular catalysts to nanoparticles and quantum dots [71, 72], or thin films. Examples of such systems have incorporated catalysts such as [Ni(terpy)$_2$]$^{2+}$ and [Ni(cyclam)]$^{2+}$ (Figure 30.21), or porphyrin-based ruthenium complexes that were then attached onto TiO$_2$ films [73]. An increase in catalytic efficiency has been achieved by greater porosity of the semiconductor to increase the surface area, as for example in submicron-sized mesoporous tantalum oxide spheres [74].

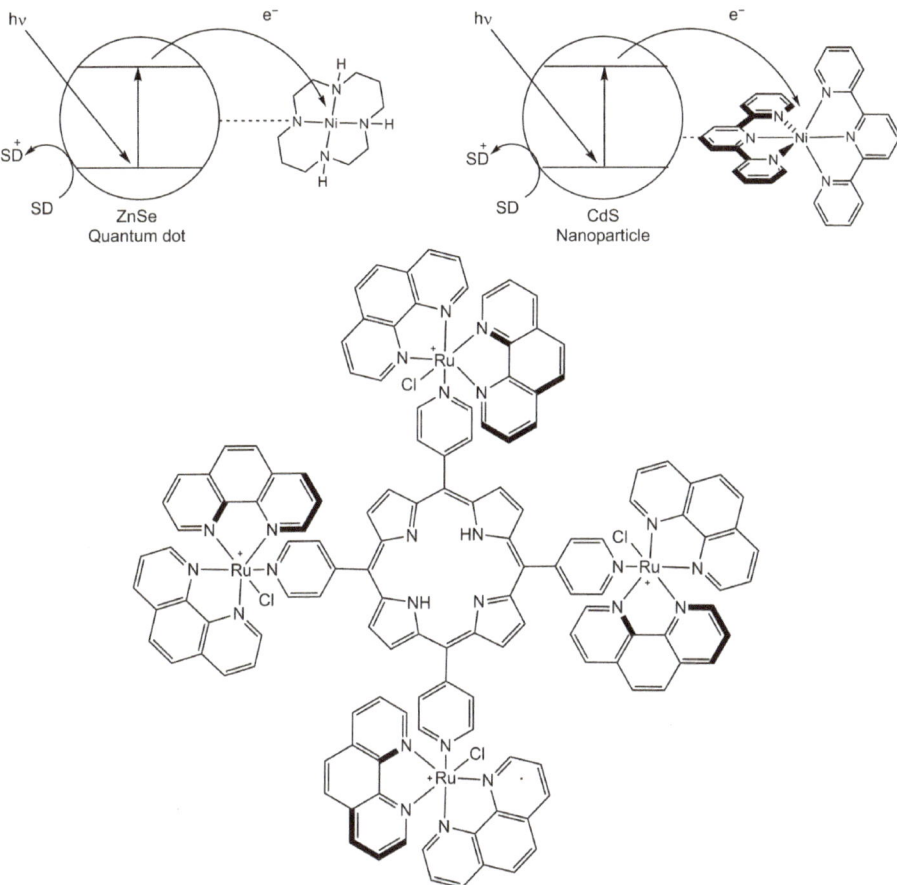

Figure 30.21: The photosensitisation of a CO$_2$ reducing electrocatalyst with a nanoparticle or quantum dot (SD – sacrificial donor), and an assembly of four Ru(II)-based units coordinated to meso-tetrapyridyl porphyrin [73].

Carbon-based electrodes, such as functionalised graphene have also been used as photosensitisers. For example, graphitic carbon nitride (gC$_3$N$_4$) was used to photosensitise a Ru–Ru dyad CO$_2$ reduction catalyst. The ruthenium catalyst was bound to the electrode via a bipyridine ligand functionalised with –CH$_2$PO$_3$$^{2-}$ groups, the –CH$_2$-linker decouples the anchor from the ligand, thus the electron transfer takes place through space, and not through a delocalised system [75]. gC$_3$N$_4$ has the benefit of being easily modifiable to allow the surface to be optimised for efficient electron transfer and increase the surface area available for catalyst loading. The catalytic system was over 80% efficient and had the highest CO turnover for any metal complex/semiconductor hybrid system as of 2017. gC$_3$N$_4$ electrodes have also been used to photosensitise an earth-abundant cobalt–porphyrin complex bound through an amide linkage formed from a pendant amine group on the carbon nitride and a ketone on one of the porphyrin pyrrole rings. This hybrid catalyst was able to reduce CO_2 to CO [76]. Various other nanostructures and materials have been used as supports – such as "nanoscrolls," [77] carbon nanotube electrodes [78], or p-type silicon [79].

It is also possible to use functionalised silica as a solid support for both photosensitisers and catalysts. Adsorption of a ruthenium–rhenium dyad complex to the silica surface gave a catalytic system capable of reducing CO_2. The use of mesoporous silica with wider channels throughout the solid resulted in an increase in catalytic performance because of faster substrate and product diffusion [80].

Another type of hybrid catalyst is based on supramolecular materials, such as metal-organic frameworks (MOFs). In one example, $[Mn(dcbpy)(CO)_3Br]$ (dcbpy = 5,5'-dicarboxylate-2,2'-bipyridyl) was bound to a robust Zr(IV) MOF. $[Ru(dmbpy)_3]^{2+}$ was used to photosensitise the catalyst. It was found that the functionalised MOF had an improved turnover number compared to free $[Mn(dcbpy)(CO)_3Br]$. This was thought to be because of the MOF providing isolated active sites, where the Mn complexes cannot dimerise, thus the catalyst remained in the active $[Mn(dcbpy)(CO)_3]^-$ form, stabilised by coordination to the framework. The coordination of the Mn catalyst to the MOF was performed under mild conditions at room temperature by post-synthetic modification methods. The ruthenium photosensitiser was not bound to the catalyst; it was in solution within the pores of the MOF. A problem was encountered when trying to recycle the MOF for successive CO_2 reductions as the turnover of CO decreased rapidly with each cycle [81]. It has been suggested that incorporating metalloporphyrin units into the MOF may lead to a viable CO_2 to CO electrocatalyst, which does not require an external dopant [82].

30.8 Photoelectrochemical cells

Hybrid catalysts are an important aspect in the development of photoelectrochemical cells, where the reductive CO_2 reduction process is coupled to an oxidation process. The realisation of this approach is highly desirable, as current catalytic systems that employ sacrificial donors are not renewable. An example of such a system is a combination of $[Re(bpy)(CO)_3Cl]$, $[Cr(Cp)(CO)_3]$ (hydrogen oxidation catalyst), and a tetraphenylporphyrin zinc(II) chromophore, used simultaneously as a model of natural photosynthesis. The chromophore possessed the required reduction potentials to donate an electron to the CO_2 catalyst, and be regenerated by the hydrogen oxidation catalyst. This allowed for a truly renewable system where the only reagents required were CO_2 and H_2. However, this did involve the use of rare-earth elements [83].

Most photoelectrochemical cells are based on hybrid catalysts, where the metal complexes are bound to electrodes connected through a circuit. The electrodes are immersed in an aqueous solution saturated with CO_2, and are separated by a semipermeable membrane, which allows proton diffusion from the water oxidation compartment to the CO_2 reduction reaction (Figure 30.22) [84]. The catalysis is

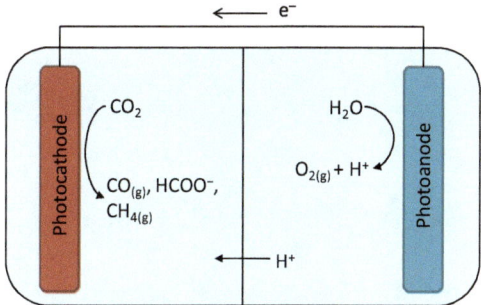

Figure 30.22: A schematic for a photoelectrochemical cell, showing how CO$_2$ reduction and water oxidation reactions can be coupled (stoichiometry not considered).

initiated through donation of an electron from the photocathode to the CO$_2$ reduction catalyst. This electron is then replenished by the water oxidation reaction, completing the cycle.

30.9 Modern spectroscopic methods of studying catalytic mechanisms

The study of ultrafast processes, such as charge separation by electron transfer, is of great importance to photocatalysis. These processes take place on a range of timescales, from femtoseconds to microseconds. One of the most important methods of studying these processes, hence understanding the catalytic mechanisms at hand, is time-resolved spectroscopy [85]. Absorption techniques, such as transient absorption (TA) and time-resolved infrared (TRIR) spectroscopy have been used to study the photophysics of many CO$_2$ reducing systems. An example is the study of charge separation within a rhenium–porphyrin dyad (Figure 30.23), which provided information about the electron transfer between the porphyrin and the catalyst, the lifetime of the charge-separated state and enabled comparison of the performances of the unimolecular versus bimolecular chromophore/catalyst combination [86, 87].

More advanced spectroscopic techniques, such as two-dimensional infrared (2D-IR) spectroscopy can be used to obtain information about the structures of excited state compounds and intermediates whilst extracting the timescales of their interconversion on the ultrafast timescale. This is possible as 2D-IR allows for the observation of spectral diffusion, which is the change in a vibrational frequency with time because of solvent–solute interactions. This allows 2D-IR study of a catalyst to obtain vibrational, spectral and solvent dynamics in multiple excited states [88–90]. Two-dimensional sum-frequency [91] generation spectroscopy has recently emerged as

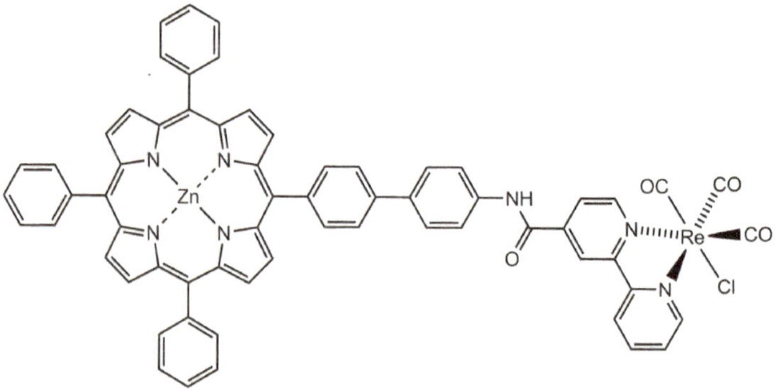

Figure 30.23: A Zn–Re dyad used to study formation of charge-separated states in CO_2 reduction catalysts using time-resolved infrared spectroscopy (TRIR) [87].

a tool to study surface attachment of catalysts, as has been demonstrated on the example of a Re(I) catalyst on single-crystalline TiO_2 surface. An application of another sophisticated method, pulsed EPR, allowed identifying a key intermediate in the catalytic reduction of CO_2 by a manganese bipyridine derivative [92].

One of the latest additions to the suit of time-resolved methods to study CO_2 reduction mechanisms is a combination of pulse-radiolytic generation and infrared detection, recently achieved by Grills et al. [34], which permits investigations of transient species of light-sensitive catalysts.

Overall, large progress has been achieved in the past decade, with the exciting prospect of light-driven CO_2 reduction using Earth-abundant elements now being within reach.

References

[1]   E. E. Benson, C. P. Kubiak, A. J. Sathrum and J. M. Smieja, Chem. Soc. Rev., 2009, 38, 89–99.
[2]   A. J. Morris, G. J. Meyer and E. Fujita, Acc. Chem. Res., 2009, 42, 1983–1994.
[3]   A. de Klerk and E. Furimsky, Catalysis in the Refining of Fischer-Tropsch Syncrude, RSC Publishing, 2010.
[4]   C. D. Windle and R. N. Perutz, Coord. Chem. Rev., 2012, 256, 2562–2570.
[5]   E. W. Abel and G. Wilkinson, J. Chem. Soc., 1959, 0, 1501.
[6]   M. S. Wrighton and D. L. Morse, J. Am. Chem. Soc., 1974, **96**, 998–1003.
[7]   J. Hawecker, J.-M. Lehn and R. Ziessel, J. Chem. Soc., Chem. Commun., 1983, 536–538.
[8]   J. Hawecker, J.-M. Lehn and R. Ziessel, J. Chem. Soc., Chem. Commun., 1984, 328–330.
[9]   A. Coleman, C. Brennan, J. G. Vos and M. T. Pryce, Coord. Chem. Rev., 2008, 252, 2585–2595.
[10] P. Kurz, B. Probst, B. Spingler and R. Alberto, Eur. J. Inorg. Chem., 2006, 2966–2974.
[11] G. J. Stor, F. Hartl, J. W. M. M. Van Outersterp and D. J. Stufkens, *Organometallics*, 1995, 14, 1115–1131.
[12] J. K. Nganga, C. R. Samanamu, J. M. Tanski, C. Pacheco, C. Saucedo, V. S. Batista, K. A. Grice, M. Z. Ertem and A. M. Angeles-Boza, Inorg. Chem., 2017, 56, 3214–3226.

[13] A. Wilting, T. Stolper, R. A. Mata and I. Siewert, Inorg. Chem., 2017, 56, 4176–4185.

[14] J. Agarwal, R. P. Johnson and G. Li, *J. Phys. Chem. A*, 2011, 115, 2877–2881.

[15] J. Hawecker, J.-M. Lehn and R. Ziessel, Helv. Chim. Acta, 1986, 69, 1990–2012.

[16] S. Ranjan, S.-Y. Lin, K.-C. Hwang, Y. Chi, W.-L. Ching, C.-S. Liu, Y.-T. Tao, C.-H. Chien, S.-M. Peng and G.-H. Lee, Inorg. Chem., 2003, 42, 1248–1255.

[17] J. R. Wagner and D. G. Hendricker, J. Inorg. Nucl. Chem., 1975, 37, 1375–1379.

[18] K. Kalyanasundaram, J. Chem. Soc., Faraday Trans., 2, 1986, 82, 2401–2415.

[19] M. S. Wrighton, D. L. Morse and L. Pdungsap, J. Am. Chem. Soc., 1975, 97, 2073–2079.

[20] H. Takeda and O. Ishitani, Coord. Chem. Rev., 2010, 254, 346–354.

[21] M. D. Doherty, D. C. Grills, J. T. Muckerman, D. E. Polyansky and E. Fujita, Coord. Chem. Rev., 2010, 254, 2472–2482.

[22] B. Kumar, M. Llorente, J. Froehlich, T. Dang, A. Sathrum and C. P. Kubiak, Annu. Rev. Phys. Chem., 2012, 63, 541–569.

[23] Y. Kou, Y. Nabetani, D. Masui, T. Shimada, S. Takagi, H. Tachibana and H. Inoue, J. Am. Chem. Soc., 2014, 136, 6021–6030.

[24] E. Portenkirchner, E. Kianfar, N. S. Sariciftci and G. Knör, ChemSusChem, 2014, 7, 1347–1351.

[25] F. P. A. Johnson, M. W. George, F. Hartl and J. J. Turner, Organometallics, 1996, 15, 3374–3387.

[26] J. Agarwal, E. Fujita, H. F. Schaefer III and J. T. Muckerman, J. Am. Chem. Soc., 2012, 134, 5180–5186.

[27] Y. Hayashi, S. Kita, B. S. Brunschwig and E. Fujita, J. Am. Chem. Soc., 2003, 125, 11976–11987.

[28] C. D. Windle, E. Pastor, A. Reynal, A. C. Whitwood, Y. Vaynzof, J. R. Durrant, R. N. Perutz and E. Reisner, Chem. – Eur. J., 2015, 21, 3746–3754.

[29] G. Handwerk and J. Gary, *Petroleum Refining*, CRC Press, 4th edn., 2001.

[30] M. J. Winter, *d-block Chemistry*, Oxford University Press, Oxford, 2011.

[31] M. Bourrez, F. Molton, S. Chardon-Noblat and A. Deronzier, Angew. Chem. Int. Ed., 2011, 50, 9903–9906.

[32] C. Riplinger, M. D. Sampson, A. M. Ritzmann, C. P. Kubiak and E. A. Carter, J. Am. Chem. Soc., 2014, 136, 16285–16298.

[33] H. Takeda, H. Koizumi, K. Okamoto and O. Ishitani, Chem. Commun., 2014, 50, 1491–1493.

[34] D. C. Grills, J. A. Farrington, B. H. Layne, S. V. Lymar, B. A. Mello, J. M. Preses and J. F. Wishart, J. Am. Chem. Soc., 2014, 136, 5563–5566.

[35] M. D. Sampson, A. D. Nguyen, K. A. Grice, C. E. Moore, A. L. Rheingold and C. P. Kubiak, J. Am. Chem. Soc., 2014, 136, 5460–5471.

[36] M. D. Sampson and C. P. Kubiak, J. Am. Chem. Soc., 2016, 138, 1386–1393.

[37] S. J. P. Spall, T. Keane, J. Tory, D. C. Cocker, H. Adams, H. Fowler, A. J. H. M. Meijer, F. Hartl and J. A. Weinstein, Inorg. Chem., 2016, 55, 12568–12582.

[38] D. C. Grills, M. Z. Ertem, M. McKinnon, K. T. Ngo and J. Rochford, Coord. Chem. Rev., 2018, 374, 173–217.

[39] J.-M. Lehn and R. Ziessel, Proc. Natl. Acad. Sci. U. S. A., 1982, 79, 701–704.

[40] P. L. Cheung, C. W. Machan, A. Y. S. Malkhasian, J. Agarwal and C. P. Kubiak, Inorg. Chem., 2016, 55, 3192–3198.

[41] C. W. Machan, C. J. Stanton III, J. E. Vandezande, G. F. Majetich, H. F. Schaefer III, C. P. Kubiak and J. Agarwal, Inorg. Chem., 2015, 54, 8849–8856.

[42] C. O. Dietrich-Buchecker, P. A. Marnot, J.-P. Sauvage, J. R. Kirchhoff and D. R. McMillin, J. Chem. Soc., Chem. Commun., 1983, 513–515.

[43] M. Sandroni, Y. Pellegrin and F. Odobel, Comptes Rendus Chim., 2016, 19, 79–93.

[44] J.-X. Zhang, C.-Y. Hu, W. Wang, H. Wang and Z.-Y. Bian, Appl. Catal., A, 2016, 522, 145–151.

[45] H. Takeda, H. Kamiyama, K. Okamoto, M. Irimajiri, T. Mizutani, K. Koike, A. Sekine and O. Ishitani, J. Am. Chem. Soc., 2018, 140, 17241–17254.

[46] J. F. Martinez, N. T. La Porte, C. M. Mauck and M. R. Wasielewski, Faraday Discuss., 2017, 198, 235–249.

[47] J. Grodkowski, P. Neta, E. Fujita, A. Mahammed, L. Simkhovich and Z. Gross, J. Phys. Chem. A, 2002, 106, 4772–4778.

[48] P. Chábera, K. S. Kjaer, O. Prakash, A. Honarfar, Y. Liu, L. A. Fredin, T. C. B. Harlang, S. Lidin, J. Uhlig, V. Sundström, R. Lomoth, P. Persson and K. Wärnmark, J. Phys. Chem. Lett., 2018, 9, 459–463.

[49] Y. Liu, K. S. Kjær, L. A. Fredin, P. Chábera, T. Harlang, S. E. Canton, S. Lidin, J. Zhang, R. Lomoth, K.-E. Bergquist, P. Persson, K. Wärnmark and V. Sundström, Chem. – Eur. J., 2015, 21, 3628–3639.

[50] P. Chábera, Y. Liu, O. Prakash, E. Thyrhaug, A. El Nahhas, A. Honarfar, S. Essén, L. A. Fredin, T. C. B. Harlang, K. S. Kjær, K. Handrup, F. Ericson, H. Tatsuno, K. Morgan, J. Schnadt, L. Häggström, T. Ericsson, A. Sobkowiak, S. Lidin, P. Huang, S. Styring, J. Uhlig, J. Bendix, R. Lomoth, V. Sundström, P. Persson and K. Wärnmark, Nature, 2017, 543, 695–699.

[51] J.-M. Savéant, Chem. Rev., 2008, 108, 2348–2378.

[52] C. Costentin, S. Drouet, M. Robert and J.-M. Savéant, Science (80-.)., 2012, 338, 90–94.

[53] H. Rao, J. Bonin and M. Robert, Chem. Commun., 2017, 53, 2830–2833.

[54] Y. Pellegrin and F. Odobel, C. R. Chim., 2017, 20, 283–295.

[55] J. Bonin, M. Chaussemier, M. Robert and M. Routier, ChemCatChem, 2014, 6, 3200–3207.

[56] J. Bonin, M. Robert and M. Routier, J. Am. Chem. Soc., 2014, 136, 16768–16771.

[57] H. Rao, L. C. Schmidt, J. Bonin and M. Robert, Nature, 2017, 548, 74–77.

[58] Y. Wu, J. Jiang, Z. Weng, M. Wang, D. L. J. Broere, Y. Zhong, G. W. Brudvig, Z. Feng and H. Wang, ACS Cent. Sci., 2017, 3, 847–852.

[59] L. Chen, Z. Guo, X.-G. Wei, C. Gallenkamp, J. Bonin, E. Anxolabéhère-Mallart, K.-C. Lau, T.-C. Lau and M. Robert, J. Am. Chem. Soc., 2015, 137, 10918–10921.

[60] B. Fisher and R. Eisenberg, J. Am. Chem. Soc., 1980, 102, 7361–7363.

[61] M. Beley, J.-P. Collin, R. Ruppert and J.-P. Sauvage, J. Chem. Soc., Chem. Commun., 1984, 2, 1315–1316.

[62] A. H. A. Tinnemans, T. P. M. Koster, D. H. M. W. Thewissen and A. Mackor, Recl. Trav. Chim. Pays-Bas, 1984, 103, 288–295.

[63] Y. Zhao and Z. Liu, Chin. J. Chem. 2018, 36, 455–460.

[64] J. Zhao, X. Wang, Z. Xu and J. S. C. Loo, J. Mater. Chem. A, 2014, 2, 15228–15233.

[65] T. W. Woolerton, S. Sheard, E. Reisner, E. Pierce, S. W. Ragsdale and F. A. Armstrong, J. Am. Chem. Soc., 2010, 132, 2132–2133.

[66] A. Grabulosa, M. Beley, P. C. Gros, S. Cazzanti, S. Caramori and C. A. Bignozzi, Inorg. Chem., 2009, 48, 8030–8036.

[67] S. Sato, T. Morikawa, S. Saeki, T. Kajino and T. Motohiro, Angew. Chem. Int. Ed., 2010, 49, 5101–5105.

[68] T. Morikawa, S. Saeki, T. Suzuki, T. Kajino and T. Motohiro, Appl. Phys. Lett., 2010, 96, 142111–142113.

[69] T. M. Suzuki, H. Tanaka, T. Morikawa, M. Iwaki, S. Sato, S. Saeki, M. Inoue, T. Kajino and T. Motohiro, Chem. Commun., 2011, 47, 8673–8675.

[70] T. E. Rosser, C. D. Windle and E. Reisner, Angew. Chem. Int. Ed., 2016, 55, 7388–7392.

[71] M. F. Kuehnel, K. L. Orchard, K. E. Dalle and E. Reisner, J. Am. Chem. Soc., 2017, 139, 7217–7223.

[72] M. F. Kuehnel, C. D. Sahm, G. Neri, J. R. Lee, K. L. Orchard, A. J. Cowan and E. Reisner, Chem. Sci., 2018, 9, 2501–2509.

[73] A. F. Nogueira, A. L. B. Formiga, H. Winnischofer, M. Nakamura, F. M. Engelmann, K. Araki and H. E. Toma, Photochem. Photobiol. Sci., 2004, 3, 56–62.

[74] T. M. Suzuki, T. Nakamura, S. Saeki, Y. Matsuoka, H. Tanaka, K. Yano, T. Kajino and T. Morikawa, J. Mater. Chem., 2012, 22, 24584–24590.

[75] R. Kuriki and K. Maeda, Phys. Chem. Chem. Phys., 2017, 19, 4938–4950.

[76] G. Zhao, H. Pang, G. Liu, P. Li, H. Liu, H. Zhang, L. Shi and J. Ye, Appl. Catal., B, 2017, 200, 141–149.

[77] L. A. Faustino, B. L. Souza, B. N. Nunes, A.-T. Duong, F. Sieland, D. W. Bahnemann and A. O. T. Patrocinio, ACS Sustain. Chem. Eng., 2018, 6, 6073–6083.

[78] B. Reuillard, K. H. Ly, T. E. Rosser, M. F. Kuehnel, I. Zebger and E. Reisner, J. Am. Chem. Soc., 2017, 139, 14425–14435.

[79] K. Alenezi, S. K. Ibrahim, P. Li and C. J. Pickett, Chem. – Eur. J., 2013, 19, 13522–13527.

[80] Y. Ueda, H. Takeda, T. Yui, K. Koike, Y. Goto, S. Inagaki and O. Ishitani, ChemSusChem, 2015, 8, 439–442.

[81] H. Fei, M. D. Sampson, Y. Lee, C. P. Kubiak and S. M. Cohen, Inorg. Chem., 2015, 54, 6821–6828.

[82] S. Wannakao, W. Jumpathong and K. Kongpatpanich, Inorg. Chem., 2017, 56, 7200–7209.

[83] S. Muramulla, H. D. Arman, C. G. Zhao and E. R. T. Tiekink, Acta Crystallogr., Sect. E Struct. Rep. Online, 2009, 65, 31–42.

[84] N. Zhang, R. Long, C. Gao and Y. Xiong, Sci. China Mater., 2018, 61, 771–805.

[85] R. Berera, R. van Grondelle and J. T. M. Kennis, Photosynth. Res., 2009, 101, 105–118.

[86] C. D. Windle, M. W. George, R. N. Perutz, P. A. Summers, X. Z. Sun and A. C. Whitwood, Chem. Sci., 2015, 6, 6847–6864.

[87] A. Gabrielsson, F. Hartl, H. Zhang, J. R. L. Smith, M. Towrie, A. Viček and R. N. Perutz, J. Am. Chem. Soc., 2006, 128, 4253–4266.

[88] L. M. Kiefer, J. T. King and K. J. Kubarych, Acc. Chem. Res., 2015, 48, 1123–1130.

[89] M. Delor, I. V Sazanovich, M. Towrie, S. J. Spall, T. Keane, A. J. Blake, C. Wilson, A. J. H. M. Meijer and J. A. Weinstein, J. Phys. Chem. B, 2014, 118, 11781–11791.

[90] W. Xiong, J. E. Laaser, P. Paoprasert, R. A. Franking, R. J. Hamers, P. Gopalan and M. T. Zanni, J. Am. Chem. Soc., 2009, 131, 18040–18041.

[91] H. Vanselous, P. E. Videla, V. S. Batista and P. B. Petersen, J. Phys. Chem. C, 2018, 122, 26018–26031.

[92] M. Bourrez, M. Orio, F. Molton, H. Vezin, C. Duboc, A. Deronzier and S. Chardon-Noblat, Angew. Chem. Int. Ed., 2014, 53, 240–243.

Index

https://doi.org/10.1515/9783110665147-031